Sustainable Environmental Waste Management Strategies

Series Editors

Alok Prasad Das, Department of Life Sciences, Rama Devi Women's University, Bhubaneswar, Odisha, India

Eric van Hullebusch, Équipe ACE, Institut de Physique du Globe de Paris, PARIS CEDEX 05, France

Sustainable waste management is an absolute imperative in a world that is already dealing with the effects of climate change. This book series seeks to publish an extensive collection of scientific books targeting researchers, scientists, professionals, academics, students, and anyone else involved in the wide, diverse field of sustainable waste management. The series accepts monographs, edited volumes, textbooks, and conference proceedings. Topics covered include but are not limited to environmental pollution and remediation, sources of environmental waste, environmental and human health, sustainable waste management, circular economy, nature-inspired environmental solutions, pollution prevention, and energy-efficient technologies for waste management. The series puts a particular emphasis on developing countries/the majority world but accepts proposals globally as long as the topic is in scope.

Eepsita Priyadarshini · Angeles Sanroman ·
Alok Prasad Das
Editors

Advanced Nanotechnology Applications for Environmental Sustainability

Springer

Editors
Eepsita Priyadarshini
Department of Biomedical Sciences
School of Biosciences and Technology
Galgotias University
Greater Noida, Uttar Pradesh, India

Angeles Sanroman
CINTECX, University of Vigo
BIOSUV, Department of Chemical Engineering
Vigo, Spain

Alok Prasad Das
Department of Life Sciences
Rama Devi Women's University
Bhubaneswar, Odisha, India

ISSN 3005-1622 ISSN 3005-1630 (electronic)
Sustainable Environmental Waste Management Strategies
ISBN 978-3-032-12932-1 ISBN 978-3-032-12933-8 (eBook)
https://doi.org/10.1007/978-3-032-12933-8

This Springer imprint is published by the registered company Springer Nature Switzerland AG
The registered company address is: Gewerbestrasse 11, 6330 Cham, Switzerland

Preface

Climate change, environmental pollution, and depletion of natural resources are the some of challenges that necessitates requirement of innovative and sustainable solutions across all sectors. Nanotechnology, is one such promising scientific advancement that has offered transformative potential in environmental protection, remediation, and monitoring. The amalgamation of nanotechnology with environmental science marks a crucial shift in our approach to sustainability—one that this book seeks to explore.

The book titled *Advanced Nanotechnology Applications for Environmental Sustainability* aims to serve as an inclusive and interdisciplinary resource that bridges the gap between emerging nanotechnological innovations and their real-world applications in environmental contexts. In this book, contributions from leading researchers and practitioners in the fields of nanoscience, environmental engineering, and sustainability studies are brought together to provide a well-rounded understanding of how nanomaterials and nano-enabled technologies are being developed and deployed for water purification, air and soil remediation, waste management and environmental sensing.

The book has multifaceted utilization such as to educate students, academicians, and policy-makers on the potential and limitations of nanotechnology in addressing environmental challenges; to encourage interdisciplinary collaboration between nanotechnologists and environmental scientists; and present real-world case studies and discussions on ethical, regulatory, and safety considerations. The broad scope, interdisciplinary approach, and emphasis on practical application makes the book different from other available books. The book is catered to a wide audience from researchers and educators to policy-makers and industry professionals making it a valuable tool for both academic study and applied research. Importantly, the book addresses challenges such as nanotoxicity, environmental risks, and regulatory uncertainties, thereby providing a holistic understanding of the field.

As editors, we consider and understand that knowledge-sharing and collaboration are necessary to comprehending the full prospective of nanotechnology in building a sustainable future. We hope this book will encourage its readers to explore, innovate, and contribute meaningfully to this evolving and impactful field.

Greater Noida, India — Eepsita Priyadarshini
Vigo, Spain — Angeles Sanroman
Bhubaneswar, India — Alok Prasad Das

Acknowledgments

Coming together is a beginning, staying together is progress, and working together is success

—*Henry Ford*

The essence of collaboration articulated in these words effortlessly reflects the journey of this book. The book *Advanced Nanotechnology Applications for Environmental Sustainability* has been a collective effort, made possible by the dedication, expertise, and support of many individuals and institutions.

We extend our heartfelt gratitude to all the contributors whose dedication, enthusiasm, and expertise have made this book possible. The insightful chapters the authors contributed reflect the depth and diversity of current research in the domain of nanotechnology for environmental sustainability. Our gratitude also extends to the institutions and organizations supporting the authors' research endeavours.

We are thankful for the support and encouragement of our parent institution, for providing an intellectually vibrant environment and the resources necessary for the successful completion of this project. In addition, we also gratefully acknowledge the support and guidance provided by our publisher, whose editorial and production teams have been immensely helpful throughout the preparation of this book.

Lastly, we extend our appreciation to our families for their unwavering support and patience during this work.

Eepsita Priyadarshini
Angeles Sanroman
Alok Prasad Das

Contents

About the Editors

Dr. Eepsita Priyadarshini is an Assistant Professor at the School of Biosciences and Technology, Galgotias University. She holds a Ph.D. in Biological Sciences from CSIR-IMMT (AcSIR), with expertise in nanoparticle synthesis, biosensors, and bionanotechnology. She has completed prestigious fellowships including DST-Women Scientist (WOS-B), DST-NPDF, and DS Kothari PDF. She has received prestigious awards, including the Chancellor's Gold Medal and Newton-Bhaba Ph.D. Placement Fellowship, and has collaborated with leading institutions globally, University College London, UK being one of those.

Her research focuses on nanomaterials for biomedical and environmental applications and has earned her a place in the Top 2% Scientists of the World (2023, 2024, Elsevier). Dr. Priyadarshini has authored over 40 peer-reviewed publications with an h-index 24 and over 2400 citations. She also serves as an editor/reviewer for several international journals, and has been invited to speak at several scientific platforms.

Angeles Sanroman is a Professor of Chemical Engineering at the University of Vigo and currently serves as Director of the CINTECX Research Center. With a Ph.D. in Chemistry and a degree in Chemical Engineering from the University of Santiago de Compostela, she leads the BIOSUV Group, focusing on sustainable processes for environmental remediation, advanced oxidation technologies, and green chemistry.

Professor Sanromán has authored more than 400 scientific publications, achieving over 15,500 citations and an h-index of 63 (Scopus), with a strong international impact (41% of co-authored papers with researchers abroad). She has been included in the Top 2% Scientists of the World for the last five consecutive years. In addition, she has led over 50 competitive research and innovation projects and transfer contracts, both national and international.

She has held key evaluation roles within Spain's national R&D system, including as a member of the Spanish Research Agency (AEI) expert panels in Chemical Technology (2007–2011), Technology Transfer (2015–2018) and Materials, Energy and Environment (2019–2022).

Recognized with the Parga Pondal Research Medal (2023) and other distinctions, Prof. Sanromán combines scientific leadership with a strong commitment to mentoring young researchers with the supervisión of supervised 28 Ph.D. theses and science outreach, particularly through initiatives promoting women in science and sustainability education.

Dr. Alok Prasad Das is presently serving as an Assistant Professor in the Department of Life Sciences, at Rama Devi Women's University, Odisha India. Before that, Dr. Alok was working as an Assistant Professor in the Department of Chemical and Polymer Engineering, at Tripura Central University, Agartala, India. He is an academician, researcher, and author of journal articles, several books, and book chapters. He is an Editorial member of several reputed journals. With his 15 years of research experience, Dr. Das has been supported through various research grants from the Department of Science and Technology and the Department of Biotechnology, Ministry of Mines, Government of India.

His research and innovation work, done in national and international collaborations, has appeared in over 120 publications; and his contributions have been recognized by several awards. His area of expertise includes Biological Wastewater treatment, Bioremediation bioleaching of mining pollutants, Bioconversion of environmental waste to value-based products, Bioremediation of Microplastics and Microfiber Pollutants, and Geomicrobiology of extreme habitats.

He has guided several postgraduate and Ph.D. scholars. His current Google Scholar: Citations is 5580 and his h-index is 43. Dr. Das is enlisted in the world's top 2% influential scientists as per the Stanford University list for 2023 and 2024.

Nanomaterial's: Types, Synthesis and Application Environment Remediation

Sateesh Yadav, Supriya Tripathy, and Navinchandra Gopal Shimpi

Abstract In recognition of their special qualities and wide range of uses, nanomaterials have grown in importance in environmental science. This chapter discusses the various types of nanomaterials, how they are produced, and their use in environmental remediation. It draws attention to the several types of nanomaterials, including those based on metal, carbon, and polymers, and their functional properties and environmental applications. The influence of various synthesis methods on the properties and efficacy of nanomaterials is analysed, encompassing sol–gel, hydrothermal, and chemical vapor deposition techniques. Furthermore, the use of nanomaterials for soil, water and air pollution remediation is investigated, demonstrating their ability to address environmental issues through catalysis, adsorption and degradation. The chapter emphasizes how promising nanomaterials are for improving the efficiency of technologies for environmental cleanup and encouraging long-term approaches to pollution control.

Keywords Nanomaterials · Synthesis techniques · Environmental remediation · Sustainable nanotechnology

1 Introduction

1.1 Overview of Nanomaterials

Nanomaterials are a special class of materials that have at least one dimension in the nanoscale range (1–100 nm). Their remarkable physical, chemical, and biological capabilities set them apart from their bulk counterparts (Khin et al., 2012). Based on their dimensional properties, nanomaterials can be broadly categorized

S. Yadav · S. Tripathy · N. G. Shimpi (✉)
Laboratory of Materials Science, Department of Chemistry, University of Mumbai, Mumbai, India
e-mail: navin_shimpi@rediffmail.com; navinshimpi@chemistry.mu.ac.in

E. Priyadarshini et al. (eds.), *Advanced Nanotechnology Applications for Environmental Sustainability*, Sustainable Environmental Waste Management Strategies,
https://doi.org/10.1007/978-3-032-12933-8_1

as zero-dimensional (0D), one-dimensional (1D), two-dimensional (2D), or three-dimensional (3D). Because of their distinct qualities that distinguish them from traditional bulk materials, nanomaterials are essentially significant (Guerra et al., 2018a). Quantum mechanical events significantly alter the chemical, physical and biological features of materials at the nanoscale. These changes result from a significant increase in the surface area-to-volume ratio and the dominance of quantum confinement processes at nanometric levels. Amazing surface reactivity, increased mechanical strength, special optical qualities, higher electrical and thermal conductivity, and quantum mechanical behaviour are some of the key characteristics that set nanomaterials apart (Kalita & Baruah, 2020). Building on theoretical underpinnings in quantum physics, surface chemistry, and materials science, the idea of nanomaterials arose from interdisciplinary scientific advancements in the latter half of the twentieth century. Many people believe that Richard Feynman's groundbreaking 1959 speech "There's Plenty of Room at the Bottom" served as the conceptual forerunner of nanotechnology, suggesting the possibility of atomic and molecular-scale matter manipulation (De Luca & Ferrer, 2017). The intersection of physics, chemistry, biology, and materials science is represented by nanomaterials, which present hitherto unheard-of chances for scientific advancement and technological innovation. With the potential to revolutionize numerous technological fields, the study of nanomaterials keeps pushing the limits of scientific knowledge (Tripathy et al., 2023).

1.2 Importance in Environmental Remediation

In addition to their distinct physicochemical characteristics, nanomaterials have become a ground-breaking technology with enormous potential for environmental remediation. They provide creative answers to challenging environmental problems (Thorat et al., 2024). In order to handle the growing complexity of pollution, environmental remediation is a crucial worldwide issue that calls for creative technology solutions. Due to their outstanding performance qualities and distinct physicochemical properties, nanomaterials have become a game-changer, providing previously unheard-of capabilities for tackling contamination of the environment (Tripathy et al., 2024a). Undoubtedly, one of the biggest issues facing society today is environmental pollution. Research into new technology for cleaning up soil, water, and air pollutants is ongoing. Among the many alarming risks are particulate matter, heavy metals, fertilizers, pesticides, herbicides, oil spills, sulfurous gasses, industrial effluents, sewage, and organic compounds (Tripathy et al., 2024b). Many different methods can be employed to accomplish this goal because environmental restoration can include a wide variety of materials. Since it may be challenging to capture and degrade environmental pollutants due to their complex mixture of many compounds, low reactivity, and high volatility have been the focus of current studies using nanomaterials for the development of innovative environmental remediation methods (Singh et al., 2018).

2 Types of Nanomaterials

2.1 Carbon-Based Nanomaterials

One of the most revolutionary advances in materials science over the last three decades is the invention of carbon-based nanomaterials, which has completely changed how we approach technological innovation in a variety of fields. The physical, chemical, and electronic characteristics of these remarkable structures—which are distinguished by their nanoscale size and carbon-dominant composition—differ significantly from those of their bulk counterparts, providing previously unheard-of chances for both scientific and industrial development (Vaseashta et al., 2007). The family of carbon nanomaterials includes a wide range of allotropes, such as fluorescent carbon quantum dots, atomically thin graphene sheets, cylindrical carbon nanotubes, soccer ball-shaped fullerenes, and crystalline nanodiamonds. Each of these allotropes has unique properties that can be precisely engineered for specific functionalities. From energy conversion and nanoelectronics to medicine and environmental remediation, their remarkable mechanical strength, remarkable electrical conductivity, remarkable thermal properties, controllable optical behaviours, and inherent biocompatibility have made them revolutionary materials (Yunus et al., 2012). Following the discovery of fullerenes in 1985, carbon nanotubes in 1991, and graphene in 2004, there were consecutive waves of scientific excitement that led to intense research efforts all around the world. The adaptability of carbon nanomaterials—their ability to be functionalized with different chemical groups, integrated into composite materials, structured in intricate designs, and controlled at the atomic level to obtain desired properties—makes them especially intriguing. Applications range from very sensitive biosensors and targeted medication delivery systems to flexible electronics and ultra strong composite materials (Chou et al., 2005). Even though our knowledge and application of these materials has advanced significantly, problems still exist with scalable production, accurate structural control, systematic characterization, and thorough toxicity evaluation. However, carbon nanomaterials continue to move from lab experiments to commercial realities as synthesis methods advance and interdisciplinary partnerships grow, with new uses continuously redefining technical possibilities. In addition to enhancing our basic knowledge of nanoscale phenomena, the continued investigation of these materials holds promise for resolving important worldwide issues in sustainability, healthcare, and energy, establishing carbon nanomaterials as the mainstays of technological innovation in the twenty-first century (Gupta & Silver, 1998).

Graphene: One subcaste of carbon granules organized in a two-dimensional honeycomb cage is famous for possesses remarkable thermal, electrical, and mechanical characteristics. Carbon granules make up graphene, a planar hexagonal honeycomb network in two dimensions structures. In utmost cases, a graphene distance is about 1 nm thick (Bosetti et al., 2002).

Fullerenes: Molecules made entirely of carbon, generally in the form of concave spheres, ellipsoids or tubes. Carbon-grounded nanomaterials are extensively used in electronics, energy storehouse, and detectors. Having compasses that range from 8.2 nm for single layers to 4–36 nm for multilayered fullerenes, the globular structure is composed of from 28 to 1500 carbon tittles. Fullerenes contain nanomaterials made of spherical concave coops that resemble allotropic forms of carbon. Due to their great strength, structure, electron affinity, stiffness, and electrical conductivity, they have attracted marketable interest (Xiu et al., 2012).

2.2 *Metal-Based Nanomaterials*

These nanomaterials are composed of metal or metal alloys at the nanoscale, and they frequently have unique parcels compared to their bulk counterparts. Examples include:

Gold nanoparticles: Known for their use in biomedical applications like drug delivery, diagnostics, and imaging.

Silver nanoparticles: Commonly used for their antimicrobial properties in medical and consumer products.

Titanium dioxide (TiO_2) nanoparticles: Used in applications like photocatalysis, self-cleaning surfaces, and sunscreens. Metal-based nanomaterials offer high electrical conductivity, catalytic activity, and enhanced optical properties.

2.3 *Polymeric Nanomaterials*

A novel class of sophisticated materials known as polymeric nanomaterials combines the special qualities that arise at the nanoscale dimension with the adaptability of polymers. These complex structures, which usually have at least one dimension between 1 and 100 nm, are made of natural or synthetic polymers that have been precisely created to have exact control over their molecular architecture, content, and functioning. By adjusting their molecular weight, chemical composition, crystallinity, hydrophobicity, biodegradability, and surface properties, polymeric nanomaterials—which are distinguished by their remarkable tunability—can be engineered to incorporate an astounding range of physical, chemical, and biological properties (Pal et al., 2007). Customized applications in a wide range of technological fields are made possible by the variety of morphologies that can be achieved within this materials category, such as nanoparticles, nanogels, nanospheres, nanocapsules, nanofibrous structures, and polymer brushes. They have important uses in drug delivery systems, scaffolds for tissue engineering, environmental cleanup technologies, smart coatings, sensors, flexible electronics, and energy storage devices, demonstrating their importance in contemporary science. Polymeric nanomaterials are at the forefront of innovation in biomedicine, electronics, and environmental science because of their unique advantages, which include biocompatibility, stimuli-responsiveness, improved mechanical properties, and the ability to release encapsulated substances under controlled conditions (Lu & Chiu, 2006). Polymeric nanomaterials are becoming more and more platforms for tackling complex issues in technology development, sustainability, and healthcare as interdisciplinary research advances synthesis methodologies, characterization techniques, and functional understanding of these materials. Through precisely engineered nanoscale architectures, these materials promise transformative solutions for enduring global challenges. These include:

Polymeric nanoparticles: Tiny particles that can encapsulate drugs or other active ingredients for controlled release in drug delivery systems.

Nanofibers: Polymer fibers with diameters in the nanometer range, often used for filtration, tissue engineering, and sensors.

Nanocomposites: Polymers that have been combined with nanofillers to enhance their mechanical, thermal, and electrical parcels. Polymeric nanomaterials are largely versatile, with operations in drug, electronics, and environmental remediation.

Nanocomposites: Polymers that have been combined with nanofillers to enhance their mechanical, thermal, and electrical parcels. Polymeric nanomaterials are largely versatile, with operations in drug, electronics, and environmental remediation (Li et al., 2001).

2.4 Composite and Hybrid Nanomaterials

Composite and hybrid nanomaterials combine two or more different types of nanomaterials to exploit their unique properties. Water quality has been monitored using intercross nanocomposite materials, which are comparable to rGO mixed with metal nanoparticles or metal oxides. These sensors are able to identify contaminants in water sources, such as heavy metals. Heavy metal ion detection continues to face obstacles in reaching peak performance in terms of speed, sensitivity, and broad identification of heavy metal ions. Biosensors for environmental monitoring have used hybrid nanocomposites, which are composed of biomolecules and nanomaterials. Certain pollutants, toxics, or diseases that are prevalent in the atmosphere can be found by these biosensors. For precise and quick environmental investigation, combining nanomaterials and biomolecules offers great perceptivity, selectivity, and biocompatibility (Li et al., 2014). Examples include:

Composite nanomaterials: Materials that integrate different nanomaterials, such as carbon nanotubes with polymers or metals, to achieve enhanced mechanical strength, conductivity, or durability.

Hybrid nanomaterials: A combination of organic and inorganic materials at the nanoscale, such as graphene wrapped in a polymer shell, which provides benefits like increased stability and functionality. These materials are often tailored for specific applications like sensors, catalysts, and energy storage devices.

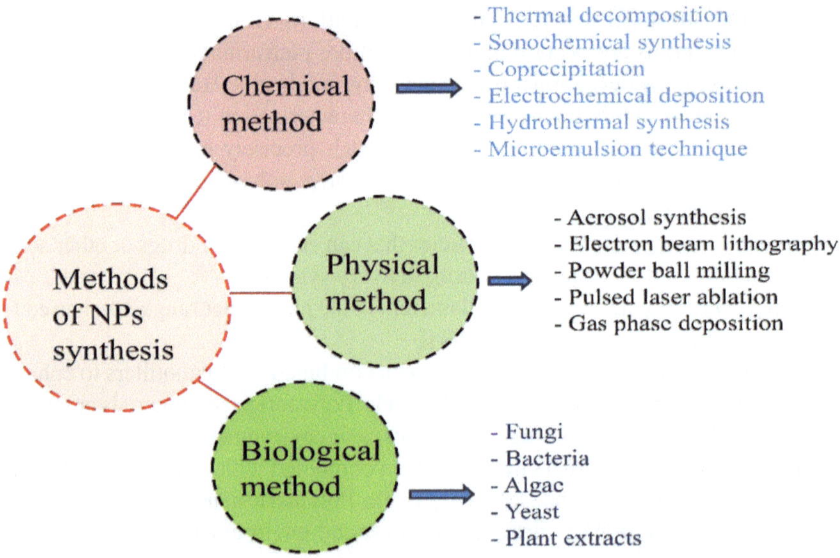

Fig. 1 Diagrammatic illustration of the production of nanoparticles using various techniques

3 Synthesis Techniques

See Fig. 1.

4 Physical Methods for Synthesis

4.1 Mechanical Milling

In mechanical milling, high-energy forces including impact, shear, and compression are used to grind or mechanically break-bulk materials into nanoscale particles.

Powders or fine particles for use in materials science, electronics, and energy storage are frequently produced using this technique. It's easy and scalable but may produce contamination due to friction between materials.

4.2 Physical Vapor Deposition (PVD)

In material science, physical vapor deposition, or PVD, is a commonly used method for applying coatings or thin films to a variety of surfaces. This method creates a thin, consistent coating by vaporizing material from a solid or liquid source and then condensing it onto a substrate. Techniques including thermal evaporation, sputtering, or laser ablation are used to accomplish the vaporization. In order to avoid contamination and guarantee the purity of the material deposited, PVD procedures usually take place in a vacuum atmosphere. To vaporize the material to be deposited—typically metals, alloys, or ceramics—it must first be heated or subjected to an intensive particle bombardment. After passing through the vacuum, this vapor condenses on the substrate's surface, which is often made of glass, plastic, or metal. PVD coatings provide increased wear resistance, corrosion resistance, and durability. These coatings are frequently utilized in sectors where accuracy and material performance are crucial, including manufacturing, electronics, automotive, and aerospace (Park & Lee, 2014).

5 Chemical Methods for Synthesis

5.1 Sol–Gel Process

The sol–gel technique produces nanomaterials, usually metal oxides, by turning a liquid "sol" (a colloidal solution) into a solid gel. The gel is created by hydrolysis and condensation processes, with the precursor sol often being a metal alkoxide. With this technique, thin films, coatings, and nanoparticles with exact control over size and shape can be produced. Low temperatures are often used in all sol–gel production processes. As a result, there will be less pollution and energy use. Some advantages of the sol–gel technique include the production of special materials like zeolites, aerogels, and ordered porous solids through organic–inorganic hybridization. Additionally, the sol–gel process can be used to create nanoparticles, nanorods, nanotubes, etc.

5.2 Chemical Vapor Deposition (CVD)

The method of CVD uses gaseous reactants are chemically reacted on a heated substrate surface, leading to the deposition of solid materials (often in nanostructured form). It's widely used to create thin films, nanotubes, and nanowires, and is important in the semiconductor industry, particularly for fabricating materials like graphene and carbon nanotubes. Because of its economic feasibility, ease of processing, capacity to deposit a variety of materials, and relatively simple equipment, CVD is frequently employed in industry. Single-crystalline or nanocrystalline films may form under specific deposition circumstances.

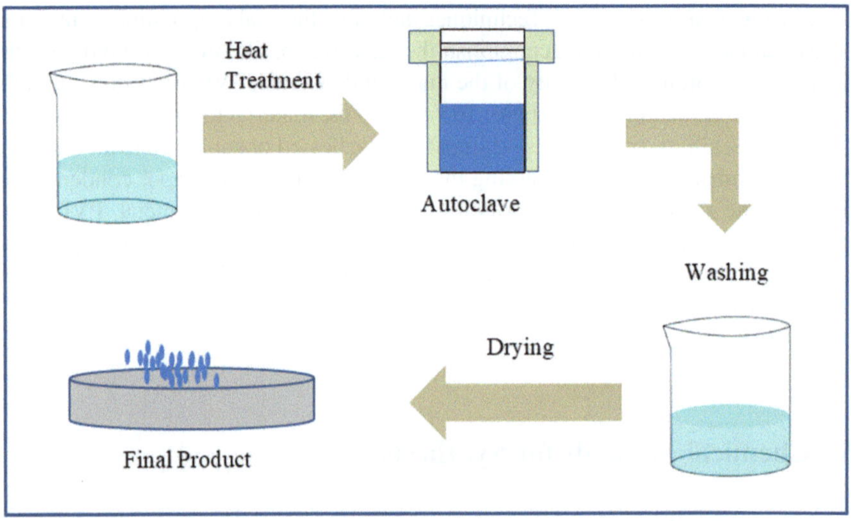

Fig. 2 Schematic representation of hydrothermal process

5.3 Hydrothermal-Solvothermal Method

These methods involve the synthesis of nanomaterials under high temperature and pressure in a sealed vessel (autoclave) with water or an organic solvent as the medium. In hydrothermal synthesis, water is the solvent, while solvothermal uses other solvents like alcohol or acids. These methods are often used for producing metal oxide nanoparticles, quantum dots, and catalysts (Fig. 2) (Bao et al., 2011).

6 Biological Synthesis Methods

6.1 Plant-Based Synthesis

Plant-based synthesis involves the use of plant extracts, parts, or metabolites to produce nanoparticles. The plants contain various bioactive compounds like polyphenols, flavonoids, and alkaloids, which have reducing and stabilizing properties. These substances stabilize metal ions to avoid agglomeration while assisting in their reduction into nanoparticles (Guerra et al., 2018b). Among the often-utilized plants are:

Tea: Frequently utilized in the production of silver and gold nanoparticles, it is well-known for having a high antioxidant content.

Cactus: Certain cactus species are used due to their capacity to reduce metal salts and stabilize nanoparticles.

Mint and Aloe Vera: These plants also have potential for nanoparticle synthesis due to their natural reducing agents.

Compared to chemical processes, plant-based synthesis is less harmful, more economical, and environmentally benign. Additionally, it is scalable and suitable for large-scale production. Applications for the generated nanoparticles can be found in a various industry, including electronics, environmental remediation (water purification), and medicine (drug delivery, antibacterial agents).

6.2 Microbial Synthesis

Microorganisms including bacteria, fungus, and algae are used in microbial synthesis to create nanoparticles. Through metabolic processes, these microbes can convert metal ions into nanomaterials (Sreeja & Vidya Shetty, 2016). For instance:

Bacteria like Escherichia coli and Bacillus species can reduce metal ions such as gold, silver and copper to form nanoparticles. These bacteria also help in stabilizing the nanoparticles by producing extracellular biopolymers.

Fungi especially species like Aspergillus and Fusarium, are used to synthesize nanoparticles through the reduction of metal salts. It is well known that fungi may create nanoparticles with a good size distribution and great stability.

Algae can also be employed to produce nanoparticles because of their capacity to uptake and reduce metal ions through their cell walls and surface.

7 Properties of Nanomaterials Relevant to Environmental Remediation

7.1 Surface Area and Reactivity

Given their small size and extensive surface exposure, nanomaterials usually have a high surface area-to-volume ratio. This characteristic is crucial for environmental remediation, as it enhances their reactivity. A larger surface area means more active sites are available for reactions, making nanomaterials highly effective in processes such as:

Adsorption: Nanomaterials can adsorb contaminants (like heavy metals, organic pollutants, or toxins) onto their surface, helping to remove these harmful substances from the environment.

Catalysis: Nanomaterials can act as catalysts in breaking down pollutants (e.g., degradation of organic compounds, degradation of dyes in wastewater).

Chemical Reactivity: The increased surface reactivity of nanoparticles makes them highly effective for chemical reactions, such as reducing toxic chemicals in contaminated soil or water.

8 Optical and Electronic Properties

The optical and electronic properties of nanomaterials change as their size decreases, often showing unique behaviours compared to bulk materials. This is due to quantum effects that become more significant at the nanoscale (Gao et al., 2013). These properties are particularly useful for environmental remediation in areas such as:

Sensors: Nanomaterials with enhanced optical properties (like fluorescence or absorption) can be used for detecting pollutants in water or air, making them useful in environmental monitoring and early detection of contaminants.

Photocatalysis: Nanomaterials like titanium dioxide (TiO_2) are frequently employed for photocatalytic degradation of pollutants when exposed to UV light. This is applicable to degrade organic contaminants in the air or water, such as pesticides or volatile organic compounds (VOCs).

Electronics for Remediation Systems: The electronic characteristics of nanoparticles enable them to be used in advanced remediation technologies, such as electrochemical sensors or energy-efficient filtration systems.

9 Stability and Biocompatibility

Stability and biocompatibility are crucial factors to take into account while using nanomaterials for environmental applications, especially for long-term or sustainable remediation efforts. These properties are important for ensuring the safety and effectiveness of the materials:

Stability: Nanomaterials need to be stable under environmental conditions (e.g., varying pH, temperature, moisture) to insure their long-term effectiveness in remediation. Instability may lead to the breakdown or loss of efficacity over time. Materials like carbon nanotubes or metal oxides are often chosen for their durability and resistance to degradation.

Biocompatibility: Nanomaterials used for environmental remediation should be non-toxic to the environment, plants, and animals. They should not introduce additional hazards during the cleanup process. For example, biocompatible nanomaterials like certain types of biosynthesized nanoparticles (from plants or microbes) have a minimal ecological footprint, creating them safer for extended use in sensitive ecosystems (Poguberović et al., 2016).

10 Applications of Nanomaterials in Environmental Remediation

10.1 Water Purification

Nanomaterials are gradually used in water treatment proceedings because of their larger surface area, reactivity, and versatility.

10.2 Adsorption of Heavy Metals and Pollutants

Heavy metals (including lead, mercury, and arsenic) and other harmful contaminants can be removed from contaminated water by using nanoparticles (such carbon nanotubes, metal oxides, and activated carbon nanoparticles). Many of the adulterants found in water aqueducts been discovered to be hazardous to both environment and human health. The finding and removal of arsenic from aqueducts and watersheds must take precedence in pollution management efforts due to its severe toxicity. In nature, arsenic is present in the Earth's crust in some quantity, as well as in trace levels in gemstones, air, water and soil, and to a lesser degree in living things including plants, animals and microorganisms (Hooshyar et al., 2013).

Nanomaterials provide a high surface area for adsorption, leading to more effective elimination of pollutants, even at low levels. Nanoparticles like zero-valent iron (ZVI) can chemically reduce heavy metals and contaminants, rendering them less toxic. Polymer nanoparticles have a variety of applications, such as sunscreen and water purification. Polymeric nanoparticles have amphiphilic plats, which are based on the same idea as surfactant micelles and have hydrophobic and hydrophilic regions in each molecule (Xiong et al., 2011).

10.3 Degradation of Organic Contaminants

Photocatalytic nanomaterials (e.g., titanium dioxide (TiO_2), zinc oxide (ZnO)) can break down organic pollutants like pesticides, dyes, and pharmaceuticals under UV or visible light. Nano catalysts can also degrade toxic organic compounds in wastewater through processes like oxidation or reduction, offering an environmentally friendly approach to water treatment (Younas et al., 2014).

11 Air Pollution Control

Nano catalysts can be used to convert harmful gaseous pollutants (e.g., NOx, CO, VOCs) into less harmful substances. Metal oxide nanomaterials (e.g., TiO_2, SnO_2) are employed in catalytic converters and air purifiers for efficient removal of gases. Nanomaterial-based filters can capture gases from industrial emissions, improving air quality and reducing toxic air pollution. Nanotechnology has the ability to destroy harmful gasses found in the air (Rasalingam et al., 2014). The adsorption of CNTs and gold particles are two instances of how hazardous gasses can be cleaned using nanotechnology. Hexagonally organized carbon atoms surround the tube axis in graphene sheets in CNTs. The face of CNTs and the two benzene rings of dioxin have a robust trade relationship. Furthermore, dioxin molecules interact with the whole surface of nanotubes with a pervious wall, or 2.9 nm, as well as the potential for overlaying events that boost the internal adsorption capacity. The renewal of the adsorbent at high temperatures has also benefited from CNTs' strong oxidation resistance (Chen et al., 2013).

12 Soil Remediation

Nanomaterials offer effective solutions for the cleanup of contaminated soils, providing more efficient and less disruptive alternatives to traditional methods (Seredych & Bandosz, 2007).

12.1 Contaminant Extraction

Nanoparticles like iron oxide nanoparticles can be used to extract contaminants (heavy metals, hydrocarbons) from soil via mechanisms such as adsorption and dissolution. They can also break down or immobilize contaminants, reducing their bioavailability and toxicity to plants and animals. Like polynuclear sweet hydrocarbons (PAHs), hydrophobic organic groundwater contaminants bind explosively to soils and are challenging to get rid of. Amphiphilic polyurethane (APU) nanoparticles have been created to be used in the rehabilitation of PAH-polluted soil (Ebrahim et al., 2016).

12.2 Soil Stabilization and Conditioning

Nanomaterials can enhance soil structure and stability by promoting the aggregation of soil particles. Nano-silica and nano-clays are used to strengthen soil and reduce erosion, improving agricultural productivity and mitigating soil degradation (Patel et al., 2019).

13 Challenges and Limitations

Although nanomaterials have great environmental applications, they nonetheless confront several obstacles. Scalability is still an issue; larger implementations of laboratory achievements frequently fail. Because nanoparticles have the potential to inflict unanticipated ecological harm, toxicity concerns remain. Aggregation decreases efficacy in practical settings, and post-deployment recovery poses technological challenges (Dutta et al., 2023). Widespread adoption is hampered by cost, especially in underdeveloped nations where environmental remediation is most urgently needed. Around the world, regulatory systems continue to vary. Long-term performance is impacted by material stability, which varies depending on the environment (Xu et al., 2019). Problems with standardization make reproducibility and quality control more difficult. For nanoparticles to reach their full potential in environmental remediation applications, these constraints need to be overcome. The potential toxicity of nanomaterials to ecosystems and human health remains a major concern. Some nanoparticles may exhibit harmful biological effects, including bioaccumulation and toxicity to aquatic life, plants, and animals (Mallikarjunaiah et al., 2020). Research on how nanoparticles move and behave in the environment are needed to understand their long-term impact. The price of mass-producing nanomaterials remains high compared to traditional materials. This poses challenges for widespread application, particularly in developing countries (Bandyopadhyay & Jana, 2020).

Scalability and reproducibility of the synthesis methods need to be improved to make nanomaterials economically viable for industrial applications in environmental remediation. There are uncertainties in regulation due to the novelty and complexity of nanomaterials. Proper standards and guidelines need to be set up for the safe production, handling, and disposal of nanomaterials. Public perception and concerns about the unknown risks associated with nanomaterials may affect their acceptance in environmental applications (Jana et al., 2021).

14 Future Perspectives

The future of nanomaterials in environmental remediation looks promising, with several areas showing great potential for innovation. The creation of innovative and sophisticated sets of materials and equipment is necessary for the global pollution remediation industry to succeed in the future. Some of the most pressing issues of the twenty-first century are addressed by nanoscience and nanotechnology, which provide avenues for material design and link the structure and function of biomolecules to human physiology (Song et al., 2018).

14.1 Advancements in Green Synthesis

Green synthesis methods for producing nanomaterials using sustainable, eco-friendly processes are being developed. This would address concerns regarding the environmental impact of nanomaterial production. Biologically synthesized nanoparticles (e.g., using plants, bacteria, or fungi) could provide a more economical and environmentally friendly substitute to conventional chemical methods (Koech et al., 2023).

14.2 Potential for Nano-Bio Hybrid Systems in Remediation

Nano-bio hybrid systems—combining nanomaterials with biological systems (e.g., bacteria, enzymes, plant roots) have shown great promise in improving the efficiency and selectivity of environmental remediation processes. These systems could offer more biodegradable, efficient, and selective approaches to removing pollutants, harnessing both nanomaterial and biological properties for synergistic effects (Tripathy et al., 2025).

References

Bandyopadhyay, A., & Jana, D. (2020). A review on role of tetra-rings in graphene systems and their possible applications. *Reports on Progress in Physics, 83*(5), Article 056501. https://doi.org/10.1088/0034-4885/83/5/056501

Bao, Q., Zhang, D., & Qi, P. (2011). Synthesis and characterization of silver nanoparticle and graphene oxide nanosheet composites as a bactericidal agent for water disinfection. *Journal of Colloid and Interface Science, 360*(2), 463–470. https://doi.org/10.1016/j.jcis.2011.05.009

Bosetti, M., Masse, A., Tobin, E., & Cannas, M. (2002). Silver coated materials for external fixation devices: In vitro biocompatibility and genotoxicity. *Biomaterials, 23*(3), 887–892. https://doi.org/10.1016/S0142-9612(01)00198-3

Chen, X., Cen, C., Tang, Z., Zeng, W., Chen, D., Fang, P., & Chen, Z. (2013). The key role of pH value in the synthesis of titanate nanotubes-loaded manganese oxides as a superior catalyst for the selective catalytic reduction of NO with NH_3. *Journal of Nanomaterials, 2013*, Article 871528. https://doi.org/10.1155/2013/871528

Chou, K.-S., Lu, Y.-C., & Lee, H.-H. (2005). Effect of alkaline ion on the mechanism and kinetics of chemical reduction of silver. *Materials Chemistry and Physics, 94*(2–3), 429–433. https://doi.org/10.1016/j.matchemphys.2005.05.029

De Luca, A., & Ferrer, B. B. (2017). Nanomaterials for water remediation: Synthesis, application and environmental fate. In C. M. Hussain (Ed.), *Nanotechnologies for environmental remediation* (pp. 25–60). Springer. https://doi.org/10.1007/978-3-319-53162-5_2

Dutta, S., Kumar, S., Yu, Y., Rubahn, H.-G., Kumar, Y., & Awasthi, K. (2023). Functional nanomaterials in flexible gas sensors: Recent progress and future prospects. *Materials Today Chemistry, 29*, Article 101428. https://doi.org/10.1016/j.mtchem.2023.101428

Ebrahim, S. E., Sulaymon, A. H., & Saad Alhares, H. (2016). Competitive removal of Cu^{2+}, Cd^{2+}, Zn^{2+}, and Ni^{2+} ions onto iron oxide nanoparticles from wastewater. *Desalination and Water Treatment, 57*(10), 20915–20929. https://doi.org/10.1080/19443994.2015.1112310

Gao, P., Ng, K., & Sun, D. D. (2013). Sulfonated graphene oxide-ZnO–Ag photocatalyst for fast photodegradation and disinfection under visible light. *Journal of Hazardous Materials, 262*, 826–835. https://doi.org/10.1016/j.jhazmat.2013.08.026

Guerra, F., Attia, M. F., Whitehead, D. C., & Alexis, F. (2018a). Nanotechnology for environmental remediation: Materials and applications. *Molecules, 23*(7), 1760. https://doi.org/10.3390/molecules23071760

Guerra, F. D., Attia, M. F., Whitehead, D. C., & Alexis, F. (2018b). Nanotechnology for environmental remediation: Materials and applications. *Molecules, 23*(7), 1760. https://doi.org/10.3390/molecules23071760

Gupta, A., & Silver, S. (1998). Molecular genetics: Silver as a biocide: Will resistance become a problem? *Nature Biotechnology, 16*(9), 888. https://doi.org/10.1038/nbt1098-888

Hooshyar, Z., Rezanejade Bardajee, G., & Ghayeb, Y. (2013). Sonication enhanced removal of nickel and cobalt ions from polluted water using an iron-based sorbent. *Journal of Chemistry, 2013*, Article 786954. https://doi.org/10.1155/2013/786954

Jana, S., Bandyopadhyay, A., Datta, S., Bhattacharya, D., & Debnarayan, J. (2021). Emerging properties of carbon based 2D material beyond graphene. *Journal of Physics: Condensed Matter*. Advance online publication. https://doi.org/10.1088/1361-648X/ac3075

Kalita, E., & Baruah, J. (2020). Environmental remediation. In V. M. Khot & A. B. Salunkhe (Eds.), *Colloidal metal oxide nanoparticles* (pp. 525–576). Elsevier. https://doi.org/10.1016/b978-0-12-813357-6.00014-0

Khin, M. M., Nair, A. S., Babu, V. J., Murugan, R., & Ramakrishna, S. (2012). A review on nanomaterials for environmental remediation. *Energy and Environmental Science, 5*(8), 8075–8109. https://doi.org/10.1039/c2ee21818f

Koech, L., Rutto, H., & Seodigeng, T. (2023). Experimental evaluation of lime spray drying for SO_2 absorption. *AIMS Environmental Science, 10*(3), 325–340. https://doi.org/10.3934/environsci.2023019

Li, Y.-H., Wang, S., Cao, A., Zhao, D., Zhang, X., Xu, C., Luan, Z., Ruan, D., Liang, J., Wu, D., et al. (2001). Adsorption of fluoride from water by amorphous alumina supported on carbon nanotubes. *Chemical Physics Letters, 350*(5–6), 412–416. https://doi.org/10.1016/S0009-2614(01)01351-3

Li, D., Cui, F., Zhao, Z., Liu, D., Xu, Y., Li, H., & Yang, X. (2014). The impact of titanium dioxide nanoparticles on biological nitrogen removal from wastewater and bacterial community shifts in activated sludge. *Biodegradation, 25*(2), 167–177. https://doi.org/10.1007/s10532-013-9648-z

Lu, C., & Chiu, H. (2006). Adsorption of zinc (II) from water with purified carbon nanotubes. *Chemical Engineering Science, 61*(4), 1138–1145. https://doi.org/10.1016/j.ces.2005.08.007

Mallikarjunaiah, S., Pattabhiramaiah, M., & Metikurki, B. (2020). Application of nanotechnology in the bioremediation of heavy metals and wastewater management. In *Nanotechnology for food, agriculture, and environment* (pp. 297–321). Springer. https://doi.org/10.1007/978-3-030-31938-0

Pal, S., Tak, Y. K., & Song, J. M. (2007). Does the antibacterial activity of silver nanoparticles depend on the shape of the nanoparticle? A study of the gram-negative bacterium *Escherichia coli*. *Applied and Environmental Microbiology, 73*(6), 1712–1720. https://doi.org/10.1128/AEM.02218-06

Park, J.-Y., & Lee, I.-H. (2014). Photocatalytic degradation of 2-chlorophenol using Ag-doped TiO_2 nanofibers and a near-UV light-emitting diode system. *Journal of Nanomaterials, 2014*, Article 250803. https://doi.org/10.1155/2014/250803

Patel, K. D., Singh, R. K., & Kim, H.-W. (2019). Carbon-based nanomaterials as an emerging platform for theranostics. *Materials Horizons, 6*(3), 434–469. https://doi.org/10.1039/C8MH00966J

Poguberović, S. S., Krčmar, D. M., Maletić, S. P., Kónya, Z., Pilipović, D. D. T., Kerkez, D. V., & Rončević, S. D. (2016). Removal of As(III) and Cr(VI) from aqueous solutions using "green" zero-valent iron nanoparticles produced by oak, mulberry and cherry leaf extracts. *Ecological Engineering, 90*, 42–49. https://doi.org/10.1016/j.ecoleng.2016.01.083

Rasalingam, S., Peng, R., & Koodali, R. T. (2014). Removal of hazardous pollutants from wastewaters: Applications of TiO_2–SiO_2 mixed oxide materials. *Journal of Nanomaterials, 2014*, Article 617405. https://doi.org/10.1155/2014/617405

Seredych, M., & Bandosz, T. J. (2007). Removal of ammonia by graphite oxide via its intercalation and reactive adsorption. *Carbon, 45*(10), 2130–2132. https://doi.org/10.1016/j.carbon.2007.06.007

Singh, J., Dutta, T., Kim, K. H., Rawat, M., Samddar, P., & Kumar, P. (2018). Green synthesis of metals and their oxide nanoparticles: Applications for environmental remediation. *Journal of Nanobiotechnology, 16*(1), 84. https://doi.org/10.1186/s12951-018-0408-4

Song, R., Wang, Q., Mao, B., Wang, Z., Tang, D., Zhang, B., Zhang, J., Liu, C., He, D., & Wu, Z. (2018). Flexible graphite films with high conductivity for radio-frequency antennas. *Carbon, 130*, 164–169. https://doi.org/10.1016/j.carbon.2018.01.019

Sreeja, S., & Vidya Shetty, K. (2016). Microbial disinfection of water with endotoxin degradation by photocatalysis using Ag@TiO_2 core shell nanoparticles. *Environmental Science and Pollution Research, 23*(18), 18154–18164. https://doi.org/10.1007/s11356-016-6841-8

Thorat, D., Tripathy, S., & Shimpi, N. G. (2024). Graphene: A wonder material of the twenty-first century. In D. P. Das & N. G. Shimpi (Eds.), *Graphene-based photocatalysts* (pp. 3–23). Springer; *Advanced Structured Materials*, 217. https://doi.org/10.1007/978-3-031-66260-7_1

Tripathy, S., Rodrigues, J., & Shimpi, N. G. (2023). Nano-biomaterials: Perspectives for medical applications in the diagnosis and treatment of diseases. In *Biomedical applications of nanostructured materials* (Vol. 145, pp. 92–130). Materials Research Forum LLC. https://doi.org/10.21741/9781644902370-4

Tripathy, S., Rodrigues, J., & Shimpi, N. G. (2024a). Fabrication of nanofillers: The future research. In S. Thomas, N. G. Shimpi, & P. Mishra (Eds.), *Handbook of nanofillers* (pp. 1–29). Springer Nature. https://doi.org/10.1007/978-981-99-3516-1_22-2

Tripathy, S., Rodrigues, J., & Shimpi, N. G. (2024b). Silks. In *Reference module in materials science and materials engineering*. Elsevier. https://doi.org/10.1016/B978-0-323-95486-0.00121-6

Tripathy, S., Rodrigues, J., & Shimpi, N. G. (2025). Hydrothermal-ultrasonication synthesis of $NiFe_2O_4$ nanospheres for temperature-controlled xylene gas sensor. *ChemistrySelect*, e202500146. https://doi.org/10.1002/slct.202500146

Vaseashta, A., Dimova-Malinovska, D., Berman, A., Walton, J., Snow, E., & Langley, D. (2007). Nanostructures in environmental pollution detection, monitoring, and remediation. *Science and Technology of Advanced Materials, 8*(1), 47–59. https://doi.org/10.1016/j.stam.2006.11.003

Xiong, Z., Ma, J., Ng, W. J., Waite, T. D., & Zhao, X. S. (2011). Silver-modified mesoporous TiO_2 photocatalyst for water purification. *Water Research, 45*(7), 2095–2103. https://doi.org/10.1016/j.watres.2010.12.019

Xiu, Z., Zhang, Q., Puppala, H. L., Colvin, V. L., & Alvarez, P. J. J. (2012). Negligible particle-specific antibacterial activity of silver nanoparticles. *Nano Letters, 12*(8), 4271–4275. https://doi.org/10.1021/nl301934w

Xu, Y., Zhang, W., Huang, X., Shi, J., Zou, X., Li, Z., & Cui, X. (2019). Adsorptive stripping voltammetry determination of hexavalent chromium by a pyridine functionalized gold nanoparticles/three-dimensional graphene electrode. *Microchemical Journal, 149*, Article 104022. https://doi.org/10.1016/j.microc.2019.104022

Younas, H., Qazi, I. A., Hashmi, I., Ali Awan, M., Mahmood, A., & Qayyum, H. A. (2014). Visible light photocatalytic water disinfection and its kinetics using Ag-doped titania nanoparticles. *Environmental Science and Pollution Research, 21*(1), 740–752. https://doi.org/10.1007/s11356-013-1980-7

Yunus, I. S., Wang, J., Wang, Y., & He, W. (2012). Nanotechnologies in water and air pollution treatment. *Environmental Technology Reviews, 1*(1), 136–148. https://doi.org/10.1080/21622515.2012.733966

Bacterial Exopolymeric Substances (EPS): Promising Biomaterials for Synthesis of Biogenic Nanoparticles for Environmental Bioremediation

Yostnamayee Panda, Alaka Sanghamitra, Sonali Sandipta Dash, and Dhananjay Soren

Abstract Bacterial exopolymeric substances (EPS) is primarily composed of exopolysaccharides, truncated peptides, and fragmented small nucleic acids with some unknown polyanionic substances. The high biodegradable and biocompatible nature of biofilms provide a potential platform for nanoparticle synthesis. Currently the microbial exopolysaccharides are widely being investigated as biomaterials for biogenic synthesis of nanoparticles. Metal ions undergo a sequence of transformation on an EPS matrix to reduce its polymerisation, well known as metal chelators. Nano-biotechnology is making significant strides across various scientific applications like drug delivery, biosensors, and vaccine development. Among these the most challenging and emerging application includes wastewater treatment, where biogenic nanoparticles in the form of biopolymeric metallic nanomaterials are gaining attention for their potential in bioremediation. These nanoparticles benefit from a high surface area to volume ratio, enhancing their catalytic activity for treating contaminated water. Biogenic nanotechnology merges biology and material science, utilizing nanoparticles that possess unique qualities due to their size similarity to biomolecules like proteins and nucleic acids. This method of nanoparticle synthesis yields particles with favourable characteristics, including good polydispersity, size variability, and stability. However, challenges such as lengthy reduction times, complex processing, and aggregation of biogenic metal nanoparticles are inevitable phenomenon. To mitigate these issues bacterial EPS can serve dual roles as reducing and capping agents during nanoparticle synthesis. The charged functional groups in EPS can effectively bind metal ions, facilitating entrapment of various environmental contaminants, including those from the pharmaceutical and textile industries. This makes bacterial EPS a promising alternative for enhancing wastewater treatment efficiency.

Y. Panda · A. Sanghamitra · S. S. Dash · D. Soren (✉)
Enzymology and Cancer Biology Laboratory, Department of Zoology, Ravenshaw University, Cuttack, Odisha, India
e-mail: dsoren@ravenshawuniversity.ac.in

Centre of Excellence in Environmental and Public Health, Department of Zoology, Ravenshaw University, Cuttack, Odisha, India

E. Priyadarshini et al. (eds.), *Advanced Nanotechnology Applications for Environmental Sustainability*, Sustainable Environmental Waste Management Strategies,
https://doi.org/10.1007/978-3-032-12933-8_2

Keywords ExoPolymeric Substances (EPS) · Biogenic nanoparticle · Bioremediation

1 Introduction

One of the most challenging and deteriorating worldwide issue in this twenty-first century is environmental pollution. Contaminating air, water, soil of our ecosystem, it affects human health and biodiversity. Some of it contains dangerous pollutants like heavy-metal including lead which affect central nervous system, arsenic and cadmium known for their carcinogenic effect can causes pulmonary system disorder, where polycyclic aromatic hydrocarbons (PAHs) function as mutagen and can act as immune system suppressors (Awewomom et al., 2024). Traditional clean-up techniques include chemical treatments; physical procedures like expensive engineering solutions with energy-intensive process may occasionally lead to formation of environmentally hazardous by products or substances (Mishra et al., 2024; Priyadarshini et al., 2024; Reddy Koteswara et al., 2025). Consequently, there is a rising interest in pollution mitigation strategies that are more ecological friendly, efficient and sustainable. A more natural and economical method of cleaning up the environment is bioremediation which uses microorganisms or their by-products to break down, neutralize or change its toxic effects (Kuppan et al., 2024). Nanotechnology has appeared as an alternative way to remove toxins from water bodies. Conventional methods of producing nanoparticles like oxidation and reduction are known as chemical processes; likewise condensation and evaporation are the physical processes have shown to be highly effective. However, these technologies do have certain drawbacks which include high energy need and usage of expensive and dangerous compounds that may cause nanoparticles to stick to surfaces. Scientists are constantly developing accessible, affordable and environmentally safe alternatives to these limitations (Nkosi et al., 2024). Bacterial EPS acts as a good bioremediation agent among various other agents due to its biocompatibility and versatile properties. Bacterial EPS are extracellular polymeric secretions of bacteria that form biofilms and present themselves inside it, to hide themselves from external challenges. Microbial biofilm exhibits both structural and physiological complexity, functioning as a cohesive entity composed of several microbial communities. The biofilm secreted by microbes typically grows on a substrate and is submerged in an extracellularly secrets polymeric substance that the bacteria produce. In order to survive in hostile environments, both autotrophic and heterotrophic microorganisms create stable biofilms (Singh et al., 2021). EPS matrix primarily composed of polysaccharides along with protein, lipid and nucleic acid. Additionally, EPS are essential for improving microbial communication, boosting nutrient uptake and allowing bacteria to flourish under harsh conditions (Karygianni et al., 2020). In addition to its biological roles, EPS has attracted more interest recently due to its potential as a crucial component in the production of biogenic nanoparticles (NPs). Many bacterial species generate EPSs, which have been utilized as a possible model for the quick manufacture of metallic nanoparticles. Since the reduction of metal through bacterial secreted EPS are easy to employ, safe and environmentally benign, they become more popular. Metal ions, organic substances and contaminants could all be readily interacting with bacterial EPS. Therefore, it is utilized in the production of biogenic nanoparticles which are essential for environmental bioremediation (Sathiyanarayanan et al., 2017a).

EPS helps in reducing, stabilizing and capping of the metals when they undergo the process of synthetization into nanoparticles like silver, iron oxide, zinc oxide, copper, gold nanoparticles etc. Green chemistry approach delivers the biosynthesis of nanoparticles utilizing EPS with a significant advantage over conventional chemical synthesis methods. The resultant nanoparticles are non-toxic, biodegradable and more sustainable since they are produced using low-energy, ecologically friendly techniques that do not call for severe conditions or hazardous chemicals (Escárcega-González et al., 2018).

Bacterial EPS-synthesised biogenic nanoparticles have characteristics nature such a large surface area, strong reactivity and ease of function for its particular uses. These characteristics make them very useful for adsorption, breaking down and transformation of contaminants to simpler molecules in environmental bioremediation. Likewise, by binding to organic contaminants, heavy metals and other hazardous elements, the biogenic nanoparticles can reduce their harmfulness or remove them from polluted settings. Additionally, by enhancing the activity of microorganisms engaged in pollution breakdown, these biogenic nanoparticles can foster synergistic interactions between chemical and biological remediation processes (Borah et al., 2025). Despite the great potential of nanoparticle mediated bio-remediation, the use of bacterial EPS for environmental bioremediation and nanoparticle production is still in its initial stages. The aim of this book chapter is to provide detail exploration towards role of bacterial EPS as a biomaterial for synthesis of biogenic nanoparticles with particular focus on their applications in environmental bioremediation.

2 Bacterial EPS

Biofilm architecture in EPS is maintained with extracellular secretion by the bacteria under aqueous condition. EPS is primarily composed of proteins or peptides, polysaccharides, nucleic acids, lipids and water. EPS acts like a glue to adhere the bacteria with its substratum and also provides communication with other microbes for associative aggregation, antioxidants defence, and maintenance of carbon storage as resources through nutrients entrapment etc. (Flemming et al., 2025). Bacterial EPS displays extensive antimicrobial resistance towards antibiotics and antimicrobial agents due to its shielding architecture design above single or mixed bacterial species. However, the antibacterial properties of EPS depend on its molecular weight and chain length. The bigger EPS molecules transform from double helical forms to macroscopic gels, which results in high viscosity. The viscosity-related biological activity will thus enhance the antibacterial action (Zhou et al., 2019). The extracellular polymeric substance usually constituted with polysaccharides, proteins and

DNA. It also contains certain extent of humic acids and lipids. The investigation about composition of EPS in numerous strains of microorganism demonstrated that there is variable extent of biopolymer present. These overall components found in EPS collectively acting as a protective barrier for encapsulated bacterial cells (Olaimat et al., 2024).

Bacterial exopolysaccharides are mainly two types; homo-polysaccharides which contain same repeating unit of monosaccharides and hetero-polysaccharides containing two or more units of different monosaccharides. Homo-polysaccharides are mainly cellulose, dextran, crudlan, inulin and levan. Hetro-polysaccharides are xanthan, hyaluronic acid, alginate, kefiran etc. Based on its function bacteria EPS can be constructive, absorptive, surface active, redox active, informative and nutritive types (Kaur & Dey, 2023). Along with these exopolysaccharides, the alteration of EPS by group substitution to modify the structure of polysaccharides and promote specific biological activity has been highlighted as an emerging trend. Carboxymethylation, acetylation, phosphorylation and sulfonation are among the current strategies for polysaccharide modification (Wang et al., 2023).

3 Bacterial EPS Mediated Biogenic Nanoparticle Synthesis

The name "EPS" refers to extracellular materials generated by bacteria, which are mostly contain polysaccharides, proteins, lipids and nucleic acids. These complex biomolecules serve important roles in the synthesis, stability and reduction of metal ions for formation of nanoparticles. The use of bacterial EPS for nanoparticle synthesis has various benefits, including eco-friendliness, cost-effectiveness and the possibility of ecologically sustainable operations. Bacterial EPS has proved a good capping and reducing agent for metal nanoparticle synthesis over other biological synthesizer (Ahmed et al., 2014). Metal nanoparticles have various applications in the fields like medicine, environmental science and agriculture. Biofilm research on their synthesis methods has quickly grown over recent years (Kato & Suzuki, 2020). Gold and silver metal nanoparticles have mostly been used in sensing and biomedical application like imaging processing using magnetic nanoparticles, where silver nanoparticles are used for photo detective purpose. In past pandemic years gold nanoparticles have been used in various purposes like detection of COVID-19 virus and light induced magnetisim (Cheng et al., 2020).

4 Mechanism Behind EPS Mediated Nanoparticle Synthesis

Synthesis of metal nanoparticles through bacterial exopolysaccharides usually involves two major steps such as (i) Reduction and (ii) Stabilization (Sharma et al., 2015). A mechanism of nanoparticle synthesis has been elaborated (Fig. 1).

Reduction is the process of controlled synthesis of metal nanoparticle with variable size and structural assembly. This process of reduction requires reducing agents where bacterial exopolysaccharides like organic macromolecules show an excellent material for reduction of nanoparticles from metal ions (Mohanty et al., 2025; Priyadarsini & Das, 2024; Sharma et al., 2019).

Bacterial exopolysaccharides contain various negatively charged groups known as poly-anionic groups. So, they have the capacity to bind the metals due to their functional moieties and can form polysaccharide-metal complex. There is a strong electrostatic attraction develops between metal cationic ions and anionic groups of EPS such as carboxylic, phosphoric groups, respectively (Raveendran et al., 2015). In the process of reduction from metal ions to metal nanoparticles, hydroxyl group present in EPS oxidized to carbonyl groups and alcoholic groups; and aldehyde groups oxidized to carboxylic groups which are known to be the most important chemical process undergo during the process of nanoparticles synthesis. During nanoparticles synthesis through a size reduction procedure, the EPS matrix provides a thin sheet that serves as a stabilizing and capping agent. So that they form a tight size distribution and prevent clump with each other. Similarly, the mechanism of EPS attachment

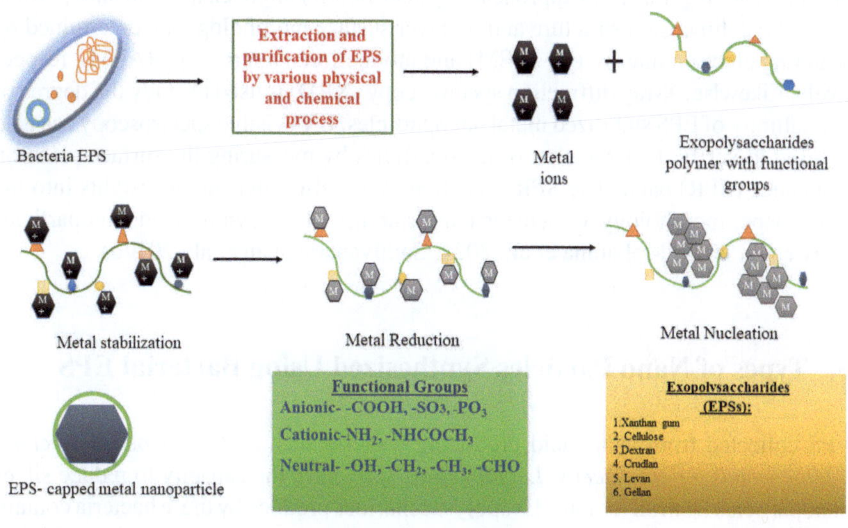

Fig. 1 General Illustration of Bacterial EPS as a reducing, stabilizing and nucleating agent to produce metal nanoparticles. Metal ions can be coordinated by the functional groups of the exopolysaccharide

to the surface of the nanoparticles occurs through a steric repulsion between them which impede their aggregation (Escárcega-González et al., 2018). The aggregation step is also known as nucleation steps where metals ions comes together to form nucleus. Bacterial EPS play a crucial role in the nucleation process where it provides a template for metal ion nucleation because it allows metal ions to bind to specific sites on the EPS molecules. In a study of gold nanoparticles formation different EPS like surfactants with varying hydrophobicity and hydrophilicity are tested to study their effect on shape, size and their distribution. EPS with stronger binding affinity to the nanoparticles surfaces tend to produce more stable well defined nanoparticles (Suárez-López et al., 2022). Since EPS has a muco-adhesion properties metal nanoparticles in presence of EPS promotes a neutral low surface energy and low level of non-specific protein receptor recognition capping which results synthetization of biogenic nanoparticles for versatile applications (Kanmani & Lim, 2013). Some study also states that monosaccharides like glucose, galactose, rhamnose have hemiacetyl reducing ends which perform reduction of metals and formation of nanoparticles. Synthesis of biogenic nanoparticles occurs using EPS are characterized through various microscopic and spectroscopic analysis. Polysaccharides functional group present in nanoparticles is detected through Fourier transform infrared spectroscopy (FTIR), because metal nanoparticles get functional groups during metal reduction and subsequent stabilization process. Though bacteria EPS contains mostly polyanionic functional groups, EPS stabilized nanoparticles exhibit net negative charges which can be measures by zeta potential analyser. Similarly, the average value of nanoparticle size and homogeneity of colloidal solution is determined by dynamic light scattering (DLS) principle, known as photon correlation spectroscopy. Due to polymer capping, this DLS approach may also identify slight changes in nanoparticle size. Three dimension structure and outlayer surface morphology are determined by scanning electron microscopy (SEM) and atomic force microscopy (AFM), respectively. Likewise, X-ray diffraction spectroscopy (XRD) is used to study the degree of crystallinity of EPS stabilized metal nanoparticles. UV–visible spectroscopy can also be used to verify the formation of nanoparticles by measuring the surface plasmon resonance (SPR) band. The SPR transition might also offer minor insights into the size, shape, morphology, dimensions and stability of the synthesized nanoparticles (Dey et al., 2024; Moharana et al., 2024; Sathiyanarayanan et al., 2017b).

5 Types of Nano Particles Synthesized Using Bacterial EPS

EPS collected from lactic acid producing bacteria such as *Enterococcus faecium, Pediococcudszs pentosaceus, Lactobacillus spp.*, have the capacity to reduce silver nanoparticles from silver ions. Exopolysaccharides produce by these bacteria contain different monomers like manose, glucose, galactose and fructose which participate in the process of oxidation and reduction reaction during synthesis of silver nanoparticles. In a study when partly purified EPS (10 mg) had been utilized to form a homogenous solution in MilliQ water (10 ml) with $AgNO_3$ (9 mM) in constant stirring for overnight in dark, the dispersion of solution resulted with change in colour from colourless to yellow, suggesting the synthesis of polymeric silver nanoparticles where EPS contain only glucose monomer (Saravanan et al., 2017). Similar research was reported where gold nanoparticles synthetization occurs using EPS extracted from bacterium *Bacillus megaterium* MSBN04, where the EPS act as a good reducing and stabilizing agent for synthesis of gold nanoparticle (Au–NP). Transmission Electron Microscopic image showed that the crystalline spherical structure of Au–NPs capped inside the EPS and its synthesis depends on the concentration of EPS and synthesized nanoparticles has been shown for antimicrobial activity against various disease forming pathogenic bacteria (Sathiyanarayanan et al., 2014). The EPS extracted from one of the spore forming strain *Bacillus subtilis* VT03 present inside the intestine of fresh water fish *Oreochromis mossambicus* contain glucan as per FTIR and NMR report which provided a good surface area and act as an excellent stabilizer and reducing agent for ecofriendly iron oxide nanoparticles (FeONPs) formulation. This EPS stabilized FeONPs had shown inhibitory activity against bacteria infecting both human and fish (Vignesh et al., 2015). Actinobacterial strain EPS was also used to produce ZnO–NPs (zinc oxide nanoparticles) in a more environmentally friendly manner. The crystalline ZnO–NPs nanoparticles that were created have a size range of 18–40 nm. The resulting zinc oxide nanoparticles have demonstrated efficacy against inflammation and significant antioxidant activity. Additionally, ZnO–NPs demonstrated excellent bactericidal activity against clinically harmful bacterial strains (Rajeswaran et al., 2023). Another study where *Enterococcus faecium* MC-5 bacterial exopolysaccharides help in production of selenium nanoparticles. The –OH group of EPS interact with selenium nanoparticle which resulted into an excellent C–O–Se bond formation for SeNP and its dispersal in EPS matrix. This works reveals that bacterial EPS has a critical role as a biologically suitable polymer template for SeNP for dimension and dispersion management where the selenium nanoparticle conjugate with EPS has been observed to scatter in the range of 76 nm size (Tilwani et al., 2023).

The usage of silver nanoparticles (AgNPs) made from exopolysaccharides (EPS) extracted from two alkaliphilic bacteria, *Bacillus krulwichiae* M2.5 and *Bacillus cellulosilyticus* M4.1, which were isolated from a hyper alkaline spring in the Philippines. The synthesized AgNPs were characterized using various biophysical methods like UV–Vis spectroscopy, FTIR (Fourier Transform Infrared Spectroscopy), SEM (Scanning electron Microscope) and EDX (Energy Dispersive X-ray Spectroscopy) were used to characterize the AgNPs and measured around 25.88 ± 10.49 nm and 23.99 ± 8.43 nm, respectively. The AgNPs remarkably affect the formation of biofilm in *Klebsiella pneumoniae, Pseudomonas aeruginosa* and *Staphylococcus aureus* in a dose-dependent manner, where their increased efficacy towards inhibition of existing biofilms had been observed with higher AgNP concentrations. The work emphasizes the potential of alkaliphilic bacteria's EPS for biosynthesizing AgNPs with biological applications, notably in the treatment of biofilm-related illnesses (Galvez et al., 2019). The exopolysaccharide EPS B3 15, EPS T14 and poly-γ-glutamic acid γ-PGA-APA isolated from thermophilic bacteria of the Mediterranean Eolian Islands (Italy) hydrothermal vent has been used as both stabilizing and reducing agents for synthesis of gold, silver and silver chloride nanoparticles (AuNPs and Ag/AgCl NPs) (Scala et al., 2019). The exopolysaccharide collected from *Lactobacillus fermentum* help to aggregate gold nanoparticles of different characters (spherical, rods and prisms). The bacteria *L. fermentum* exopolysachharides helps to aggregate gold from Au (lll) solution because gold nanoparticles have high affinity towards bacterial EPS (González et al., 2019). The demand of new synthetic strategies in biomedicine is increasing gradually since last decades through application of eco-friendly, non-toxic green technology. The application of magnesium oxide based nanoparticles (MgO NPs) in healthcare system have recently gained an enormous attention for its unique and versatile biochemical features like high stability with redox potentials, wide range of cationic capacity, biocompatibility and biodegradability. Bacteria like *Lactobacillus plantarum* and *Lactobacillus sporogenes* extract were recently been used as bio-template for synthesis of magnesium oxide nanoparticles (MgONPs) (Ramezani Farani et al., 2023). According to this study nanoparticie are produce not in the presence of bacterial exopolysaccharides but inside the bacterial extra-cellular surface. A bacterium *Bacillus megaterium* was utilized to produce silver (Ag), lead (Pb) and cadmium (Cd) nanoparticles extracellularly. After being exposure to various differential concentrations of inorganic solvents like silver nitrate, lead nitrate and cadmium nitrate solutions, nanoparticles have been accumulated on bacterial cell wall. The cell lysate collected after centrifugation was analyzed by UV–Vis spectroscopy, X-ray diffraction (XRD) analysis, transmission electron microscopy (TEM) and energy dispersive spectroscopy (EDS). The nanoparticles had absorption maxima at around 435 nm (AgNPs), 330 nm (CdNPs) and 410 nm (PbNPs). XRD confirms the production of Ag, PbS and CdS particles. TEM examination indicated that the size of nanoparticles varied within 10 and 20 nm (Prakash et al., 2010).

6 Nanoparticles Synthesis from Some Selected Bacterial Exopolysaccharides

Some known exopolysaccharide producing from bacteria are Gellan, Dextran, Xanthan, Crudlan, Hyaluronic acid, cellulose, starch, xylan, alginate etc. (Table 1). These carbohydrate polymers are homo and hetro-polysaccharides which are mostly composed glucose, galactose and mannose as monomeric units. In addition, neutral rare sugars like fucose, rhamnose, uronic acids and amino sugars were also observed in bacterial EPSs as monomeric units.

Table 1 Bacterial EPS synthesized nanoparticles

Bacterial EPS	Metallic nanoparticle	Application	References
Gellan gum	Au–NP	Drug delivery and biomedical application	Dhar et al. (2008), Dhar et al. (2011)
	Ag–NP	Biomedical application	Dhar et al. (2012)
	Fe_3O_4–NP	Drug delivery carrier, anticancer	Sivakumar et al. (2014)
Dextran	Ag–NP	Antibacterial efficacy	Hussain et al. (2014)
	Fe_3O_4–NP	Biocompatibility	Shaterabadi et al. (2017)
	Au–NP	Biocompatibility	Wang et al. (2010)
Hyaluronic acid	Au–NP	Drug delivery	Kim et al. (2023)
	Fe_3O_4–NP	Biomedical application	
Cellulose	Au–NP	Catalysts, environmental remediation	Yan et al. (2016)
	Ag–NP	Antibacterial, antioxidant and photocatalytic activities	Visagamani et al. (2023)
	CdS–NPs	Nanofibers, thermal stability	Li et al. (2009)
Alginate	Ag–NP	Sensing low concentration of pollutant at water	Bhagyaraj and Krupa (2020)
	Au–NP	Biomedical Application	Liu et al. (2022)
Lactic Acid producing bacterial (EPS)	Ag–NP	Degradation of organic azo-dyes molecules	Saravanan et al. (2017)
Bacillus subtilis VT03 EPS	FeO–NP	Antibacterial activity against pathogenic strain	Vignesh et al. (2015)
Actinobacterial strain EPS	ZnO–NP	Bactericidal action against clinical pathogenic bacterial strains	Rajeswaran et al. (2023)
Enterococcus faecium MC-5 EPS	Se–NP	Antimicrobial activities, biomedical and cosmetics industry application	Tilwani et al. (2023)

(continued)

Table 1 (continued)

Bacterial EPS	Metallic nanoparticle	Application	References
Bacillus cellulosilyticus M4.1 and *Bacillus krulwichiae* M2.5 EPS	Ag–NP	Anti-biofilm activity against *Pseudomonas aeruginosa, Klebsiella pneumoniae* and *Staphylococcus aureus*	Galvez et al. (2019)
Lactobacillus plantarum and Lactobacillus sporogenes	MgO–NP	Biocompatible and biodegradable	Ramezani Farani et al. (2023)

6.1 Gellan

Gellan gum is a microbial exopolysaccharide act as reducing and simultaneously as stabilizing agent for the synthesis of gold nanoparticles used for drug delivery agents and in many other biomedical uses. During assessment of its toxicity effect through oral administration route, it was found to be non-toxic for oral use. The results suggest that Au nanoparticles reduced by gellan gum could specifically target cancer cells since they are non toxic in nature to animals when administered orally (Dhar et al., 2011).

In the present situation among all metallic nanoparticles, gold nanoparticles have high non cytotoxicity and biocompatibility nature with unique dimension and surface dependent properties. It has wide biomedical application. These findings explore the green synthesis and stabilization of gold nanoparticles (AuNPs) with the utilization of natural bacterial exopolysaccharide gellan gum. These stabilized AuNPs were used to load the chemotherapeutic drug doxorubicin hydrochloride (an anthracycline antibiotic), which expressed cytotoxic properties on glioma cell lines LN-18 and LN-229 (Human). This study highlights the gellan gum based reduction of gold nanoparticles as potentially effective medication delivery vehicle, offering improved stability and efficacy for cancer treatment (Dhar et al., 2008).

The gellan gum is also used to synthesize silver nanoparticles as reducing and capping agent. The gel forming characteristics of gellan gum is exploited to prepare hydrogels that incorporate silver nanoparticles. These hydrogels could have applications in wound healing or controlled release of silver for antimicrobial purposes. After synthesized AgNPs are characterized to various spectroscopic analysis, its cytotoxic effects was also investigates in mouse embryonic fibroblast cells (NIH3T3), since cytotoxicity is a crucial determinants of the safety of nanomaterials for biological applications (Dhar et al., 2012).

6.2 Dextran

Dextran is an exopolysaccharide secreted by bacteria which is a hydrophilic and water soluble substance. This exopolysaccharide is synthesized by the bacterial cell depending on its requirement during variable environmental conditions. Dextran is a natural macromolecule made up of α-(1–6) connected D-glucose units with different branching. This works explored about biosynthesis of dextran (Dx) mediated silver nanoparticles (AgNPs) as a stabilizer. The AgNPs were prepared through a chemical reduction method, where three different concentrations of dextran such as 0.4, 0.8 and 1.2% (w/v) were used under controlled temperature (80 °C) conditions. Silver nitrate ($AgNO_3$) served as the precursor for silver nanoparticles formation where dextran (Dx) has been utilized as both capping and steady agent under diffused sunlight conditions (Kavitha et al., 2015). Furthermore, after drying, the dextran-stabilized Ag NPs solution produced thin coatings that remained stable for months without altering the pure AgNPs' plasmon band. Thereby synthesised AgNPs show very impressive efficacy in the antibacterial experiment (Hussain et al., 2014).

Dextran as EPS mediated synthesis of gold nanoparticle and its toxicological effect in mammalian cells has been investigated. They developed room temperature strategies for preparation of biocompatible AuNPs on dextran based polysaccharide cap for acting as both reducing agent and stabilizer. Surface chemistry of nanoparticle is the primary contributor of its toxicity to the cell. However, the dextran, which act as a stabilizer and reducing agent for AuNPs synthesis provide an excellent biocompatibility to the cell as compared to citrate capped and CTAB capped AuNPs (Wang et al., 2010).

6.3 Hyaluronic Acid

Hyaluronic acid is a natural linear polymer secret by bacteria *Streptococcus* sp. and *Bacillus* sp. It has wide application in cosmetics and pharmaceuticals industry primarily for drug deliver due to its biocompatible nature (Waoo et al., 2023). Extensive studies have been conducted on creation of gold nanoparticles mediated by hyaluronic acid and their usage in medical administration. Additionally, this work highlights that HA-coated gold nanoparticles provide a potential oral, intestinally targeted drug delivery system with controlled release properties by choosing Sulfasalazine (SSZ), a poorly water-soluble drug (Kim et al., 2023). Due to their high r2 relaxivity, use of iron oxide nanoparticles, utilized in magnetic resonance imaging (MRI) have garnered much attention. This iron oxide nanoparticles (Fe_3O_4) is stabilized by hyaluronic acid and synthesized by electrostatically adsorbed onto the surface of the nanoparticles after hydrothermal co-precipitation. The water-soluble HA with more affinity for CD44 receptor that is over expressed in many tumors and serves as a targeted ligand in addition to being a used as stabilizer. Transmission electron microscopy (TEM) imaging revealed that the resultant HA-stabilized Fe_3O_4

nanoparticles were sub-20 nm in size and demonstrated long-term colloidal stability in aqueous solution (Zhang et al., 2022).

6.4 Cellulose

Cellulose is naturally occurring eco-friendly biopolymer which is environmentally benign and biodegradable. Cellulose is made up of linear repeating units of 1,4-β glucan chains. It is structurally characterized by its strong intramolecular and intermolecular hydrogen bonding with the neighbouring glucose rings covalently linked in C1 and C4 intermolecular configuration. This research focuses on green synthesis of cellulose-mediated silver nanoparticles (cellulose/Ag NPs), that exhibit remarkable antibacterial, antioxidant, and photocatalytic activities. The generation of silver nanoparticles in this process has crystallinity and size range of 20–30 nm (Visagamani et al., 2023). Cellulose nano-crystals also supported by gold nanoparticles synthesized through simple and environmental friendly green synthesis process. The gold nanoparticles are used in cellulose nanocrystal formation to avoid use of toxic chemicals or through complex procedures, respectively. Some of the study demonstrates that cellulose nanocrystals not only act as stabilizing support for the gold nanoparticles but also enhance their catalytic activity. The CNCs/Au NPs exhibited high stability and reusability making them suitable for practical applications in catalysis and environmental remediation (Yan et al., 2016).

6.5 Alginate

Alginate is an exopolysaccharide produced by bacteria that is made up of straight chains α-l-guluronate (G) and β-d-mannuronate (M) units. It is a simple method in which natural polymer sodium alginate has been used as capping agent and glucose as reducing agent to synthesize hetero structured alginate-capped AgNPs.

It has been reported that the synthesised Ag-NPs were small (30 ± 18 to 45 ± 12 nm) and spherical during the initial (1–10) hours of the reaction and evolved to become hetero structured after 48 h with particle size reaching 60 ± 25 nm. The alginate-capped Ag-NPs showed excellent sensing activity towards detecting pollutants at low concentrations in water (Bhagyaraj & Krupa, 2020). This review highlights the synthesis process of gold nanoparticle which has shown potential biomedical application through alginate based reduction and stabilization process. Alginate, a biocompatible and natural polymer, was used to stabilize the gold nanoparticles. This is important for ensuring the colloidal stability of the nanoparticles, which is crucial for their potential biomedical applications. The introduction of alginate accelerates the rate of reaction which prevents aggregation and precipitation due to long time discharge properties through microwave-induced plasma-in-liquid process (MWPLP) and provided longterm colloidal stability (Liu et al., 2022).

7 Role of EPS Mediated Biogenic Nano Particle in Environmental Bioremedition

Bioremediation is a method that uses biological processes to reduce pollutant concentrations to pre-contamination levels. Green and sustainable nano technological approaches towards the production of nanoparticles and their use in environmental bioremediation have achieved significant concern nowadays (Azubuike et al., 2016). The application of microbial resources for the synthesis of cheap and cost effective, efficient and environment friendly nanoparticles with significant applications in controlling environmental pollution has been described briefly in this review. The recent research on biogenic nanomaterials for bioremediation has enormous potential towards sustainable, efficient and successful removal of environmental pollutions (Table 2).

A recent research provides a comprehensive idea about the green synthesis of comprehensive nanoparticle like iron oxide (Fe_2O_3) and zero-valent iron (ZVI) and their usage in controlling ecological pollution, precisely in water and soil remediation. Iron nanoparticles have been found to be used in de-chlorination procedures. Iron nanoparticles have also been employed in dye degradation. Several researchers have successfully used iron nanoparticles in the treatment of waste water remediation. Considering the use of iron nanoparticles in environmental pollution remediation, a synergetic link between nanotechnology and bioremediation (Nano-bioremediation) has been proposed (Anyanwu et al., 2021). The biogenic nanoparticles mainly function by adsorbing heavy metals, effectively removing them from the soil. These heavy metals like lead, arsenic and cadmium are the primary causative agents impeding significant risks towards plant growth and human health. So the adsorption process helps reduce soil toxicity, thus improving its quality for agricultural use (Jha et al., 2024).

A research on polychlorinated biphenyls (PCBs) toxicity illustrated that due to their long term persistent nature it poses as a threat for contamination of both soil and water system. Bioremediation of this chemical contaminated soil had been carried out by novel strategy, where electro-kinetic technology had been used in combination with xanthan gum which stabilized Pd/Fe nano-suspension provided highest removal efficacy of PCB (Fan et al., 2013). Since the size of nanoparticles formed was very tiny in nature it is easy to penetrate into the contaminated sites (Galdames et al., 2022). A study claims that new exopolysaccharides called EPS-605 self-aggregate to produce 88 nm diameter nanoparticles with spherical form where mannose, glucose and galactose contribute majority of this new EPS, undergone with a number of changes like phosphorylation, acylation, sulfation and carboxylation, as well as a strong negative charge. This EPS demonstrates exceptional bio-sorption capabilities, efficiently adsorbing heavy metals (Pb^2, Cu^2, Cd^2) and the dye methylene blue. Simultaneously, EPS has also been used as reductant for creation of silver and gold nanoparticles (Li et al., 2017). *Lactobacillus plantarum* R503 a bacteria secret an EPS-R503 which has been used for biogenesis of nanosheets made up of from

Table 2 Biogenic nanoparticle in environmental bioremediation

Biological source	Synthesized nanoparticles	Size	Applications	References
Aspergillus niger BSC-1, (Fungi)	Iron oxide Nanoparticles	20–40 nm	Removal of chromium like heavy metal through adsorption	Chatterjee et al. (2020)
Saccharomyces cerevisiae (Fungi)	Titania Nano powder	<21 nm	Removal of Chromium (VI) heavy metal	Choudhury et al. (2017)
D. radiodurans R1 strain (Polyextremophilic bacteria)	Magnetic Iron Nanoparticles	141.8 nm to 164.2 nm	Removal of Arsenic	Kim et al. (2019)
Chlorococcum sp. MM11 (Algae)	Iron Nanoparticles	20 to 50 nm	Hexavalent Chromium reduction and remediation to Trivalent Chromium	Subramaniyam et al. (2014)
Citrobacter freundii Y9 (Bacterial EPS)	Selenium Nanoparticles	–	Removal of mercury from ground water	Wang et al. (2018)
E. coli (Bacteria)	Zinc Oxide Nanoparticles	10 nm	Reduces the levels of harmful Bacteria, Dyes, and Heavy metals	Somu and Paul (2018)
Bacillus subtilis NCIB 3610, (EPS)	Carbon Nanoparticles	100 nm	Scavenging hazardous materials, pharmacophoric chemicals	Savadiya et al. (2024)
Endophytes	Iron oxide and Titanium dioxide	1-100 nm	Adsorb and neutralize contaminants in contaminated water	Manganyi et al. (2024)
Novel EPS 605	Silver and Gold Nanoparticles	88 nm	Bio-sorption capabilities, efficiently adsorbing heavy metals (Pb^{2+}, Cu^{2+}, Cd^{2+}) and the dye methylene blue	Li et al. (2017)

(continued)

Table 2 (continued)

Biological source	Synthesized nanoparticles	Size	Applications	References
Pseudomonas aeruginosa JP-11 (EPS)	Cadmium Sulfate Nanoparticles	–	Removal of toxic metal pollutant Cadmium from contaminated water bodies	Raj et al. (2016)
Bacillus stratosphericus (A07) (EPS)	Silver Nanoparticles	87 nm	Biodegradation of harmful Azo dyes like Methyl Orange and Congo Red	Anil et al. (2023)

manganese dioxide nanoparticles (MnO_2 NSs), where EPS-R503 can act as stabilizing and reducing agents, resulting in the EPS-MnO_2 nanocomposite. Similarly, with the addition of fluorescein this nanocomposite can be is converted into EPS-MnO_2–FL which possess extensive biocompatibility nature and could be used to detect glutathione (GSH) (Li et al., 2021). The marine bacteria *Pseudomonas aeruginosa* JP-11 which produces extracellular polymeric substances (EPS) and forms biofilms, is used to create cadmium sulfide nanoparticles (CdS–NPs). The biochemical process involves interaction of sulfur and cadmium ions in order to produce EPS-capped CdS–NPS, with electron-deficient cadmium atoms serving as requisite places to affix EPS as organic capping agent. The combined effects of EPS and sulfur produced CdS nanocrystals and by stabilizing their surface energy, EPS could prevent growth of CdS–NPs crystals grains from growing further. Additionally, the creation of cadmium sulphide nanoparticles mediated by bacterial EPS aids in the dismissal of the detrimental effect of cadmium metal pollution from aqueous solution (Raj et al., 2016). Bacteria *Bacillus stratosphericus* (A07), produces EPS which used in synthesis of silver nanoparticles is sized about 87 nm. Additionally, these EPS stabilizes and biosynthesis the metal nanoparticles for appropriate, affordable and environmental friendly benign option towards biodegradation of hazardous Azo based dyes like Congo Red and Methyl Orange, among others. These biodegradation techniques primarily employ the transfer of electron from the reducing agent to dye molecule, which is aided by nanoparticles and results in the destruction of the dye chromophore (Anil et al., 2023). Nanoparticles are applied in bioremediation of organic as well as inorganic pollutants of aquatic ecosystems as adsorbents, sensors, membranes, nano-filters, nano-catalysts etc., and the efficiency of nanomaterial for bioremediation is dependent on various environmental factors like pH, temperature, particle size and shape, contact duration and absorbent dosage (Kaul & Sharma, 2022). The remediation of pollutants has been greatly affected by nanotechnology. The effectiveness and potential of the decontamination process are significantly enhanced by nano-bioremediation. This green synthesized nanoparticle mediated nano-bioremediation process has attracted the attention of scientists in the past few years and has demonstrated encouraging outcomes in managing detrimental contaminants from soil and water through the use of nanoparticles. The interaction between

EPS-mediated nanoparticles and pollutants is central to their role in bioremediation. The nanoparticles can enhance the process of pollutant removal through various mechanisms, including adsorption, reduction and catalytic degradation (Torimiro et al., 2021).

The extracellularly secreted polysaccharide by bacteria recently been identified as a viable bio-ingredients for production of silver nano-particles (AgNP) production. EPS contains poly-negatively charged functional groups (phosphate, hydroxyl, sulfate and carboxylic), which efficiently decrease and maintain AgNPs. EPS-facilitated AgNPs have wide spectrum of antimicrobial action in contradiction of a several harmful pathogens, such as fungus and along with bacteria both from Gram-positive and Gram-negative strains bacteria (Dey et al., 2023). EPS-mediated nanoparticles primarily aid bioremediation by adsorbing deleterious heavy metals from soil or water. The chain of exopolysaccharides carry functional groups which helps to bind ionic group of heavy metals like arsenic (As), mercury (Hg), cadmium (Cd), effectively eliminating them from the environment. Similarly, due to their larger surface area, EPS-synthesised nanoparticles usually bind more metal ions (Ghosh et al., 2024). Carbon based nano-materials like graphene and activated carbon, metal oxides and composite materials are being developed to target metal ions in contaminated water sources. Microstructure of nanoparticles such as size, surface functional group which is mostly provided by EPS and surface charge greatly affects the adsorption performance of nano-materials. Smaller particles typically have a higher surface area, leading to improved adsorption efficiency (Luo et al., 2021). One of the recent studies emphasizes vital role of EPS in the process of bioremediation through metal nanoparticles. EPS function as natural stabilizers for metal nanoparticles, averting agglomeration and regulating their size. This increases the bioavailability of metal nanoparticles, facilitating a more effective interaction with the target contaminants. EPS improves the elimination of hazardous metal ions from polluted settings by increasing their absorption and reduction. Function of EPS produced by *Halomonas* sp. has been demonstrated through in vivo exposure to Cd/Pb and polyvinylpyrrolidone (PVP) coated iron oxide nanoparticles (FeO–NPs, 20 mg L^{-1}) for 72 h. This discovery provides a better understanding on the utilization of nanoparticle enhanced bioremediation by indicating the functional implications of EPS in nanoparticle-metal-EPS interactions (Cao et al., 2022).

8 Conclusion and Future Prospective

The extensive requirement of nanoparticles in industrial sectors and environmental remediation purposes has created the primary focus on synthesis of green nanoparticles over chemical mediated nanoparticles synthesis. Nanoparticles synthesized from extracted bacterial EPS is considered as most promising green chemistry approach as microbial EPS not only act as reducing and stabilizing agent for nanoparticle synthesis, it also creates specific chemical alterations, in order to improve the solubility, reducibility and availability of polysaccharide functional groups.

Nevertheless, there are numerous disadvantages to microbial synthesis of nanoparticles, it include challenges related to achieving uniformity in shape, symmetry, composition and size of nanoparticles as these factors are influenced by environmental conditions like the medium's temperature and pH. Further, application of EPS mediated synthesizing nanoparticles in biomedical health setups sometimes have shown to elicit immune response by acting as antigenic determinants. However, the diverse functional groups present in EPS along with their ability to stabilize and reduce nanoparticles make them ideal candidates with enhanced efficiency and specificity towards environmental clean-up processes.

Microbial EPS-based metal nanoparticles are useful for bioremediation, wastewater treatment, and sustainable agricultural practices, respectively. Their biocompatibility and biodegradability property pose fewer environmental risks than those associated with conventional nanomaterial which make them suitable for prolonged use in the environment. Future research on biofilm should focus on understanding the interactions between EPS-based nanoparticles and various contaminants at the molecular level will provide deeper insights into their mechanisms of action, allowing for the design of tailored nanoparticles for specific environmental pollutants.

Acknowledgements All the authors duly acknowledge the Department of Science and Technology, Government of India, New Delhi for infrastructure development by DST-FIST (First cycle, Year 2012, and Second Cycle, Year 2024, File No.SR/FST/LS-I/2024/1270). The Centre of Excellence in Environment and Public Health, Odisha State Higher Education Council, Government of Odisha (File No. 26913/HED/HE-PTC-WB-02-17/OHEPEE) by the World Bank Fund is duly acknowledged for the Infrastructure development for the Department of Zoology.

Author DS duly acknowledge the Science and Engineering Research Board (SERB), Government of India, New Delhi for Laboratory infrastructure facility created by the PI through Extramural Research Grants (File No. EEQ/2016/000808).

Author AS and DS (Principal Investigator) duly acknowledge the Science and Technology Department, Government of Odisha for financial assistant to execute the Extramural Research Project on Advanced area of Biotechnology (File No. ST-BT-MISC-0009-2022-2744/ST).

Author YP duly acknowledge the Council for Scientific and Industrial Research (CSIR), New Delhi towards Junior Research Fellowship for financial assistant to pursue the Ph.D. program.

Author Contribution YP: Draft writing and manuscript preparation; AS: manuscript editing; SSD: data collection and design; DS: conceptualization, data analysis, review and editing. All authors contributed to the book chapter.

Conflict of Interest The authors declare no conflict of interest.

References

Ahmed, K. B. A., Kalla, D., Uppuluri, K. B., & Anbazhagan, V. (2014). Green synthesis of silver angold nanoparticles employing levan, a biopolymer from Acetobacter xylinum NCIM 2526, as a reducing agent and capping agent. *Carbohydrate Polymers, 112*, 539–545.

Anil, A., Sanjeev, K. G., Kamarudheen, N., Sebastian, P. M., & Rao, K. V. B. (2023). EPS-mediated biosynthesis of nanoparticles by Bacillus stratosphericus A07, their characterization and potential application in azo dye degradation. *Archives of Microbiology, 205*(2), 72. https://doi.org/10.1007/s00203-023-03415-0

Anyanwu, N. G., Stanley, H. O., Okpokwasili, G. C., & Akaranta, O. (2021). Biogenic nanoparticles and their environmental applications in bioremediation and pollution control. *International Journal of Innovative Science and Research Technology, 6*(10), 113–121.

Awewomom, J., Dzeble, F., Takyi, Y. D., Ashie, W. B., Ettey, E. N. Y. O., Afua, P. E., et al. (2024). Addressing global environmental pollution using environmental control techniques: A focus on environmental policy and preventive environmental management. *Discover Environment, 2*(1), 8.

Azubuike, C. C., Chikere, C. B., & Okpokwasili, G. C. (2016). Bioremediation techniques–classification based on site of application: Principles, advantages, limitations and prospects. *World Journal of Microbiology and Biotechnology, 32*, 1–18.

Bhagyaraj, S., & Krupa, I. (2020). Alginate-mediated synthesis of hetero-shaped silver nanoparticles and their hydrogen peroxide sensing ability. *Molecules (Basel, Switzerland), 25*(3), 435. https://doi.org/10.3390/molecules25030435

Borah, G., Thakur, A., Kumar, A., Komal, K., Kumar, M., Singh, P., & Borah, P. (2025). Microbial fabrication of biogenic nanoparticles and their applications in water remediation. *Total Environment Engineering*, 100006.

Cao, X., Xu, L., Chen, Y. P., Decho, A. W., Cui, Z., & Lead, J. R. (2022). Contribution, compositions structure of EPS by in vivo exposure to elucidate the mechanisms of nanoparticle-enhanced bioremediation to metals. *Environmental Science and Technology*, *56*(2), 896–906.

Chatterjee, S., Mahanty, S., Das, P., Chaudhuri, P., & Das, S. (2020). Biofabrication of iron oxide nanoparticles using manglicolous fungus Aspergillus niger BSC-1 and removal of Cr (VI) from aqueous solution. *Chemical Engineering Journal, 385*, Article 123790.

Cheng, O. H.-C., Son, D. H., & Sheldon, M. (2020). Light-induced magnetism in plasmonic gold nanoparticles. *Nature Photonics, 14*(6), 365–368.

Choudhury, P. R., Bhattacharya, P., Ghosh, S., Majumdar, S., Saha, S., & Sahoo, G. C. (2017). Removal of Cr (VI) by synthesized titania embedded dead yeast nanocomposite: Optimization and modeling by response surface methodology. *Journal of Environmental Chemical Engineering, 5*(1), 214–221.

Dey, S., Dash, A., & Das, A. P. (2024). Bioremediation as a Sustainable and Circular Solution for Mining Waste management and Bioenergy Generation. *Sustainable Management of Mining Waste and tailings: A Circular Economy Approach* (pp. 107–119). https://doi.org/10.1201/9781003442455-6

Dey, G., Patil, M. P., Banerjee, A., Sharma, R. K., Banerjee, P., & Maity, J. P., et al. (2023). The role of bacterial exopolysaccharides (EPS) in the synthesis of antimicrobial silver nanomaterials: A state-of-the-art review. *Journal of Microbiological Methods*, 106809.

Dhar, S., Mali, V., Bodhankar, S., Shiras, A., Prasad, B. L. V., & Pokharkar, V. (2011). Biocompatible gellan gum-reduced gold nanoparticles: Cellular uptake and subacute oral toxicity studies. *Journal of Applied Toxicology, 31*(5), 411–420.

Dhar, S., Murawala, P., Shiras, A., Pokharkar, V., & Prasad, B. L. V. (2012). Gellan gum capped silver nanoparticle dispersions and hydrogels: Cytotoxicity and in vitro diffusion studies. *Nanoscale, 4*(2), 563–567.

Dhar, S., Reddy, E. M., Shiras, A., Pokharkar, V., & Prasad, B. E. E. (2008). Natural gum reduced/stabilized gold nanoparticles for drug delivery formulations. *Chemistry–A European Journal, 14*(33), 10244–10250.

Escárcega-González, C. E., Garza-Cervantes, J. A., Vázquez-Rodríguez, A., & Morones-Ramírez, J. R. (2018). Bacterial exopolysaccharides as reducing and/or stabilizing agents during synthesis of metal nanoparticles with biomedical applications. *International Journal of Polymer Science, 2018*(1), 7045852.

Fan, G., Cang, L., Qin, W., Zhou, C., Gomes, H. I., & Zhou, D. (2013). Surfactants-enhanced electrokinetic transport of xanthan gum stabilized nanoPd/Fe for the remediation of PCBs contaminated soils. *Separation and Purification Technology, 114*, 64–72.

Flemming, H. C., van Hullebusch, E. D., Little, B. J, Neu, T. R., Nielsen, P. H., Seviour, T., Stoodley, P., Wingender, J., & Wuertz, S. (2025). Microbial extracellular polymeric substances in the environment, technology and medicine. *Nature Review in Microbiology, 23*(2):87. https://doi.org/10.1038/s41579-024-01098-y

Galdames, A., Ruiz-Rubio, L., Brettes, P., Etxebarria, J., & Vilela, J. L. V. (2022). Nanobioremediation for soil remediation: An introduction. In *Nano-bioremediation: Fundamentals and applications* (pp. 479–500). Elsevier.

Galvez, A. M., Ramos, K. M., Alexis, J. T., & Baculi, R. (2019). Bacterial exopolysaccharide-mediated synthesis of silver nanoparticles and their application on bacterial biofilms. *The Journal of Microbiology, Biotechnology and Food Sciences, 8*(4), 970.

Ghosh, A., et al. (2024). Mechanism and application of bacterial exopolysaccharides: An advanced approach for sustainable heavy metal abolition from soil. *Carbohydrate Research, 544*, 109247.

González, A., Garcés, V., Sabio, L., Velando, F., López-Haro, M., & Gálvez, N., et al. (2019). Optical and tomography studies of water-soluble gold nanoparticles on bacterial exopolysaccharides. *Journal of Applied Physics, 126*(5).

Hussain, M. A., Shah, A., Jantan, I., Tahir, M. N., Shah, M. R., Ahmed, R., & Bukhari, S. N. A. (2014). One pot light assisted green synthesis, storage and antimicrobial activity of dextran stabilized silver nanoparticles. *Journal of Nanobiotechnology, 12*, 1–6.

Jha, A., Barsola, B., Pathania, D., Raizada, P., Thakur, P., Singh, P., et al. (2024). Nano-biogenic heavy metals adsorptive remediation for enhanced soil health and sustainable agricultural production. *Environmental Research*, 118926.

Karygianni, L., Ren, Z., Koo, H., & Thurnheer, T. (2020). Biofilm matrixome: extracellular components in structured microbial communities. *Trends in Microbiology, 28*(8), 668–681. https://doi.org/10.1016/j.tim.2020.03.016

Kanmani, P., & Lim, S. T. (2013). Synthesis and structural characterization of silver nanoparticles using bacterial exopolysaccharide and its antimicrobial activity against food and multidrug resistant pathogens. *Process Biochemistry, 48*(7), 1099–1106.

Kato, Y., & Suzuki, M. (2020). Synthesis of metal nanoparticles by microorganisms. *Crystals, 10*(7), 589.

Kaul, I., & Sharma, J. G. (2022). Nanotechnology for the bioremediation of organic and inorganic compounds in aquatic ecosystem/marine ecosystem. *Journal of Applied Biology and Biotechnology, 10*, 22–33.

Kaur, N., & Dey, P. (2023). Bacterial exopolysaccharides as emerging bioactive macromolecules: From fundamentals to applications. *Research in Microbiology, 174*(4), Article 104024.

Kavitha, S., Geetha, D., & Ramesh, P. S. (2015). Synthesis and characterizations of silver colloid nanoparticles stabilized by dextran. *Journal of Environmental Nanotechnology, 4*(1), 50–55.

Kim, H. K., Jeong, S. W., Yang, J. E., & Choi, Y. J. (2019). Highly efficient and stable removal of arsenic by live cell fabricated magnetic nanoparticles. *International Journal of Molecular Sciences, 20*(14), 3566.

Kim, H. M., Park, J. H., Choi, Y. J., Oh, J. M., & Park, J. (2023). Hyaluronic acid-coated gold nanoparticles as a controlled drug delivery system for poorly water-soluble drugs. *RSC Advances, 13*(8), 5529–5537.

Kuppan, N., Padman, M., Mahadeva, M., Srinivasan, S., & Devarajan, R. (2024). A comprehensive review of sustainable bioremediation techniques: Eco friendly solutions for waste and pollution management. *Waste Management Bulletin.*

Li, X., Chen, S., Hu, W., Shi, S., Shen, W., Zhang, X., & Wang, H. (2009). In situ synthesis of CdS nanoparticles on bacterial cellulose nanofibers. *Carbohydrate Polymers, 76*(4), 509–512.

Li, C., Zhou, L., Yang, H., Lv, R., Tian, P., Li, X., et al. (2017). Self-assembled exopolysaccharide nanoparticles for bioremediation and green synthesis of noble metal nanoparticles. *ACS Applied Materials and Interfaces, 9*(27), 22808–22818.

Li, C., Zhang, X., Guo, Y., Seidi, F., Shi, X., & Xiao, H. (2021). Naturally occurring exopolysaccharide nanoparticles: Formation process and their application in glutathione detection. *ACS Applied Materials and Interfaces, 13*(17), 19756–19767.

Liu, H., Ikeda, K., Nguyen, M. T., Sato, S., Matsuda, N., Tsukamoto, H., et al. (2022). Alginate-stabilized gold nanoparticles prepared using the microwave-induced plasma-in-liquid process with long-term storage stability for potential biomedical applications. *ACS Omega, 7*(7), 6238–6247.

Luo, J., Fu, K., Yu, D., Hristovski, K. D., Westerhoff, P., & Crittenden, J. C. (2021). Review of advances in engineering nanomaterial adsorbents for metal removal and recovery from water: Synthesis and microstructure impacts. *ACS ES&T Engineering, 1*(4), 623–661.

Manganyi, M. C., Dikobe, T. B., & Maseme, M. R. (2024). Exploring the potential of endophytic microorganisms and nanoparticles for enhanced water remediation. *Molecules, 29*(12), 2858. https://doi.org/10.3390/molecules29122858

Mishra, P., Tiwari, A., Dubey, V. K., Kushwaha, Singh, S. K., Singh, V. K., & Singh, N. K. Arbuscular Mycorrhizal Fungi (AMF): A natural tool for phytoremediation of heavy metals (HMs). In: Madhav S, Gupta GP, Yadav RK, Mishra R, Hullebusch EV (Edn) (2024). *Phytoremediation.* Springer, Cham. https://doi.org/10.1007/978-3-031-60761-5_11

Mohanty, M., Mohanty, J., Priyadarsini, S., Rodriguez-Diaz, J. M., Yadav, S., Carbal-Pinto, M. M. S., Nath, H., Ram, D. K., Reddy Koteswara, G., & Das A. P. (2025). Environmental Hydrocarbon Pollutants: Sources, Transport, and Effect on Human Health. In *Environmental Hydrocarbon Pollutants: Sources, Transport, and Effect on Human Health.* Springer Nature Book Chapter, (p. 21–44). https://doi.org/10.1007/978-3-031-81615-4_2

Nkosi, N. C., Basson, A. K., Ntombela, Z. G., Dlamini, N. G., & Pullabhotla, R. V. S. R. (2024). Green synthesis, characterization and application of silver nanoparticles using bioflocculant: A review. *Bioengineering (Basel, Switzerland), 11*(5), 492. https://doi.org/10.3390/bioengineering11050492

Olaimat, A. N., Ababneh, A. M., Al-Holy, M., Al-Nabulsi, A., Osaili, T., Abughoush, M., Ayyash, M., & Holley, R. A. (2024). A review of bacterial biofilm components and formation, detection methods, and their prevention and control on food contact surfaces. *Microbiology Research, 15*(4), 1973–1992. https://doi.org/10.3390/microbiolres15040132

Prakash, A., Sharma, S., Ahmad, N., Ghosh, A., & Sinha, P. (2010). Bacteria mediated extracellular synthesis of metallic nanoparticles. *International Research Journal of Biotechnology, 1*(5), 071–079.

Priyadarsini, A., & Das, A. P. (2024). Analytical and structural characterization of waste lithium-ion batteries for their effective recycling strategy. *Environmental Chemistry and Ecotoxicology.* https://doi.org/10.1016/j.enceco.2014.12.004

Priyadarshini, E., Kumar, R., Balakrishnan, K., Pandit, S., Kumar, R., Jha, N. K., & Gupta, P. K. (2024) *ACS Applied Bio Materials, 7*(5), 2604–2619. https://doi.org/10.1021/acsabm.4c00024

Raj, R., Dalei, K., Chakraborty, J., & Das, S. (2016). Extracellular polymeric substances of a marine bacterium mediated synthesis of CdS nanoparticles for removal of cadmium from aqueous solution. *Journal of Colloid and Interface Science, 462*, 166–175.

Rajeswaran, S., Veeraiyan, B., Saravanan, V., Manickam, E., & Danaraj, J. (2023). Biogenic facile green synthesis of actinobacterium exopolysaccharide-fabricated zinc oxide nanoparticles for the diverse biomedical applications. *Biomass Conversion and Biorefinery*, 1–15.

Ramezani Farani, M., Farsadrooh, M., Zare, I., Gholami, A., & Akhavan, O. (2023). Green synthesis of magnesium oxide nanoparticles and nanocomposites for photocatalytic antimicrobial, antibiofilm and antifungal applications. *Catalysts, 13*(4), 642.

Raveendran, S., Chauhan, N., Palaninathan, V., Nagaoka, Y., Yoshida, Y., Maekawa, T., & Kumar, D. S. (2015). Extremophilic polysaccharide for biosynthesis and passivation of gold nanoparticles and photothermal ablation of cancer cells. *Particle and Particle Systems Characterization, 32*(1), 54–64.

Reddy Koteswara, G., Thanay Allam, S. S., Gasem, A., Tummalacharla, P., Ramakrishnan, G., Yadav, K. K., Nath, H., Ram, D. K., & Das, A. (2025) Zero Waste Initiatives in Industrial

Hydrocarbons and Waste Management. In Environmental Hydrocarbon Pollution and Zero Waste Approach Towards a Sustainable Waste Management (pp. 105–126). https://doi.org/10.1007/978-3-031-81615-4

Saravanan, C., Rajesh, R., Kaviarasan, T., Muthukumar, K., Kavitake, D., & Shetty, P. H. (2017). Synthesis of silver nanoparticles using bacterial exopolysaccharide and its application for degradation of azo-dyes. *Biotechnology Reports, 15*, 33–40.

Sathiyanarayanan, G., Vignesh, V., Saibaba, G., Vinothkanna, A., Dineshkumar, K., Viswanathan, M. B., & Selvin, J. (2014). Synthesis of carbohydrate polymer encrusted gold nanoparticles using bacterial exopolysaccharide: A novel and greener approach. *RSC Advances, 4*(43), 22817–22827.

Sathiyanarayanan, G., Dineshkumar, K., & Yang, Y. H. (2017a). Microbial exopolysaccharide-mediated synthesis and stabilization of metal nanoparticles. *Critical Reviews in Microbiology, 43*(6), 731–752. https://doi.org/10.1080/1040841X.2017.1306689

Sathiyanarayanan, G., Dineshkumar, K., & Yang, Y.-H. (2017b). Microbial exopolysaccharide-mediated synthesis and stabilization of metal nanoparticles. *Critical Reviews in Microbiology, 43*(6), 731–752.

Savadiya, B., Pandey, G., & Misra, S. K. (2024). Remediation of pharmacophoric laboratory waste by using biodegradable carbon nanoparticles of bacterial biofilm origin. *Environmental Research, 252*, Article 118969.

Scala, A., Piperno, A., Hada, A., Astilean, S., Vulpoi, A., Ginestra, G., Marino, A., Nostro, A., Zammuto, V., & Gugliandolo, C. (2019). Marine bacterial exopolymers-mediated green synthesis of noble metal nanoparticles with antimicrobial properties. *Polymers, 11*(7), 1157. https://doi.org/10.3390/polym11071157

Sharma, D., Kanchi, S., & Bisetty, K. (2015). Biogenic synthesis of nanoparticles: A review. *Arabian Journal of Chemistry*. https://doi.org/10.1016/j.arabjc.2015.11.002

Sharma, D., Kanchi, S., & Bisetty, K. (2019). Biogenic synthesis of nanoparticles: A review. *Arabian Journal of Chemistry, 12*(8), 3576–3600.

Shaterabadi, Z., Nabiyouni, G., & Soleymani, M. (2017). High impact of in situ dextran coating on biocompatibility, stability and magnetic properties of iron oxide nanoparticles. *Materials Science and Engineering: C, 75*, 947–956.

Singh, S., Datta, S., Narayanan, K. B., & Rajnish, K. N. (2021). Bacterial exo-polysaccharides in biofilms: Role in antimicrobial resistance and treatments. *Journal, Genetic Engineering and Biotechnology, 19*(1), 140. https://doi.org/10.1186/s43141-021-00242-y

Sivakumar, B., Aswathy, R. G., Sreejith, R., et al. (2014). Bacterial exopolysaccharide based magnetic nanoparticles: A versatile nanotool for cancer cell imaging, targeted drug delivery and synergistic effect of drug and hyperthermia mediated cancer therapy. *Journal of Biomedical Nanotechnology, 10*(6), 885–899.

Somu, P., & Paul, S. (2018). Casein based biogenic-synthesized zinc oxide nanoparticles simultaneously decontaminate heavy metals, dyes, and pathogenic microbes: A rational strategy for wastewater treatment. *Journal of Chemical Technology and Biotechnology, 93*(10), 2962–2976.

Suárez-López, R., Puntes, V. F., Bastús, N. G., Hervés, C., & Jaime, C. (2022). Nucleation and growth of gold nanoparticles in the presence of different surfactants. A dissipative particle dynamics study. *Scientific Reports, 12*(1), 13926. https://doi.org/10.1038/s41598-022-18155-2

Subramaniyam, V., Subashchandrabose, S. R., Thavamani, P., Megharaj, M., Chen, Z., & Naidu, R. (2014). Chlorococcum sp. MM11-a novel phyco-nanofactory for the synthesis of iron nanoparticles. *Journal of Applied Phycology, 27*(2).

Tilwani, Y. M., Lakra, A. K., Domdi, L., Jha, N., & Arul, V. (2023). Preparation, physicochemical characterization, and in vitro biological properties of selenium nanoparticle synthesized from exopolysaccharide of enterococcus faecium MC-5. *BioNanoScience, 13*(2), 413–425.

Torimiro, N., Daramola, O. B., Oshibanjo, O. D., Otuyelu, F. O., Akinsanola, B. A., Yusuf, O. O., et al. (2021). Ecorestoration of heavy metals and toxic chemicals in polluted environment using microbe-mediated nanomaterials. *International Journal of Environmental Bioremediation and Biodegradation, 9*(1), 8–21.

Vignesh, V., Sathiyanarayanan, G., Sathishkumar, G., Parthiban, K., Sathish-Kumar, K., & Thirumurugan, R. (2015). Formulation of iron oxide nanoparticles using exopolysaccharide: Evaluation of their antibacterial and anticancer activities. *RSC Advances, 5*(35), 27794–27804.

Visagamani, A. M., Shanthi, D., Muthukrishnaraj, A., Venkatadri, B., Ahamed, J. I., & Kaviyarasu, K. (2023). Innovative preparation of cellulose-mediated silver nanoparticles for multipurpose applications: Experiment and molecular docking studies. *ACS Omega, 8*(42), 38860–38870. https://doi.org/10.1021/acsomega.3c02432

Wang, Y., Zhan, L., & Huang, C. Z. (2010). One-pot preparation of dextran-capped gold nanoparticles at room temperature and colorimetric detection of dihydralazine sulfate in uric samples. *Analytical Methods, 2*(12), 1982–1988.

Wang, X., Zhang, D., Qian, H., Liang, Y., Pan, X., & Gadd, G. M. (2018). Interactions between biogenic selenium nanoparticles and goethite colloids and consequence for remediation of elemental mercury contaminated groundwater. *Science of the Total Environment, 613*, 672–678.

Wang, W., Yuhao, J., Liu, N., Shi, S., & Hao, L. (2023). Structural characteristics of microbial exopolysaccharides in association with their biological activities: A review. *Chemical and Biological Technologies in Agriculture, 10*(1), 137.

Waoo, A. A., Singh, S., Pandey, A., Kant, G., Choure, K., Amesho, K. T., & Srivastava, S. (2023). Microbial exopolysaccharides in the biomedical and pharmaceutical industries. *Heliyon, 9*(8).

Yan, W., Chen, C., Wang, L., Zhang, D., Li, A. J., Yao, Z., & Shi, L. Y. (2016). Facile and green synthesis of cellulose nanocrystal-supported gold nanoparticles with superior catalytic activity. *Carbohydrate Polymers, 140*, 66–73.

Zhang, W., Zhang, Z., Lou, S., Chang, Z., Wen, B., & Zhang, T. (2022). Hyaluronic acid-stabilized Fe3O4 Nanoparticles for promoting In vivo Magnetic Resonance Imaging of Tumors. *Frontiers in Pharmacology, Experimental Pharmacology and Drug Discovery Section, 13*. https://doi.org/10.3389/fphar.2022.918819

Zhou, Y., Cui, Y., Qu, X. (2019) Exopolysaccharides of lactic acid bacteria: Structure, bioactivity and associations: A review. *Carbohydrate Polymers, 207*, 317–332. https://doi.org/10.1016/j.carbpol.2018.11.093

Diatoms Nanotechnology: Prospects and Challenges for Sustainable Environment

Rinku, Stanzen Angmo, Ayushmaan Srivastava, Khundrakpam Merina, Sneha Sindhwal, and Archana Tiwari

Abstract Diatom presents a promising frontier in nanotechnology for modern biomedical applications and material science. Diatoms possess a unique feature in their intricate silica-based cell walls, frustules, which shows high surface area, porosity and nanoscale precision which closely resemble artificially synthesized porous silica and offers a lucrative and sustainable alternative for diverse technological applications. Diatom biosilica holds distinctive properties such as chemical stability, thermal resistance, biocompatibility and tailored surface chemistry which makes it exceptionally suitable for nanomedicine, microfluidics, biosensing and photonics. Furthermore, nanoparticles synthesized using diatoms serve as efficient platforms for various applications including targeted therapeutics like drug and gene delivery and photodynamic therapy because of their distinctive morphology and biofunctionalization potential. The capability of diatoms to synthesize nanoscale architectures through biological processes serves an ideal model for energy-efficient and scalable nanofabrication which can surpass conventional synthetic methods. This chapter focuses on various aspects of diatom frustule offering significant insights for designing biomaterial, carving path for remarkable innovations in applications such as renewable energy technologies, bioimaging and molecular filtration. Harnessing the potential of diatom in formulating nanostructures which could generate new prospects in modern technologies, illustrating their importance in next-generation nanotechnology.

Keywords Biomaterials · Diatoms · Drug delivery · Nanotechnology · Silver Nanoparticles

---

Rinku · S. Angmo · A. Srivastava · K. Merina · S. Sindhwal · A. Tiwari (✉)
Diatom Research Laboratory, Amity Institute of Biotechnology, Amity University, Noida, Uttar Pradesh, India
e-mail: panarchana@gmail.com

E. Priyadarshini et al. (eds.), *Advanced Nanotechnology Applications for Environmental Sustainability*, Sustainable Environmental Waste Management Strategies,
https://doi.org/10.1007/978-3-032-12933-8_3

1 Introduction

Diatoms are major group among phytoplanktons, encased in a porous micro-shell (also known as frustule) (Rea & De Stefano, 2022) composed of opaline silica which provides intricate and glass-like patterns on the cell wall (Zobi, 2022). They come under the class Bacillariophyceae, Bacillariophyta division and are classified into centrales and pennates based on their morphological structure (Parikh, 2021). They are ubiquitous and inhabit all aquatic ecosystems including freshwater, marine and brackish water (Hamed, 2023).

Diatoms are unicellular and photosynthetic (Roychoudhury et al., 2022) and play a crucial role in the ocean's cycle of nutrients and organic matter, accounts for approximately 40% of the carbon fixation globally (Roychoudhury et al., 2022) and 20–50% of production of the oxygen (Zobi, 2022). They are also responsible for the production of biosilica and are the primary producers in the aquatic food chain, contributing to more than 25% of primary outputs (Parikh, 2021). During their reproduction cycle, they annually convert seven billion metric tons of silicon from the waters in which they thrive (Zobi, 2022). When these microorganisms die, they sediment and accumulate at the bottom of the ocean floor, resulting in the formation of siliceous oozes, which over time fossilized and forms diatomaceous earth (DE) (Roychoudhury et al., 2022).

Frustules are composed of biogenic silica, also known as hydrated silica ($SiO_2.nH_2O$). It comprises 97% of dissolved silica (Roychoudhury et al., 2022). They are characterized by their unique cellular architecture, i.e., highly ordered nanopatterned 3-D siliceous cell walls. They are a natural source of biosilica, which can be obtained in large amounts through algae replication. DE, which is also known as diatomite is a most abundant source of biosilica (Terracciano et al., 2018). Diatoms form their cell wall through the process known as biosilification. They are good at absorbing silica from their surroundings. During moist conditions, silica are present in the form orthosilicic acid ($Si(OH)_4$). Biosilification starts when the diatoms uptake silicon in the form of silicic acid through specific transporter protein. Then these silicas are polycondensed and accumulate in the specific vesicles recognized as silica deposition vesicles (SDV), where it can be accumulated upto 1000 × higher concentration as compared to the surrounding environments (Roychoudhury et al., 2022).

Synthesis of nanoparticles for commercial use is time and energy consuming and there is also production of toxic compounds in the environment. However, there is an increase demand of cost-effective, rapid and greener methods to produce nanoparticles (Rabiee et al., 2021). Therefore, diatom generates nano-silica as building blocks and are considered as a unique machinery for nano-silica biosynthesis (Roychoudhury et al., 2022). Greener biosynthesis of golden nanoparticles (~10–30 nm) was accomplished through diatom species (*Navicula atomus* and *Diadesmis gallicaat*) under laboratory conditions. Extracellular polysaccharides secreted by diatoms and their frustules can be used in the nanoparticles synthesis. It was observed that various diatom species can be employed for the nanoparticle biosynthesis including *Nitzschia sp.* which has been used in the sustainable synthesis of gold nanoparticles (~43 nm) whereas *Amphora* sp. was used in the synthesis of spherical silver nanoparticles (~20–25 nm) and they exhibited high antimicrobial against both gram negative and gram positive bacteria (Rabiee et al., 2021).

Recently, nanoscales properties of diatom frustules have attracted the interest of many scientists. Their macro, meso, and microporous silica have useful applications in the nanotechnology field due to their high surface area and physicochemical properties. Diatoms produce porous silica (including macro and meso) and it has been reported that they are a potential nanomaterial due to their pore size (1 micron) (Hamed, 2023). They are used in biosensing, material science, drug carrier (Zobi, 2022), gene delivery, bioimaging, tissue engineering, hemorrhage control (Rabiee et al., 2021) and also, they have been used for fabrication of various silica-material nano- and micro-composites (Roychoudhury et al., 2022). Diatom frustules, biosilica and diatomite nanoparticles have shown notable capabilities in the industrial applications in biomedicine and pharmaceuticals, due to their characteristic's properties and good biocompatibility. Silica nanoparticles have exhibited various features which make them the better tool for gene and drug delivery carrier. These features include enhancement of drug solubility, high biocompatibility, accumulation capabilities, high drug loading capacity (Rabiee et al., 2021).

2 Green Biosynthesis of Nanoparticles

Nanoparticles (NPs), with at least one dimension ranging from 1 to 100 nm, act as a bridge between bulk materials and atomic or molecular structures (Thakkar et al., 2010). Due to their small size, increased reactivity and large surface area with free dangling bonds, they show remarkable and unique properties compared to their bulk counterparts (Hussain et al., 2016). Since the nineteenth century, scientists have recognized the ability of biological organisms to reduce metal precursors, but the underlying mechanisms remain largely unexplored. The advancement of efficient green synthesis methods, which utilize capping, natural reducing and stabilizing agents, without relying on expensive, toxic chemicals or high energy consumption, has drawn interest to biological approaches (Mohanpuria et al., 2008).

Urbanization is driving significant changes in environmental dynamics, necessitating innovative approaches to address the challenges it presents, particularly in waste management and pollution. The green synthesis of nanoparticles emerges as a promising solution, leveraging biogenic materials to produce eco-friendly nanomaterials. This method not only mitigates the environmental impact associated with traditional chemical synthesis but also promotes the recycling of organic waste, aligning with sustainable urban development goals. By utilizing renewable resources and reducing reliance on hazardous chemicals, green synthesis contributes to cleaner production processes and the development of advanced materials that can be applied in various sectors, including healthcare and environmental remediation

(Chopra et al., 2022; Mohanty et al., 2024). Biologically synthesized nanoparticles (b-NPs) are gaining significant importance due to their eco-friendly production methods, cost-effectiveness, and multifunctional applications in various fields, particularly in biomedicine. Unlike conventional chemical synthesis, which often involves hazardous materials and harsh conditions, the green synthesis of b-NPs utilizes natural biological precursors such as plants, fungi, algae and bacteria, resulting in minimal environmental impact and enhanced biocompatibility. This approach not only simplifies the synthesis process by eliminating the need for toxic reducing agents and capping agents but also allows for the incorporation of bioactive compounds that enhance the therapeutic efficacy of the nanoparticles (Bhardwaj et al., 2020; Li et al., 2011). Nanoparticle synthesis (Fig. 1) is primarily categorized into two approaches: (a) top-down and (b) bottom-up methods. The top-down approach involves breaking down bulk materials into nanoscale particles through techniques like mechanical milling, laser ablation, or spark ablation, which can produce uniform nanoparticles but may require extensive post-processing. In contrast, the bottom-up approach assembles nanoparticles from atomic or molecular precursors using methods such as chemical vapor deposition (CVD), sol–gel processes, or biological synthesis, which allows for greater control over particle size and morphology. While top-down methods are effective for creating specific shapes and sizes, bottom-up methods often provide a eco-friendlier and more versatile route to nanoparticle production (Patra & Baek, 2014).

Biologically derived nanoparticle synthesis employs a bottom-up approach, where the process is facilitated by stabilizing and reducing agents. The synthesis typically involves three key steps: (a) selecting an appropriate solvent medium, (b) choosing an ecofriendly and environmentally safe reducing agent, and (c) opting for a non-toxic

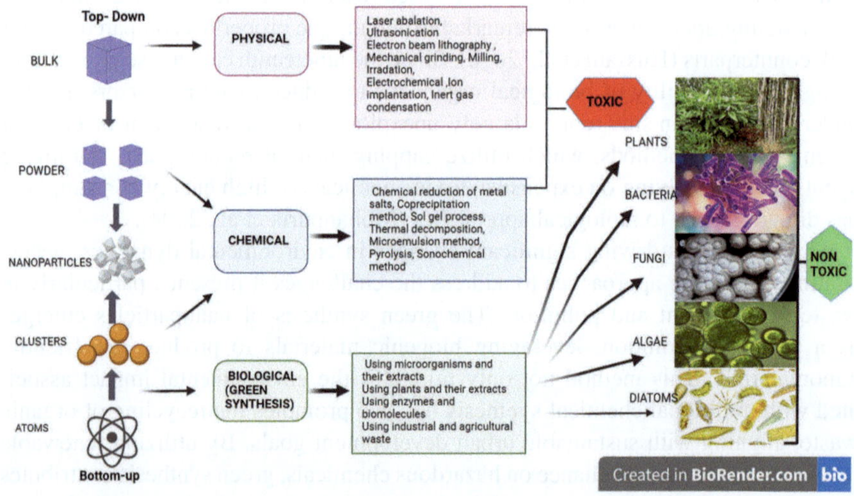

Fig. 1 Nanoparticle synthesis methods

material to serve as a capping agent to stabilize the synthesized nanoparticles (Mohanpuria et al., 2008). Nanotechnology offers remarakable advantages over conventional methods due to the various range of components available from biological systems for nanoparticle formation. The rich biodiversity of these biological components has been utilized for the synthesis of bionanomaterials which are environmentally friendly and can be applied in various medical fields. Physicochemical synthesis of nanoparticles for commercial applications is typically time-consuming, energy-intensive and requires high pressures, temperatures and the use of harmful or toxic chemicals, all of which are hazardous to the environment (Farjadian et al., 2019). As indicated, controlling hydrolysis time and temperature significantly impacts the average particle size and distribution shape. Consequently, there is a growing need for more sustainable methods, rapid, safer and cost-effective for producing nanoparticles (Bala et al., 2024; Das et al., 2024). Renewable, biological and sustainable resources offer ecofriendly strategies for synthesizing nanoparticles while reducing the need for complex procedures which can be labor intensive (Table 1). Among these sustainable resources, diatoms can be used as natural resources for the ecofriendly synthesis of various nanoparticles through metabolic insertion and in situ coating or fabrication (Rabiee et al., 2021).

Diatoms are unicellular microalgae belonging to the class Bacillariophyceae, characterized by their unique silica cell walls known as frustules. These organisms thrive in diverse aquatic environments and play a vital role in biogeochemical cycles globally. Their ability to produce biogenic silica has garnered significant interest in various fields, particularly in nanotechnology and biotechnology, where they serve as eco-friendly platforms for synthesizing nanoparticles.

Silicon is a valuable material with various range of applications in nanotechnology including nanomedicine, synthetic production processes and pharmaceutical sciences. Through the deposition of silica, Diatoms are capable of forming micro- or nanoscale structures that can be utilized for nanomedicine, especially for drug delivery systems (Wang et al., 2013). Moreover, diatoms are favorable candidates for the phytoremediation of heavy metals with remarakable commercial potential (Dey et al., 2024; Mishra et al., 2024). Their significant ability to tolerate pollutants and toxic metals makes them an interesting alternative for both the synthesis of metallic nanoparticles and toxic metal phytoremediation (Kiran Marella et al., 2020). Typically, the synthesis of silica (nano)materials on commercial scale through chemical methods requires extreme pH conditions, pressures and temperatures along with the use of toxic and harmful chemicalsmeanwhile, in contrast to the above, the biosynthesis of silica occurs at moderate neutral pH, low pressures and temperatures, emphasizing the advantages of biological based synthetic systems.

The commonly used procedure for in vivo NP biosynthesis by diatoms is demonstrated in Fig. 2. After cell growth in 'normal' culture media, the diatom biomass is extracted and is added in a metal salt solution. During the incubation period, reduction reaction takes place between the meta solution and diatoms leading to nanoparticles synthesis.

Table 1 Synthesis of nanoparticles

Factor	Chemical method	Biological method	References
Synthesis mechanism	Involves reduction of metal salts using chemical reducing agent	Utilizes living organisms or extracts to reduce metal ions	Banjara et al. (2024)
Reaction conditions	Often requires harsh conditions (high temperature, extreme pH)	Typically occurs under milder conditions (ambient temperature, neutral pH)	Anu Mary Ealia and Saravanakumar (2017)
Environmental impact	Can produce toxic byproducts and waste, raising pollution concern	Generally, more environmentally friendly with minimal harmful byproducts	Ijaz et al. (2020)
Cost efficiency	Can be cost-effective for large scale production but involves expensive reagents	May have higher initial setup costs but lower operational costs over time	Zhang et al. (2011)
Size and morphology control	Provides precise control over size and shape through adjustable parameters	Can produce unique shapes and sizes but may have variability in characteristics	Abuzeid et al. (2023)
Toxicity and biocompatibility	May leave residual toxic chemicals, raising biocompatibility concerns	Typically yields more biocompatible nanoparticles suitable for medical applications	Buzea et al. (2007)
Scalability	Easily scalable with established protocols for large-scale production	Scalability can be more complex due to biological growth conditions and variability	Yu et al. (2008)
Reproducibility	High reproducibility with consistent results when parameters are controlled	Reproducibility can be affected by biological variability and environmental factors	Ijaz et al. (2020)
Material variety	Can synthesize a wide range of metals and alloys with specific properties	Limited to the types of metals that can be reduced biologically; often produces pure metals or oxides	Saba Hasan (2015)

(continued)

Table 1 (continued)

Factor	Chemical method	Biological method	References
Safety concerns	Involves handling of hazardous chemicals, posing risks to health and safety	Generally safer as it involves natural processes, though handling of biological materials still requires precautions	Vinita Vishwakarma (2010)
Applications	Widely used in electronics, catalysis, and various industrial applications	Particularly suited for biomedical applications like drug delivery, imaging, and biosensing	Khan et al. (2019)

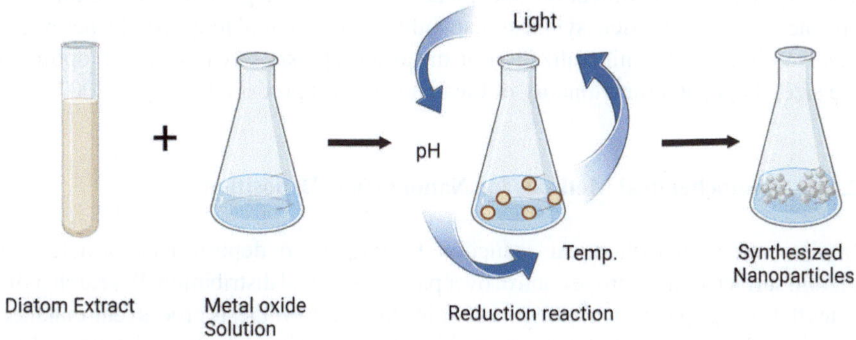

Fig. 2 Green nanoparticle synthesis by using diatoms

2.1 Scalability of Diatom Nanoparticle Production

While chemical methods benefit from established industrial protocols, scaling biological methods requires innovation in optimization of diatom cultivation.

2.1.1 Rapid Cultivation

Diatoms can be cultivated in diverse aquatic environments and under controlled conditions, allowing for rapid biomass production. Their growth rates can be optimized through nutrient management and light exposure, facilitating continuous production cycles.

2.1.2 Cost-Effectiveness

The use of diatoms reduces reliance on expensive chemical reagents typically required in traditional nanoparticle synthesis methods. Diatom frustules, which are abundant and can be sourced from natural deposits (diatomite), provide a low-cost alternative for producing silica-based nanoparticles (Rabiee et al., 2021).

2.1.3 Modular Production

The ability to manipulate diatom cultures to produce specific types of nanoparticles allows for tailored applications across various industries, enhancing their market viability. Genetic modification and metabolic engineering of diatom strains can enhance their nanoparticle synthesis capabilities. By manipulating specific pathways involved in silica biomineralization or metal ion uptake, researchers can optimize the size, shape, and functionality of the nanoparticles produced (Kröger, 2007).

2.1.4 Photochemical Methods for Nanoparticle Deposition

Implementing photochemical reduction techniques to deposit nanoparticles on diatom surfaces can improve control over particle size and distribution. Research indicates that using specific wavelengths of light during the synthesis process can enhance nanoparticle growth and uniformity, making it suitable for scalable applications (Jara et al., 2021).

2.1.5 Integration with Other Technologies

Combining diatom nanoparticle production with other nanotechnology methods, such as hybridization with other nanomaterials or incorporating them into composite materials, can expand their functionality and application range. This integration can lead to innovative products that meet diverse industrial needs while leveraging the unique properties of diatoms (Roychoudhury et al., 2021).

3 Modifications to the Frustule

Diatom frustule alterations are changes to diatoms' silica-based cell walls that can be caused by environmental conditions, genetic factors, or specialized growth factors. These frustules, composed of two overlapping thecae, have complicated and patterned structures that can change size, shape, or symmetry, as well as pore density and ornamentation. Changes in the chemical makeup of silica, such as the introduction of metals like titanium or aluminium, or modifications in related organic molecules, are also possible. Diatoms may change the thickness or ornamentation of their frustules in response to factors such as silica availability, light, pH, or nutritional circumstances (Falasco & Badino, 2011). Functionally, these changes can increase buoyancy, reduce predation, improve light collection for photosynthesis, or provide resistance to mechanical stress and water currents. Such modifications are not only environmentally beneficial, but they also have uses in nanotechnology, biosensors, and materials science (Bok et al., 2022; De Tommasi & De Luca, 2022; Falasco & Badino, 2011).

3.1 Different Modification Techniques

3.1.1 Physical Method

Physical approaches for changing diatom frustules include manipulating their growth and morphology mechanically, optically, or environmentally. Nanopatterning is an important process in which mechanical forces or Molds influence the arrangement of silica nanopores and ridges, resulting in unique shapes used in photonics and biosensors. Light manipulation, which involves varying intensity, duration, or wavelength, can increase silica uptake and change ornamentation, making it useful for improving frustules for solar energy acquisition (De Tommasi & De Luca, 2022). Ultrasound therapies use high-frequency sound waves to cause stress, disrupt patterns, or improve nutrient intake, and they are commonly used in material science for surface modification (Gholami et al., 2020). Temperature modulation affects the metabolic processes involved in silica deposition, with lower temperatures resulting in thinner frustules and higher temperatures increasing their density (Javaheri et al., 2015). Pressure fluctuations that imitate deep-sea conditions cause structural adaptations, whereas controlled water currents increase frustule resilience to hydrodynamic stress. Post-growth procedures such as laser etching and abrasion enable for exact changes in nanotechnology applications (Tamburini et al., 2006). These methods are useful because to their precision, scalability, and environmentally benign approach, making them essential for designing diatom frustules in sophisticated scientific and industrial applications.

3.1.2 Chemical Method

Chemical methods for manipulating diatom frustules involve changing the composition and deposition process of silica and other components within the frustules. One important approach is silica precipitation, which involves adjusting the concentration of dissolved silica in the growth media. This affects the pace at which silica is deposited, resulting in differences in the frustule's size, thickness, and overall shape (Soleimani et al., 2020). Another important technique is metal doping, which involves injecting metals like titanium, aluminium, or zinc into the growth media.

These metals are integrated into the silica structure, which alters the mechanical and chemical properties of the frustules. For example, titanium doping can improve the frustule's susceptibility to oxidative stress (Brzozowska et al., 2020), whereas aluminium inclusion can change its light-absorbing properties, which may be advantageous in solar energy harvesting applications (Vicente-Garcia et al., 2024). Additionally, pH changes are frequently utilized to alter frustule production. Changing the pH of the culture medium influences silica solubility and, as a result, frustule structure (Vrieling et al., 1999). A more acidic or basic environment can promote various types of silica deposition, resulting in changes in surface texture and pore size (Hervé et al., 2012). Finally, chemical etching is the process of selectively dissolving sections of the frustule using acids or alkalis, resulting in patterned surfaces or the extraction of certain components for use in nanotechnology applications (Townley et al., 2008). These chemical approaches enable precise control over the shape and activity of diatom frustules, making them useful in a variety of industrial, environmental, and therapeutic applications.

3.2 Recent Methods for Functionalizing Diatom Frustules

3.2.1 Genetic Engineering

Diatom genetic engineering entails modifying their genetic material such that specific proteins or functional molecules interact with their silica-based frustules. This process normally begins by inserting foreign genes into the diatom's genome, either through stable transformation or transient expression approaches. The inserted genes are frequently intended to create proteins with distinct characteristics, such as enzymes, antibodies, or peptides, which can bind to or coat the diatom frustules (Delalat et al., 2015). Once produced, these proteins can either stick to the frustule's surface or incorporate into the silica matrix as it forms. Researchers can, for example, modify diatoms to create enzymes that degrade pollutants or antibodies that preferentially bind to pathogens or poisons, effectively transforming the diatoms into living bio-filters or sensors. For example, one study found that genetically modifying the diatom *P. tricornutum* allowed it to produce and secrete human IgG antibodies against the Hepatitis B Virus surface protein (Hempel & Maier, 2012). This approach allows for the in vivo inclusion of functional molecules, which means that the frustules are transformed in real time as the diatom grows, giving precise control over the amount and placement of functional molecules. Genetic engineering also allows for the development of new materials, such as diatom-based biosensors, biocatalysts, and drug delivery systems, by utilizing the unique structural and chemical features of diatom frustules. This technique allows diatoms to be modified for a variety of applications in biotechnology, environmental science, and nanotechnology (Ford et al., 2020).

3.2.2 Chemical Doping

Chemical doping is the technique of incorporating various metals into the silica-based structure of diatom frustules during their growth. Examples include titanium, aluminium, and zinc. This is usually accomplished by putting these elements into the culture medium, where they are absorbed by the diatoms as they grow and form their frustules. The presence of these dopants alters the physical, chemical, and structural features of the frustules in a way that is not seen in ordinary diatom growth. For example, titanium doping can improve the mechanical strength and resistance of the frustules to external stress (Brzozowska et al., 2020, 2022) whilst aluminium can affect their optical properties, making them suitable in light-harvesting applications (Vicente-Garcia et al., 2024). Zinc, on the other hand, is commonly employed to modify the electronic properties of frustules, perhaps making them acceptable for use in electronic devices or sensors (Lamastra et al., 2017). Researchers can use these features to customize diatom frustules for specific uses, such as enhancing their functionality in nanotechnology, environmental sensing, or biological applications. The ability to adjust dopant quantities and kinds also allows for fine customisation of frustule properties, allowing the fabrication of new materials with desired qualities for a wide range of industrial and scientific applications (Vicente-Garcia et al., 2024).

3.2.3 Surface Functionalization

Surface functionalization of diatom frustules entails changing the frustule's exterior structure or chemical composition to improve its capacity to interact with certain molecules, making it more useful for a variety of applications, including biosensing and environmental remediation. One typical technique is Chemical etching, which uses acids or alkalis to selectively dissolve sections of the silica frustule (Townley et al., 2008). This method can produce micro- and nanoscale surface patterns such as holes, ridges, and cavities, increasing surface area and improving target molecule binding affinity (Losic et al., 2007). Researchers may manage the etching process to change the porosity and roughness of the frustule surface, making it more sensitive to specific environmental conditions or chemical interactions. Surface coating is another approach that involves depositing functional groups, polymers, or nanoparticles onto the surface of the frustule. These coatings can be engineered to have additional qualities such as greater hydrophobicity, improved charge interactions, or the capacity to bind specific biomolecules like proteins, antibodies, or enzymes (Villani et al., 2019). Coating the frustules can help prevent degradation in hostile settings, making them more durable for long-term use. These surface changes are especially useful in applications like as biosensing, where functionalized frustules may detect certain diseases or environmental contaminants, and environmental remediation, where they can absorb pollutants or heavy metals from water and air. The ability to adjust the surface features of diatom frustules makes them extremely adaptable and successful in many applications.

3.2.4 Nano Chemical Property Enhancement

Nanomechanical property enhancement of diatom frustules entail investigating the static and dynamic mechanical behaviours of these silica-based structures, which indicate their potential for usage in advanced nanotechnological and materials science applications (Cvjetinovic et al., 2023). Diatoms have frustules with complicated and resilient nanoscale patterns, which contribute to their exceptional mechanical qualities, including great strength and flexibility. These features are mostly attributable to the hierarchical structure and the frustules' distinct micro- and nanoporous architectures. Researchers evaluate diatom frustules' ability to tolerate mechanical stress by measuring their static properties, which include stiffness, hardness, and fracture resistance (Cvjetinovic et al., 2023). The dynamic properties are concerned with the behaviour of the frustules under changing forces, such as their reaction to vibration, oscillation, and deformation under variable loads.

The nanomechanical study of diatom frustules is significant because it lights on how these natural structures might be copied or modified for technological applications. For example, their large surface area and mechanical flexibility make them excellent for application in sensors, drug delivery devices, and lightweight but strong materials (Bhat et al., 2024). Diatoms' capacity to preserve structural integrity while exposed to a variety of environmental conditions is being investigated for use in nanoelectronics and biomaterials, where durability and usefulness at the nanoscale are critical (Romann et al., 2016). These investigations allow researchers to tune the features of diatom frustules through functionalization or surface modification, customizing them to specific applications like biosensing, drug delivery, and nanoengineering. In essence, knowing the nanomechanical characteristics of diatom frustules is critical to their application in cutting-edge nanotechnology and materials research.

3.3 *Engineering Diatom Frustules for Drug Delivery and Biosensing*

To alter the features of diatom frustules, researchers used chemical doping and surface functionalization. By integrating metal ions such as gold and silver during diatom growth, scientists were able to improve the optical properties of the frustules, making them appropriate for biosensing applications that detect changes in light emission or absorption (Rabiee et al., 2021).

Furthermore, surface modification techniques such as chemical etching were used to develop specific surface characteristics capable of interacting with target molecules. These alterations enhanced the frustules' adsorption capability, making them suitable candidates for transporting therapeutic chemicals in drug delivery systems (Ojo et al., 2021). Researchers also functionalized the frustules' surfaces with biomolecules like antibodies or peptides to boost specificity when targeting certain

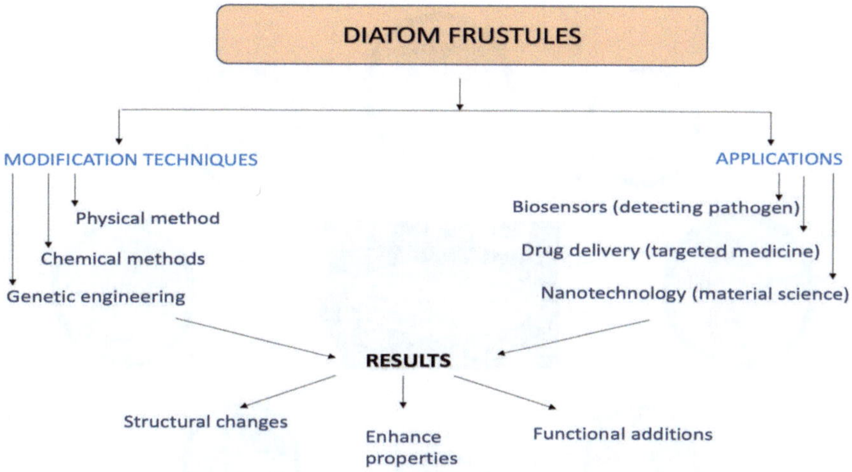

Fig. 3 Diatom frustule engineering and applications

cells or organs. The modified diatom frustules revealed increased ability to transport medicinal molecules such as chemotherapeutics and antibiotics. Their large surface area allowed for efficient drug loading, while surface changes guaranteed regulated and targeted drug release (Terracciano et al., 2018). In vitro investigations revealed that diatom frustules could release medications in response to specific environmental stimuli, such as pH shifts, which are found in certain disease environments (e.g., tumors or infection sites). By incorporating metal nanoparticles into diatom frustules, the modified structures developed plasmonic capabilities, allowing them to detect changes in the nearby environment, such as the presence of specific biomolecules (e.g., pathogens or cancer biomarkers) (Moharana et al., 2024). These biosensors demonstrated good sensitivity and specificity, making them useful for quick diagnosis and environmental monitoring (Fig. 3). Such engineering approaches emphasize the potential of biomaterials like diatoms to address healthcare concerns while also delivering long-term answers in nanotechnology and environmental research (Zobi, 2022).

4 Applications of Diatom-Based Nanoparticles

Due to the exclusive properties of diatomaceous earth and its silica-based structures, diatom nanotechnology has gained its place among the promising biomedical sciences. Unicellular algae known as diatoms have extremely versatile nanoporous cell walls named frustules. This document reviews four critical biomedical applications of diatom nanotechnology: drug delivery, gene delivery, anticancer treatments, and tissue engineering, with the recent developments in these areas (Fig. 4).

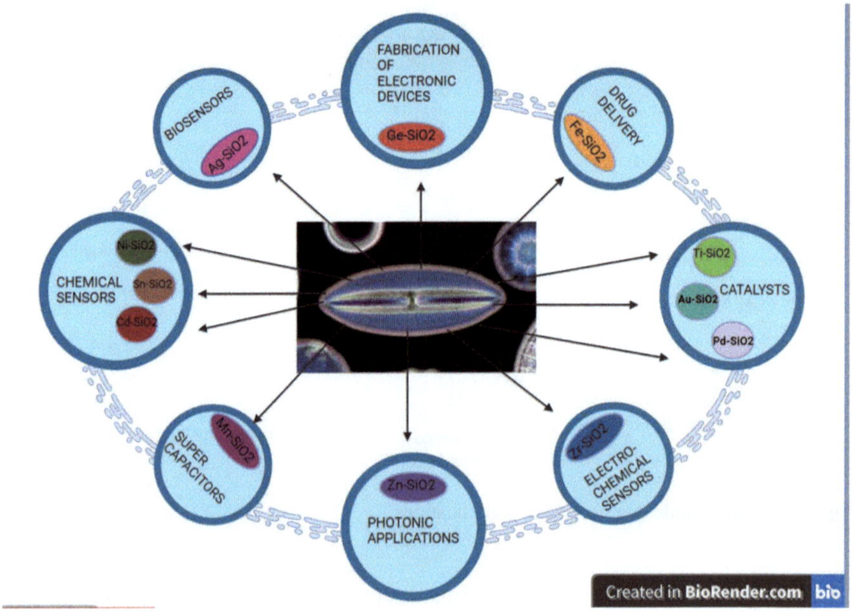

Fig. 4 Applications of diatom based nanoparticles

4.1 Drug Delivery

Diatom nanotechnology has a great advantage in drug delivery systems because of the high surface area, porosity, biocompatibility, and functionalizability of diatom biosilica. These properties allow for the loading, protection, and controlled release of therapeutic agents.

4.2 High Drug Loading Capacity

The nanoporous structure of diatom frustules makes it possible to encapsulate drugs efficiently, whether hydrophilic or hydrophobic. This will lead to a higher drug loading than with conventional carriers. Kang et al. (2024) showed that biosilica from diatoms could encapsulate chemotherapeutic agents such as doxorubicin, leading to sustained release at the target site (Lim et al., 2023).

4.3 *Controlled Release and Stimuli-Responsive Systems*

Functionalization of diatom biosilica with pH-sensitive or enzyme-responsive coatings enables controlled drug release. For example, in cancer therapy, diatom carriers can release drugs selectively in the acidic microenvironment of tumors, reducing off-target effects and minimizing systemic toxicity. Delalat et al. (2015) have pointed out the use of antibody-functionalized diatom biosilica for site-specific drug delivery, which enhances therapeutic efficacy and reduces side effects (Delalat et al., 2015).

4.4 *Combination Therapy*

Diatom-based drug carriers can be filled with multiple drugs or therapeutic agents, thus making it possible to apply combination therapy. For instance, combining chemotherapeutic agents with anti-inflammatory drugs may expedite treatment in diseases such as cancer and chronic inflammation.

4.5 *Gene Delivery*

Gene delivery is another key area where diatom nanotechnology has proved to be promising. The delivery of genetic materials, such as DNA, RNA, or CRISPR-associated systems, to target cells can revolutionize treatments for genetic disorders and cancer (Delalat et al., 2015).

4.5.1 Stable Encapsulation

Biosilica is formed by diatoms shields genetic molecules from enzymatic degradation in the systemic circulation. Incorporation within frustules of diatoms allows nucleic acids to stay stable and available in biological form for enhanced efficacy in treatment. The researches have proven the delivery potential of siRNA and plasmid DNA through diatom-based carriers, for gene therapy purposes (Rabiee et al., 2021).

4.5.2 Targeted Gene Delivery

The cell-targeting ligands functionalized on diatom surfaces can guide gene delivery to specific cell types. This targeted approach reduces off-target effects and improves therapeutic index. For example, antibody-functionalized diatoms have been explored to deliver CRISPR-Cas9 components to cancer cells and thereby suppress oncogenes (Rabiee et al., 2021).

4.5.3 Applications in Immunotherapy

Immunotherapy is also targeted through diatom biosilica-based gene delivery systems. These systems are able to transport genetic materials encoding cytokines or tumor-specific antigens and, thus, can be able to activate immune responses against cancer cells (Tramontano et al., 2023).

4.6 Anti-cancer Treatment

The development of diatom nanotechnology has paved the way for an innovative anti-cancer treatment involving drug delivery, photodynamic therapy, and combinations.

4.6.1 Targeted Chemotherapy

Diatom biosilica has a high drug-loading capacity and can be functionalized with targeting ligands, making it an excellent carrier for anticancer drugs. Functionalized diatom nanoparticles deliver drugs directly to tumor cells, thus causing minimal damage to healthy tissues. Tramontano et al. (2023) showed that antibody-functionalized diatom biosilica selectively targeted breast cancer cells, increasing the cytotoxic effects of chemotherapeutic drugs (Delalat et al., 2015).

4.6.2 Photodynamic Therapy (PDT)

In PDT, photosensitizing agents are activated by light to generate ROS, which destroy cancer cells. Diatom biosilica can efficiently carry photosensitizers, ensuring their accumulation at tumor sites. Terracciano et al. (2023) reviewed the use of diatom-based carriers for colorectal cancer PDT, highlighting their ability to enhance therapeutic outcomes (Tramontano et al., 2023).

4.6.3 Photothermal Therapy (PTT)

The photothermal therapy is possible with nanomaterial-coated diatoms, like gold nanoparticles, which produce localized heat through laser irradiation for killing cancer cells. The PDT/Pošt combined platform using the diatom biosilica has further addressed the challenge in treatment efficacy improvement.

4.6.4 Overcoming Drug Resistance

MDR represents one of the most important current challenges in cancer drug treatment. The concept of drug delivery using diatom nanocarriers can combine MDR modulators and chemotherapeutic agents to avoid resistance mechanisms within the cancer cells (Tramontano et al., 2023).

4.7 Tissue Engineering

The biocompatibility and porous structure of diatom biosilica make it an attractive material for tissue engineering. Its ability to support cell growth and differentiation offers immense potential for regenerating damaged tissues.

4.7.1 Bone Tissue Engineering

Diatom biosilica can mimic the structure of bone tissue, making it an ideal scaffold for bone regeneration. When combined with bioactive molecules, such as growth factors or osteogenic agents, diatom scaffolds promote bone formation and healing (Kang et al., 2024).

4.7.2 Soft Tissue Engineering

The surface of diatoms is also presently being investigated in soft tissue engineering. Hydrogels or bioactive compounds may be attached to the surface for the purpose of making scaffolds that would ensure epithelial or endothelial cells' growth (Tramontano et al., 2023).

4.7.3 Vascularization and Angiogenesis

In tissue engineering, vascularization is a critical feature for tissue survival. Diatom scaffolds functionalized with angiogenic factors, such as vascular endothelial growth factor (VEGF), can promote new blood vessel formation, thereby facilitating tissue integration and repair (Kang et al., 2024).

4.7.4 3D Bioprinting Applications

Diatom-based inks are being developed for 3D bioprinting. Those inks can be used to fabricate complex structures intended to work as scaffolds for different tissue types, such as cartilage, skin, and muscle (Tramontano et al., 2023).

4.8 Environmental Applications

Due to the exceptional characteristics of diatomaceous earth, such as a porous structure, large surface area, biocompatibility, and functional versatility, diatom nanotechnology has vast potential for applications in solving environmental problems. It allows diatoms to perform multifaceted functions in environmental applications, such as phytoremediation, surface-enhanced Raman spectroscopy (SERS), biocontrol, and nutraceuticals. The following are explorations on these applications, emphasizing the latest research and findings.

4.8.1 Phycoremediation

Phytoremediation, an idea to use plants and algae for an effective removal or neutralization of pollutants from the environment, has been greatly enhanced with diatom nanotechnology application. Diatoms biosilica, a natural nanomaterial, may serve as excellent adsorbent, and thus, heavy metal and pesticide carriers.

4.8.2 Heavy Metals Removal

Diatom frustules have a large surface area and hydroxyl groups that attract heavy metals such as lead (Pb), cadmium (Cd), and mercury (Hg). The functionalized biosilica enhances the adsorption efficiency. For instance, Ojo et al. (2021) modified the amine groups on the diatom biosilica, and they observed high adsorption capacities compared to traditional adsorbents in arsenic removal from wastewater (Ojo et al., 2021).

4.8.3 Degradation of Pesticide

Agricultural runoffs containing pesticides can be decomposed by the use of diatoms. Combinations of the diatom frustules and photocatalytic material, such as titanium dioxide, TiO_2, improve their light absorption property and the photodegradation ability of organic polluters. Recently, Wu et al. used TiO_2-coated diatoms to photodegrade atrazine under ultraviolet light in a study they published in 2021 (Delalat et al., 2015).

4.8.4 Oil Spill Cleanup

Oil spill remediation has been considered for diatomaceous earth through its hydrophobic properties and high oil absorption capacity. In this regard, modified diatom frustules developed with nanotechnology can selectively adsorb hydrocarbons from water, providing a cheap and green approach to treating marine oil pollution.

4.9 Surface-Enhanced Raman Spectroscopy (SERS)

Surface-enhanced Raman spectroscopy is a powerful analytical tool for detecting low concentrations of pollutants and environmental contaminants. The integration of diatom biosilica into SERS substrates has improved sensitivity and signal enhancement due to the unique structural and optical properties of diatom frustules. The periodic nanoporous structures of diatoms act as natural photonic crystals, which enhance light scattering and improve SERS signals. Functionalization of diatom surfaces with noble metals such as gold or silver creates highly sensitive SERS platforms. Liu et al. (2024) reported the fabrication of gold-coated diatom biosilica for detecting trace amounts of heavy metals in water, achieving detection limits as low as 10 parts per billion (ppb) (Rabiee et al., 2021). Diatom-based SERS platforms are being used for real-time environmental monitoring. For example, they can detect polycyclic aromatic hydrocarbons (PAHs) in polluted water sources or monitor pesticide residues in agricultural fields. The portability and cost-effectiveness of diatom-based SERS sensors make them practical for widespread environmental applications (Min et al., 2024). SERS coupled with other detection methods, for example, fluorescence or electrochemical sensing, expands the capabilities of diatom-based sensors. The systems can simultaneously detect several contaminants and provide holistic environmental assessments (Ahmed & Mohamed, 2024).

4.10 Biocontrol

Diatoms have significant potential as biocontrol agents in agriculture and aquaculture. Their natural properties, in conjunction with nanotechnology, allow them to serve as biological pesticides, fungicides, and antibacterials (De Tommasi & De Luca, 2022). Diatomaceous earth has been an ideal natural pesticide for many years. Its abrasive natures smash into the exoskeletons of insects, leading to dehydration and death. Through nanotechnology enhancement, diatom biosilica can be functionalized with essential oils or biocidal compounds for enhanced effectiveness in pest kill. Goswami et al. (2022) successfully showed that diatom-based nanopesticides could effectively control aphids and whiteflies, while chemical pesticide use could be reduced (Mariyam et al., 2024). Diatom biosilica can be used as an antifungal agent carrier in the fight against plant pathogens. For instance, composites containing diatoms, chitosan, or copper nanoparticles have been demonstrated to limit the growth of Fusarium and Aspergillus fungal species, which account for most of the crop damage (Zobi, 2022). Diatom-based biocontrol agents reduce bacterial and algal contamination in water systems of aquaculture. Functionalized diatom biosilica with antimicrobial peptides or silver nanoparticles inhibits the growth of pathogenic bacteria, thereby creating a healthier aquatic environment (Kang et al., 2024).

4.11 Nutraceuticals

Being an excellent source of bioactive compounds, such as omega-3 fatty acids, antioxidants, and polysaccharides, makes diatoms extremely valuable in the nutraceutical industry. In addition, the applications of these microorganisms were further expanded using nanotechnology for enhancing the bioavailability and stability of these compounds (Singla & Chakkaravarthi, 2017). Diatoms are the main producers of eicosapentaenoic acid (EPA) and docosahexaenoic acid (DHA), two essential omega-3 fatty acids with various health benefits. Encapsulation of these compounds in diatom biosilica improves their stability by preventing oxidation and improving delivery in functional foods and supplements (Kang et al., 2024). Diatoms contain antioxidants such as carotenoids and polyphenols, which fight oxidative stress and inflammation. Through functionalized diatom biosilica, the delivery of such compounds in targeted therapies is improved, enhancing their bioavailability and therapeutic efficacy (Singla & Chakkaravarthi, 2017). Silica in diatoms, for example, forms a prebiotic that fosters a healthy gut by supporting the growth of beneficial microorganisms. Further, diatomaceous earth is currently being investigated as a dietary supplement that could enhance digestion and mineral absorption (Singla & Chakkaravarthi, 2017). Diatoms are being used to create value-added functional foods and beverages. Diatom-based nanocarriers can be used to encapsulate vitamins or probiotics for delivery, ensuring controlled release and improved bioavailability (Singla & Chakkaravarthi, 2017).

4.12 Technological Applications

Diatoms are microscopic algae with silica-based nanoporous frustules that have attracted the attention of researchers in nanotechnology due to their unique structural, optical, and chemical properties. Their intricate and highly organized silica architectures are biological in origin and offer cost-effective and sustainable templates for advanced technological applications. The present document examines three critical technological applications of diatom nanotechnology: biochips, electroluminescent devices, and biosensing, along with recent advancements and supporting research (Cai et al., 2019).

4.12.1 Biochips

Biochips are miniaturized devices that have multiple functionalities for diagnostics, genomics, and proteomics. The fabrication and performance of these biochips have been significantly boosted in recent times by properties intrinsically associated with diatom biosilica including their high surface area, porosity, and biocompatibility (Singh et al., 2023).

4.12.2 Diatom Bio-Silica as a Template for Microarrays

Diatom frustules provide natural nanostructured surfaces that can be used as templates for immobilizing biomolecules such as DNA, RNA, or proteins (Jeffryes et al., 2008). The high surface-to-volume ratio of diatom frustules increases the density of immobilized biomolecules, improving the sensitivity and throughput of biochips. Hussein et al. (2022) demonstrated the use of functionalized diatom biosilica in creating DNA microarrays with enhanced hybridization efficiency, enabling rapid genetic screening (Delalat et al., 2015).

4.12.3 Combination with Photonic and Plasmonic Structures

Diatom-based biochips may integrate photonic and plasmonic structures for enhanced signal detection. The light-manipulation property of photonic crystals, such as those in diatom frustules, enhances the amplification of optical signals. For example, Liu et al. (2024) demonstrated a real-time protein detection assay by combining biosilica derived from diatoms with gold nanoparticles in a plasmonic biochip, allowing for femtomolar sensitivity (Rabiee et al., 2021).

4.12.4 Point-of-Care Diagnostics

Diatom-based biochips are being designed for portable, cost-effective diagnostic tools. The devices can be used in resource-limited settings for detecting diseases like tuberculosis, HIV, and cancers. Functionalization with specific antibodies or aptamers enables rapid biomarker detection with minimal sample preparation (Rabiee et al., 2021).

4.12.5 Electroluminescent Devices

The optical and electrical properties presented by diatoms make them unique for the formation of optoelectronic biosilica materials that are highly beneficial in electroluminescent devices and display technologies. The new paradigm of nanoscale manipulation in light makes diatoms potentially useful for enhancing the performance and efficiency of electroluminescent devices (Wojtczak et al., 2025).

4.12.6 Diatom Biosilica as Light-Scattering Medium

The frustules of diatoms contain periodic nanostructures that have an excellent scattering and diffracting ability to increase the brightness and uniformity of electroluminescent devices. Sun et al. (2024) found that diatom biosilica added into LED structures enhanced the light extraction efficiency by 25% while decreasing energy consumption (Tramontano et al., 2023).

4.12.7 Hybrid Nanocomposites for Improved Luminescence

The hybrids can be fabricated by combining diatoms with quantum dots, rare earth elements, or organic dyes to generate hybrid nanocomposites with controlled optical properties. Such composites have been explored for electroluminescent displays to render more vibrant and intense colors. For instance, the composites of diatom-frustules doped with europium ions resulted in improved red emission in OLEDs (organic light-emitting diodes) (Kang et al., 2024).

4.12.8 Flexible and Wearable Devices

Diatom-based nanocomposites have attracted increasing interest as potential alternative methods for flexible and wearable electroluminescent devices. Modified diatom frustules offer superior mechanical strength and flexibility for the formation of bendable displays and lighting systems, hence opening up opportunities for future consumer electronics (De Tommasi & De Luca, 2022).

4.13 Biosensing

Biosensing technologies rely on the detection of biological molecules or processes, and diatom nanotechnology offers innovative solutions for creating highly sensitive and specific biosensors. The natural nanostructure, biocompatibility, and ability to functionalize diatom frustules make them ideal platforms for biosensing applications

(Singh et al., 2023). Diatom frustules have unique optical properties, including photoluminescence and interference effects, which can be used for label-free biosensing. Functionalized diatom biosilica can detect biomolecules by measuring changes in refractive index or light scattering. Bhat et al. (2024) developed a diatom-based optical biosensor for detecting glucose in blood, achieving high sensitivity and specificity (Appiah et al., 2024). Combination of diatom biosilica with conductive materials like graphene or metal nanoparticles allows the production of electrochemical biosensors that will allow detecting trace amounts of analytes, such as heavy metals, toxins, or disease biomarkers with high precision. For example, diatom-graphene hybrids were used to obtain a biosensor detecting dopamine, an important neurotransmitter at nanomolar concentration (Kang et al., 2024). The increasing application of diatom-based biosensors in environmental monitoring has attracted attention in the field. Diatoms can sense pesticides, heavy metals, and organic contaminants in water and soil. Functionalization of diatom frustules with enzymes or antibodies acts as selective sensing elements to improve detection accuracy (Singla & Chakkaravarthi, 2017). This breakthrough development in nanotechnology of diatoms would allow portable biosensing devices of real-time performances. Medical diagnosis and monitoring applications in environmental engineering are important users of rapid-on-site analyses using these biosensors (De Tommasi & De Luca, 2022).

4.14 Emerging Applications

Diatoms are single-celled algae that have an intricate silica-based exoskeleton. These microorganisms have gained attention as versatile nanomaterials with the potential for revolutionary applications in nanotechnology. Their natural ability to fabricate nanoscale structures with high precision and biocompatibility opens the door to innovative solutions in various fields, including programmable nanostructures and energy storage (Lim et al., 2023). The unique properties of diatomaceous materials, such as their porous structure, large surface area, and versatility in chemical modifications, make them ideal candidates for these emerging applications (Tramontano et al., 2023). Among some of the very exciting areas is the development of programmable nanostructures in the field of diatom nanotechnology. Diatoms have an innate ability to engineer the biosynthesis of highly ordered complex nanostructures in itself. The main advantage is always the precise structures mostly formed at nanoscale offer a basis on which advanced functionalities can be established in materials customized for desired objectives. Through the genetic and metabolic pathways of diatoms, scientists can control the biosynthesis of their silica frustules to develop nanostructures with tailored characteristics for a wide range of applications such as drug delivery, biosensing, and tissue engineering (Aye & Sato, 2022).

The unique pore structure of diatom shells provides an opportunity for incorporating various biomolecules or nanoparticles into these systems to design multifunctional platforms for targeted therapeutic applications. By designing diatoms to express certain proteins or enzymes within their silica matrix, one can program the particles to respond to external stimuli such as pH, temperature, or electromagnetic fields. This kind of programmability opens up a new avenue of nanocarriers that are capable of delivering therapeutic agents in a controlled and site-specific manner, thus increasing the efficacy of treatments (Kang et al., 2024). The potential to form composite materials also provides the use of diatoms for templates to develop novel materials. Highly ordered geometries and chemistry at their surface allow guiding nanoparticles or functional groups in the self-assembly, opening the avenue to more sophisticated nanomaterials built by defining all of its properties with unprecedented precision. Programmable structures might eventually become breakthrough points for optoelectronics, sensors, or catalysis applications (Lugo et al., 2024).

4.14.1 Energy Storage

Energy storage is another promising application of diatom nanotechnology. With the world slowly shifting towards renewable energy sources, efficient energy storage systems are becoming increasingly critical. Diatomaceous materials, because of their high surface area and tunable porosity, have significant potential for use in batteries, supercapacitors, and other energy storage devices (Mishra et al., 2017). Engineers may improve the lithium-ion battery's performance through the porous structure of diatom silica by enhancing conductivity and anode or cathode materials' capacity. Researchers can develop electrodes with improved electrochemical properties by infusing nanoparticles into the diatom shell or using diatom-derived materials as templates (Lugo et al., 2024). Moreover, diatom-based materials can be engineered to increase their charge storage and release capacity, which would enhance the energy storage efficiency (Appiah et al., 2024). In the case of supercapacitors, diatom-based nanostructures form an ideal framework for enhancing the charge storage capacity. The hierarchical porosity and high surface area of diatom frustules may be used in designing electrodes that have improved charge–discharge cycles, thus furthering the development of more efficient energy storage solutions. In addition, by combining diatom-derived materials with other nanomaterials, such as graphene or carbon nanotubes, hybrid electrodes can be made that exhibit much higher energy storage capacity and stability (Mishra et al., 2017). Diatoms are also becoming increasingly used in energy storage systems, which aligns with the growing demand for sustainable, eco-friendly materials. Diatomaceous materials are available in abundance, biodegradable, and nontoxic, making them a good option for the development of green technologies. By using natural processes of diatom growth, researchers can develop energy storage solutions that are both efficient and environmentally friendly, thus aligning with global efforts to combat climate change (Lugo et al., 2024). The future development of programmable nanostructures and energy storage devices through diatom nanotechnology holds significant promise. Biosynthesis of diatom silica can be manipulated, and a platform can be established to produce complex functional nanomaterials that may be applied in diverse applications ranging from drug delivery systems to advanced energy storage devices.

Advancements in research in the areas described above are likely to eventually position diatom-based materials at the center of efforts to confront several of the biggest challenges facing nanotechnology, energy, and environmental sustainability (Aye & Sato, 2022).

5 Conclusion

Diatoms are promising candidates for environmentally friendly source for the biosynthesis of nanoparticles because of their capability to synthesize biogenic silica through the process known as biosilification. Their natural capability to uptake and accumulate silica makes them ideal for the sustainable synthesis of nanomaterials with various applications including in nanotechnology and biomedicine. The unique characteristics of diatom frustules including high porosity, surface area and biocompatibility enables the development of advanced materials which further finds their use in biosensing, drug delivery and other biomedical applications. As the demand for greener and cost-effective methods of synthesis of nanoparticle increases, diatoms can be employed as a viable alternative to traditional chemical methods while reducing hazardous environmental effects of chemical counterparts and promoting sustainable practices of green biosynthesis of nanoparticles. The continued exploration of diatom-based nanomaterials will likely result to further innovations in biotechnological and industrial fields.

Acknowledgements The authors would like to acknowledge the University for providing the necessary infrastructure support. Financial support for this study had been provided by the ASEAN-India Science, Technology & Innovation Cooperation (AISTDF Secretariat), India [Grant number CRD/2022/000595].

References

Abuzeid, H. M., Julien, C. M., Zhu, L., & Hashem, A. M. (2023). Green synthesis of nanoparticles and their energy storage, environmental, and biomedical applications. *Crystals, 13*(11), 1576. https://doi.org/10.3390/cryst13111576

Ahmed, M. A., & Mohamed, A. A. (2024). Advances in ultrasound-assisted synthesis of photocatalysts and sonophotocatalytic processes: A review. *Iscience, 27*(1), Article 108583. https://doi.org/10.1016/j.isci.2023.108583

Anu Mary Ealia, S., & Saravanakumar, M. P. (2017). A review on the classification, characterisation, synthesis of nanoparticles and their application. *IOP Conference Series: Materials Science and Engineering, 263*, 032019. https://doi.org/10.1088/1757-899X/263/3/032019

Appiah, E. S., Dzikunu, P., Akinwamide, S. O., Fangnon, E. A. K., Mensah-Darkwa, K., Andrews, A., Agyemang, F. O., Nartey, M. A., Makgopa, K., & Bossuyt, S. (2024). A review on progress and prospects of diatomaceous earth as a bio-template material for electrochemical energy storage: Synthesis, characterization, and applications. *Ionics, 30*(12), 7809–7860. https://doi.org/10.1007/s11581-024-05825-6

Aye, S., & Sato, Y. (2022). Therapeutic applications of programmable DNA nanostructures. *Micromachines, 13*(2), 315. https://doi.org/10.3390/mi13020315

Bala, K., Verma, S., Dogra, A., Das, A. P., Kumar, R., & Nandini. (2024). Sustainable management for urban plastic waste generation. *Sustainable Management of Urban Plastic Waste Through Circular Economic Approaches*, 211–236. https://doi.org/10.1201/9781003465447-13

Banjara, R. A., Kumar, A., Aneshwari, R. K., Satnami, M. L., & Sinha, S. K. (2024). A comparative analysis of chemical versus green synthesis of nanoparticles and their various applications. *Environmental Nanotechnology, Monitoring and Management, 22*, Article 100988. https://doi.org/10.1016/j.enmm.2024.100988

Bhardwaj, B., Singh, P., Kumar, A., Kumar, S., & Budhwar, V. (2020). Eco-friendly greener synthesis of nanoparticles. *Advanced Pharmaceutical Bulletin, 10*(4), 566–576. https://doi.org/10.34172/apb.2020.067

Bhat, K., Ajees, M. A., Kumar, P., Vibha, Bhat, V. G., Nayak, R., & Mazumder, N. (2024). Diatoms: harnessing nature's microscopic marvels for biosensing and multifaceted applications. *Biophysical Reviews, 17*(1), 103–125. https://doi.org/10.1007/S12551-024-01249-8

Bok, M. K., Chin, C. H., Choi, H. J., Ham, J. H., & Chang, B. S. (2022). Analysis of composition and microstructure of diatom frustules in mud on the coast of Boryeong- city, South Korea. *Applied Microscopy, 52*(1). https://doi.org/10.1186/s42649-022-00082-1

Brzozowska, W., Sprynskyy, M., Wojtczak, I., Dabek, P., Witkowski, A., & Buszewski, B. (2020). "Outsourcing" diatoms in fabrication of metal-doped 3D biosilica. *Materials, 13*(11). MDPI AG. https://doi.org/10.3390/ma13112576

Brzozowska, W., Sprynskyy, M., Wojtczak, I., Dąbek, P., Markuszewski, M. J., Witkowski, A., & Buszewski, B. (2022). Metabolically doping of 3D diatomaceous biosilica with titanium. *Materials, 15*(15). https://doi.org/10.3390/ma15155210

Buzea, C., Pacheco, I. I., & Robbie, K. (2007). Nanomaterials and nanoparticles: Sources and toxicity. *Biointerphases, 2*(4), MR17–MR71. https://doi.org/10.1116/1.2815690

Cai, Z., Ye, Y., Wan, X., Liu, J., Yang, S., Xia, Y., Li, G., & He, Q. (2019). Morphology-dependent electrochemical sensing properties of iron oxide-graphene oxide nanohybrids for dopamine and uric acid. *Nanomaterials, 9*(6), 835. https://doi.org/10.3390/nano9060835

Chopra, H., Bibi, S., Singh, I., Hasan, M. M., Khan, M. S., Yousafi, Q., Baig, A. A., Rahman, M. M., Islam, F., Emran, T. Bin, & Cavalu, S. (2022). Green metallic nanoparticles: Biosynthesis to applications. *Frontiers in Bioengineering and Biotechnology, 10*. https://doi.org/10.3389/fbioe.2022.874742

Cvjetinovic, J., Luchkin, S. Y., Statnik, E. S., Davidovich, N. A., Somov, P. A., Salimon, A. I., Korsunsky, A. M., & Gorin, D. A. (2023). Revealing the static and dynamic nanomechanical properties of diatom frustules—Nature's glass lace. *Scientific Reports, 13*(1). https://doi.org/10.1038/s41598-023-31487-x

Das, A., Dey, S., Bala, K., Gnanasekaran, R., & Das, A. P. (2024). Microbial bioremediation technology for sustainable treatment and management of synthetic microfiber waste. *Sustainable Microbial Technology for Synthetic and Cellulosic Microfiber Bioremediation*, 221–238. https://doi.org/10.1007/978-3-031-62600-5_12

De Tommasi, E., & De Luca, A. C. (2022). Diatom biosilica in plasmonics: Applications in sensing, diagnostics and therapeutics [Invited]. *Biomedical Optics Express, 13*(5), 3080. https://doi.org/10.1364/BOE.457483

Delalat, B., Sheppard, V. C., Rasi Ghaemi, S., Rao, S., Prestidge, C. A., McPhee, G., Rogers, M.-L., Donoghue, J. F., Pillay, V., Johns, T. G., Kröger, N., & Voelcker, N. H. (2015). Targeted drug delivery using genetically engineered diatom biosilica. *Nature Communications, 6*(1), 8791. https://doi.org/10.1038/ncomms9791

Dey, S., Dash, N., Ray, S. K., Priyadarsini, S., Chhotaray, C., & Das, A. P. (2024). Occurrence and quantification of tire wear plastics pollutants from highway city dust of Bhubaneswar City of Odisha. *Sustainable Microbial Technology for Synthetic and Cellulosic Microfiber Bioremediation*, 77–89. https://doi.org/10.1007/978-3-031-62600-5_5

Falasco, E., & Badino, G. (2011). *The role of environmental factors in shaping diatom frustule: Morphological plasticity and teratological forms*. https://www.researchgate.net/publication/233816694

Farjadian, F., Roointan, A., Mohammadi-Samani, S., & Hosseini, M. (2019). Mesoporous silica nanoparticles: Synthesis, pharmaceutical applications, biodistribution, and biosafety assessment. *Chemical Engineering Journal, 359*, 684–705. https://doi.org/10.1016/j.cej.2018.11.156

Ford, N. R., Xiong, Y., Hecht, K. A., Squier, T. C., Rorrer, G. L., & Roesijadi, G. (2020). Optimizing the design of diatom biosilica-targeted fusion proteins in biosensor construction for bacillus anthracis detection. *Biology, 9*(1). https://doi.org/10.3390/biology9010014

Gholami, P., Khataee, A., & Bhatnagar, A. (2020). Environmentally superior cleaning of diatom frustules using Sono-Fenton process: Facile fabrication of nanoporous silica with homogeneous morphology and controlled size. *Ultrasonics Sonochemistry, 64*. https://doi.org/10.1016/j.ultsonch.2020.105044

Goswami, P., Mathur, J., & Srivastava, N. (2022). Silica nanoparticles as novel sustainable approach for plant growth and crop protection. *Heliyon, 8*(7), e09908. https://doi.org/10.1016/J.HELIYON.2022.E09908

Hamed, A. F. (2023). Review article a brief and general overview on diatoms and their applications. *The Egyptian Journal of Phycology, 24*.

Hasan, S. (2015). A review on nanoparticles: Their synthesis and types. *Research Journal of Recent Sciences*.

Hempel, F., & Maier, U. G. (2012). An engineered diatom acting like a plasma cell secreting human IgG antibodies with high efficiency. *Microbial Cell Factories, 11*. https://doi.org/10.1186/1475-2859-11-126

Hervé, V., Derr, J., Douady, S., Quinet, M., Moisan, L., & Lopez, P. J. (2012). Multiparametric analyses reveal the pH-dependence of silicon biomineralization in diatoms. *PLoS ONE, 7*(10). https://doi.org/10.1371/journal.pone.0046722

Hussain, I., Singh, N. B., Singh, A., Singh, H., & Singh, S. C. (2016). Green synthesis of nanoparticles and its potential application. *Biotechnology Letters, 38*(4), 545–560. https://doi.org/10.1007/s10529-015-2026-7

Hussein, H. A., Nazir, M. S., Azra, N., Qamar, Z., Seeni, A., Din, T. A. D. A.-A. T., Abdullah, M. A., Hussein, H. A., Nazir, M. S., Azra, N., Qamar, Z., Seeni, A., Din, T. A. D. A.-A. T., & Abdullah, M. A. (2022). Novel Drug and Gene Delivery System and Imaging Agent Based on Marine Diatom Biosilica Nanoparticles. *Marine Drugs, 20*(8). https://doi.org/10.3390/MD20080480

Ijaz, I., Gilani, E., Nazir, A., & Bukhari, A. (2020). Detail review on chemical, physical and green synthesis, classification, characterizations and applications of nanoparticles. *Green Chemistry Letters and Reviews, 13*(3), 223–245. https://doi.org/10.1080/17518253.2020.1802517

Jara, N., Milán, N. S., Rahman, A., Mouheb, L., Boffito, D. C., Jeffryes, C., & Dahoumane, S. A. (2021). Photochemical synthesis of gold and silver nanoparticles—a review. *Molecules, 26*(15), 4585. https://doi.org/10.3390/molecules26154585

Javaheri, N., Dries, R., Burson, A., Stal, L. J., Sloot, P. M. A., & Kaandorp, J. A. (2015). Temperature affects the silicate morphology in a diatom. *Scientific Reports, 5*. https://doi.org/10.1038/srep11652

Jeffryes, C., Solanki, R., Rangineni, Y., Wang, W., Chang, C., & Rorrer, G. L. (2008). Electroluminescence and photoluminescence from nanostructured diatom frustules containing metabolically inserted germanium. *Advanced Materials, 20*(13), 2633–2637. https://doi.org/10.1002/adma.200800292

Kang, S., Woo, Y., Seo, Y., Yoo, D., Kwon, D., Park, H., Lee, S. D., Yoo, H. Y., & Lee, T. (2024). A descriptive review on the potential use of diatom biosilica as a powerful functional biomaterial: A natural drug delivery system. *Pharmaceutics, 16*(9), 1171. https://doi.org/10.3390/pharmaceutics16091171

Khan, I., Saeed, K., & Khan, I. (2019). Nanoparticles: Properties, applications and toxicities. *Arabian Journal of Chemistry, 12*(7), 908–931. https://doi.org/10.1016/j.arabjc.2017.05.011

Kiran Marella, T., Saxena, A., & Tiwari, A. (2020). Diatom mediated heavy metal remediation: A review. *Bioresource Technology, 305*, Article 123068. https://doi.org/10.1016/j.biortech.2020.123068

Kröger, N. (2007). Prescribing diatom morphology: Toward genetic engineering of biological nanomaterials. *Current Opinion in Chemical Biology, 11*(6), 662–669. https://doi.org/10.1016/j.cbpa.2007.10.009

Lamastra, F. R., Grilli, M. L., Leahu, G., Belardini, A., Li Voti, R., Sibilia, C., Salvatori, D., Cacciotti, I., & Nanni, F. (2017). Diatom frustules decorated with zinc oxide nanoparticles for enhanced optical properties. *Nanotechnology, 28*(37). https://doi.org/10.1088/1361-6528/aa7d6f

Li, X., Xu, H., Chen, Z.-S., & Chen, G. (2011). Biosynthesis of nanoparticles by microorganisms and their applications. *Journal of Nanomaterials, 2011*, 1–16. https://doi.org/10.1155/2011/270974

Lim, H., Seo, Y., Kwon, D., Kang, S., Yu, J., Park, H., Lee, S. D., & Lee, T. (2023). Recent progress in diatom biosilica: A natural nanoporous silica material as sustained release carrier. *Pharmaceutics, 15*(10), 2434. https://doi.org/10.3390/pharmaceutics15102434

Liu, Y., Liu, L., Sun, M., Xu, X., Kuang, H., Xu, C., & Xu, L. (2024). Development of a gold nanoparticles-based visual immuno-sensing platform for the rapid detection of gibberellins. *Food Control, 156*, 110113. https://doi.org/10.1016/J.FOODCONT.2023.110113

Losic, D., Mitchell, J. G., Lal, R., & Voelcker, N. H. (2007). Rapid fabrication of micro- and nanoscale patterns by replica molding from diatom biosilica. *Advanced Functional Materials, 17*(14), 2439–2446. https://doi.org/10.1002/adfm.200600872

Lugo, M. C., Saito, M., Kitamura, M., Ide, Y., Koide, S., & Mayama, S. (2024). Imaging and quantitative analysis on the etching of diatom frustules via digital holographic microscopy. *ACS Biomaterials Science and Engineering, 10*(2), 1106–1111. https://doi.org/10.1021/acsbiomaterials.3c01349

Mariyam, S., Upadhyay, S. K., Chakraborty, K., Verma, K. K., Duhan, J. S., Muneer, S., Meena, M., Sharma, R. K., Ghodake, G., & Seth, C. S. (2024). Nanotechnology, a frontier in agricultural science, a novel approach in abiotic stress management and convergence with new age medicine-a review. *Science of the Total Environment, 912*, Article 169097. https://doi.org/10.1016/j.scitotenv.2023.169097

Min, K. H., Kim, D. H., Youn, S., & Pack, S. P. (2024). Biomimetic diatom biosilica and its potential for biomedical applications and prospects: A review. *International Journal of Molecular Sciences, 25*(4), 2023. https://doi.org/10.3390/ijms25042023

Mishra, S., Dash, D., & Das, A. P. (2024). Isolation and molecular characterization of polyester-degrading bacterial strains from municipal wastewater and their application in microplastic degradation in contaminated wastewater. *Geomicrobiology Journal*. https://doi.org/10.1080/01490451.2024.2376101

Mishra, M., Arukha, A. P., Bashir, T., Yadav, D., & Prasad, G. B. K. S. (2017). All new faces of diatoms: Potential source of nanomaterials and beyond. *Frontiers in Microbiology, 8*. https://doi.org/10.3389/fmicb.2017.01239

Mohanpuria, P., Rana, N. K., & Yadav, S. K. (2008). Biosynthesis of nanoparticles: Technological concepts and future applications. *Journal of Nanoparticle Research, 10*(3), 507–517. https://doi.org/10.1007/s11051-007-9275-x

Mohanty, M., Tripathy, B., Koteswara Reddy, G., & Das, A. P. (2024). Greener technologies for establishing restraint over microplastic pollution. *Sustainable Microbial Technology for Synthetic and Cellulosic Microfiber Bioremediation*, 239–258. https://doi.org/10.1007/978-3-031-62600-5_13

Moharana, S., Dey, S., Priyadarsini, S., Kumar, M. S., & Das, A. P. (2024). Identification and characterization of microplastic pollutants from the marine sediments of paradeep coast of Bay of Bengal, India for their sustainable management. *Environmental Science and Engineering, Part, F2333*, 57–73. https://doi.org/10.1007/978-3-031-51792-1_4

Ojo, O. A., Olayide, I. I., Chidinma Akalabu, M., Olaitan Ajiboye, B., Ojo, A. B., Oyinloye, B. E., & Ramalingam, M. (2021). Nanoparticles and their biomedical applications. *Biointerface Research in Applied Chemistry, 11*(1), 8431–8445. https://doi.org/10.33263/BRIAC111.84318445

Parikh, H. (2021). Diatom biosilica as a source of nanomaterials. *International Journal of All Research Education and Scientific Methods (IJARESM), 9*(11).

Patra, J. K., & Baek, K.-H. (2014). Green nanobiotechnology: Factors affecting synthesis and characterization techniques. *Journal of Nanomaterials, 2014*(1). https://doi.org/10.1155/2014/417305

Rabiee, N., Khatami, M., Jamalipour Soufi, G., Fatahi, Y., Iravani, S., & Varma, R. S. (2021). Diatoms with invaluable applications in nanotechnology, biotechnology, and biomedicine: Recent advances. *ACS Biomaterials Science and Engineering, 7*(7), 3053–3068. https://doi.org/10.1021/acsbiomaterials.1c00475

Rea, I., & De Stefano, L. (2022). Special issue on new frontiers in diatom nanotechnology. *Applied Sciences (Switzerland), 12*(20). MDPI. https://doi.org/10.3390/app122010332

Romann, J., Chauton, M. S., Hanetho, S. M., Vebner, M., Heldal, M., Thaulow, C., Vadstein, O., Tranell, G., & Einarsrud, M. A. (2016). Diatom frustules as a biomaterial: Effects of chemical treatment on organic material removal and mechanical properties in cleaned frustules from two Coscinodiscus species. *Journal of Porous Materials, 23*(4), 905–910. https://doi.org/10.1007/s10934-016-0147-6

Roychoudhury, P., Golubeva, A., Dąbek, P., Gloc, M., Dobrucka, R., Kurzydłowski, K., & Witkowski, A. (2021). Diatom mediated production of fluorescent flower shaped silver-silica nanohybrid. *Materials, 14*(23), 7284. https://doi.org/10.3390/ma14237284

Roychoudhury, P., Bose, R., Dąbek, P., & Witkowski, A. (2022). Photonic nano-/microstructured diatom based biosilica in metal modification and removal—A review. *Materials, 15*(19). MDPI. https://doi.org/10.3390/ma15196597

Singh, A. K., Mittal, S., Das, M., Saharia, A., & Tiwari, M. (2023). Optical biosensors: A decade in review. *Alexandria Engineering Journal, 67*, 673–691. https://doi.org/10.1016/j.aej.2022.12.040

Singla, V., & Chakkaravarthi, S. (2017). Applications of prebiotics in food industry: A review. *Food Science and Technology International, 23*(8), 649–667. https://doi.org/10.1177/1082013217721769

Soleimani, M., Rutten, L., Maddala, S. P., Wu, H., Eren, E. D., Mezari, B., Schreur-Piet, I., Friedrich, H., & van Benthem, R. A. T. M. (2020). Modifying the thickness, pore size, and composition of diatom frustule in Pinnularia sp. with Al3+ ions. *Scientific Reports, 10*(1). https://doi.org/10.1038/s41598-020-76318-5

Sun, X., Zhang, M., Liu, J., Hui, G., Chen, X., & Feng, C. (2024). The art of exploring diatom biosilica biomaterials: from biofabrication perspective. *Advanced Science, 11*(6), 2304695. https://doi.org/10.1002/ADVS.202304695; PAGEGROUP:STRING:PUBLICATION

Tamburini, C., Garcin, J., Gregori, G., Leblanc, K., Rimmelin, P., & Kirchman, L. (2006). Pressure effects on surface mediterranean prokaryotes and biogenic silica dissolution during a diatom sinking experiment. *Aquatic Microbial Ecology, 43*, 267–276. https://doi.org/10.3354/ame043267i

Terracciano, M., De Stefano, L., & Rea, I. (2018). Diatoms green nanotechnology for biosilica-based drug delivery systems. *Pharmaceutics, 10*(4), 242. https://doi.org/10.3390/pharmaceutics10040242

Thakkar, K. N., Mhatre, S. S., & Parikh, R. Y. (2010). Biological synthesis of metallic nanoparticles. *Nanomedicine: Nanotechnology, Biology and Medicine, 6*(2), 257–262. https://doi.org/10.1016/j.nano.2009.07.002

Townley, H. E., Parker, A. R., & White-Cooper, H. (2008). Exploitation of diatom frustules for nanotechnology: Tethering active biomolecules. *Advanced Functional Materials, 18*(2), 369–374. https://doi.org/10.1002/adfm.200700609

Tramontano, C., De Stefano, L., & Rea, I. (2023). Diatom-based nanomedicine for colorectal cancer treatment: New approaches for old challenges. *Marine Drugs, 21*(5), 266. https://doi.org/10.3390/md21050266

Vicente-Garcia, C., Vona, D., Flemma, A., Cicco, S. R., & Farinola, G. M. (2024). Diatoms in focus: Chemically doped biosilica for customized nanomaterials. *ChemPlusChem.* https://doi.org/10.1002/cplu.202400462

Villani, M., Onesto, V., Coluccio, M. L., Valpapuram, I., Majewska, R., Alabastri, A., Battista, E., Schirato, A., Calestani, D., Coppedé, N., Zappettini, A., Amato, F., Di Fabrizio, E., & Gentile, F. (2019). Transforming diatomaceous earth into sensing devices by surface modification with gold nanoparticles. *Micro and Nano Engineering, 2*, 29–34. https://doi.org/10.1016/j.mne.2018.11.006

Vishwakarma, V., Samal, S. S., & Manoharan, N. (2010). Safety and risk associated with nanoparticles—a review. *Journal of Minerals and Materials Characterization and Engineering.*

Vrieling, E. G., Gieskes, W. W. C., & Beelen, T. P. M. (1999). Silicon deposition in diatoms: Control by the pH inside the silicon deposition vesicle 1. *Journal of Phycology, 35.*

Wang, Y., Cai, J., Jiang, Y., Jiang, X., & Zhang, D. (2013). Preparation of biosilica structures from frustules of diatoms and their applications: Current state and perspectives. *Applied Microbiology and Biotechnology, 97*(2), 453–460. https://doi.org/10.1007/s00253-012-4568-0

Wojtczak, I., Brzozowska, W., Bekissanova, Z., Trykowski, G., Rybczyński, P., Ośmiałowski, B., & Sprynskyy, M. (2025). Diatom biosilica modified with Ce–Tb mixed oxide twinning nanoparticles and with polyphases quasi-crystalline Tb oxide nanoparticles. *Colloids and Surfaces a: Physicochemical and Engineering Aspects, 704*, Article 135463. https://doi.org/10.1016/j.colsurfa.2024.135463

Yu, C.-H., Tam, K., & Tsang, E. S. C. (2008). *Chapter 5 Chemical methods for preparation of nanoparticles in solution* (pp. 113–141). https://doi.org/10.1016/S1570-002X(08)00205-X

Zhang, X., Yan, S., Tyagi, R. D., & Surampalli, R. Y. (2011). Synthesis of nanoparticles by microorganisms and their application in enhancing microbiological reaction rates. *Chemosphere, 82*(4), 489–494. https://doi.org/10.1016/j.chemosphere.2010.10.023

Zobi, F. (2022). Diatom biosilica in targeted drug delivery and biosensing applications: Recent studies. *Micro, 2*(2), 342–360. https://doi.org/10.3390/micro2020023

Fabrication of Phyto-Nanoparticles and Their Possible Outcomes: With a Special Reference to Iron Oxide Nanoparticles

Alisha Nayak, Suman Jagatee, Dillip Kumar Bishi, and Sakti Kanta Rath

Abstract Nanotechnology is a field of research that is constantly progressing and exploring the world of nanoparticles, which are minute particles with dimensions ranging from 1 to 100 nm. Its impact extends to various areas of study, including basic physics and medicinal chemistry. The minuscule size, rotational phase, and physical properties of nanoparticles are frequently used because they have been shown to modify the behavior of every other material that comes in contact with them. Various physical, elemental, and biological approaches can be employed to readily produce these particles. However, biological techniques are the most demanded preparation method because it is faster, more environmentally friendly and easier than other methods. Given the growing threat that climate change poses to the environment, it is much desired to develop sustainable and eco-friendly methods for producing metallic nanoparticles (Nps). This review gives a thorough analysis of the most recent studies on the production, uses, effects on the environment, and prospects for iron oxide nanoparticles are also provided. Also this review discuss briefly the latest advancements and prospective uses for environmentally friendly Silver Nanoparticles (AgNps), Gold Nanopaticles (AuNps), Copper/Copper Oxide Nanoparticles (CuO/CuNps), Zinc Oxide Nanoparticles (ZnO Nps).

Keywords Nanoparticles · Plant extracts · Green synthesis · Ecofriendly · Iron oxide nanoparticles

A. Nayak · S. Jagatee · S. K. Rath (✉)
Department of Life Sciences, Rama Devi Women's University, Bhubaneswar, Odisha, India
e-mail: saktikantarath@rdwu.ac.in

D. K. Bishi
Department of Biotechnology, Rama Devi Women's University, Bhubaneswar, Odisha, India

E. Priyadarshini et al. (eds.), *Advanced Nanotechnology Applications for Environmental Sustainability*, Sustainable Environmental Waste Management Strategies,
https://doi.org/10.1007/978-3-032-12933-8_4

1 Introduction

In recent days, a rapidly expanding topic in Science and nanotechnology is the creation of remarkable materials known as nanoparticles (Algebaly et al., 2022). The technological progress of the twenty-first century was creation of nanotechnology. This multidisciplinary field deals with the creation, application, and management of tiny materials with an average size below 100 nm (Jadoun et al., 2021). Nanomaterials are applied in a wide range of fields, such as bio-sciences, materials science, microbial biology, the field of engineering, computational imaging, appliances, environmental remediation, medical science, and various sectors of industry. Their primary characteristic is that they have a huge surface area to volume ratio (Saif et al., 2016). Nanoparticles can be made of a single ingredient or mixture of several materials. Depending on their electromagnetism and chemical structure, they can exist as dispersed aerosols, colloidal material, or suspensions of materials (Nair et al., 2022). The methodologies followed for synthesis of metallic nanoparticles is illustrated in (Fig. 1).

Nanomaterials can be created using a variety of techniques, including the co-precipitation, solvothermal, thermal decomposition, sol–gel, and hydrothermal synthesis. The physical and chemical approaches are cost effective and hazardous. Hazardous chemicals like hydrazine, hydrogen peroxide, sodium borohydride and dimethyl hydrazine are employed in chemical synthesis processes. All of these approaches employ chemicals that are considered to be environmentally toxic. To produce nanoparticles without using dangerous chemicals, researchers are looking

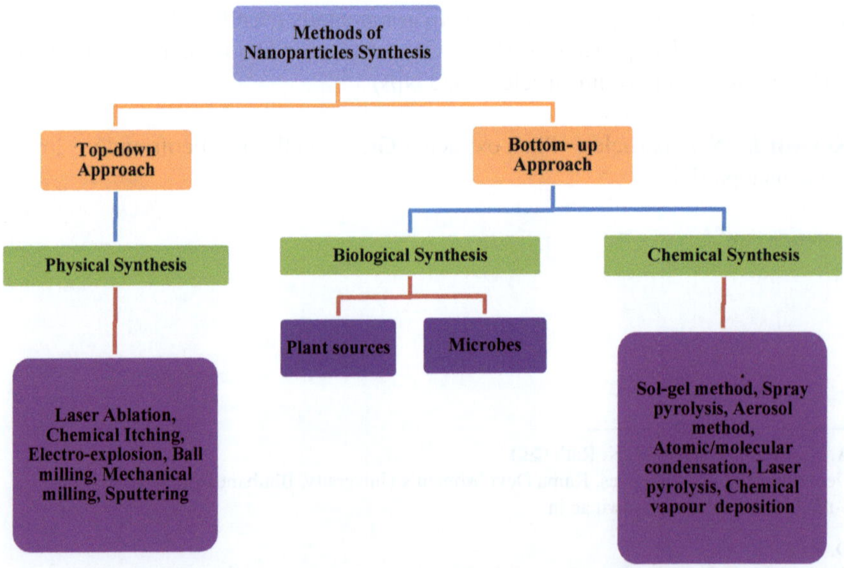

Fig. 1 Schematic representation of methods for synthesis of metallic nanoparticles

at a green synthesis alternative. Green nanomaterial development has so acquired importance recently due to its minimal cost, harmless, biocompatibility, and favorable ecological feature (Kirdat et al., 2021). The metallic nanoparticles produced through the green chemistry methodology exhibit enhanced stability and desirability. Due to their greater stability and nontoxic in nature this green synthesized metallic nanoparticles are found to have wide range of therapeutic applications which is depicted in (Fig. 2). This green approach primarily employs microorganisms, plants, algae, bacteria, yeast, and fungi. Three major factors influence the green method: the use of ecologically friendly solvents, the use of effective reducing agents, and the incorporation of non-toxic components for stabilization (Nair et al., 2022). A diversity of response factors, containing solvent, temperature, pressure and pH (acidic, basic or neutral) conditions influence green synthesis techniques using biological precursors (Singh et al., 2018). The eco-friendly production of nanoparticles makes use of plant extract, which is rich in plant derived chemicals such as phenol, terpenoids, amides, flavonoids, aldehydes, and ketones—all of which play a crucial role in the stabilization and reduction of nanoparticles (Kirdat et al., 2021). It has been hypothesized that plants are better appropriate for green nanoparticle production than microbes since they are not infectious and do not require the time-consuming procedure of maintaining cell cultures (Ahmad et al., 2021).

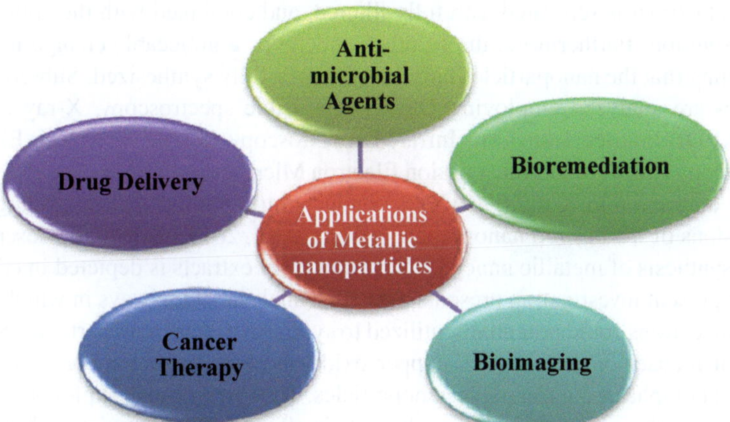

Fig. 2 Metallic nanoparticles application

2 Synthesis of Nanoparticles Using Plant-Mediated Green Approaches

Recent publications have extensively explored the utilization of microorganisms, such as fungi, bacteria, and plants, in the fabrication of diverse nanoparticles. The microorganisms exhibit significant functionality in the reduction of metal compounds to nanoparticles, due to their inherent minimizing or antioxidant properties. However, because very aseptic conditions must be maintained, microbe-mediated synthesis is not the most efficient biological method for producing nanoparticles for commercial use. Therefore, using plant extracts instead of microorganisms for this purpose may be preferable due to their ease of improvement, lesser biological hazards, and more involved process of maintaining cell cultures (Roy & Das, 2015). Biological techniques are utilized to synthesize metallic nanoparticles in various forms and sizes from plant components such as leaves, stems, roots, calluses, flowers, peels, seeds. Plant contains different phytochemicals which has high potential that can convert metal salts in to respective nanoparticles (Kumar et al., 2021). The illustration in (Fig. 3) demonstrates the involvement of many phytochemicals in the green production of metallic nanoparticles. Typically, botanical specimens undergo a thorough rinsing process using distilled water, followed by desiccation at ambient temperature to remove any residual moisture. Next, the desiccated sample is heated and then the plant extract is separated, carefully filtered, and combined with the appropriate metal solution. Furthermore, the solution undergoes a noticeable change in color, suggesting that the nanoparticles have been successfully synthesized. Subsequently, the Nps are analyzed employing Ultraviolet–visible spectroscopy, X-ray diffraction (XRD), Fourier Transform Infrared spectroscopy (FTIR), Scanning Electron Microscopy (SEM), and Transmission Electron Microscopy (TEM). The analysis of energy dispersion in X-ray (EDAX) is conducted to optimize the morphology and dimensions of the desired nanoparticles (Wang et al., 2019). Schematic description of the synthesis of metallic nanoparticles from plant extracts is depicted in (Fig. 4).

The present investigation presents a detailed analysis of the ways in which plant-based inventions could potentially utilized to synthesize metal nanoparticles. Specifically, it focuses on silver, gold, copper oxide and zinc oxide nanoparticles, with particular emphasis on iron oxide nanoparticles. The research also explores the practical uses of these nanoparticles, such as their effectiveness in antimicrobial treatment, drug delivery, cytotoxicity, dye degradation, waste water treatment, larvicidal activity, and antioxidant properties.

2.1 *Silver (Ag) Nanoparticles*

Nanotechnology is a fast growing area with many potential uses in biological sciences. Furthermore, silver nanoparticles (AgNps) have been used as an antibiotic and decontaminator that is mostly devoid of side effects. Ag Nps have extensive

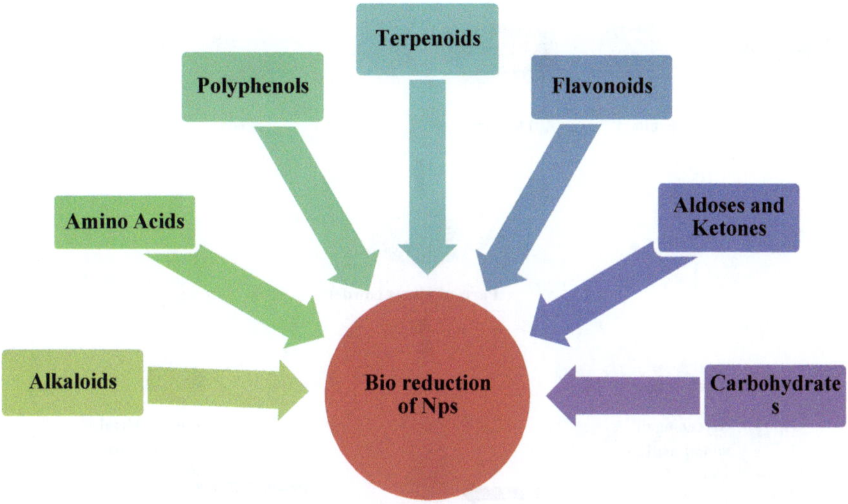

Fig. 3 Phytochemicals for bio-reduction and fabrication of Nps

antifungal, antibacterial, and antiviral activities (Salleh et al., 2020). Silver nanoparticles are capable of entering the walls of bacterial cells, distorting the structure of cell membranes and potentially inducing the death of cells. The efficacy of these nanoparticles can be ascribed to their substantial surface area to volume ratio, in conjunction with their remarkably diminutive size. The release of silver ions facilitates the generation of reactive oxygen species, hinder DNA replication, and enhance the permeability of cell membranes (Yin et al., 2020). Ag Nps can be synthesized in various sizes and forms, depending on the extraction technique used. Ag nanoparticles are available in various shapes, such as spheres, rods, wires, and plates (Shenashen et al., 2014). A study was conducted using the leaf extract of *Catharanthus roseus* to create Ag Nps. These nanoparticles showed promising results in removing chromium and cadmium, and also demonstrated antibacterial capabilities against the pathogenic bacteria *S. aureus* (Verma & Bharadvaja, 2022). The synthesis of silver nanoparticles was conducted utilizing an extract derived from the leaves of *Polyalthia longifolia.* The dimensions of the particles were observed within a range of 10 to 40 nm. The antifungal capabilities were assessed through experimentation involving *Alternaria alternata,* a species of plant pathogenic fungi (Dashora et al., 2022). Researchers have endeavored to employ fruit waste in an innovative way by employing *Mangifera indica* fruit waste to produce silver nanoparticles. Subsequently, the characterization of these nanoparticles was conducted utilizing various spectroscopic methods, including UV–Vis, Zeta potential analysis, FTIR, XRD, EDX, and TEM. The evaluation of the antibacterial, antifungal, and cytotoxic properties of spherical silver nanoparticles was conducted using HeLa, MCF-7, and normal fibroblast cell lines. The dimensions of the particles were quantified to be 26.85 nm in both dimension and morphology (Donga & Chanda, 2021). Ag Nps were made from *Azadirachta indica*

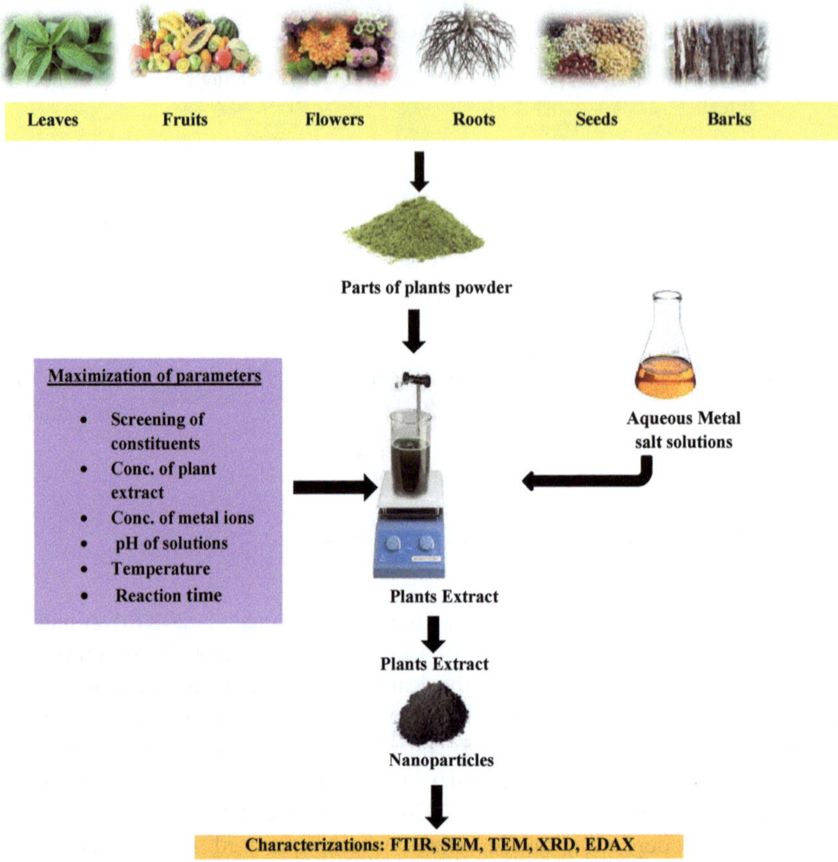

Fig. 4 Flow chart demonstrating the procedure for creating nanoparticles with the assistance of parts of plants

leaf extract, which has a particle size range of 67.94 ± 0.72 nm. A plant disease caused by *Xanthomonas oryzae pv. oryzae* was used to test the effectiveness of the Ag Nps that were developed as antimicrobial agents. The findings indicated a significant level of antibacterial activity, which was found to be superior to that of 200 mg/l of streptocycline (Mankad et al., 2020). *Annona senegalensis* seed bark extract was used to create spherical Ag Nps, which demonstrated potent antibacterial activity against *Pseudomonas aeuginosa* and *Kleriella oxytoxa* (Olabemiwo et al., 2021). The process of synthesizing silver nanoparticles (Ag Nps) incorporated the use of *Ocimum canum* leaf extract as a capping agent. The morphological shapes were validated through scanning electron microscopy (SEM) and X-ray diffraction (XRD), indicating an average particle size of 15.72 nm. The nanoparticles demonstrated the most significant Zone of Inhibition (ZOI) in relation to *E. coli.* (Tailor et al., 2020). An eco-friendly approach was employed to extract silver nanoparticles (Ag Nps) from the roots of *Berberis asiatica.* In order to complete the extraction process, it was necessary to take an aqueous root extract. Silver nanoparticles were found to have a spherical morphology and an average dimension of 9.8 nm, according to research that was carried out with the assistance of a transmission electron microscope (TEM) (Dangi et al., 2020). Additionally, floral extract of *Jatropha integerrima Jacq.* was used to synthesize AgNps, which had a mean dimension of 50.07 nm. In addition, they revealed the powerful antibacterial activity of AgNps against various bacterial strains (Suriyakala et al., 2022). A synthesis of Green AgNps was conducted using the root extract of *Premna integrifolia. L.* AgNps indicated effectiveness against clinically isolated strains of *P. aeruginosa, S. aureus, and E. coli*, as well as HepG2 cells, which exhibited substantial cytotoxicity against the AgNps. *P. aeruginosa*, in particular, had the highest level of sensitivity to these nanoparticles (Singh et al., 2023). A research investigation was carried out to explore the efficacy of *Lallemantia royleana* leaf extract in the synthesis of silver nanoparticles (AgNps). The results showed that these nanoparticles possess remarkable antioxidant and antibacterial properties, effectively combating *Candida*, Gram positive and negative bacteria. The resultant AgNps showed a dose-dependent anti-inflammatory activity and In vitro anti-arthritic efficacy. The newly generated AgNps had a greater cytotoxic impact on human breast (MCF-7) and hepatoma (Hep-G2) cancer cells. Methylene blue, an organic pollution dye, was used to test the biocatalytic activity of the produced AgNps, and they were shown to be able to catalyze the MB (Methylene blue) dye (Sharifi-Rad et al., 2024).

2.2 Gold (Au) Nanoparticle

Gold nanoparticles (Au Nps) are a crucial component in nanometal engineering, particularly in terms of their antibacterial properties. A comprehensive investigation has been conducted on the antibacterial efficacy of Au Nps, focusing on their diameter, structure, concentration, and capping agent (Rabiee et al., 2022). Furthermore, the study indicated that green produced AuNps, which were manufactured utilizing *Scrophularia striata* extract, had antibacterial activity against Gram-positive and Gram-negative bacterial strains, as well as demonstrating that they have efficient antileishmanial capabilities. In addition, gold nanoparticles were utilized as a noble nanocatalyst for the purpose of degrading the color methylene blue (Roya Alizadeh et al., 2023). Synthesis of Au Nps using *Citrullus colocynthis* seed extract has also been reported. SEM image showed spherical structure with range from 7 to 33 nm (Mubeen et al., 2022). Gold nanoparticles derived from *Physalis minima* demonstrate antibacterial, antioxidant, and antidiabetic properties, which may significantly augment their applicability within the biopharmaceutical and nano-biopharmaceutical regions (Sekar et al., 2022). This study explored the synthesis of gold nanoparticles using *Combretum erythrophyllum* leaf extract, as well as the effectiveness of these nanoparticles against bacteria. Nanoparticles, which had an average size of 13.2 nm, demonstrated antibacterial activity against Gram-positive and Gram-negative bacteria that are pathogenic, as well as cell viability against normal and malignant cells (Fanoro et al., 2021). An extract of the leaves of *Pongammia pinnata*, which is also referred to as pongam, was employed in order to demonstrate the ecologically friendly synthesis of gold nanoparticles that were mediated by plants and their propensity to inhibit the growth of bacteria. The size of the gold nanoparticles that were created was around 16 nm in size. It has been demonstrated that gold nanoparticles has a powerful antimycobacterial activity against *M. tuberculosis.* Furthermore, these nanoparticles have the ability to prevent the proliferation of breast cancer cell lines (Govindaraju et al., 2020). *Colus scutellarioides* (L.) Benth leaf extract was used to successfully synthesize AuNps using a plant-mediated, environmentally friendly method. Researchers compared the antioxidant and anticancer effects of AuNps on two human cell lines: MDA-MB-231 and Hs-27. They found that the inhibition of MDA-MB-231 cells increased proportionally with increasing AuNPs content, suggesting that these nanoparticles may have a role in the treatment of breast cancer. However, when it came to the normal Hs-27 cells, which are not malignant, the produced AuNps showed no signs of toxicity (Al-Mafarjy et al., 2024).

2.3 Copper Oxide (CuO) Nanoparticles

One of the elements that are most prevalent on Earth is copper. Due to its various characteristics, including great malleability, strong corrosion resistance, and superior electrical and thermal conductivity, it has played a significant role in history. Due to their application in heat transfer systems, water purification, and surgical equipment antimicrobial coating, copper nanoparticles (CuNps) have gained attention from the wider community (Crisan et al., 2022). Copper nanoparticle synthesis is substantially more difficult than noble metal preparation due to the possibility of copper oxidation with air (Tamilvanan et al., 2014). During preparation, copper changes into CuO and Cu_2O, and then to Cu^{2+}, causing difficulties in continued synthesis of CuNps in a normal condition (Edis et al., 2019). To avoid this problem, copper nanoparticles are produced in an inert gas environment. To limit oxidation, protective polymers or surfactants are used in some circumstances (Tamilvanan et al., 2014). The study utilized *Eucalyptus Globoulus* leaf extract to synthesize 88 nm spherical Copper oxide (CuO Nps) particles, as demonstrated in SEM Images, which effectively eliminated methyl orange from aqueous solution (Alhalili et al., 2022). Spherical shaped copper oxide nanoparticles were produced from *Averrhoa carambola* leaf extract. To examine the antibacterial activity of CuO Nps, the Agar well diffusion technique was utilized. It was determined by them that the ability of green produced CuO Nps to kill two gram (+)ve and gram (−)ve bacterial strains can open up new possibilities for the development of antibacterial (Saha et al., 2023). The research suggests that copper nanoparticles (CuNps) might be produced by employing the leaf extract of *Camelia sinensis.* This extract demonstrated a high level of effectiveness in the process of photo degrading bromophenol blue dye when exposed to sunlight. Through the utilization of FT-IR, SEM, TEM, and EDX examination, the CuNps were characterized, revealing that they were composed of agglomerated particles that had an average size of 60 ± 6 nm (Ahmed et al., 2019). The utilization of phytochemicals that are taken from an extract of a plant known as *Cissus arnotiana* has resulted in the generation of copper nanoparticles, which have garnered a great deal of interest in the field of nanotechnology. The potential of biosynthesized copper nanoparticles (CuNps) in nanomedicine and biomedical applications was demonstrated by their strong antibacterial activity against *Escherichia coli* and their antioxidant capabilities that were equivalent to those of ascorbic acid (Rajeshkumar et al., 2019). CuNps were synthesized from *Celastrus paniculatus* willd leaf extract, with particle sizes ranging from 2 to 10 nm. An investigation was conducted on the antifungal activity of a synthetic copper nanoparticle against the plant pathogenic fungus *Fusarium oxysporum*. Moreover, the deterioration of the dye known as methylene blue demonstrated the CuNps' photocatalytic potential (Mali et al., 2020). The *Cinnamomum zelanicum* aqueous extract was used to create CuNps with a spherical morphology, and its antioxidant properties were further examined. This Nps was also administered to people as a therapy for many forms of lung cancer (Liu et al., 2021). The study investigated the antibacterial properties of copper nanoparticles (CuNps) generated from garlic husk extracts. The findings showed CuNps effectively inhibited both Gram-positive and Gram-negative bacteria, with an average crystal size of 14 nm, suggesting potential applications in nanomedicine (Senthilkumar et al., 2024).

2.4 Zinc Oxide (ZnO) Nanoparticles

Zinc oxide nanoparticles (ZnO Nps) are a crucial metal oxide material that have been extensively employed in materials research because of its special chemical, physical, and biological advantages, which include their non-toxicity, affordability, and ecological friendliness. Hexagonal zinc oxide can occasionally be found mixed with cubic zinc or as cubic rock salt (Shaba et al., 2021). Zinc oxide, also known as ZnO, is a semiconducting metal oxide that has rapidly acquired popularity in recent years due to the extensive variety of uses it offers in the fields of electrical products, biomolecular research, optical science, medicinal platforms, and the production of antimicrobial medications (Fahimmunisha et al., 2020). The research demonstrated that ZnO nanoparticles synthesized from *Ocimum americanum* display notable antibacterial activity against a range of Gram-positive, Gram-negative bacteria and fungal pathogens, indicating their potential for use in antimicrobial therapies (Narendra kumar et al., 2019). According to the study ZnO nanoparticles that were made with *Crotalaria verrucosa* leaf extract had antibacterial action against a range of microorganisms. The nanoparticles also showed that they caused cellular damage, apoptotic DNA damage, and anti-proliferative effects on carcinoma cells, indicating that the pharmaceutical, food processing, and biomedical sectors may find use for them as antibacterial and anti-cancer agents (Sana et al., 2020). Zinc oxide nanoparticles derived from *Berberis aristata* leaf extract, demonstrated antibacterial activity against various bacteria strains and were found to prevent urinary tract infections (Chandra et al., 2019). The ZnO nanoparticles that were created by the extract of *Ziziphus jujuba* leaves were utilized for the purpose of adsorbing and removing harmful Pb(II) ions as well as two unique azo dyes, namely methyl orange (MO) and methylene blue (MB) (Alharthi et al., 2021). The study investigated the antibacterial properties of green-synthesized ZnO–Nps from *Dysphania ambrosioides* extract, emphasizing the importance of nanoparticle size in enhancing their effectiveness against various strains (Álvarez-Chimal et al., 2022). According to the study, methylene blue dye was possibly efficiently destroyed by zinc oxide nanoparticles made from neem and eucalyptus leaf extracts, which have an impressive photocatalytic activity of 97.98% (Supin, 2023). Green ZnO–Nps from *Moringa oleifera* leaves showed strong hepatoprotective effects against CCl_4, a substance that causes liver damage in rats. These nanoparticles could potentially prevent liver damage by boosting antioxidant enzyme activity and restoring liver and kidney function (El-Beltagi et al., 2024)

2.5 Iron Oxide Nanoparticles

The size of iron oxide nanoparticles can range from 1 to 100 nm (Nair et al., 2022). The fields of medicine, biosensors, and drug delivery have gained new opportunities as a result of the development of microscopic iron oxide nanoparticles with remarkable properties like low toxicity, size-dependent magnetic behavior, superparamagnetism, biological compatibility, durability in physiological and natural environments, and high saturation magnetization (Zambri et al., 2019). Iron (Fe) and the oxygen molecule (O) combine to make iron oxide in its native state. In nature, IONps are most frequently found in three different chemical forms: magnetite (Fe_3O_4), maghemite (γ-Fe_2O_3), and hematite (α-Fe_2O_3). Magnetite is currently attracting a lot of scientific attention when compared to other kinds of IONps. They offer significant advantages such as superparamagnetism, a greater surface area, a high surface-to-volume ratio, and a quick and simple approach for separating the target from the solution because the external magnetic field modifications. As a result, IONps (magnetite) have been widely studied and employed in biomedicine (Gambhir et al., 2022). Iron oxide nanoparticles, or IONps, are one of the many different kinds of nanomaterials that have the benefit of being easier to separate from other nanomaterials and have potent catalytic and reductive capabilities for use in treating waste water. IONps are widely employed in a number of scientific applications, including the removal of antibiotics, colored pigments, and hazardous chemicals from water sources, as well as target site drug delivery and tumor destruction (Bhuiyan et al., 2020). Iron-based nanoparticles have once again been shown to be effective against a broad variety of pathogenic strains of bacteria and fungi. This is primarily attributable to the fact that these nanoparticles are able to produce highly reactive oxygen species (ROS) (Muthukumar et al., 2019). In vivo cell monitoring is being performed with the use of high-sensitivity magnetic resonance imaging (MRI) probes that are present in iron oxide nanoparticles (IONps). These probes have also made it feasible to develop a variety of diagnostic devices that are capable of detecting and tracking diseases (Masthoff et al., 2019). The bark, leaves, roots, shoots, and stems of plants, as well as the entire plant, can be used to obtain extracts for the synthesis of IONps. After thoroughly cleaning the plant components with tap water, they are cleaned to remove any impurities with double-distilled water and allowed to air dry. The sample is dried, weighed, and then crushed. Subsequently, plant powder and distilled water are combined and boiled, stirring constantly. Afterwards, the mixture is preserved for future use by filtering it via Whatman filter paper. When iron precursors (iron chloride, iron nitrate, and iron sulphate) are reacted with, plant biomolecules serve as reducing agents and produce iron oxide nanoparticles (IONps). Several methodologies have been utilized to assess the characteristics of synthetic IONps. UV–vis (Ultraviolet visible) spectroscopy is employed to confirm the synthesis of IONps. In order to ascertain the morphological properties, several microscopy techniques, such as atomic force microscopy (AFM), scanning electron microscopy (SEM), and transmission electron microscopy (TEM), are utilized. It is possible to determine which functional groups are present by the use of Fourier transform infrared spectroscopy, often known as FTIR. While EDX and total reflection X-ray fluorescence (TXRF) are two techniques that might be used to determine the element's purity and composition, X-ray diffraction is the technique that is applied to evaluate the crystallinity of IONps that have been engineered (Akintelu et al., 2021). When compared to other nanoparticles, iron oxide nanoparticles are affordable, effective, and simple to use, which makes their magnetic properties commonly used. Because of their strong magnetic properties, iron oxide nanoparticles were initially used in biology and then in medicine for the magnetic separation of cells and biological products. They were also used for magnetic particle system guidance in order to transport drugs to specified sites in the body (Ali et al., 2016).

3 Biological Applications of Iron Oxide Nanoparticles

3.1 Antimicrobial Activity

A major challenge in the area of biomedicine is examining the interactions between microbes and magnetic IONps. These interactions harm the cellular components of bacteria and mediate the formation of ROS. This results in the coordinated targeting and eradication of bacteria or other microorganisms (Gambhir et al., 2022). Numerous studies have tested the antibacterial properties of IONps produced from plants. Iron oxide nanoparticles were generated in the study by using an aqueous extract of *Laurus nobilis L.* leaves. These nanoparticles had shown a moderate level of effectiveness against the microorganisms *Aspergillus flavus, Penicillium spinulosum,* and *Listeria monocytogenes*. Semiconductor devices, gas sensors, light photocatalysis, and antibacterial and antifungal medications are all potential applications for these nanoparticles (Jamzad & Kamari, 2020). The round-shaped Fe_3O_4 nanoparticles (Nps) were synthesized using plant metabolites from *Bombax malabaricum*. These Nps, with a mean size of 09 $\pm$ 2 nm, exhibited strong antibacterial properties and moderate antifungal effects against different strains (Awais et al., 2023). The research found that Fe_2O_3 nanoparticles made from *Achyranthes aspera* were highly effective in killing bacteria including *Klebsiella pneumonia*, *Escherichia coli*, and *Bacillus*. Researchers determined that 50 mg/mL was the amount that effectively inhibited *Klebsiella pneumonia* the most. The study found that Fe_2O_3 nanoparticles may efficiently inhibit bacterial growth, which is surprising given their small size. Because of their ability to pass through bacterial cell walls, they are being considered as a possible antibacterial agent (Ahmad et al., 2022). According to the findings of the research, Fe_3O_4 nanoparticles derived from *Mikania mikrantha* leaf extract were able to suppress the development of five different bacterial strains and four different fungus strains. The strain of *Candida albicans* was found to have the highest inhibition zone recorded. Their ability to penetrate bacterial cell walls, despite their very small size, suggests they may have medicinal use as an antibacterial agent (Biswas et al., 2021). According to the findings of the research, iron oxide nanoparticles that were bioreduced from the water-soluble extract of avocado rind exhibited powerful antibacterial activity against the pathogens that were examined. IO-Nps exhibited the greatest inhibition against *Proteus and Staphylococcus sp.* (Kamaraj et al., 2019). The synthesis of green FeO nanoparticles was accomplished by the use of the leaf extract of *Nicotiana plumbaginifolia*. Experiments conducted with the Muller-Hinton agar disk diffusion technique, which can be applied to Gram-positive as well as Gram-negative bacterial strains, revealed that these nanoparticles have powerful antibacterial characteristics. This gives more evidence that the nanoparticles may be effective in the treatment of infection caused by microorganisms (Duan et al., 2024).

3.2 Drug Delivery for Cancer Therapy

Drug-loaded nanoparticles ought to carry medications to the intended place for an appropriate period without damaging healthy cells. Only a few strategies exist for loading drugs onto nanoparticles, such as incorporating medication components onto the nanoparticles' exterior coating or importing the molecules that contain the drug in the encapsulating material. It has been established that drug-loaded Fe_3O_4–Nps are injected into blood vessels and subsequently located in the appropriate location (cancer cells or tumor) using a magnetic field that surrounds them. This can aid in medication retention and site-specific drug release, increasing the efficiency of treating cancer cells without killing neighboring healthy cells (Yew et al., 2020).

An environmentally sustainable technique for producing IONps with a size of up to 50 nm by utilizing an aqueous extract derived from green tea leaves. Drug loading experiments revealed that iron oxide nanoparticles derived from liquids had a 94% drug concentration on their active surface. Based on in vitro drug release studies, doxorubicin (DOX) was shown to be released more effectively in a slightly acidic (pH 4.5) environment compared to a neutral (pH 7.4) one (Raoof et al., 2023). Additionally, a rabbit model in vivo bio distribution investigation of radiolabeled IONps revealed fast organ accumulation in the heart, liver, and kidneys. This study made the case for green synthetic iron oxide nanoparticles as promising nanocarriers for imaging diagnostics and effectively delivering DOX (Doxorubicin) to certain organs. The most prevalent hazard today is cancer, a group of illnesses defined by uncontrollable cell growth and division. Even in the present day, this remains a significant cause of death in humans, ranking second only to heart disorders. It is a critical health concern that demands attention. Iron oxide nanoparticles generate reactive oxygen species (ROS), which impede the growth of cancer cells. Iron oxide is a significant nanoparticle in the realm of transition metal oxides. It is valued for its diverse oxidative state, magnetic properties, and affordability. By applying a magnetic field externally, it is simple to drive magnetic iron oxide nanoparticles toward the tumor cell. Iron oxide nanoparticles, which included *Annona muricata* as a bioreductant, were investigated for the possibility that they may inhibit the growth of cancer cells. Researchers employed Hela cancer cells and HEK 293 cells, which are representative of human embryonic kidneys, in order to determine whether or not the nanoparticles were detrimental. As a consequence of the nanoparticles, both cell lines experienced the death of their cells, as indicated by the measurements. A concentration of 48.97 μg/mL was found for Hela cells, whereas the concentration of 63.05 μg/mL was found for HEK 293 cells, which led to a 50% reduction in cell death. Considering that the iron oxide nanoparticles exhibited a greater degree of cytotoxicity towards the cancerous Hela cells in contrast to the normal HEK 293 cells, it would suggest that they have the ability to inhibit the progress of cancer (Elemike et al., 2024).

Recent research on biosynthesized iron oxide nanoparticles from *Pimenta dioica* leaf extract demonstrated a strong cytotoxic impact on DLD-1 human colon cancer cells, indicating their safety and effectiveness in treating cancer (Pillai et al., 2022). The IONps derived from *Nigella sativa* seeds were cytotoxic to lung and colorectal cancer cell lines, inducing cell death and inhibiting proliferation. They induced programmed cell death by activating apoptotic genes and decreasing anti-apoptotic gene expression, pointing to potential cancer treatment targets (Aliya et al., 2023). Researchers found that iron oxide nanoparticles might kill MCF-7 breast cancer cells by functionalizing with the bioactive components of stem bark extract from *Cinchona officinalis*. One potential application for iron nanoparticles in cancer therapy delivery systems is their capacity to kill MCF-7 cells (Al-Harbi et al., 2022). The MTT (3-(4, 5-dimethylthiazolyl-2)-2, 5-diphenyltetrazolium bromide) assay was used in a research to examine the potential anticancer properties of hematite nanoparticles derived from a Piper betel leaf extract. The nanoparticles also included hematite. The research demonstrated that there was a cytotoxic effect on A549 (Lung Cancer) cells, which was concentration dependant. The cytotoxic response that occurred differed depending on the concentration (Yoonus et al., 2021). Iron oxide nanoparticles derived from *Carica papaya* leaf extract have been shown to be toxic at maximal doses against Hela, BHK-21, and Vero cell lines; yet, their remarkable activity raises the possibility of using them to destroy tumor cells (Bhuiyan et al., 2020).

3.3 Environmental Remediation

To make a variety of goods more appealing to consumers, dyes are frequently utilized in industrial settings. The majority of the time, however, typical wastewater treatments does not eliminate this class of chemicals, leading to a pollutant load. Using iron oxide nanorods derived from *Withania coagulans*, the study investigates the photocatalytic degradation of Safranin dye and how well the nanorods reduce the dye when exposed to sunlight. Biologically produced nanorods have been shown to have stronger photocatalytic activity compared to those that were manufactured chemically. The presence of biomolecules derived from the plant extract is responsible for the enhanced efficiency of the nanorods (Qasim et al., 2020). *Psidium guajava* and *Moringa oleifera* combined to produce iron oxide nanoparticles. The degradation of Methylene blue dyes by photocatalysis employing these nanoparticles under solar radiation was demonstrated to lead to a gradual reduction in dye fluorescence. The dye solution's color shift demonstrated how this Nps aided in the degradation process and emphasized their possible application in ecological remediation and water purification (Madubuonu et al., 2019). The researchers synthesized α-Fe_2O_3 nanoparticles using *Vernonia amygdalina* extract and examined their capacity to degrade dyes by photocatalysis. The results demonstrated the considerable effects of experimental factors such as pH and light source on degradation efficiency, suggesting their potential application as catalysts for wastewater treatment (Weldegebrieal & Sibhatu, 2021). The goal of research was to create iron oxide nanoparticles from beet stalks (*Beta vulgaris*) and pumpkin leaves (*Cucurbita moschata*) in order to eliminate food colorings. The study demonstrated that these nanoparticles had superior adsorption capabilities towards tertrazine and bordeaux red dyes (Lima Barizão et al., 2020). The efficacy of iron oxide nanoparticles in adsorbing arsenic was investigated by synthesizing them using green leaf extracts from five distinct plant species. Eucalyptus leaf nanoparticles demonstrated the highest level of adsorption efficiency. Research has shown that certain non-agricultural plants, like oak and eucalyptus, are more effective at producing iron nanoparticles for environmentally friendly arsenic removal compared to green tea or black tea leaves (Kamath et al., 2020).

The decolorization of the dyes from dark blue/yellow to colorless indicated that the Fe_2O_3 nanoparticles derived from *Achyranthes aspera* efficiently broke down the indigo carmine and bromothymol blue dyes. Over time, the dye solutions' absorbance steadily dropped, indicating the effectiveness of Fe_2O_3 nanoparticles in the degradation of dyes (Awais et al., 2023). In comparison to other reported nanoparticles, iron oxide nanoparticles synthesized using Amla seed extract demonstrated exceptional performance in removing Methylene Blue dye. The nanoparticles achieved maximum dye removal in less time and dose, and they did so without the need to control parameters like pH or temperature. The optimal elimination of Methylene Blue dye was achieved by the combination of iron oxide nanoparticles and $NaBH_4$, demonstrating the complementary effects of both elements (Ashraf et al., 2023). There were a variety of dyes that were applied in order to test the biocatalytic activity of FeNps that were generated from *Polygonum plebeium*. As a result of the findings of the investigation, it was revealed that FeNps are efficient in the degradation of dyes. This indicates the potential for the materials to be exploited in the process of environmental restoration (Begum et al., 2024).

3.4 Larvicidal Effect

Mosquito-borne illnesses are a serious problem for society that contribute to a number of problems, including those that are highly frequent in India and include, malaria,, Japanese encephalitis, lymphatic filariasis, chikungunya fever, and dengue fever. Plant-derived larvicides are a promising class of natural pesticides because they are environmentally friendly, have low levels of toxicity for organisms other than the target, and contain a variety of chemical compounds that cooperate to target various biological processes, lowering the risk of resistance (Arasu et al., 2019). Exploring the application of green synthesised iron oxide nanoparticles (IONps) derived from the leaf extract of *Grevillea robusta* as a larvicide for *Aedes aegypti* and *Anopheles stephenti* mosquito larvae. High death rates for both mosquito species were seen in larvicidal bioassays at varying dosages (80–400 ppm). The findings showed notable patterns in larval mortality that was dependent on both time and dose, and there was no noticeable distinction in the two species of mosquitoes or in the methods used to handle mosquito larvae (Zargham et al., 2023). The *Cassia auriculata* flower's aqueous extract was used to green synthesizes iron oxide nanoparticles. IONps had high larvicidal ability against the *Aedes aegypti* mosquito, the vector of dengue fever. In the research that was carried out on *Anopheles subpictus* mosquito larvae, the larvicidal activity of iron oxide nanoparticles that were produced by employing the leaf extract of *Celosia argentea* was proved effectively. Through extensive testing, it was found that higher concentrations of the nanoparticles exhibited a greater ability to eliminate larvae, resulting in a higher mortality rate. This discovery opens up the potential for using these nanoparticles as effective larvicides to effectively control mosquito populations (Sudhakar et al., 2022).

3.5 Antioxidant Activity

The presence of antioxidants is essential for ensuring that our cells continue to function properly and that they continue to be healthy. Free radicals have the potential to create serious health problems. By evaluating the DPPH (1, 1-diphenyl-2-picrylhydrazyl) free radical scavenging activity of iron oxide nanoparticles, it is possible to establish the degree to which these particles are successful in neutralizing free radicals.

This assay, which utilizes ascorbic acid as a positive control, is designed to quantify the different amounts of iron oxide nanoparticles that are dependent on dose. Percentages are used to illustrate the findings about the nanoparticles and the ascorbic acid. Reactive oxygen species (ROS) are crucial for the capacity to scavenge free radicals. The IC_{50} value of iron oxide nanoparticles, or the quantity of material needed to produce 50% scavenging activity indicates the potential of these particles. The iron oxide nanoparticles that were produced with the help of *Celosia argentea* leaf extract shown strong antioxidant activity; the greatest percentage was seen at 97% in the 80 μg/mL concentration. Ascorbic acid, which served as a reference over the whole range of dosages, was shown to have less antioxidant activity than the nanoparticles. Through the use of DPPH free-radical scavenging, the antioxidant activity was evaluated, and the results indicated that the nanoparticles offered potential protection against the damage caused by free radicals (Velsankar et al., 2022). The research indicates that the iron oxide nanoparticles that were synthesized from the leaf extract of *Ficus carica* have a high level of antioxidant activity. The DPPH free radical scavenging activity was equivalent to 5.14 mg of ascorbic acid, and the ORAC tests were equivalent to 7.360 mg of Trolox. About 50% of the DPPH radicals in the environment might be eliminated by these nanoparticles (Ustun et al., 2022). Several tests were conducted to assess the antioxidant activities of several plant extracts, including those of *Moringa oleifera, Terminalia bellirica,* and Fe Nps prepared from these extracts. The antioxidant assay findings showed that the plant extracts and the Fe Nps made from them have antioxidant properties, which might lead to their use in treating oxidative damage and associated diseases (Jegadeesan et al., 2019). A study used the leaf extract of Cilantro (*Coriandrum sativum L.*) to produce iron oxide nanoparticles. IONps have an IC_{50} value of 86.68 mg/ml on the DPPH radical, which suggests that they are just as efficient at scavenging radicals as regular ascorbic acid (Singh et al., 2020). According to the study, iron oxide nanoparticles may be used as antioxidant agents in biological applications because of their potential antioxidant qualities. In addition, an extract of the *Madhuca indica* plant was employed in the process of producing FeONps. In this study, a comparison was done between the plant extract, the control, and many different sources of plant extract. According to the findings, nanoparticles exhibited the maximum antioxidant activity level, measuring 81.5% when the concentration was set at 1000 μg/mL during the experiment. There was a trend toward antioxidant activity with rising FeONps concentration (Shabbir et al., 2023). The study revealed that FeONps, synthesized from *Nicotiana plumbaginifolia*, showed strong radical trapping activity, with a noticeable color change from violet to yellow. Increased FeONps concentration increased antioxidant activity, making them promising for further research in antioxidant applications (Duan et al., 2024).

In addition to the previously mentioned biological applications of iron oxide nanoparticles, Table 1 presents a summarized data regarding the utilization of various plant species regarding the synthesis of iron oxide nanoparticles and their associated bioactivity.

4 Conclusion

Nanotechnology has the potential to revolutionize the world through its diverse applications across various disciplines. At present, scientists recommend for the production of nanoparticles through the green chemistry method. This method allows for the creation of metallic nanoparticles with precise attributes, including dimensions, form, and morphology, by utilizing reducing, capping, and stabilizing agents. These nanoparticles have a multitude of applications in addressing environmental concerns and ensuring human health and safety. The plant-based nanoparticles have a broad variety of uses within the realm of medicine, including antibacterial activities, therapeutic medicines, and potential antiviral activities. The synthesis of nanoparticles through biological means is considered to be a more dependable approach when compared to chemical agents. This is because plant extracts and physiologically active biomolecules can be cultivated in various regions. The nanoparticles synthesized through a green chemistry approach have a vast array of potential applications. These nanoparticles can be conveniently characterized using various spectroscopic methods, allowing for more precise modeling studies. This review study attempts to offer a complete outline of the many elements of IONps and their applications in the field of biomedicine. This review focuses on the recent advancements in IONps-based research, covering the entire process from synthesis to application. Furthermore, this source provides a brief description of different categories of nanoparticles that are specifically designed for biomedical applications. Further investigation is necessary to conduct clinical trials in order to determine the suitability of multifunctional iron oxide nanoparticles with magnetic characteristics for in vivo usage on an industrial scale.

Table 1 Different plant extracts used for synthesizing Iron oxide nanoparticles

Sl no.	Plants name	Plants part	Size and morphology	Applications	References
1	*Platanus orientalis*	Leaves	38 nm and spherical	Antifungal activity	Devi et al. (2019)
2	*Plantago major*	Leaves	4.6–30.6 nm and spherical	Dye removal activity	Lohrasbi et al. (2019)
3	*Pheonix dactylifera*	Leaves	12.61 nm and spherical	Antioxidant activity	Abdullah et al. (2020)
4	*PunicaGranatum*	Fruits peel	26.52–158.44 nm and spherical	Anticancer activity against nasopharyngeal carcinoma cell line	Yusefi et al. (2020)
5	*Euphorbia herita*	Leaves	25–80 nm and rough surface with cavity like structure	Antimicrobial activity against bacterial and fungal pathogen	Ahmad et al. (2021)
6	*Canthium coromandelicum*	Leaves	19.25 nm and spherical	Antibacterial and Photocatalytic activity	Sudhakar et al. (2021)
7	*Psidiumguajava L*	Leaves	12.64 nm and mostly cubic and partly spherical	Antibacterial, antifungal, dye degradation activity	Adhikari et al. (2022)
8	*Spondias pinnata*	Fruits	5.38 nm, and Needle shaped	Anticancer activity against lung and breast cancer cell lines	Prabhu et al. (2022)
9	*Melia azedarach*	Flowers	231.43 ± 5.21 nm and Spherical	Antimicrobial and antioxidant activity	Muzafar et al. (2022)
10	*Nephrolepis exaltata*	Whole plant	30 to 70 nm and roughly spherical	Antiplasmodial and Cytotoxic activity	Nadeem et al. (2022)
11	*Terminalia catappa*	Leaves	19.14 nm and spherical and polygonal	Antimicrobial and Cytotoxic activity	Elemike et al. (2023)
13	*Anastatica hierochuntica*	Whole plant	30–70 nm and spherical	Antibacterial, antioxidant and in vitro cytotoxic activity against MCF-7 cell lines	Vahini et al. (2024)
14	*Andrographis serpyllifolia*	Leaves	20.11 nm and triclinic crystalline structure	Larvicidal activity against larvae of *Culex quinquefasciatus* mosquito	Soundarya and Karmegam (2023)
15	*Calotropis procera*	Latex	3 and 5 nm and spherical	Nanocatalyst for biodiesel production	Kumar et al. (2024)

(continued)

Table 1 (continued)

Sl no.	Plants name	Plants part	Size and morphology	Applications	References
16	*Hibiscus sabdariffa*	Calyces	61 nm and Spherical	Antibacterial activity	Saod et al. (2024)
17	*Aegle marmelos*	Leaves	18.78 nm and spherical	Antibacterial, antifungal, antioxidant, cytotoxic and Catalytic activity of Basic Fuchsin dye	Rama et al. (2024)

Author Contributions SKR conceptualized and designed the methodology of paper, AN investigated and wrote the paper, SJ reviewed and edited the paper, SKR and DKB gave the final correction of the manuscript. All authors read and approved the manuscript.

Funding Preparation of this review article received no funding from any organization.

Declarations

Ethical Approval Ethical approval is not applicable here.

References

Abdullah, J. A. A., Salah Eddine, L., Abderrhmane, B., Alonso-González, M., Guerrero, A., & Romero, A. (2020). Green synthesis and characterization of iron oxide nanoparticles by *Pheonix dactylifera* leaf extract and evaluation of their antioxidant activity. *Sustainable Chemistry and Pharmacy, 17*, 1–7. https://doi.org/10.1016/j.scp.2020.100280

Adhikari, A., Chhetri, K., Acharya, D., Pant, B., & Adhikari, A. (2022). Green synthesis of iron oxide nanoparticles using *Psidium guajava* L. Leaves extract for degradation of organic dyes and anti-microbial applications. *Catalysts, 12*, 1–14. https://doi.org/10.3390/catal12101188

Ahmad, W., Kumar Jaiswal, K., & Amjad, M. (2021). *Euphorbia herita* leaf extract as a reducing agent in a facile green synthesis of iron oxide nanoparticles and antimicrobial activity evaluation. *Inorganic and Nano-Metal Chemistry, 51*, 1147–1154. https://doi.org/10.1080/24701556.2020.1815062

Ahmad, W., Singh, V., Ahmed, S., & Nur-e-Alam, M. (2022). A comprehensive study on antibacterial antioxidant and photocatalytic activity of *Achyranthes aspera* mediated biosynthesized Fe_2O_3 nanoparticles. *Results in Engineering, 14*, 1–7. https://doi.org/10.1016/j.rineng.2022.100450

Ahmad, B., Shireen, F., Rauf, A., Shariati, M. A., Bashir, S., Patel, S., Khan, A., Rebezov, M., Khan, M. U., Mubarak, M. S., & Zhang, H. (2021). Phyto-fabrication, purification, characterisation, optimisation, and biological competence of nano-silver. *IET Nanobiotechnology, 15*(1), 1–18. https://doi.org/10.1049/nbt2.12007

Ahmed, A., Usman, M., Liu, Q. Y., Shen, Y. Q., Yu, B., & Cong, H. L. (2019). Plant mediated synthesis of copper nanoparticles by using *Camelia sinensis* leaves extract and their applications in dye degradation. *Ferroelectrics, 549*, 61–69. https://doi.org/10.1080/00150193.2019.1592544

Akintelu, S. A., Oyebamiji, A. K., Olugbeko, S. C., Folorunso, A. S. (2021). Green synthesis of iron oxide nanoparticles for biomedical application and environmental remediation: A review. *Ecletica Quimica Journal, 46*, 17–37. https://doi.org/10.26850/1678-4618eqj.v46.4.2021.p17-37

Algebaly, A. S., Mohammed, A. E., Abutaha, N., & Elobeid, M. M. (2022). Biogenic synthesis of silver nanoparticles: Antibacterial and cytotoxic potential. *Saudi Journal of Biological Sciences, 27*, 1340–1351. https://doi.org/10.1016/j.sjbs.2019.12.014

Alhalili, Z. (2022). Green synthesis of copper oxide nanoparticles CuO Nps from *Eucalyptus Globoulus* leaf extract: Adsorption and design of experiments. *Arabian Journal of Chemistry, 15*, 1–17. https://doi.org/10.1016/j.arabjc.2022.103739

Al-Harbi, L. N., Al-Shammari, G. M., Subash-Babu, P., Mohammed, M. A., Alkreadees, R. A., & Yagoub, A. E. A. (2022). *Cinchona officinalis* phytochemicals-loaded iron oxide nanoparticles induce cytotoxicity and stimulate apoptosis in MCF-7 human breast cancer cells. *Nanomaterials, 12*, 1–20. https://doi.org/10.3390/nano12193393

Alharthi, M. N., Ismail, I., Bellucci, S., & Salam, M. A. (2021). Green synthesis of zinc oxide nanoparticles by *Ziziphus jujuba* leaves extract: Environmental application, kinetic and thermodynamic studies. *Journal of Physics and Chemistry of Solids, 2021*(158), 1–14. https://doi.org/10.1016/j.jpcs.2021.110237

Ali, A., Zafar, H., Zia, M., ul Haq, I., Phull, A. R., Ali, J. S., Hussain, A. (2016). Synthesis, characterization, applications, and challenges of iron oxide nanoparticles. *Nanotechnology, Science and Applications, 9*, 49–67. https://doi.org/10.2147/NSA.S99986

Aliya, S., Rethinasabapathy, M., Yoo, J., Kim, E., Chung, J. Y., & Cha, J. H. S. (2023). Phytogenic fabrication of iron oxide nanoparticles and evaluation of their in vitro antibacterial and cytotoxic activity. *Arabian Journal of Chemistry, 16*, 1–15. https://doi.org/10.1016/j.arabjc.2023.104703

Al-Mafarjy, S. S., Suardi, N., Ahmed, N. M., Kernain, D., Hisham Alkatib, H., Dheyab, M. A. (2024). Green synthesis of gold nanoparticles from *Coleus scutellarioides* (L.) Benth leaves and assessment of anticancer and antioxidant properties. *Inorganic Chemistry Communications, 161*, 1–11. https://doi.org/10.1016/j.inoche.2024.112052

Álvarez-Chimal, R., García-Pérez, V. I., Álvarez-Pérez, M. A., Tavera-Hernández, R., Reyes-Carmona, L., Martínez-Hernández, M., & Arenas-Alatorre, J. Á. (2022). Influence of the particle size on the antibacterial activity of green synthesized zinc oxide nanoparticles using *Dysphania ambrosioides* extract, supported by molecular docking analysis. *Arabian Journal of Chemistry, 15*, 1–19. https://doi.org/10.1016/j.arabjc.2022.103804

Arasu, M. V., Arokiyaraj, S., Viayaraghavan, P., Kumar, T. S. J., Duraipandiyan, V., Al-Dhabi, N. A., & Kaviyarasu, K. (2019). One step green synthesis of larvicidal, and azo dye degrading antibacterial nanoparticles by response surface methodology. *Journal of Photochemistry and Photobiology B: Biology, 190*, 154–162. https://doi.org/10.1016/j.jphotobiol.2018.11.020

Ashraf, I., Singh, N. B., & Agarwal, A. (2023). Green synthesis of iron oxide nanoparticles using Amla seed for methylene blue dye removal from water. *Materials Today: Proceedings, 72*, 311–316. https://doi.org/10.1016/j.matpr.2022.07.404

Awais, S., Munir, H., Najeeb, J., Anjum, F., Naseem, K., Kausar, N., Shahid, M., Irfan, M., & Najeeb, N. (2023). Green synthesis of iron oxide nanoparticles using *Bombax malabaricum* for antioxidant, antimicrobial and photocatalytic applications. *Journal of Cleaner Production, 406*, 1–13. https://doi.org/10.1016/j.jclepro.2023.136916

Begum, S., Sahoo, T., Swain, S., Nayak, A., Das, S., Patel, S., Rath, S. K., & Rath, C. C. (2024). Fabrication of iron nanoparticles using different bioactive precursors, their characterization and bioactivity evaluation. *Sustainable Chemistry for the Environment, 6*, 1–11. https://doi.org/10.1016/j.scenv.2024.100100

Bhuiyan, M. S. H., Miah, M. Y., Paul, S. C., Aka, T. D., Saha, O., Rahaman, M. M., Sharif, M. J. I., Habiba, O., & Ashaduzzaman, M. (2020). Green synthesis of iron oxide nanoparticle using *Carica papaya* leaf extract: Application for photocatalytic degradation of remazol yellow RR dye and antibacterial activity. *Heliyon, 6*, 1–13. https://doi.org/10.1016/j.heliyon.2020.e04603

Biswas, A., Vanlalveni, C., Lalfakzuala, R., Nath, S., & Rokhum, S. L. (2021). *Mikania mikrantha* leaf extract mediated biogenic synthesis of magnetic iron oxide nanoparticles: Characterization and its antimicrobial activity study. *Materials Today: Proceedings, 42*, 1366–1373. https://doi.org/10.1016/j.matpr.2021.01.108

Chandra, H., Patel, D., Kumari, P., Jangwan, J. S., & Yadav, S. (2019). Phyto-mediated synthesis of zinc oxide nanoparticles of *Berberis aristata*: Characterization, antioxidant activity and antibacterial activity with special reference to urinary tract pathogens. *Materials Science and Engineering: C, 102*, 212–220. https://doi.org/10.1016/j.msec.2019.04.035

Crisan, M. C., Teodora, M., Lucian, M. (2022). Copper nanoparticles: Synthesis and characterization, physiology, toxicity and antimicrobial applications. *Applied Sciences, 12*, 1–18. https://doi.org/10.3390/app12010141

Dangi, S., Gupta, A., Gupta, D. K., Singh, S., & Parajuli, N. (2020). Green synthesis of silver nanoparticles using aqueous root extract of *Berberis asiatica* and evaluation of their antibacterial activity. *Chemical Data Collections, 28*, 1–10. https://doi.org/10.1016/j.cdc.2020.100411

Dashora, A., Rathore, K., Raj, S., & Sharma, K. (2022). Synthesis of silver nanoparticles employing *Polyalthia longifolia* leaf extract and their in vitro antifungal activity against phytopathogen. *Biochemistry and Biophysics Reports, 31*, 1–9. https://doi.org/10.1016/j.bbrep.2022.101320

de Lima Barizão, A. C., Silva, M. F., Andrade, M., Brito, F. C., Gomes, R. G., & Bergamasco, R. (2020). Green synthesis of iron oxide nanoparticles for tartrazine and bordeaux red dye removal. *Journal of Environmental Chemical Engineering, 8*, 1–10. https://doi.org/10.1016/j.jece.2019.103618

Devi, H. S., Boda, M. A., Shah, M. A., Parveen, S., & Wani, A. H. (2019). Green synthesis of iron oxide nanoparticles using *Platanus orientalis* leaf extract for antifungal activity. *Green Processing and Synthesis, 8*, 38–45. https://doi.org/10.1515/gps-2017-0145

Donga, S., & Chanda, S. (2021). Facile green synthesis of silver nanoparticles using *Mangifera indica* seed aqueous extract and its antimicrobial, antioxidant and cytotoxic potential (3-in-1 system). *Artificial Cells, Nanomedicine, and Biotechnology, 49*, 292–302. https://doi.org/10.1080/21691401.2021.1899193

Duan, Y. T., Soni, K., Patel, D., Choksi, H., Sangani, C. B., Sharaf, S. W., Lalit, A. K., & Kumar, A. R. (2024). Green synthesis of iron oxide nanoparticles using *Nicotiana plumbaginifolia* and their biological evaluation. *Journal of Molecular Liquids, 396*, 1–10. https://doi.org/10.1016/j.molliq.2024.123985

Edis, Z., Haj Bloukh, S., Ashames, A., Ibrahim, M. (2019). Copper-based nanoparticles, their chemistry and antibacterial properties: A review. *Chemistry for a Clean and Healthy Planet*, 401–428. https://doi.org/10.1007/978-3-030-20283-5_24

El-Beltagi, H. S., Rageb, M., El-Saber, M. M., El-Masry, R. A., Ramadan, K. M. A., Kandeel, M., Alhajri, A. S., & Osman, A. (2024). Green synthesis, characterization, and hepatoprotective effect of zinc oxide nanoparticles from *Moringa oleifera* leaves in CCl4-treated albino rats. *Heliyon, 10*, 1–21. https://doi.org/10.1016/j.heliyon.2024.e30627

Elemike, E. E., Nna, P. J., Ikenweke, C., Onwudiwe, D., Omotade, E. T., & Singh, M. (2023). Synthesis, characterization, anti-cancer and antimicrobial studies of iron oxide nanoparticles mediated by *Terminalia catappa* (Indian almond) leaf extract. *Inorganic Chemistry Communications, 155*, 1–6. https://doi.org/10.1016/j.inoche.2023.111048

Elemike, E. E., Nna, P. J., Nzete, S. C., Onwudiwe, D. C., Oghenekohwiroro, E., Singh, M., & Hossain, I. (2024). Green Synthesis, characterization, antimicrobial and anticancer properties of iron oxide nanoparticles mediated by *Annona muricata* (sour sop) leaf extract. *Nanotechnology for Environmental Engineering, 9*, 437–444. https://doi.org/10.1007/s41204-024-00374-9

Fahimmunisha, B. A., Ishwarya, R., AlSalhi, M. S., Devanesan, S., Govindarajan, M., & Vaseeharan, B. (2020). Green fabrication, characterization and antibacterial potential of zinc oxide nanoparticles using *Aloe socotrina* leaf extract: A novel drug delivery approach. *The Journal of Drug Delivery Science and Technology, 55*, Article 101465. https://doi.org/10.1016/j.jddst.2019.101465

Fanoro, O. T., Parani, S., Maluleke, R., Lebepe, T. C., Varghese, J. R., Mavumengwana, V., & Oluwafemi, O. S. (2021). Facile green, room-temperature synthesis of gold nanoparticles using

Combretum erythrophyllum leaf extract: Antibacterial and cell viability studies against normal and cancerous cells. *Antibiotics, 10*, 1–14. https://doi.org/10.3390/antibiotics10080893

Gambhir, R. P., Rohiwal, S. S., Tiwari, A. P. (2022). Multifunctional surface functionalized magnetic iron oxide nanoparticles for biomedical applications: A review *Applied Surface Science, 11*, 1–14. https://doi.org/10.1016/j.apsadv.2022.100303

Govindaraju, K., Vasantharaja, R., Uma Suganya, K. S., Anbarasu, S., Revathy, K., Pugazhendhi, A., Karthickeyan, D., & Singaravelu, G. (2020). Unveiling the anticancer and antimycobacterial potentials of bioengineered gold nanoparticles. *Process Biochemistry, 96*, 213–219. https://doi.org/10.1016/j.procbio.2020.06.016

Jadoun, S., Arif, R., Jangid, N. K., & Meena, R. K. (2021). Green synthesis of nanoparticles using plant extracts: A review. *Environmental Chemistry Letters, 19*, 355–374. https://doi.org/10.1007/s10311-020-01074-x

Jamzad, M., & Kamari, B. M. (2020). Green synthesis of iron oxide nanoparticles by the aqueous extract of *Laurus nobilis* L. leaves and evaluation of the antimicrobial activity. *Journal of Nanostructure in Chemistry, 10*, 193–201. https://doi.org/10.1007/s40097-020-00341-1

Jegadeesan, G. B., Srimathi, K., Santosh, S. N., Manishkanna, S., & Vignesh, D. (2019). Green synthesis of iron oxide nanoparticles using *Terminalia bellirica* and *Moringa oleifera* fruit and leaf extracts: Antioxidant, antibacterial and thermoacoustic properties. *Biocatalysis and Agricultural Biotechnology, 21*, 1–14. https://doi.org/10.1016/j.bcab.2019.101354

Kamaraj, M., Kidane, T., Muluken, K. U., & Aravind, J. (2019). Biofabrication of iron oxide nanoparticles as a potential photocatalyst for dye degradation with antimicrobial activity. *International Journal of Environmental Science and Technology, 16*, 8305–8314. https://doi.org/10.1007/s13762-019-02402-7

Kamath, V., Chandra, P., & Jeppu, G. P. (2020). Comparative study of using five different leaf extracts in the green synthesis of iron oxide nanoparticles for removal of arsenic from water. *International Journal of Phytoremediation, 22*, 1278–1294. https://doi.org/10.1080/15226514.2020.1765139

Kirdat, P. N., Dandge, P. B., Hagwane, R. M., Nikam, A. S., Mahadik, S. P., & Jirange, S. T. (2021). Synthesis and characterization of ginger *(Z. officinale)* extract mediated iron oxide nanoparticles and its antibacterial activity. *Materials Today: Proceedings, 43*, 2826–2831. https://doi.org/10.1016/j.matpr.2020.11.422

Kumar, P. P., Bhatlu, M. L. D., Sukanya, K., Karthikeyan, S., & Jayan, N. (2021). Synthesis of magnesium oxide nanoparticle by eco-friendly method (green synthesis)–a review. *Materials Today: Proceedings, 37*, 3028–3030. https://doi.org/10.1016/j.matpr.2020.08.726

Kumar, S., Kumari, S., Kumari, A., Kondal, N., Choudhary, P., Dhiman, V., & Sharma, R. (2024). Green synthesis of iron oxide nanoparticles from *Calotropis procera* latex: An eco-friendly catalyst for biodiesel production from *Calotropis procera* seed oil. *Clean Technologies and Environmental Policy*, 1–13. https://doi.org/10.1007/s10098-023-02737-3

Liu, H., Wang, G. N., Liu, J., Nan, K., Zhang, J., Guo, L., & Liu, Y. (2021). Green synthesis of copper nanoparticles using *Cinnamomum zelanicum* extract and its applications as a highly efficient antioxidant and anti-human lung carcinoma. *Journal of Experimental Nanoscience, 16*, 410–423. https://doi.org/10.1080/17458080.2021.1991577

Lohrasbi, S., Kouhbanani, M. A. J., Beheshtkhoo, N., Ghasemi, Y., Amani, A. M., & Taghizadeh, S. (2019). Green Synthesis of iron nanoparticles using *Plantago major* leaf extract and their application as a catalyst for the decolorization of azo dye. *BioNanoScience, 9*, 317–322. https://doi.org/10.1007/s12668-019-0596-x

Madubuonu, N., Aisida, S. O., Ali, A., Ahmad, I., Zhao, T., Botha, S., Maaza, M., & Ezema, F. I. (2019). Biosynthesis of iron oxide nanoparticles via a composite of *Psidium guavaja-Moringa oleifera* and their antibacterial and photocatalytic study. *Journal of Photochemistry and Photobiology B: Biology, 199*, 1–9. https://doi.org/10.1016/j.jphotobiol.2019.111601

Mali, S. C., Dhaka, A., Githala, C. K., Trivedi, R. (2020). Green synthesis of copper nanoparticles using *Celastrus paniculatus* Willd. Leaf extract and their photocatalytic and antifungal properties. *Biotechnology Reports, 27*, 1–9. https://doi.org/10.1016/j.btre.2020.e00518

Mankad, M., Patil, G., Patel, D., Patel, P., & Patel, A. (2020). Comparative studies of sunlight mediated green synthesis of silver nanoparaticles from *Azadirachta indica* leaf extract and its antibacterial effect on *Xanthomonas oryzae* pv. *Oryzae*. *Arabian Journal of Chemistry, 13*, 2865–2872. https://doi.org/10.1016/j.arabjc.2018.07.016

Masthoff, M., Buchholz, R., Beuker, A., Wachsmuth, L., Kraupner, A., Albers, F., Freppon, F., Helfen, A., Gerwing, M., Höltke, C., Hansen, U., Rehkämper, J., Vielhaber, T., Heindel, W., Eisenblätter, M., Karst, U., Wildgruber, M., & Faber, C. (2019). Introducing specificity to iron oxide nanoparticle imaging by combining 57Fe-based MRI and mass spectrometry. *Nano Letters, 19*, 7908–7917. https://doi.org/10.1021/acs.nanolett.9b03016

Mubeen, B., Rasool, M. G., Ullah, I., Rasool, R., Imam, S. S., Alshehri, S., Ghoneim, M. M., Alzarea, S. I., Nadeem, M. S., & Kazmi, I. (2022). Phytochemicals mediated synthesis of AuNps from *Citrullus colocynthis* and their characterization. *Molecules, 27*, 1–13. https://doi.org/10.3390/molecules27041300

Muthukumar, H., Mohammed, S. N., Chandrasekaran, N., Sekar, A. D., Pugazhendhi, A., & Matheswaran, M. (2019). Effect of iron doped zinc oxide nanoparticles coating in the anode on current generation in microbial electrochemical cells. *International Journal of Hydrogen Energy, 44*, 2407–2416. https://doi.org/10.1016/j.ijhydene.2018.06.046

Muzafar, W., Kanwal, T., Rehman, K., Perveen, S., Jabri, T., Qamar, F., Faizi, S., & Shah, M. R. (2022). Green synthesis of iron oxide nanoparticles using *Melia azedarach* flowers extract and evaluation of their antimicrobial and antioxidant activities. *Journal of Molecular Structure, 1269*, 1–11. https://doi.org/10.1016/j.molstruc.2022.133824

Nadeem, F., Fozia, F., Aslam, M., Ahmad, I., Ahmad, S., Ullah, R., Almutairi, M. H., Aleya, L., & Abdel-Daim, M. M. (2022). Characterization, antiplasmodial and cytotoxic activities of green synthesized iron oxide nanoparticles using *Nephrolepis exaltata* aqueous extract. *Molecules, 27*, 1–14. https://doi.org/10.3390/molecules27154931

Nair, G. M., Sajini, T., & Mathew, B. (2022). Advanced green approaches for metal and metal oxide nanoparticles synthesis and their environmental applications. *Talanta Open, 5*, 1–11. https://doi.org/10.1016/j.talo.2021.100080

Narendra Kumar, H. K., Chandra Mohana, N., Nuthan, B. R., Ramesha, K. P., Rakshith, D., Geetha, N., & Satish, S. (2019). Phyto-mediated synthesis of zinc oxide nanoparticles using aqueous plant extract of *Ocimum americanum* and evaluation of its bioactivity. *SN Applied Sciences, 1*, 1–9. https://doi.org/10.1007/s42452-019-0671-5

Olabemiwo, O. M., Akintelu, S. A., Waheed, A. A., Okunlola, D. S., Akinwale, D. R., Adeyinka, G. C., Adeniji, T. W., & Adebisi, S. A. (2021). Green synthesis of silver nanoparticles using stem bark extract of *Annona senegalensis*: Characterization and its antibacterial potency. *Current Research in Green and Sustainable Chemistry, 4*, 1–6. https://doi.org/10.1016/j.crgsc.2021.100219

Pillai, R. R., Sreelekshmi, P. B., Meera, A. P., & Thomas, S. (2022). Biosynthesized iron oxide nanoparticles: Cytotoxic evaluation against human colorectal cancer cell lines. *Materials Today: Proceedings, 50*, 187–195. https://doi.org/10.1016/j.matpr.2022.01.151

Prabhu, P., Rao, M., Murugesan, G., Narasimhan, M. K., Varadavenkatesan, T., Vinayagam, R., Lan Chi, N. T., Pugazhendhi, A., & Selvaraj, R. (2022). Synthesis, characterization and anticancer activity of the green-synthesized hematite nanoparticles. *Environmental Research, 214*, 1–8. https://doi.org/10.1016/j.envres.2022.113864

Qasim, S., Zafar, A., Saif, M. S., Ali, Z., Nazar, M., Waqas, M., Haq, A. U., Tariq, T., Hassan, S. G., Iqbal, F., Shu, X. G., & Hasan, M. (2020). Green synthesis of iron oxide nanorods using *Withania coagulans* extract improved photocatalytic degradation and antimicrobial activity. *Journal of Photochemistry and Photobiology B: Biology, 204*, 1–11. https://doi.org/10.1016/j.jphotobiol.2020.111784

Rabiee, N., Ahmadi, S., Akhavan, O., & Luque, R. (2022). Silver and gold nanoparticles for antimicrobial purposes against multi-drug resistance bacteria. *Materials, 15*, 1–25. https://doi.org/10.3390/ma15051799

Rajeshkumar, S., Menon, S., Venkat Kumar, S., Tambuwala, M. M., Bakshi, H. A., Mehta, M., Satija, S., Gupta, G., Chellappan, D. K., Thangavelu, L., & Dua, K. (2019). Antibacterial and antioxidant potential of biosynthesized copper nanoparticles mediated through *Cissus arnotiana* plant extract. *Journal of Photochemistry and Photobiology B: Biology, 197*, 1–6. https://doi.org/10.1016/j.jphotobiol.2019.111531

Rama, P., Thangapushbam, V., Sivakami, S., Jothika, M., Mariselvi, P., Sundaram, R., & Muthu, K. (2024). Preparation, characterization of green synthesis FeO nanoparticles and their photocatalytic activity towards Basic Fuschin dye. *Journal of the Indian Chemical Society, 101*, 1–8. https://doi.org/10.1016/j.jics.2024.10114

Raoof, F., Munawar, A., Ahmad, M., Rizvi, S. F. A., Ali, Z., & Shahid, A. B. (2023). Multifunctional iron oxide nanocarriers synthesis for drug delivery, diagnostic imaging, and biodistribution study. *Applied Biochemistry and Biotechnology, 195*, 4469–4484. https://doi.org/10.1007/s12010-023-04345-9

Roy, S., & Das, T. K. (2015). Synthesis of silver nanoparticles-a review. *International Journal of Plant Biology, 3*, 1044–1055.

Roya Alizadeh, S., Nemati Seyghalan, H., Hashemi, Z., & Ali, E. M. (2023). *Scrophularia striata* extract mediated synthesis of gold nanoparticles; their antibacterial, antileishmanial, antioxidant, and photocatalytic activities. *Inorganic Chemistry Communications, 156*, 1–11. https://doi.org/10.1016/j.inoche.2023.111138

Saha, T., Bin Mobarak, M., Uddin, M. N., Quddus, M. S., Naim, M. R., & Pinky, N. S. (2023). Biogenic synthesis of copper oxide (CuO) Nps exploiting *Averrhoa carambola* leaf extract and its potential antibacterial activity. *Materials Chemistry and Physics, 305*, 1–12. https://doi.org/10.1016/j.matchemphys.2023.127979

Saif, S., Tahir, A., & Chen, Y. (2016). Green synthesis of iron nanoparticles and their environmental applications and implications. *Nanomaterials, 6*, 1–26. https://doi.org/10.3390/nano6110209

Salleh, A., Naomi, R., Utami, N. D., Mohammad, A. W., Mahmoudi, E., Mustafa, N., & Fauzi, M. B. (2020). The potential of silver nanoparticles for antiviral and antibacterial applications: A mechanism of action. *Nanomaterials, 10*, 1–20. https://doi.org/10.3390/nano10081566

Sana, S. S., Kumbhakar, D. V., Pasha, A., Pawar, S. C., Grace, A. N., Singh, R. P., Nguyen, V. H., Le, Q. V., & Peng, W. (2020). *Crotalaria verrucosa* leaf extract mediated synthesis of zinc oxide nanoparticles: Assessment of antimicrobial and anticancer activity. *Molecules, 25*, 1–21. https://doi.org/10.3390/molecules25214896

Saod, W. M., Al-Janaby, M. S., Gayadh, E. W., Ramizy, A., & Hamid, L. L. (2024). Biogenic synthesis of iron oxide nanoparticles using *Hibiscus sabdariffa* extract: Potential for antibiotic development and antibacterial activity against multidrug-resistant bacteria. *Current Research in Green and Sustainable Chemistry., 8*, 1–8. https://doi.org/10.1016/j.crgsc.2024.100397

Sekar, V., Al-Ansari, M. M., Narenkumar, J., Al-Humaid, L., Arunkumar, P., & Santhanam, A. (2022). Synthesis of gold nanoparticles (AuNps) with improved anti-diabetic, antioxidant and anti-microbial activity from *Physalis minima. The Journal of King Saud University–science, 34*, 1–8. https://doi.org/10.1016/j.jksus.2022.102197

Senthilkumar, A., Muthuswamy, R., Nallal, U. M., Ramaiyan, S., Kannan, P., Muthupandi, S., Lakshminarayanan, S. P., Sambasivam, S., & Ayyar, M. (2024). Green synthesis of copper nanoparticles from agro-waste garlic husk. *Zeitschrift Für Physikalische Chemie, 238*, 75–88. https://doi.org/10.1515/zpch-2023-0291

Shaba, E. Y., Jacob, J. O., Tijani, J. O., & Suleiman, M. A. T. (2021). A critical review of synthesis parameters affecting the properties of zinc oxide nanoparticle and its application in wastewater treatment. *Applied Water Science, 11*, 1–41. https://doi.org/10.1007/s13201-021-01370-z

Shabbir, M. A., Naveed, M., Rehman, S. U., Ain, N. U., Aziz, T., Alharbi, M., Alsahammari, A., & Alasmari, A. F. (2023). Synthesis of iron oxide nanoparticles from madhuca indica plant extract and assessment of their cytotoxic, antioxidant, anti-inflammatory, and anti-diabetic properties via different nanoinformatics approaches. *ACS Omega, 8*, 33358–33366. https://doi.org/10.1021/acsomega.3c02744

Sharifi-Rad, M., Elshafie, H. S., & Pohl, P. (2024). Green synthesis of silver nanoparticles (AgNps) by *Lallemantia royleana* leaf extract: Their bio-pharmaceutical and catalytic properties. *Journal of Photochemistry and Photobiology, a: Chemistry, 448*, 1–14. https://doi.org/10.1016/j.jphotochem.2023.115318

Shenashen, M. A., El-Safty, S. A., & Elshehy, E. A. (2014). Synthesis, morphological control, and properties of silver nanoparticles in potential applications. *Particle and Particle Systems Characterization, 31*, 293–316. https://doi.org/10.1002/ppsc.201300181

Singh, J., Dutta, T., Kim, K. H., Rawat, M., Samddar, P., & Kumar, P. (2018). Green synthesis of metals and their oxide nanoparticles: Applications for environmental remediation. *Journal of Nanobiotechnology, 16*, 1–24. https://doi.org/10.1186/s12951-018-0408-4

Singh, K., Chopra, D. S., Singh, D., & Singh, N. (2020). Optimization and ecofriendly synthesis of iron oxide nanoparticles as potential antioxidant. *Arabian Journal of Chemistry, 13*, 9034–9046. https://doi.org/10.1016/j.arabjc.2020.10.025

Singh, C., Anand, S. K., Upadhyay, R., Pandey, N., Kumar, P., Singh, D., Tiwari, P., Saini, R., Tiwari, K. N., Mishra, S. K., & Tilak, R. (2023). Green synthesis of silver nanoparticles by root extract of *Premna integrifolia* L. and evaluation of its cytotoxic and antibacterial activity. *Materials Chemistry and Physics, 297*, 1–14. https://doi.org/10.1016/j.matchemphys.2023.127413

Soundarya, V., & Karmegam, N. (2023). Antimicrobial and mosquito larvicidal activity of iron oxide nanoparticles phytosynthesized from the medicinal plant *Andrographis serpyllifolia. Notulae Scientia Biologicae, 15*, 1–12. https://doi.org/10.55779/nsb15411668

Sudhakar, C., Poonkothai, M., Selvankmuar, T., Selvam, K., Rajivgandhi, G., Siddiqi, M. Z., Alharbi, N. S., Kadaikunnan, S., & Vijayakumar, N. (2021). Biomimetic synthesis of iron oxide nanoparticles using *Canthium coromandelicum* leaf extract and its antibacterial and catalytic degradation of Janus green. *Inorganic Chemistry Communications, 133*, 1–6. https://doi.org/10.1016/j.inoche.2021.108977

Sudhakar, C., Poonkothai, M., Selvankumar, T., & Selvam, K. (2022). Facile synthesis of iron oxide nanoparticles using *Cassia auriculata* flower extract and accessing their photocatalytic degradation and larvicidal effect. *Journal of Materials Science: Materials in Electronics, 31*, 1434–11445. https://doi.org/10.1007/s10854-022-08116-w

Supin, K. K., Vasundhara, M. (2023). Green synthesis of ZnO nanoparticles from Neem and Eucalyptus leaves extract for photocatalytic applications. *Materials Today: Proceedings, 92*, 787–795. https://doi.org/10.1016/j.matpr.2023.04.343

Suriyakala, G., Sathiyaraj, S., Devanesan, S., AlSalhi, M. S., Rajasekar, A., Maruthamuthu, M. K., Babujanarthanam, R. (2022). Phytosynthesis of silver nanoparticles from *Jatropha integerrima* Jacq. Flower extract and their possible applications as antibacterial and antioxidant agent. *Saudi Journal of Biological Sciences, 29*, 680–688. https://doi.org/10.1016/j.sjbs.2021.12.007

Tailor, G., Yadav, B. L., Chaudhary, J., Joshi, M., & Suvalka, C. (2020). Green synthesis of silver nanoparticles using *Ocimum canum* and their anti-bacterial activity. *Biochemistry and Biophysics Reports, 24*, 1–5. https://doi.org/10.1016/j.bbrep.2020.100848

Tamilvanan, A., Balamurugan, K., Ponappa, K., & Kumar, B. M. (2014). Copper nanoparticles: Synthetic strategies, properties and multifunctional application. *International Journal of Nanoscience, 13*, 1–22. https://doi.org/10.1142/S0219581X14300016

Ustun, E., Onbas, S. C., Celik, S. K., Ayvaz, M. C., Sahin, N. (2022). Green synthesis of iron oxide nanoparticles by using *Ficus Carica* Leaf extract and its antioxidant activity. *Biointerface Research in Applied Chemistry, 12*, 2108–2116. https://doi.org/10.33263/BRIAC122.21082116

Vahini, M., Rakesh, S. S., Subashini, R., Loganathan, S., & Prakash, D. G. (2024). In vitro biological assessment of green synthesized iron oxide nanoparticles using *Anastatica hierochuntica* (Rose of Jericho). *Biomass Conversion and Biorefinery, 14*, 19005–19015. https://doi.org/10.1007/s13399-023-04018-x

Velsankar, K., Parvathy, G., Mohandoss, S., Krishna Kumar, M., & Sudhahar, S. (2022). *Celosia argentea* leaf extract-mediated green synthesized iron oxide nanoparticles for bio-applications. *Journal of Nanostructure in Chemistry, 12*, 625–640. https://doi.org/10.1007/s40097-021-00434-5

Verma, A., & Bharadvaja, N. (2022). Plant-mediated synthesis and characterization of silver and copper oxide nanoparticles: Antibacterial and heavy metal removal activity. *Journal of Cluster Science, 33*, 1697–1712. https://doi.org/10.1007/s10876-021-02091-8

Wang, Y., O'Connor, D., Shen, Z., Lo, I. M. C., Tsang, D. C. W., Pehkonen, S., Pu, S., & Hou, D. (2019). Green synthesis of nanoparticles for the remediation of contaminated waters and soils: Constituents, synthesizing methods, and influencing factors. *Journal of Cleaner Production, 226*, 540–549. https://doi.org/10.1016/j.jclepro.2019.04.128

Weldegebrieal, G. K., & Sibhatu, A. K. (2021). Photocatalytic activity of biosynthesized α-Fe_2O_3 nanoparticles for the degradation of methylene blue and methyl orange dyes. *Optik, 241*, 1–15. https://doi.org/10.1016/j.ijleo.2021.167226

Yew, Y. P., Shameli, K., Miyake, M., Ahmad Khairudin, N. B. B., Mohamad, S. E. B., Naiki, T., & Lee, K. X. (2020). Green biosynthesis of superparamagnetic magnetite Fe_3O_4 nanoparticles and biomedical applications in targeted anticancer drug delivery system: A review. *Arabian Journal of Chemistry, 13*, 2287–2308. https://doi.org/10.1016/j.arabjc.2018.04.013

Yin, I. X., Zhang, J., Zhao, I. S., Mei, M. L., Li, Q., & Chu, C. H. (2020). The antibacterial mechanism of silver nanoparticles and its application in dentistry. *International Journal of Nanomedicine, 15*, 2555–2562. https://doi.org/10.2147/IJN.S246764

Yoonus, J., Resmi, R., & Beena, B. (2021). Evaluation of antibacterial and anticancer activity of green synthesized iron oxide (α-Fe_2O_3) nanoparticles. *Materials Today: Proceedings, 46*, 2969–2974. https://doi.org/10.1016/j.matpr.2020.12.426

Yusefi, M., Shameli, K., Ali, R. R., Pang, S. W., & Teow, S. Y. (2020). Evaluating anticancer activity of plant-mediated synthesized iron oxide nanoparticles using *Punica Granatum* fruit peel extract. *Journal of Molecular Structure, 1204*, 1–11. https://doi.org/10.1016/j.molstruc.2019.127539

Zambri, N. D. S., Taib, N. I., Abdul Latif, F., & Mohamed, Z. (2019). Utilization of neem leaf extract on biosynthesis of iron oxide nanoparticles. *Molecules, 24*, 1–12. https://doi.org/10.3390/molecules24203803

Zargham, F., Afzal, M., Rasool, K., Manzoor, S., & Qureshi, N. A. (2023). Larvicidal activity of green synthesized iron oxide nanoparticles using *Grevillea robusta* Cunn. Leaf extract against vector mosquitoes and their characterization. *Experimental Parasitology, 252*, 108586. https://doi.org/10.1016/j.exppara.2023.108586

Nanomaterials in Environmental Remediation: Opportunities and Challenges

Jaya Dayal, Preeti Chaurasia, and Neha Nagar

Abstract The use of nanomaterials is growing day by day, and by virtue of their unique properties, they serve as an efficient treatment of environmental contaminants. With respect to their bulk forms, nanomaterials have distinct physicochemical characteristics, including high surface area-to-volume ratio, high reactivity, and the ability to interact at the molecular level, that enable them to be specifically effective for environmental applications such as air, water, and soil remediation. Despite the several potential environmental applications that have shown promise in the laboratory scale, the integration of nanoparticles into environmental technologies faces significant challenges to a successful large-scale usage of nanotechnology, including technical, regulatory, economic, health, environmental, and regulatory challenges. This chapter provides an introduction to nanotechnology, discussing its characteristics as well as its potential applications in the field of pollution removal. It discusses the major challenges that define the use of nanomaterials in the environment, providing a thorough overview of the hazards and limitations of nanomaterials in environmental applications. Furthermore, it also highlights the recent developments and emerging research initiatives that are aimed at addressing these challenges. Nanomaterials would, under the right circumstances and mitigating existing challenges to their usage, contribute towards achieving wonders for environment sustainability and effectively addressing the global environment problems.

Keywords Nanomaterials · Environmental contaminants · Challenges · Physicochemical parameters · Pollutants removal

J. Dayal (✉) · P. Chaurasia · N. Nagar
Department of Microbiology, S.S. Jain Subodh P.G. College, Jaipur, India
e-mail: jaya22subodh@gmail.com

P. Chaurasia
e-mail: preetimicro@gmail.com

N. Nagar
e-mail: neha.subodhmicro@gmail.com

E. Priyadarshini et al. (eds.), *Advanced Nanotechnology Applications for Environmental Sustainability*, Sustainable Environmental Waste Management Strategies,
https://doi.org/10.1007/978-3-032-12933-8_5

1 Introduction

Rapid worldwide industrialization and exponential population growth result in significant increases in pollution levels, posing serious threats to environmental sustainability and public health. Environmental pollution represents a critical global issue, as even minimum amounts of hazardous substances can lead to severe negative impacts on both human and animal health (Priyadarshini & Das, 2024; Priyadarshini et al., 2024).

These pollutants, ranging from heavy metals and toxic chemicals to greenhouse gases and microplastics/nanoplastics, have widespread ecological and health consequences.

The produced wastes are emitted into the environment in several forms. Air pollutants include hazardous gases, notably carbon oxides. ozone, nitrogen oxides, sulfur oxides, along with volatile organic compounds (VOCs) and suspended airborne particles. Water and soil contaminants include organic matters such as hydrocarbons, phenols, pesticides, and insecticides, in addition to microbial pathogens and heavy metals (Behera & Das, 2025; Mohanty et al., 2025; Satapathy et al., 2024).

According to the World Health Organization (WHO), approximately 91% of the worldwide population lives in regions with pollution levels higher than recommended air quality. Around 4.2 million early deaths worldwide are linked to ambient air pollution (Ju et al., 2020). Similarly, industrial effluents containing heavy metals pollute water bodies, endangering aquatic life and human populations dependent on these resources. Soil and water contamination have been linked to poor health and illness outcomes. According to a WHO article, exposure to chemical pollution led to 1.3 million deaths in 2012, increasing to 2 million deaths by 2019 (WHO, 2021). Chemical pollutants that pose potential harm consist of heavy metals, organic solvents, benzene, polyfluorinated alkyl substances (PFAS), polycyclic aromatic hydrocarbons (PAHs), and pesticides (WHO, 2016). Pollution in the soil and water can lead to food contamination by heavy metals, polychlorinated biphenyls, and dioxins (Munzel et al., 2025; World Health Organization, 2022; Nagar et al., 2021).

To monitor and control hazardous environmental contaminants, it is essential to establish efficient, affordable, and sustainable technology. For centuries, humans have incinerated or buried solid waste in their surroundings to prevent its accumulation. In order to reduce environmental pollution, numerous remediation strategies have been implemented over the past few decades, including biological, chemical, and physical procedures (see Table 1). Nevertheless, many of these methods exhibit significant disadvantages, including higher costs, operational complexities, and secondary pollution (Asghar et al., 2024; Chaurasia et al., 2023; Dayal et al., 2023; Singh et al., 2022).

Bioremediation offers environmental and sustainability benefits over physical and chemical alternatives. However, there are still many limitations in the way of using biological treatments to eradicate environmental contaminants.

Nanotechnology, especially via nanoparticles, presents a promising and innovative approach to overcome these limitations. This advanced field includes various

Table 1 Common physical, chemical and biological methods for pollutants removal

Physical methods	Chemical methods	Biological methods
Adsorption on activated carbon	Ozonation	Biostimulation
Ion exchange	Fenton method	Bioaugmentation
Membrane filtration	Oxidation	Phytoremediation
Irradiation	Electrochemical destruction	Biosorption
Reverse osmosis	Photochemical	Natural attenuation
Coagulation and flocculation	Sodium hypochlorite	Composting
Electrokinetic coagulation		Bioventing

innovative technologies that function at the nanoscale (1–100 nm), facilitating the fabrication of materials, systems, and devices with different physicochemical characteristics. Nanotechnology offers exceptional opportunities for pollution reduction, resource recovery, and environmental monitoring by precise control of the shape, size, and composition of matter at the atomic and molecular levels (Ibrahim et al., 2016).

Nanoparticles, characterized by dimensions less than 100 nm, exhibit distinct features relative to bulk materials. These features include a higher surface-area-to-volume ratio and distinct chemical reactivity, making them effective in interacting with contaminants at the molecular level. Unlike traditional materials, nanoparticles can be engineered to specifically target several pollutants, such as organic compounds, metals, or pathogens, thus enhancing the precision and effectiveness of treatment systems. These nanoscale materials show potential to significantly improve treatment efficiency while reducing secondary pollution and are being actively explored for their diverse applications in the environment.

A variety of nanoparticles have demonstrated considerable potential in removing contaminants from wastewater (Dayal et al., 2019). Their mechanisms include adsorption, photocatalysis, antimicrobial action, and magnetic separation, each suited to particular types of contaminants. For example, titanium dioxide (TiO_2) nanoparticles and zinc oxide (ZnO) nanoparticles are widely known for their photocatalytic properties, as they generate reactive oxygen species (ROS) upon light exposure, facilitating the decomposition of organic contaminants (Tamanna et al., 2024). Carbon-based nanomaterials (for example, carbon nanotubes (CNTs) and graphene nanomaterials) demonstrate significant adsorption capacities attributed to their extensive surface areas and adaptable surface characteristics, making them effective for the elimination of various pollutants (Kumar, 2021).

Despite the promise nanoparticles hold, several challenges hinder their widespread implementation in environmental remediation. Concerns around nanoparticle toxicity, environmental persistence, bioaccumulation, cost of production, and scalability remain. The release of nanoparticles into natural water systems may pose ecological and health risks, as certain nanoparticles could persist in the environment and adversely affect aquatic organisms and humans (Kumar, 2021). Further research is necessary to develop and promote environmentally sustainable, cost-efficient, and scalable methods for the production and recovery of nanomaterials to fully exploit their potential in environmental remediation.

This chapter discusses the fundamental properties of nanomaterials that enable them ideal for environmental applications. It defines the present and prospective applications of these materials in pollutant removal while critically assessing the challenges to large-scale implementation. Furthermore, we examine ongoing research efforts aimed at addressing these challenges and assess the prospective impact of nanotechnology in the field of environmental management.

2 Properties of Nanomaterials Relevant to Environmental Remediation

Nanomaterials have various features that make them appropriate for environmental applications:

- **High Surface Area-to-Volume Ratio**: The reduction of the material to the nanoscale results in an increase in surface area per unit mass. This facilitates increased interaction between larger quantities of the material and the surrounding components, thereby affecting its reactivity with pollutants.
- **Quantum effect and surface plasmon resonance**: Due to their quantum effects, nanomaterials require reduced activation energy to facilitate chemical reactions. Additionally, nanoparticles exhibit a phenomenon known as surface plasmon resonance that can be employed to detect potentially harmful chemicals.
- **Molecular level interaction**: Nanoparticles can interact with contaminants at the molecular level, leading to improved specificity and efficiency in pollutant removal.
- **Smaller size**: Nanoparticles, because of their reduced dimensions and distinct surface coatings, may infiltrate into the subsurface and remain suspended in groundwater. This allows for greater dispersion compared to bigger, macro-sized particles.
- **Catalytic Activity**: Many nanomaterials, such as metal-oxide nanoparticles, act as catalysts to accelerate degradation, reducing pollutants into harmless byproducts.

3 Applications of Nanomaterial in Environmental Remediation

Nanomaterials have demonstrated promising results in several environmental applications, including air pollution management, water treatment, and soil remediation (Fig. 1). The following sections discuss these applications in detail.

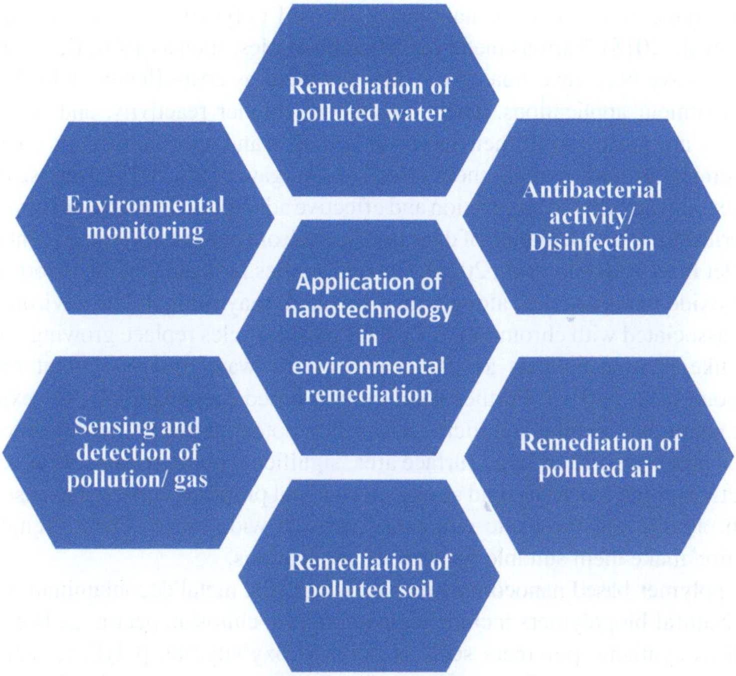

Fig. 1 Various applications of nanotechnology in environmental remediation

3.1 Nanoremediation of Waste Water

A major challenge facing the global community is providing accessible, clean, and affordable water in order to meet the continuously expanding demand. Water scarcity continues to grow in both industrialized and developing countries due to human activities that contaminate natural water resources through the release of pollutants, thereby reducing the quality of water for others. Various ions, petroleum hydrocarbons, heavy metals, pesticides, radioactive substances, and emerging contaminants, such as pharmaceuticals and personal care products, can pollute water sources. (Das & Dey et al., 2024a, 2024b; Jadhav et al., 2015; Mohanty et al., 2024a, 2024b; Priyadarsini & Das, 2025; Sahu et al., 2024a; Zamora-Ledezma et al., 2021).

Wastewater treatment utilizes physical, chemical, and biological methods. Water purification employs several methods, such as sedimentation, ion exchange, filtration, dissolved air flotation, deionization, disinfection, reverse osmosis, coagulation, and flocculation (Ibrahim et al., 2016). In contrast, these techniques of wastewater treatment may be expensive and expose the environment to secondary contaminants that are hazardous to the ecosystem.

In an effort to find efficient ways to manage contaminated water, researchers have frequently used the special qualities of nanomaterials to treat wastewater. The most common nanomaterials employed in water remediation methods are metallic nanomaterials, carbon-derived nanomaterials, and polymer-based nanocomposites (Saikia et al., 2018). Various metal oxide nanoparticles, such as TiO_2, Fe_2O_3/Fe_3O_4, and ZnO, have been investigated for their potential as cost-effective adsorbents in water treatment applications. Their dimensions, higher reactivity, and adsorption properties are attributed to their large surface area and specific affinity for particular chemical groups, making them effective (Aragaw et al., 2021). Nanoscale iron oxides exhibit stability in suspension and effective adsorption features, enabling them appropriate for the remediation of dyes in effluent from paper, textile, and paint industries (Del Prado-Audelo et al., 2021). Current studies indicate that the incorporation of iron oxide nanoparticles, along with citric acid, may mitigate the environmental effects associated with chromium (VI). TiO_2 nanoparticles replace growing contaminants like pharmaceuticals as photocatalysts for water micropollutant removal (Mahmoud et al., 2017). Furthermore, carbon based nanomaterials, for example, carbon nanotubes, exhibit significant adsorption potential due to their chemical and physical properties. Their large surface area, significant porosity, superior electronic conductivity, high reactivity, and strong mechanical properties enable the adsorption of both organic and inorganic pollutants across a wide range. Their strength and absorption make them suitable for various applications.

The polymer-based nanocomposite has remarkable metal decontamination properties. Natural biopolymers include alginate, lignin, chitosan, pectin, and cellulose, as well as synthetic polymers such as polyhydroxybutyrate, polylactic acid, and polyhydroxyalkanoates. They facilitate biopolymeric nanocomposite development (Annu et al., 2024). They help develop eco-friendly water treatment systems by using sustainable methods (Al-Hazmi et al., 2024). Water pollutants are adsorbed by polymer-based nanocomposites due to their specific functional groups and large surface area. Environmental contaminants may attach to the surface of the nanocomposite or interact with its structure, facilitating water separation. Additionally, polymeric nanocomposites serve as effective filtration agents, possessing specific properties that enable the capture or removal of microplastics, microorganisms, particulate matter, and other contaminants from water, thereby improving the quality of the water (Risch et al., 2021).

Nanotechnology also offers an effective approach for water disinfection. Numerous nanomaterials exhibit significant antimicrobial activities through various mechanisms, such as the generation of reactive oxygen species that harm cellular and acellular structures (e.g., ZnO and TiO_2), disruption of the cell envelope of bacteria (e.g., chitosan, carboxy fullerene, peptides, silver nanoparticles (AgNPs), ZnO, and carbon nanotubes), interference with energy transfer (e.g., fullerene nanoparticles and Ag nanoparticles), and inhibition of enzymes. TiO_2 is regarded as the most suitable option owing to its stability in aqueous environments, non-toxic nature when ingested, and economic (Li et al., 2008; Yunus et al., 2012).

3.2 Nanoremediation of Polluted Air

Air pollution constitutes a significant global challenge of this century, impacting both climate change and human health. Natural and anthropogenic sources both cause air pollution. Natural sources significantly impact air pollution locally, especially in arid regions prone to dust storms and wildfires. Human activities are significantly more responsible for air pollution compared to natural sources. The main anthropogenic causes of air pollution consist of automobile fuel combustion, industrial facilities, agricultural and municipal waste incineration, heat and power generation, and residential activities such as cooking, lighting, and heating. Globally, air pollution is considered the fourth most common cause of death (Das & Ghosh, 2023; Das & Ray et al., 2024a, 2024b; Dey et al., 2023). Rising levels of air pollution threaten ecosystem health (Ali et al., 2020). The most frequent hazardous air pollutants include carbon monoxide (CO), nitrogen oxides (NOx), sulphur dioxide (SO_2), lead, ground-level ozone, and particulate matter (PM_{10} and $PM_{2.5}$), (Manisalidis et al., 2020). While not considered a pollutant, CO_2 represents the primary greenhouse gas generated by anthropogenic activities.

To efficiently monitor and reduce the emissions of these gases, a variety of control and treatment systems have been established and utilized. By reducing the environmental effect of their discharge, these systems are essential to the larger initiatives of environmental sustainability. Nanotechnology is a therapeutic strategy that can effectively identify, manage, and remediate air pollution through the application of nanoparticles as adsorbents, membranes, catalysts, and sensors.

Nanotechnology suggests that nanoscale adsorbents (nanoadsorbents) can effectively address several contemporary challenges, including poor air quality. Nanoparticles possess the ability to adsorb various airborne pollutants. Nanoadsorbents have been produced and defined to detect and adsorb pollutants, including metal oxides, volatile organic compounds (VOCs), and microorganisms (Saleem et al., 2022). Scientists have used several nanoadsorbent materials to eliminate and monitor harmful gases and air pollutants. Carbon-based nanomaterials exhibit significant physical properties, making them appropriate for commercial applications as nanoadsorbents with improved affinity, selectivity, and capacity (Gabal et al., 2020). CNTs modulate methane and CO_2 emissions from industrial chimneys and vehicle exhausts. These materials are capable of capturing greenhouse gases produced during coal mining and electricity production. These can also serve as nanoadsorbents to separate various airborne pollutants due to their enhanced properties, including a strong affinity for selective adsorption of pollutants.

Many nanocatalytic procedures have been established for the breakdown or transformation of nitrogen oxides (NOx) and methane (CH_4). Nickel nanomaterials were used as nanocatalysts to promote the thermal breakdown of methane into hydrogen (Wang et al., 2012). The photocatalytic oxidation of NOx using titanate nanotubes

(TNTs) and their products has received significant attention. The removal of gas-phase pollutants and the remediation of contaminated air have gained growing attention in past years, attributed to the catalytic and adsorption properties of silica-based nanomaterials (Guerra et al., 2018).

Nanostructured membranes featuring tiny pores may effectively remove particles from emissions, providing an alternative technique for controlling air pollution. Numerous studies have concentrated on developing and optimizing nano-structured membranes to trap diverse gas pollutants (Rubina et al., 2019; Samantara et al., 2019). Nanofiber-coated filters are utilized for air filtration in various industrial plants. Nano-structured membrane filtration is particularly suitable for a variety of volatile organic carbon vapors. For example, formaldehyde (HCHO) removal from the atmosphere is extremely difficult. Consequently, various methods have been developed to increase the removal of HCHO, including the production of a carbon nanofiber membrane based on electrospun polyacrylonitrile that has good microporosity and many nitrogenous functional groups that act as an effective adsorption site. (Lee et al., 2010).

A remarkable technology, nanomaterial-enabled sensors provide exquisite detection of contaminants at the nanomolar to sub-picomolar level (Saleem et al., 2022). Nanosensors are composed of three parts: a nanomaterial, a selective detecting element, and a signal transduction technology that communicates the presence of analytes. For example, carbon nanotubes and graphene, are both frequently utilized in nanosensors owing to the fact that they have a huge surface area, are mechanically strong, and have better thermal and electric conductivity (Schroeder et al., 2019). CNT sensors are engineered to identify various gases such as methane (CH_4), nitrogen dioxide (NO_2), carbon monoxide (CO), ammonia (NH_3), sulphur dioxide (SO_2), and hydrogen (H_2) (Mohamed et al., 2022). Graphene nanoparticles exhibit remarkable thermal and mechanical characteristics. Consequently, it holds considerable relevance in electrochemical nanosensors and nanobiosensors. Additionally, their implementation in the detection of gaseous contaminants has been significant. (Li et al., 2011).

3.3 Nanoremediation of Polluted Soil

Soil contamination and degradation remain significant environmental challenges, necessitating global remedial efforts. Persistent organic pollutants (POPs), heavy metals, herbicides, and pesticides pollute soil, intensifying the problem. Biomagnification of these chemicals in polluted soil enhances the risk of intoxication of the food chain.

One of the most prevalent treatments is the adsorption or immobilization technique. This remediation methodology is distinguished by its efficacy, cost-effectiveness, and ecologically sustainable approach to removing metal contaminants from soils (Arif et al., 2020). The engineered nanomaterials utilized for the immobilization of soil pollutants comprise metal oxides (such as titanium oxide and ferric oxide), carbon-based nanomaterials, and various nanocomposites (Gaur et al., 2014). Ferric oxide nanoparticles possess a remarkable capacity for the adsorption and immobilization of heavy metals, including cadmium and arsenic, from various media samples.

Effective decolorization of organic dyes can be achieved through the use of industrial procedures that are facilitated by nanoparticles. The primary mechanism underlying the catalytic effectiveness of gold and silver nanoparticles in dye degradation involves two steps: initially, electrons accumulate on the particle surface, followed by the dye diffusion for a reduction reaction (Bastus et al., 2014).

Zero-valent iron at the nanoscale (nZVI) serves as an electron donor having a negative reduction capacity. In pilot studies, nZVI is often applied. (Cheng et al., 2021), as these facilitate the elimination of organochlorine pesticides, chlorinated organic solvents, and polychlorinated biphenyls by oxidation–reduction method of transformation (Stefaniuk et al., 2016). Carbon nanotubes have currently gained prominence because of their superior adsorption capacity compared to granular activated carbon, biochar, graphene oxides, and graphene. The integration of functional groups like -SH and NH_2, employing oxidation methods, and applying nonmagnetic and magnetic iron oxides can all enhance the adsorption capability. Carbon nanotubes can eliminate heavy metal ions such as Ni^{2+}, Pb^{2+}, Zn^{2+}, and Cu^{2+}; however, the concentration of organic material, pH, and the presence of clay and silt particles are the factors that determine the heavy metals immobilization. These can also remediate soil contaminated with cadmium, chromium (VI), total petroleum hydrocarbons, hexachlorocyclohexane, dichlorodiphenyltrichloroethane (DDT), and crude oil by promoting microbial proliferation and plant development. (Shan et al., 2015).

4 Challenges in Large-Scale Implementation

Despite their promising applications, the large-scale use of nanomaterials in environmental technologies faces several significant barriers:

4.1 Technical Challenges

4.1.1 Scalability and Reproducibility

Large-scale production of nanomaterials faces challenges in both cost and consistency. Technical challenges related to the development of products based on nanotechnology for commercialization encompass the inability to exceed current products, ineffective technology integration leading to the loss of unique characteristics post-scaling, difficulties in controlling reproducibility and batch-to-batch variations during industrial-scale manufacturing, and insufficiently trained personnel to utilize, analyze, and interpret results from various analytical instruments (Kaur et al., 2014; Rambaran et al., 2022). In large-scale operations, keeping control over important parameters, such as size, shape, and surface functionalization of particles, is especially challenging. For example, inconsistent nanoparticle size distribution might cause changes in catalytic performance, thereby compromising their stability in uses including wastewater treatment. The variability in the characteristics of carbon based nanomaterials and metal oxide nanoparticles can result in considerable performance challenges (Salaudeen et al., 2024). Single walled CNTs demonstrate variation in diameter, chirality, and length based on the synthesis method employed, which can profoundly influence their mechanical strength (Liu et al., 2013).

4.1.2 Stability Under Operational Conditions

Although nanoparticles have shown potential in contaminant remediation, several limitations exist regarding their application in contaminated sites. One of these drawbacks is that nanoparticles lose their reactivity with time. Over time, certain metal nanoparticles lose their reactivity because the soil has a blocking effect, which restricts the flow of fluids through the particles (Shahi et al., 2021).

4.2 Health and Environmental Challenges

4.2.1 Toxicological Risks

The health concerns of nanotechnology include workplace nanoparticle exposure, inhalation from polluted air, water, or food, and skin contact (Mohanty et al., 2024a, 2024b; Ray et al., 2024; Sahu et al., 2024b). Research shows that nanoparticle size and morphology affect toxicity, with tiny particles posing a higher risk. Gravitational forces pull larger particles out of the airflow. Larger particles possess greater inertia and mass, resulting in a reduced probability of staying suspended in the air for an extended duration. Smaller particles, especially nanoparticles, have less mass and inertia, allowing airborne suspension. Toxic airborne particles can penetrate the respiratory system, leading to oxidative stress, tissue damage, and inflammation. Heavy metals and organic contaminants in particles can harm health when inhaled (Khadanga et al., 2024). Additionally, DNA mutations and mitochondrial damage leading to cell death have been observed. Safety concerns may restrict the widespread implementation of nanomaterials for environmental remediation (Darwesh et al., 2022).

The extensive soil microbial variety degrades organic matter, bioremediates harmful chemicals, and recycles nutrients, maintaining environmental equilibrium. Researchers found that adding CuO-NP to soil reduced the hydrolytic activity and oxidative potential of dehydrogenase in soil microbial communities (Schimel et al., 2012). A frequently used antibiotic, Ag-NP, toxically affects *Nitrosomonas* and *Nitrobacter* in soil and aquatic environments. Poor enzyme function has reduced microbial community and diversity. This loss subsequently impacts soil fertility, biogeochemical cycles, and climate regulation (Dash et al., 2020).

4.2.2 Environmental Persistence and Bioaccumulation

In contrast to lab experiments, nanoparticle use in environmental applications poses many hazards and unforeseen consequences. Toxic nanomaterials like silver nanoparticles released into terrestrial or aquatic habitats are a major environmental hazard (Darwesh et al., 2022). Nanoparticles accumulate in the food chain due to widespread use, disrupting the food web. Fish use Cu and Na transporters to absorb Ag-NP, leading to its buildup across trophic levels (Waalewijn-Kool et al., 2014). Nanoparticle accumulations aid environmental biomagnification. The duration of nanoparticle exposure and the physicochemical properties of the nanoparticle are two parameters that facilitate the transmission of nanoparticles into the food chain (Jakubiak et al., 2014; Zhu et al., 2014). *Manduca sexta* and *Nicotiana tabacum* L. cv Xanthi serve as models for the accumulation of Au-NP, which biomagnifies within terrestrial food chains and poses toxicity risks to animals (Judy et al., 2010). To mitigate the bioaccumulation of nanomaterials in primary trophic levels, further research is required in this field.

4.2.3 Secondary Environmental Risks

Certain nanoparticles function as carriers for contaminants, thus increasing their mobility in the environment. Contaminants such as radioactive elements, polychlorinated compounds, potentially toxic elements (PTE), and pesticides in sediments or suspended solids in water interact with nanoparticle surfaces through van der Waals, ion-exchange, and electrostatic adsorption. The sorption of various contaminants can deactivate the active surfaces of nanoparticles, consequently influencing the environmental behavior and toxicity of both the nanoparticles and the adsorbed contaminants by modifying their dissolution and solubility kinetics (Lin et al., 2010). Nanoparticles function as carriers for contaminants, entering organisms in the form of complexes comprising both particles and contaminants. Consequently, the complexed pollutants are liberated within the organisms, enhancing their bioavailability and toxicity through the "Trojan horse effect" (Devasena et al., 2022).

4.3 Economic Challenges

4.3.1 Cost Challenges

The financial cost frequently presents a significant challenge for numerous new applications. Applications of nanotechnology encounter considerable economic challenges. The cost associated with the synthesis and quantification of nanocomposites is comparatively high. The elevated cost poses a significant barrier for numerous researchers. Furthermore, when initiating the large-scale implementation of nanomaterial-based technologies, the economic costs are frequently overestimated by numerous countries (Darwesh et al., 2022).

4.3.2 Commercialization Barriers

The commercialization of nanotechnology innovations is challenging due to elevated research and development costs, extended development timelines, and restricted funding access. Numerous projects encounter stagnation in the "valley of death," which represents the gap between academic research and feasible commercial investment. Furthermore, there is an insufficient infrastructure and a shortage of trained personnel to facilitate scale-up (Aithal et al., 2016).

4.4 Social Challenges

4.4.1 Regulatory and Ethical Challenges

Ethics encourages responsible technology development. The current understanding suggests that nanoparticles and their applications may have inherent risks. Potential risks vary based on nanomaterials and applications. Conducting thorough evaluations on nanotechnology concepts, processes, and applications is crucial to discover nonconformities to safety norms and regulations. The primary challenges at present include the establishment and validation of methods and instruments for the detection, characterization, and analysis of nanomaterials; the assessment of exposure level; and the establishment of comprehensive risk hazards pertaining to nanomaterials and nanotechnologies (Baran et al., 2016).

4.4.2 Societal Acceptance

As with other evolving technologies, conveners of nanotechnology innovations must identify factors influencing societal responses to their development, use, and commercialization. Recognizing and comprehending these issues will promote nanotechnology development and commercialization. As the first nanoproducts like nutritional supplements, cosmetics, pharmaceuticals, and pesticides enter the market, society naturally questions their health, environmental, and safety effects. Numerous surveys suggest that demographics, age, gender, income, education, and awareness influence the perception and acceptance of applications based on nanotechnology (Oluwoye, 2015; Pillai et al., 2017). The acceptability of nanotechnologies will depend on concerns about environmental effects, safety, and sustainability (Babatunde et al., 2019).

5 Emerging Solutions and Research Initiatives

Recent studies focus on finding and removing major barriers to the application of nanoparticles in environmental cleanup. Bioinspired and green synthesis methods using biological agents, including plants, bacteria, or biodegradable polymers, are being utilized to reduce environmental and health hazards. In targeting pollutants, surface functionalization improves the stability and specificity of nanoparticles. Nanocomposite research incorporating nanomaterials with supporting matrices improves performance as well as durability. Smart nanomaterials that react to stimuli such as pH or temperature help to minimize unintentional impacts and enable regulated release. The integration of nanotechnology with circular economy strategies is experiencing increased growth. The process includes the utilization of waste-derived materials for the synthesis of nanoparticles, the formulation of recyclable nanocomposites, and the establishment of closed-loop systems that facilitate material recovery and reuse. These initiatives effectively reduce resource depletion and waste generation while simultaneously creating new opportunities for value generation. Safer nanomaterials are being designed more and more by means of lifecycle and risk assessments. Joint projects among academics, entrepreneurs, and government agencies are developing standardized safety protocols. Tools powered by data, including artificial intelligence and machine learning, are accelerating the optimization and identification of materials for environmental uses.

6 Conclusion

The utilization of nanomaterials presents a substantial opportunity for the mitigation of key environmental concerns, including the contamination of water, the pollution of air, and the degradation of soil. They are particularly useful for environmental remediation due to the fact that they possess distinctive properties such as increased reactivity, molecular interactions, catalytic activity, and increased surface area. Despite their demonstrated potential, the implementation of nanotechnology in environmental applications on a wide scale is hindered by a number of significant challenges. Overcoming technical, economic, health, and regulatory barriers is essential for the sustainable and safe implementation of nanomaterials in environmental technologies. Because of ongoing research and technological breakthroughs, it is necessary to place an emphasis on the development of solutions that make use of nanomaterials that are not only cost-effective but also safe for the environment and scalable. Effective regulations and risk management strategies facilitate the significant contribution of nanomaterials to the advancement of a cleaner and more sustainable environment.

References

Aithal, P. S., & Aithal, S. (2016). Nanotechnology innovations and commercialization-opportunities, challenges and reasons for delay. *International Journal of Engineering and Manufacturing, 6*(6), 15–25. https://doi.org/10.5815/ijem.2016.06.02

Al-Hazmi, H. E., Łuczak, J., Habibzadeh, S., Hasanin, M. S., Mohammadi, A., Esmaeili, A., Kim, S. J., Khodadadi Yazdi, M., Rabiee, N., Badawi, M., & Saeb, M. R. (2024). Polysaccharide nanocomposites in wastewater treatment: A review. *Chemosphere, 347*, Article 140578. https://doi.org/10.1016/j.chemosphere.2023.140578

Ali, F. I. M., Awwad, F., Greish, Y. E., Abu-Hani, A. F. S., & Mahmoud, S. T. (2020). Fabrication of low temperature and fast response H_2S gas sensor based on organic-metal oxide hybrid nanocomposite membrane. *Organic Electronics, 76*, Article 105486. https://doi.org/10.1016/j.orgel.2019.105486

Annu, A., Mittal, M., Tripathi, S., & Shin, D. K. (2024). Biopolymeric nanocomposites for wastewater remediation: An overview on recent progress and challenges. *Polymers, 16*(2), 294. https://doi.org/10.3390/polym16020294

Aragaw, T. A., Bogale, F. M., & Aragaw, B. A. (2021). Iron-based nanoparticles in wastewater treatment: A review on synthesis methods, applications, and removal mechanisms. *Journal of Saudi Chemical Society, 25*(8), Article 101280. https://doi.org/10.1016/j.jscs.2021.101280

Arif, Z., Sethy, N. K., Kumari, L., Mishra, P. K., & Upadhyay, S. N. (2020). Recent advances in functionalized polymer-based composite photocatalysts for wastewater treatment. *Nano-Materials as Photocatalysts for Degradation of Environmental Pollutants*, 39–64. https://doi.org/10.1016/b978-0-12-818598-8.00003-1

Asghar, N., Hussain, A., Nguyen, D. A., Ali, S., Hussain, I., Junejo, A., & Ali, A. (2024). Advancement in Nanomaterials for environmental pollutants remediation: A systematic review on Bibliometrics analysis, material types, synthesis pathways, and related mechanisms. *Journal of Nanobiotechnology, 22*(1). https://doi.org/10.1186/s12951-023-02151-3.

Babatunde, D. E., Denwigwe, I. H., Babatunde, O. M., Gbadamosi, S. L., Babalola, I. P., & Agboola, O. (2019). Environmental and societal impact of nanotechnology. *IEEE Access, 8*, 4640–4667. https://doi.org/10.1109/access.2019.2961513

Baran, A. (2016). Nanotechnology: Legal and ethical issues. *Ekonomia I Zarządzanie, 8*(1), 47–54. https://doi.org/10.1515/emj-2016-0005

Bastus, N. G., Merkoçi, F., Piella, J., & Puntes, V. (2014). Synthesis of highly monodisperse citrate-stabilized silver nanoparticles of up to 200 nm: Kinetic control and catalytic properties. *Chemistry of Materials, 26*(9), 2836–2846. https://doi.org/10.1021/cm500316k

Behera, I. D., & Das, A. P. (2025). *Environmental hydrocarbon pollution and zero waste approach towards a sustainable waste management* (p. 436).

Chaurasia, P., Dut Jasuja, N., & Kumar, S. (2023). Textile effluent treatment methods and limitations: A sustainable and ecological aspect. *Suranaree Journal of Science and Technology, 30*(6). https://doi.org/10.55766/sujst-2023-06-e01655.

Cheng, P., Zhang, S., Wang, Q., Feng, X., Zhang, S., Sun, Y., & Wang, F. (2021). Contribution of nano-zero-valent iron and arbuscular mycorrhizal fungi to phytoremediation of heavy metal-contaminated soil. *Nanomaterials, 11*(5), 1264. https://doi.org/10.3390/nano11051264

Darwesh, O., Matter, I., & Li, zhiqiang. (2021). Challenges of nanotechnology applications in addressing environmental pollution. *Egyptian Journal of Chemistry*, *0*(0), 0–0. https://doi.org/10.21608/ejchem.2021.86072.4172.

Das, A., Ray, S. K., Mohanty, M., Mohanty, J., Dey, S., & Das, A. P. (2024b). Ecotoxicity of microplastic wastes and their sustainable management: A review. *Environmental Chemistry and Ecotoxicology, 6*, 144–152. https://doi.org/10.1016/J.ENCECO.2024.05.003

Das, A. P., & Ghosh, S. (2023). Introduction. *Manganese Mining Microorganisms*, 1–4. https://doi.org/10.1016/B978-0-12-822147-1.00005-4

Das, A., Dey, S., Bala, K., Gnanasekaran, R., & Das, A. P. (2024). Microbial bioremediation technology for sustainable treatment and management of synthetic microfiber waste. *Sustainable Microbial Technology for Synthetic and Cellulosic Microfiber Bioremediation*, 221–238. https://doi.org/10.1007/978-3-031-62600-5_12

Dash, S. R., & Kundu, C. N. (2020). Promising opportunities and potential risk of nanoparticle on the society. *IET Nanobiotechnology, 14*(4), 253–260. https://doi.org/10.1049/iet-nbt.2019.0303

Dayal, J., Singh, A., & Mathur, A. (2019). Strategies of biosynthesis of nanoparticles and their potential applications. *International Journal of Scientific Research and Reviews, 8*(1), 792–806.

Dayal, J., Singh, A., & Mathur, A. (2023). Decolorization efficiency of crude laccase extract of *Cladosporium magnoliigena* LACF6 on different textile dyes. (2023). *Biological Forum—an International Journal, 15*(2), 213–219.

Del Prado-Audelo, M. L., Garcia Kerdan, I., Escutia-Guadarrama, L., Reyna-Gonzalez, J. M., Magana, J. J., & Leyva-Gomez, G. (2021). Nanoremediation: Nanomaterials and nanotechnologies for environmental cleanup. *Frontiers in Environmental Science, 9*. https://doi.org/10.3389/fenvs.2021.793765

Devasena, T., Iffath, B., Renjith Kumar, R., Muninathan, N., Baskaran, K., Srinivasan, T., & John, S. T. (2022). Insights on the dynamics and toxicity of nanoparticles in environmental matrices. *Bioinorganic Chemistry and Applications, 2022*(1). https://doi.org/10.1155/2022/4348149.

Dey, S., Tripathy, B., Kumar, M. S., & Das, A. P. (2023). Ecotoxicological consequences of manganese mining pollutants and their biological remediation. *Environmental Chemistry and Ecotoxicology*. https://doi.org/10.1016/J.ENCECO.2023.01.001

Gabal, E., Chatterjee, S., Ahmed, F. K., & Abd-Elsalam, K. A. (2020). Carbon nanomaterial applications in air pollution remediation. *Carbon Nanomaterials for Agri-Food and Environmental Applications*, 133–153. https://doi.org/10.1016/b978-0-12-819786-8.00007-4

Gaur, N., Flora, G., Yadav, M., & Tiwari, A. (2014). A review with recent advancements on bioremediation-based abolition of heavy metals. *Environment Science: Processes Impacts, 16*(2), 180–193. https://doi.org/10.1039/c3em00491k.

Guerra, F. D., Attia, M. F., Whitehead, D. C., & Alexis, F. (2018). Nanotechnology for environmental remediation: Materials and applications. *Molecules, 23*(7), 1760. https://doi.org/10.3390/molecules23071760

Ibrahim, R. K., Hayyan, M., AlSaadi, M. A., Hayyan, A., & Ibrahim, S. (2016). Environmental application of nanotechnology: Air, soil, and water. *Environmental Science and Pollution Research, 23*(14), 13754–13788. https://doi.org/10.1007/s11356-016-6457-z

Jadhav, S. V., Bringas, E., Yadav, G. D., Rathod, V. K., Ortiz, I., & Marathe, K. V. (2015). Arsenic and fluoride contaminated groundwaters: A review of current technologies for contaminants removal. *Journal of Environmental Management, 162*, 306–325. https://doi.org/10.1016/j.jenvman.2015.07.020

Jakubiak, M., Giska, I., Asztemborska, M., & Bystrzejewska-Piotrowska, G. (2013). Bioaccumulation and biosorption of inorganic nanoparticles: Factors affecting the efficiency of nanoparticle mycoextraction by liquid-grown mycelia of *Pleurotus eryngii* and *Trametes versicolor*. *Mycological Progress, 13*(3), 525–532. https://doi.org/10.1007/s11557-013-09333

Ju, M. J., Oh, J., & Choi, Y. (2020). Changes in air pollution levels after COVID-19 outbreak in Korea. *The Science of the Total Environment, 750*, Article 141521. https://doi.org/10.1016/j.scitotenv.2020.141521

Judy, J. D., Unrine, J. M., & Bertsch, P. M. (2010). Evidence for biomagnification of gold nanoparticles within a terrestrial food chain. *Environmental Science and Technology, 45*(2), 776–781. https://doi.org/10.1021/es103031a

Kaur, I. P., Kakkar, V., Deol, P. K., Yadav, M., Singh, M., & Sharma, I. (2014). Issues and concerns in nanotech product development and its commercialization. *Journal of Controlled Release: Official Journal of the Controlled Release Society, 193*, 51–62. https://doi.org/10.1016/j.jconrel.2014.06.005

Khadanga, V., & Mishra, P. C. (2024). A review on toxicity mechanism and risk factors of nanoparticles in respiratory tract. *Toxicology, 504*, Article 153781. https://doi.org/10.1016/j.tox.2024.153781

Kumar, S. (2021). Carbon based nanomaterial for removal of heavy metals from wastewater: A review. *International Journal of Environmental Analytical Chemistry, 103*(19), 7961–7978. https://doi.org/10.1080/03067319.2021.1977930

Lee, K. J., Shiratori, N., Lee, G. H., Miyawaki, J., Mochida, I., Yoon, S.-H., & Jang, J. (2010). Activated carbon nanofiber produced from electrospun polyacrylonitrile nanofiber as a highly efficient formaldehyde adsorbent. *Carbon, 48*(15), 4248–4255. https://doi.org/10.1016/j.carbon.2010.07.034

Li, Q., Mahendra, S., Lyon, D. Y., Brunet, L., Liga, M. V., Li, D., & Alvarez, P. J. J. (2008). Antimicrobial nanomaterials for water disinfection and microbial control: Potential applications and implications. *Water Research, 42*(18), 4591–4602. https://doi.org/10.1016/j.watres.2008.08.015

Li, W., Geng, X., Guo, Y., Rong, J., Gong, Y., Wu, L., Zhang, X., Li, P., Xu, J., Cheng, G., Sun, M., & Liu, L. (2011). Reduced graphene oxide electrically contacted graphene sensor for highly sensitive nitric oxide detection. *ACS Nano, 5*(9), 6955–6961. https://doi.org/10.1021/nn201433r

Lin, D., Tian, X., Wu, F., & Xing, B. (2010). Fate and transport of engineered nanomaterials in the environment. *Journal of Environmental Quality, 39*(6), 1896–1908. https://doi.org/10.2134/jeq2009.0423

Liu, Y., Dong, X., & Chen, P. (2013). Biological and chemical sensors based on graphene materials. *Chemical Society Reviews, 41*(6), 2283–2307. https://doi.org/10.1039/C1CS15270H

Mahmoud, W. M. M., Rastogi, T., & Kümmerer, K. (2017). Application of titanium dioxide nanoparticles as a photocatalyst for the removal of micropollutants such as pharmaceuticals from water. *Current Opinion in Green and Sustainable Chemistry, 6*, 1–10. https://doi.org/10.1016/j.cogsc.2017.04.001

Manisalidis, I., Stavropoulou, E., Stavropoulos, A., & Bezirtzoglou, E. (2020). Environmental and health impacts of air pollution: A review. *Frontiers in Public Health, 8*(14), 1–13. https://doi.org/10.3389/fpubh.2020.00014

Mohamed, E. F., & Awad, G. (2022). Development of nano-sensor and Biosensor as an air pollution detection technique for the foreseeable future. *Comprehensive Analytical Chemistry*, 163–188. https://doi.org/10.1016/bs.coac.2021.11.003

Mohanty, M., Mohanty, J., Priyadarsini, S., Rodríguez-Díaz, J. M., Yadav, S., Cabral-Pinto, M. M. S., Nath, H., Ram, D. K., Koteswara Reddy, G., & Das, A. P. (2025). Environmental hydrocarbon pollutants: sources, transport, and effect on human health. *Environmental Science and Engineering, Part, F218*, 21–44. https://doi.org/10.1007/978-3-031-81615-4_2

Mohanty, J., Dey, S., Bala, K., & Das, A. P. (2024). Recent advances in microplastic pollution for its sustainable management. *Sustainable Management of Urban Plastic Waste Through Circular Economic Approaches*, 27–48. https://doi.org/10.1201/9781003465447-3

Mohanty, M., Tripathy, B., Koteswara Reddy, G., & Das, A. P. (2024). Greener technologies for establishing restraint over microplastic pollution. *Sustainable Microbial Technology for Synthetic and Cellulosic Microfiber Bioremediation*, 239–258. https://doi.org/10.1007/978-3-031-62600-5_13

Munzel, T., Hahad, O., Lelieveld, J., Aschner, M., Nieuwenhuijsen, M. J., Landrigan, P. J., & Daiber, A. (2024). Soil and water pollution and cardiovascular disease. *Nature Reviews Cardiology, 22*(2), 71–89. https://doi.org/10.1038/s41569-024-01068-0

Nagar, N., Garg, H., Sharma, N., Awe, S. A., & Gahan, C. S. (2021). Effect of pulp density on the bioleaching of metals from petroleum refinery spent catalyst. *3 Biotech, 11*(3), 143. https://doi.org/10.1007/s13205-021-02686-y

Oluwoye, J. (2015). Age as a predictor of social acceptance of nanotechnology and nano-based food: a conceptual framework. *American International Journal of Contemporary Rese, 5*(6), 14–21.

Pillai, R. G., & Bezbaruah, A. N. (2017). Perceptions and attitude effects on nanotechnology acceptance: an exploratory framework. *Journal of Nanoparticle Research, 19*(2). https://doi.org/10.1007/s11051-016-3733-2

Priyadarshini, B., & Das, A. P. (2024). The Adverse impact of synthetic microplastics released from disposable face masks pollutants on aquatic microbiota, aquatic species and humans: A review. *Environmental Science and Engineering, Part, F2827*, 67–81. https://doi.org/10.1007/978-3-031-50840-0_5

Priyadarshini, S., Jagatee, S., & Das, A. P. (2024). Synthetic fabrics and microfiber pollution–an assessment of their global impact. *Environmental Science and Engineering, Part, F2333*, 137–157. https://doi.org/10.1007/978-3-031-51792-1_8

Priyadarsini, S., & Das, A. P. (2025). Lithium bioleaching: A review on microbial-assisted sustainable technology for lithium bio-circularity. *Journal of Water Process Engineering, 69*, Article 106744. https://doi.org/10.1016/j.jwpe.2024.106744

Rambaran, T., & Schirhagl, R. (2022). Nanotechnology from lab to industry—a look at current trends. *Nanoscale Advances, 4*(18), 3664–3675. https://doi.org/10.1039/d2na00439a

Ray, S. K., Dey, S., Das, A. P., & John, A. (2024). Microplastic waste management through greener technologies : A review. *Sustainable Management of Urban Plastic Waste Through Circular Economic Approaches*, 237–261. https://doi.org/10.1201/9781003465447-14

Risch, P., & Adlhart, C. (2021). A chitosan nanofiber sponge for oyster-inspired filtration of Microplastics. *ACS Applied Polymer Materials, 3*(9), 4685–4694. https://doi.org/10.1021/acsapm.1c00799

Rubina, A., Rubinova, O., & Blasinski, P. (2019). Application of nanofilters for ventilation. *International Review of Applied Sciences and Engineering, 10*(1), 51–56. https://doi.org/10.1556/1848.2019.0008

Sahu, A., Dey, S., Mallick, K., Priyadarsini, S., Behera, I. D., & Das, A. P. (2024a). A review on advanced technology for sustainable management of synthetic microplastic waste. *Sustainable Management of Urban Plastic Waste Through Circular Economic Approaches*, 262–279. https://doi.org/10.1201/9781003465447-15

Sahu, A., Dey, S., Mallick, K., Priyadarsini, S., Behera, I. D., & Das, A. P. (2024b). A review on advanced technology for sustainable management of synthetic microplastic waste. *Sustainable*

Management of Urban Plastic Waste Through Circular Economic Approaches, 262–279. https://doi.org/10.1201/9781003465447-15

Saikia, J., Gogoi, A., & Baruah, S. (2018). Nanotechnology for water remediation. *Environmental Chemistry for a Sustainable World*, 195–211. https://doi.org/10.1007/978-3-319-98708-8_7

Salaudeen, N. H. D., Akinniranye, N. R. D., Kolawole, N. M. I., & Yahaya, N. S. (2024). Nanomaterials in electronics: Advancements and challenges in high-performance devices. *World Journal of Advanced Research and Reviews, 24*(1), 830–845. https://doi.org/10.30574/wjarr.2024.24.1.3116.

Saleem, H., Zaidi, S. J., Ismail, A. F., & Goh, P. S. (2022). Advances of nanomaterials for air pollution remediation and their impacts on the environment. *Chemosphere, 287*, Article 132083. https://doi.org/10.1016/j.chemosphere.2021.132083

Samantara, A. K., Ratha, S., & Raj, S. (2019). Functionalized graphene nanocomposites in air filtration applications. *Functionalized Graphene Nanocomposites and Their Derivatives*, 65–89. https://doi.org/10.1016/b978-0-12-814548-7.00004-0

Satapathy, S., Kar, D., Priyadarsini, S., Das, A. P., & Kuanar, A. (2024). Application of microorganisms in emerging waste management strategies and remediation of synthetic and cellulosic microfibers. *Sustainable Microbial Technology for Synthetic and Cellulosic Microfiber Bioremediation*, 183–206. https://doi.org/10.1007/978-3-031-62600-5_10

Schimel, J. P., & Schaeffer, S. M. (2012). Microbial control over carbon cycling in soil. *Frontiers in Microbiology, 3*. https://doi.org/10.3389/fmicb.2012.00348

Schroeder, V., Savagatrup, S., He, M., Lin, S., & Swager, T. M. (2018). Carbon nanotube chemical sensors. *Chemical Reviews, 119*(1), 599–663. https://doi.org/10.1021/acs.chemrev.8b00340

Shahi, P.M., Kumari, P., Mahobiya, D., & Kumar Shahi, S. (2021). Nano-bioremediation of environmental contaminants: Applications, challenges, and future prospects. *Bioremediation for Environmental Sustainability*, 83–98. https://doi.org/10.1016/b978-0-12-820318-7.00004-6

Shan, J., Ji, R., Yu, Y., Xie, Z., & Yan, X. (2015). Biochar, activated carbon and carbon nanotubes have different effects on fate of 14C-catechol and microbial community in soil. *Scientific Reports, 5*(1). https://doi.org/10.1038/srep16000.

Singh, A., Pal, D. B., Mohammad, A., Alhazmi, A., Haque, S., Yoon, T., Srivastava, N., & Gupta, V. K. (2022). Biological remediation technologies for dyes and heavy metals in wastewater treatment: New insight. *Bioresource Technology, 343*, Article 126154. https://doi.org/10.1016/j.biortech.2021.126154

Stefaniuk, M., Oleszczuk, P., & Ok, Y. S. (2016). Review on nano zerovalent iron (nZVI): From synthesis to environmental applications. *Chemical Engineering Journal, 287*, 618–632. https://doi.org/10.1016/j.cej.2015.11.046

Tamanna, N. J., Hossain, M. S., Tabassum, S., Bahadur, N. M., & Ahmed, S. (2024). Easy and green synthesis of nano-ZnO and nano-TiO_2 for efficient photocatalytic degradation of organic pollutants. *Heliyon, 10*(17), Article e37469. https://doi.org/10.1016/j.heliyon.2024.e37469

Waalewijn-Kool, P. L., Klein, K., Fornies, R. M., & van Gestel, C. A. (2014). Bioaccumulation and toxicity of silver nanoparticles and silver nitrate to the soil arthropod *Folsomia Candida. Ecotoxicology, 23*(9), 1629–1637. https://doi.org/10.1007/s10646-014-1302-y

Wang, H. Y., & Lua, A. C. (2012). Development of metallic nickel nanoparticle catalyst for the decomposition of methane into hydrogen and carbon nanofibers. *The Journal of Physical Chemistry C, 116*(51), 26765–26775. https://doi.org/10.1021/jp306519t

World Health Organization (2021). The public health impact of chemicals: knowns and unknowns—data addendum for 2019. https://www.who.int/publications/i/item/WHO-HEP-ECH-EHD-21.01.

World Health Organization (2016). The public health impact of chemicals: knowns and unknowns. http://www.who.int/publications/i/item/WHO-FWC-PHE-EPE-16.01-eng

World Health Organization. Food safety (2022). http://www.who.int/news-room/fact-sheets/detail/food-safety.

Yunus, I. S., Harwin, Kurniawan, A., Adityawarman, D., & Indarto, A. (2012). Nanotechnologies in water and air pollution treatment. *Environmental Technology Reviews, 1*(1), 136–148. https://doi.org/10.1080/21622515.2012.733966.
Zamora-Ledezma, C., Negrete-Bolagay, D., Figueroa, F., Zamora-Ledezma, E., Ni, M., Alexis, F., & Guerrero, V. H. (2021). Heavy metal water pollution: A fresh look about hazards, novel and conventional remediation methods. *Environmental Technology and Innovation, 22*, Article 101504. https://doi.org/10.1016/j.eti.2021.101504
Zhu, X., Wang, J., Zhang, X., Chang, Y., & Chen, Y. (2010). Trophic transfer of TiO_2 nanoparticles from daphnia to zebrafish in a simplified freshwater food chain. *Chemosphere, 79*(9), 928–933. https://doi.org/10.1016/j.chemosphere.2010.03.022

Explorative Role of Green Synthesized Nanoparticles in Environmental Sustainability

Sushanta Kumar Barik, Akshata Sewalikar, Mulay Manjushree Vijay, Smita Sitaram Kulkarni, Sanjay N. Harke, Annasaheb S. Khemnar, Ashish S. Deshmukh, Sadanand Y. Guhe, and Sanjay J. Dhoble

Abstract Green synthesized nanoparticles (NPs) are widely used in various environmental applications due to efficiency, less cost and ecofriendly nature. Generally, the green synthesized NPs have a lot of environmental application such as waste water treatment, disinfectants, ion sensor such as heavy metal ion detection, dye removal etc. Many industries such as textile, leather, food, cosmetics, paper and pharmaceuticals industries are used several dyes and their released to the environment produces the hazardous and toxicity to the aquatic life, animals and mankind living in the ecosystem. Green synthesized nanoparticles are ecofriendly, non-toxic and promising methods for degradation of dyes. Plant sources materials act as reducing agents during synthesis of green synthesized NPs and play an important role on degradation of toxic products in reducing pollution in the environment. Photodegradation of the waste water pollutants is a simplest way to degrade pollutants or contaminants towards non-toxic substance. Green synthesized metal/metal oxide NPs are efficient photocatalyst to dyes degradation in waste water. Green synthesized NPs play an important role on ion sensation of heavy metals in the environment. Green synthesized AgNPs, CuO NPs, Au NPs, Pd NPs, PtNPs, Co NPs are widely used for detection of heavy metals such as mercury (Hg^{2+}), Lead (Pb^{2+}), Zinc (Zn^{2+}), Cadmium

S. K. Barik (✉) · A. Sewalikar · S. N. Harke · A. S. Khemnar
Institute of Biosciences and Technology, MGM University, N-6, Cidco, Chhatrapati Sambhajinagar 431003, Maharashtra, India
e-mail: sbarik@mgmu.ac.in

M. M. Vijay · S. S. Kulkarni
Department of Microbiology, MGM IHS, Chhatrapati Sambhajinagar, Maharashtra, India

A. S. Deshmukh
Department of Respiratory Medicine, MGM Medical College and Hospital, Chhatrapati Sambhajinagar, Maharashtra, India

S. Y. Guhe
Department of Chemical Engineering, MGM'S Jawaharlal Nehru Engineering College-MGM University, Chhatrapati Sambhajinagar, Maharashtra, India

S. J. Dhoble
RTM Nagpur University, Nagpur, Maharashtra, India

E. Priyadarshini et al. (eds.), *Advanced Nanotechnology Applications for Environmental Sustainability*, Sustainable Environmental Waste Management Strategies,
https://doi.org/10.1007/978-3-032-12933-8_6

(Cd^{2+}), Arsenic (As^{2+}), Copper (Cu^{2+}) etc., dye degradation in textile industries, waste water treatment in the environment. Various green synthesized nanocomposites offer an efficient and cost-effective solution in preventing water pollution and improving the water quality of waste water. Plant derived green synthesized NPs have a unique property such as biodegradability, sustainability and biocompatibility and used as a bioremediation to control environmental pollution.

Keywords Nanotechnology · Nanoparticles (NPs) · Green synthesized · Biocompatibility · Biosensing technologies · Bioremediation · Environmental sustainability

1 Introduction

Nanotechnology is the science of manipulating matter on an atomic or molecular scale, typically within the range of 1–100 nano metres' (nm), where 1 nm equals 10^{-9} m. The term "nano meter" comes from the Greek word "nano," meaning "diminutive." At this tiny scale, materials exhibit unique physical and chemical properties, such as increased surface area, enhanced reactivity, and the ability to fine-tune characteristics. These distinctive properties have catalysed the advancement of nanotechnology across various domains, including medicine, electronics, environmental science, and consumer products (Khan et al., 2019). Nanoparticles (NPs), which are particles with dimensions at the nanoscale, are of particular interest due to their exceptional electrical, optical, and magnetic properties. This has led to their application in areas like drug delivery, diagnostic imaging, data storage, and environmental remediation. Recent research has expanded into a variety of nanoparticles, including those with core/shell structures, polymer-coated surfaces, and those made from metals and metal oxides. Examples include silver (Ag), gold (Au), copper (Cu), iron oxide (FeO), zinc oxide (ZnO), titanium dioxide (TiO_2), and silica (SiO_2) nanoparticles. Each type of nanoparticle offers unique advantages and potential applications, further advancing the field of nanotechnology. Nanoparticles are extremely small particles that measure less than 100 nm in at least one dimension. To put this size into perspective, a nano meter is one-billionth of a metre, which is approximately 1/100,000th the width of a human hair. Despite their minuscule size, nanoparticles have a wide array of unique properties and applications due to their scale and surface-to-volume ratio. At the nanoscale, materials exhibit different physical and chemical properties compared to their larger-scale counterparts. This is primarily due to the increased surface area relative to their volume, which enhances their reactivity and interaction with other substances. For instance, nanoparticles can have different optical, electrical, and magnetic properties than the same material in bulk form. This is because, at the nanoscale, quantum effects become more pronounced, leading to changes in the way particles absorb and emit light or interact with electric and magnetic fields (Joudeh et al., 2022).

Nanoparticles are categorised based on their shape and dimensionality. They can be spherical (0D), cylindrical (1D), planar (2D), or even more complex structures (3D). The shape and size of nanoparticles play a crucial role in determining their specific properties and potential uses. For instance, spherical nanoparticles are often used in drug delivery systems due to their ability to encapsulate drugs and release them in a controlled manner. The synthesis of nanoparticles can be achieved through various methods, including chemical, physical, and biological approaches. Chemical methods often involve reducing metal salts in a solution to form nanoparticles, while physical methods may use techniques like laser ablation or sputtering. Biological methods leverage natural processes, such as using microorganisms or plant extracts, to produce nanoparticles (Mekuye et al., 2023). The application of nanoparticles spans numerous fields. In medicine, they are used for targeted drug delivery, allowing for more precise treatment with fewer side effects. In electronics, nanoparticles contribute to the development of more efficient and smaller devices. They are also used in environmental applications to clean up pollutants and in agriculture to improve crop yields through more efficient delivery of fertilisers and pesticides (Das & John, 2024; Mohanty et al., 2024a, 2024b; Singh et al., 2024). Despite their benefits, the use of nanoparticles also raises concerns, particularly regarding their potential impact on human health and the environment. Due to their small size, nanoparticles can potentially enter cells and tissues, which necessitates careful study and regulation to ensure their safety. In conclusion, nanoparticles are incredibly small particles with unique properties that make them valuable in various technological and scientific fields. Their ability to exhibit different behaviours compared to larger particles opens up new possibilities for innovation and application. However, as with any advanced technology, it is essential to continue researching and regulating their use to address any potential risks associated with their application. Nanotechnology emerged as a significant scientific achievement in the twenty-first century, encompassing the synthesis, management, and application of materials smaller than 100 nm. This interdisciplinary field spans various sectors, including the environment, agriculture, food, biotechnology, and medicine. Nanoparticles (NPs) are used in a variety of applications, such as treating wastewater (Zahra et al., 2020), environmental monitoring (Rassaei et al., 2011), functional food additives (Chen et al., 2023), and antimicrobial agents (Islam et al., 2022). Their unique properties, including biocompatibility, anti-inflammatory and antibacterial activities, effective drug delivery, bioactivity, and tumour targeting have driven growth in biotechnology and applied microbiology. A nanoparticle, or ultrafine particle, is typically defined as a particle with a diameter between one and one hundred nano meters. Due to their small size and large surface area, nanoparticles often display unique size-dependent properties. As particles approach the nanoscale, their physical characteristics diverge from those of bulk materials, influenced by the loss of periodic boundary conditions in crystalline structures and interactions at the de Broglie wavelength of light wavelength scale (Guo et al., 2013). These distinctive features open up a broad range of novel applications for nanoparticles (Hasan, 2015).

2 Various Plants Sources of Green Synthesized Nanoparticles

Plants possess the capacity to accumulate varying concentrations of heavy metals across their diverse anatomical structures. Consequently, biosynthesis methodologies employing plant extracts have garnered increased attention as simplistic, efficient, cost-effective, and viable techniques, serving as commendable alternatives to conventional approaches for nanoparticle synthesis. Numerous plant species have been identified for their ability to effectively reduce and stabilise metallic nanoparticles within a unified synthesis process. A multitude of researchers have embraced green synthesis methodologies for the production of metal/metal oxide nanoparticles using plant leaf extracts. Plants harbour a rich array of biomolecules, including carbohydrates, proteins, and coenzymes, which exhibit remarkable potential for the reduction of metal salts into nanoparticles (Das et al., 2024; John & Das, 2024; Priyadarsini & Das, 2025). Initial investigations into plant extract-mediated synthesis primarily focused on gold and silver nanoparticles.

Various plant species, such as datura *(metel),* peepal (*Ficus religiosa), parijat (Nyctanthes arbour-tristis),* Bermuda grass *(Cynodondactylon),* Neem (Azadirachta indica), Coriander (Coriandrum sativum), Mustard (Brassica juncea), and Lemongrass (Cymbopogon flexuosus) have been harnessed for the synthesis of silver and gold nanoparticles. The predominant body of research in this domain has explored the ex vivo synthesis of nanoparticles, although metallic nanoparticles can also be synthesised within living plants (in vivo) through the reduction of metal salt ions absorbed in soluble forms. In vivo nanoparticle synthesis, encompassing metals such as zinc, nickel, cobalt, and copper has been observed in mustard (Brassica juncea), alfalfa (Medicago sativa), and sunflower (Helianthus annuus). Furthermore, zinc oxide nanoparticles have been successfully prepared utilising a diverse array of plant leaf extracts, including coriander (Coriandrum sativum), crown flower (Calotropis gigantea), copper leaf (Acalypha indica), China rose (Hibiscus rosa-sinensis), green tea (Camellia sinensis), and aloe leaf broth extract (Aloe barbadensis Miller) (Sathiyavimal et al., 2018).

3 Types of Green Synthesised Nanoparticles

Depending on their morphology, size, and chemical characteristics, NPs are classified into several groups. Some of the most well-known classes of NPs are listed below, based on physical and chemical features.

Silver Nanoparticles (Ag NPs): Silver nanoparticles (AgNPs) are produced through green synthesis using biological agents to reduce silver ions. This involves dissolving silver metal ions and utilising natural substances like polysaccharides, proteins, and phenolics (Tolaymat et al., 2010). Medicinal plants such as Saccharum officinarum Helianthus annuus and Aloe vera are commonly used for green synthesised nanoparticles (Chandran et al., 2006; Chaudhari et al., 2012; Dubchak et al., 2010). Methods include boiling of Nelumbo nucifera leaves and treating the filtrate with silver nitrate, resulting in a brownish-yellow solution (Santhoshkumar et al., 2011). Hibiscus rosa-sinensis leaf extract mixed with silver nitrate produces light brown nanoparticles after stirring (Philip, 2010), while Jatropha Curcas seed extract in a heated silver nitrate solution yields reddish nanoparticles (Bar et al., 2009). The bark extract of Cinnamomum zeylanicum has been used to synthesise silver nanoparticles, leveraging its water-soluble organics for reduction. This method is considered green technology due to its absence of harmful chemicals (Sathishkumar et al., 2009).

Nanotechnology is an alternative technology for the environmental challenges for the treatment of industrial waste water. The waste water bodies from tannery industry were treated with silver nanoparticles (AgNPs) by the green methods. The treatment was carried out with green synthesized nanoparticles from Aloysia triphlla extract. The treated green synthesized AgNPs was generally measured through the chemical oxygen demand, biological oxygen demand, oils and gasses, total suspended solids, E. coli in waste water. Green synthesized silver nanotechnology is generally improved the physicochemical and microbiological parameters of tannery industrial waters (Benites-Alfaro et al., 2023). Green synthesized Ag NPs have multiple effect on antimicrobial properties, radical scavenging, and photo catalytic activities. Centaurea behen leaf extract mediated green synthesized nanoparticles have a photocatalytic activity suitable for degrading organic pollutants (Abdoli et al., 2024). Another innovative approach of Ag NPs acts as environmental remediation and waste water purifications in removing organic pollutants. Green synthesized Ag NPs using leaf extract from Lawsonia inermis have superior catalytic degradation of organic pollutants, such as eosin yellow, methylene blue, methyl orange, 4-nitrophenol etc. in waste water treatment (Natrayan et al., 2024). Green synthesized Ag NPs was synthesized from the Verbascum species to check the biological properties. Ag NPs was prepared from Verbascum nudatum aqueous extract and its effect was observed on quick degradation of organic pollutants such as methylene blue (Hazman et al., 2024). Ag NPs are generally synthesized from the leaves of plant species called them "ecofriendly" function as stabilizing and reducing agent during synthesis. The ecofriendly synthesis of AgNPs using Acacia raddiana acts effectively as an environmental sensor for heavy metals (Ibrahim et al., 2024). In recent years, several green synthesized metallic nanoparticles are in interest of several environmental applications. Asterarcys species (micro alga) generally used for biological reduction of AgNo3 to produce AgNPs and have a photocatalytic activity in degrading methylene blue (Choudhary et al., 2024).

Several industries release large volume of waste water containing hazardous dyes and chemicals. The growing demand for human needs to clear the pollutants from the waste water through the proper technology for sustainable development. Nanotechnology offers a better platform to improve waste water treatment/remediation. AgNPs have an effective properties and chemical activity to remove or degrade the dye material from waste water treatment. AgNPs have been used for removal or degradation of dyes as an effective strategy for waste water management (Palani et al., 2023). Water is a basic human need and removal of heavily contaminated material source is very essential. It is very essential concern for the human being to remove pollutant of waste water and monitor its quality by sensor. AgNPs are generally used as a sensor component for monitoring various waste water pollution (Serunting et al., 2023).

Gold Nanoparticles (Au NPs): Gold nanoparticles (Au NPs) are notable for their unique properties and potential in medicine and biology (Jain et al., 2006). They are highly biocompatible (Sperling et al., 2008), exhibit tunable surface plasmon resonance (Huang & El-Sayed, 2010), and have low toxicity (Jeong et al., 2011). Initial studies used Geranium leaf extract, which contains terpenoids that reduced gold ions to nanoparticles in various shapes (Shankar et al., 2003). Subsequent research with Azadirachta indica leaf extract also produced gold nanoparticles, predominantly triangular and spherical (Shankar et al., 2004). Aloe vera leaf extract allowed control over nanoparticle shape and size, with less extract yielding larger triangles and more extract producing spherical particles (Chandran et al., 2006). Gold nanoparticles synthesised with mushroom extract showed shape variations, including triangles and spheres, affected by extract concentration and temperature (Philip, 2009). Diospyros kaki and Magnolia kobus leaf extracts demonstrated that higher temperatures and extract concentrations led to smaller, spherical nanoparticles, while lower conditions resulted in larger particles with diverse shapes (Song et al., 2009). Terminalia Catappa leaf extract rapidly reduced chloruretic ions to nanoparticles with sizes of 10–35 nm (Ankamwar, 2010). Coriander leaf extract produced particles with sizes ranging from 6.75 to 57.91 nm (Kumar et al., 2019; Narayanan & Sakthivel, 2008). Trifolium leaf chloroplasts yielded highly crystalline spherical gold nanoparticles of around 20 nm, showing potential for enhanced Raman spectroscopy (Zhang et al., 2011). Recent studies also used extracts from Salix Alba (Islam et al., 2019), Coffea Arabica (Keijok et al., 2019), Croton Caudatus Bacillus Marisflavi (Nadaf & Kanase, 2019), Croton Sparsiflorus (Boomi et al., 2020), Citrus Limonum (Bhagat et al., 2020), and Aeromonas Hydrophila (Fernando & Judan Cruz, 2020) for various biomedical and environmental applications.

Green synthesis of gold nanoparticles (AuNPs) from biological resources is an ecofriendly approach (Santhosh et al., 2022). To avoid the negative effects of technological application in the environment related health issues, adaptation of green synthesis approach is a best suitable technology in modern age. The exploration of green synthesis of nanoparticles assures as a promising green chemistry to develop a greenery nanoscience and nanotechnology that reduces harmful effects of human beings (Mohanty et al., 2024a, 2024b). The efficient biogenic mechanism dramatically reduces pollutants and hazardous chemicals in an economical and green solution for environmental factors. Sustainable preparations of gold nanoparticles by green synthesis methods are a biogenic application in environmental issues (Asiya et al., 2020). Seaweed Gracilaria crassa aqueous extract was used for green synthesis of gold nanoparticles (AuNPs) and was confirmed with their ecotoxicological potential (Kamaraj et al., 2022). The unique features of green synthesized AuNPs are biocompatible, reliable and cost-effective methods for its synthesis. The ecofriendly and non-toxic AuNPs was synthesized by reducing and capping of liquid extract of stem of Lilium longiforum and had a catalytic action on reduction of methylene blue and 4-nitrophenol (Khalil et al., 2024). AuNPs were synthesized using Canthium parviflorum leaf aqueous extract. The CNP-AuNPs could effectively remove nitrophenol dyes from textiles and other industrial waste water (Kumar et al., 2024). The environment pollution management is very essential for survival of the human being in a healthy condition. Green synthesized nanotechnology offers a most effective, safety and non-toxic, non-hazardous nanomaterials for waste water management. The green synthesized AuNPs from the flowers of Clitoriaternatea potentially eliminating the organic pollutants from industrial effluent. Generally, these green synthesized AuNPs photo catalytically degraded the bromophenol blue, 4-nitrophenol, bromocresol green in the waste water (Sa et al., 2024).

Palladium (Pd) and platinum (Pt) nanoparticles: Palladium (Pd) and platinum (Pt) nanoparticles are of significant interest due to their high density and potential applications. Their green synthesis using plant extracts is an eco-friendly and cost-effective approach. Various plant extracts, including Cinnamomum Camphora, Gardenia jasminoides, Pinus Resinosa, Anogeissus Latifolia, Glycine Max, Ocimum Sanctum, Curcuma Longa, Musa Paradisica, Cinnamomum Zeylanicum, PulicariaGlutinosa, and Diospyros Kaki, have been used for this purpose (Siddiqi & Husen, 2016). For Pd nanoparticles, a methanolic extract of Catharanthus Roseus, containing compounds with –OH groups, was mixed with an aqueous solution of [Pd(OAc)2] at 60 °C, resulting in spherical nanoparticles of around 40 nm (Kalaiselvi et al., 2015). Soybean leaf extract, rich in proteins and amino acids, also produced Pd nanoparticles, which were confirmed to be 15 nm in size (Sheny et al., 2012). Black tea leaf extract was used to prepare Pd nanoparticles, which acted as effective catalysts in the Suzuki coupling reaction and demonstrated good recycling capability (Lebaschi et al., 2017). Similarly, Anogeissus Latifolia extract yielded Pd nanoparticles with an average size of 4.8 ± 1.6 nm (Kora & Rastogi, 2018), while Chlorella vulgaris extract produced 5–20 nm Pd nanoparticles in just 10 min (Arsiya et al., 2017). For Pt nanoparticles, Azadirachta Indica (Neem) leaf extract reduces Pt4 + ions to particles averaging 5–50 nm in size (Ahmed et al., 2016). Ocimum Sanctum (tulsi) leaf broth produced irregularly shaped Pt nanoparticles of about 23 nm, with terpenoids, amino acids, and proteins aiding in the reduction process (Soundarrajan et al., 2012). Dates extracted from Barni and Ajwa varieties also synthesised Pt nanoparticles, notable for their antibacterial and antifungal properties (Al-Radadi, 2019).

Metallic palladium (Pd) NPs have a variety of applications in heterogenous and homogenous catalysis and hence the greener pathways of synthesis of Pds are more valuable. Various plant extracts of PdNPs were synthesized and characterized. Green synthesized NPs exhibits as a potential reducing agent, stabilizer and the good catalytic activity in presence of flavonoids, phenols, alkaloids, reductases, ascorbic acids etc. of plant origin (Ghosh et al., 2021). The main aim of the use of green synthesized Pd NPs is to reduce the ecotoxicity effect on the life cycle of the organisms and provides a better recommendation for sustainable development in an ecosystem (Luo et al., 2021). Green synthesized Pd NPs have an effect on degradation of organic pollutants. Pimpinella tirupatiensisplant extract was used for green synthesis of Pd NPs. Pimpinella tirupatiensis Pd NPs shown the photocatalytic activity by degrading organic dye such as Congo red (Narasaiah et al., 2017). Metallic PdNPs are well recognized for their high catalytic activity in a wide range of organic transformations of industrial importance. The green synthesis of PdNPs is an ecofriendly approach for non-toxic reducing reagents, acts as biodegradable capping and stabilizing agents and is an energy efficient approach. Green synthesized PdNPs from Origanum vulgare leaf extract act as a catalyst for oxidation of alcohols (Shaik et al., 2017). Platinum nanoparticles (PtNPs) have a unique structural, optical and catalytic properties. Green synthesis of PtNPs using plant extract as an alternative approach because of its easy scalability, simplicity, easy convenience, environmental friendliness, minimum use of hazardous materials and maximized synthesis process (Jameel et al., 2020). Green synthesized PtNPs were prepared from the Sesbania polysaccharides and observed a photocatalytic activity on degradation of dyes in the environment (Nagaraja et al., 2024).

Copper nanoparticles (Cu NPs): Copper nanoparticles (Cu NPs) have been synthesised using various plant extracts. For example, Aloe vera flower extract was employed to reduce aqueous copper ions, resulting in Cu NPs with an average size of 40 nm, as confirmed by a 578-nm peak in UV-Visible spectroscopy (Karimi & Mohsenzadeh, 2015). Additionally, Cuscuta reflexa leaf extract, rich in antioxidant phytochemicals such as Myricetin, Kaempferol, and Quercetin, was used to create a $Cu/GO/MnO_2$ nanocomposite. This extract helps in converting copper ions to nanoparticles (Naghdi et al., 2018; Rahmatullah et al., 2010; Vijikumar et al., 2016).

CuNPs immobilised on graphene oxide/MnO_2 nanocomposites, synthesised using Cuscuta Reflexa leaf extract, were employed as catalysts for the reduction of various dyes and chemicals in aqueous media (Naghdi et al., 2018). Large-scale synthesis of Cu NPs was also achieved using henna leaf extract, which were incorporated into bio composites for electronic device applications (Cheirmadurai et al., 2014). Similarly, tamarind and lemon juice were used to produce 20–50 nm Cu NPs. Barberry fruit extract facilitated the in-situ synthesis of Cu NPs on reduced graphene oxide/Fe_3O_4, which proved to be an effective and reusable catalyst for the O-arylation of phenol with aryl halides (Nasrollahzadeh et al., 2015).

Copper oxide nanoparticles CuONPs gained much more attention by green synthesis methods due to availability of large number of plant species for production of reducing and stabilizing agents required for CuO NPs synthesis. CuO NPs effectively removes toxic chemicals from the environment (Akintelu et al., 2020). CuNPs were synthesized using Phragmanthera austroarabica extract in a green synthesis method. The biosynthesized CuNPs are very excellent use as a sensor for detection of hazardous metal ions, degradation of organic dyes such as congo red, and methylene blue in the environment (Alahdal et al., 2023). CuO NPs were prepared by green synthesis approach using the extracts of Corchorus olitorus leaves (Molokhaia). The biosynthesized CuO NPs were efficiently degrade the violet dye in the environmental water samples (Aljedaani et al., 2023). Therefore, CuO NPs mediated NPs synthesis is ecofriendly, economic and non-pathogenic. CuO NPs have the properties like disinfection property to remove pathogenic E.Coli to purify the water and removal of heavy metals from waste water.

Zinc oxide nanoparticles (ZnO NPs): Zinc oxide nanoparticles (ZnO NPs) have garnered significant interest due to their applications in biomedicine, optics, and electronics. They are valued for their low cost, safety, and ease of synthesis, exhibiting high exciton binding energy (60 meV) and a large bandgap (3.37 eV). These properties contribute to their semiconducting capabilities, including high catalytic activity, wound healing, anti-inflammatory effects, ultraviolet filtering, and use in cosmetics like sunscreens. ZnO NPs also find applications in antifungal, antibacterial, drug delivery, antidiabetic, and anticancer treatments (Goswami et al., 2024).

Various plant parts (flowers, roots, seeds, leaves) have been used to synthesise ZnO NPs by mixing plant extract solutions with zinc sulphate, zinc oxide, or zinc nitrate and heating the mixture. Parameters such as time, temperature, and pH can be optimised for effective synthesis. Characterization of ZnO NPs includes techniques like energy-dispersive X-ray analysis (EDAX), scanning electron microscopy (SEM), and X-ray diffraction (XRD). For instance, Azadirachta Indica leaves and Vitex Negundo flower and leaf extracts have been used, with nanoparticles sized around 38.17 nm (Ambika & Sundrarajan, 2015).

Additionally, Bacillus Licheniformis produced ZnO nanoflowers, which showed enhanced photostability and photocatalytic activity for methylene blue dye degradation, achieving 83% degradation with 74% degradation over three cycles (Auld, 2001). Lactobacillus plantarum was also used to synthesise ZnO NPs, which exhibited moderate stability with a zeta potential of −15.3 mV (Selvarajan & Mohanasrinivasan, 2013).

Zinc oxide NPs (ZnO-NPs) synthesized from M. fragrans extracts have a potential effect on environmental application. ZnO-NPs were ecofriendly, non-toxic and biocompatible in nature act as a photo catalytic agent by degrading the methylene blue, larvicidal activity in the environmental water (Faisal et al., 2021). The leaf extracts of Strobilanthes barbatus Zinc oxide NPs (ZnO-NPs) were environmental sustainable to remove the hazardous textile dyes from industrial water bodies (Gangwar et al., 2023). The sustainable green synthesis of ZnO-NPs using plant extracts gained much attention due to ecofriendly nature. ZnO-NPs have multiple roles in environmental remediation such as waste water treatment, soil remediation, effectively removing pollutants from the water bodies (Jadoun et al., 2024).

Titanium dioxide (TiO_2) nanoparticles: Titanium dioxide (TiO_2) nanoparticles are valued for their unique morphologies and surface chemistry, with applications spanning textiles, plastics, cosmetics, and pollution reduction. Green synthesis using plant extracts offers a non-toxic alternative for TiO_2 nanoparticle production. The synthesis involves reacting plant extracts with TiO_2 salts, with nanoparticle formation indicated by colour changes in the reaction mixture. Characterization methods such as UV-Visible spectroscopy, FTIR, XRD, SEM, and others confirm nanoparticle formation. For example, TiO_2 nanoparticles with spherical shapes were synthesised using Annona squamosa leaf extract (Roopan et al., 2012). Similarly, Jatropha Curcas leaf extract produced TiO_2 nanoparticles characterised by various analytical techniques (Goutam et al., 2018). Other plant extracts include Catharanthus roseus, which yielded 25–110 nm TiO_2 nanoparticles with irregular morphologies due to aliphatic amines and alcohols (Velayutham et al., 2012). Moringa Oleifera leaf extract generated 100-nm TiO_2 nanoparticles with notable wound healing properties (Sivaranjani & Philominathan, 2016). Calotropis Gigantea leaf extract produced TiO_2 nanoparticles with acaricidal activity (Marimuthu et al., 2013), while Cucurbita pepo seeds extract resulted in uniform spherical nanoparticles (Abisharani et al., 2019).

Biosynthesis of Titanium oxide (TiO_2) NPs were obtained from various plants species (Herbs, shrubs, and trees), microorganisms (bacteria, fungi, algae) and proteins, peptides and starch derivatives etc. Particularly, the TiO_2 NPs were used for treatment of waste water. Green synthesized TiO_2 NPs from cinnamon powder extracts have a strong photocatalytic activity to remove dyes pollutants from waste water (Irshad et al., 2021; Nabi et al., 2019). The biosynthesized TiO_2 NPs were prepared from leaves of Syzygiumcumini in a non-toxic, cost-effective and environmentally friendly and were employed from removal of lead from Industrial waste water (Sethy et al., 2020). Green synthesized NPs from pomegranate peel was used as a disinfectant source for waste water without any environmental damage (Abu-Dalo et al., 2019).

4 Future Perspectives of Green Synthesised Nanoparticles

The future of green-synthesised nanoparticles is promising, with the potential to transform various fields, particularly in medicine, environmental remediation, and technology. Green synthesis—using biological organisms such as plants, fungi, or bacteria—offers a sustainable alternative to traditional chemical methods. This eco-friendly approach reduces harmful byproducts, making the synthesis process more environmentally responsible. In the medical field, green-synthesised nanoparticles could significantly enhance drug delivery systems, allowing for targeted therapies with reduced toxicity. Their biocompatibility and ability to be customised for specific cellular interactions may lead to more effective treatments for conditions like cancer, with minimised side effects. Additionally, these nanoparticles could improve the accuracy of medical imaging as contrast agents. For environmental applications, these nanoparticles may serve crucial roles in pollution remediation. They can be designed to capture heavy metals, degrade toxic substances, and eliminate pathogens, thus providing valuable solutions for environmental cleanup and public health concerns. In wastewater remediation, various chemical groups are being used to functionalize nano particles so as to improve their affinity towards contaminants in the wastewater. Due to its properties, these particles can form ligands, which are recyclable, and have high selectivity towards toxicants in wastewater. Thus, achieving wastewater remediation through adsorption, as a catalyst for oxidation reaction and photocatalysis, to break down and remove harmful pollutants in wastewater. Very common examples are: disinfection and microbial control of ground water, surface water and drinking water, removal of organic and inorganic solvents, color dyes, heavy metals, biological toxins and pathogen responsible for various diseases. As nanotechnology is advancing day by day, new composite materials membranes are being synthesised embedded or encapsulated with application specific designed nanoparticles (e.g. Boron, Carbon nanotubes etc.) for water remediation. Many nanoparticles such as zero-valent iron is also being used for soil remediation.

Moreover, the integration of green-synthesised nanoparticles in renewable energy technologies, such as solar panels and batteries, can lead to more efficient energy conversion and storage systems, aiding the transition to sustainable energy sources. As research continues, the combination of green synthesis methods with artificial intelligence and machine learning may expedite the discovery of new applications and optimise the properties of nanoparticles for specific needs. Ultimately, green-synthesised nanoparticles have the potential to advance sustainability, enhance health outcomes, and foster innovation across various sectors.

5 Conclusion

Nanotechnology plays an important role in many applications due to unique physicochemical properties of nanoparticles. The green synthesis of nanoparticles represents a transformative approach in nanotechnology, leveraging plant extracts and microorganisms to produce biocompatible materials with significant applications across diverse fields, particularly healthcare, green nanotechnology is extensively being used at it is a sustainable and environmentally friendly method for wastewater and soil remediation. In these methods, nanoparticles are synthesised using microorganisms and plant extracts as reducing agents, which eliminates the toxicity associated otherwise. This eco-friendly method not only facilitates large-scale production but also enhances the therapeutic potential of nanoparticles for complex diseases, paving the way for advancements in personalised medicine and targeted therapies. The unique properties of various nanoparticles, including metal-based, carbon-based, and metal oxides, drive ongoing research into their roles in drug delivery, diagnostics, and environmental applications. As the field continues to evolve, the integration of green synthesis will likely yield sustainable solutions that address pressing health and environmental challenges, reinforcing the promise of nanotechnology in shaping future innovations. In future, nanotechnology, still need to be exploited more for water and soil remediation, production of bio hydrogen and bio-electricity from wastewater to support industrial economy.

Acknowledgements Thanks to the Obama Foundation (www.obama.org). Thanks to Governor Gretchen Whitmer and Governor Kathy Hochul, Michigan State and New York State, USA. Thanks to MGM University administration.

References

Abdoli, M., Khaledian, S., Mavaei, M., Hajmomeni, P., Ghowsi, M., Qalekhani, F., Nemati, H., Fattahi, A., & Sadrjavadi, K. (2024). Centaurea behen leaf extract mediated green synthesized

silver nanoparticles as antibacterial and removing agent of environmental pollutants with blood compatible and hemostatic effects. *Scientific Reports, 14*(1), 13941.
Abisharani, J. M., Devikala, S., Kumar, R. D., Arthanareeswari, M., & Kamaraj, P. (2019). Green synthesis of TiO_2 nanoparticles using Cucurbita pepo seeds extract. *Materials Today: Proceedings, 14*, 302–307.
Abu-Dalo, M., Jaradat, A., Albiss, B. A., & Al-Rawashdeh, N. A. (2019). Green synthesis of TiO_2 NPs/pristine pomegranate peel extract nanocomposite and its antimicrobial activity for water disinfection. *Journal of Environmental Chemical Engineering, 7*(5), Article 103370.
Ahmed, S., Ahmad, M., Swami, B. L., & Ikram, S. (2016). Green synthesis of silver nanoparticles using Azadirachta indica aqueous leaf extract. *Journal of Radiation Research and Applied Sciences, 9*(1), 1–7.
Akintelu, S. A., Folorunso, A. S., Folorunso, F. A., & Oyebamiji, A.K. (2020). Green synthesis of copper oxide nanoparticles for biomedical application and environmental remediation. *Heliyon, 6*(7).
Alahdal, F. A., Qashqoosh, M. T., Manea, Y. K., Mohammed, R. K., & Naqvi, S. (2023). Green synthesis and characterization of copper nanoparticles using Phragmanthera austroarabica extract and their biological/environmental applications. *Sustainable Materials and Technologies, 35*, Article e00540.
Aljedaani, R. O., Kosa, S. A., & Abdel Salam, M. (2022). Ecofriendly green synthesis of copper (II) oxide nanoparticles using corchorus olitorus leaves (Molokhaia) extract and their application for the environmental remediation of direct violet dye via advanced oxidation process. *Molecules, 28*(1), 16.
Al-Radadi, N. S. (2019). Green synthesis of platinum nanoparticles using Saudi's dates extract and their usage on the cancer cell treatment. *Arabian Journal of Chemistry, 12*(3), 330–349.
Ambika, S., & Sundrarajan, M. (2015). Antibacterial behaviour of Vitex negundo extract assisted ZnO nanoparticles against pathogenic bacteria. *Journal of Photochemistry and Photobiology B: Biology, 146*, 52–57.
Ankamwar, B. (2010). Biosynthesis of gold nanoparticles (green-gold) using leaf extract of Terminalia catappa. *Journal of Chemistry, 7*(4), 1334–1339.
Arsiya, F., Sayadi, M. H., & Sobhani, S. (2017). Green synthesis of palladium nanoparticles using Chlorella vulgaris. *Materials Letters, 186*, 113–115.
Asiya, S. I., Pal, K., Kralj, S., El-Sayyad, G. S., De Souza, F. G., & Narayanan, T. (2020). Sustainable preparation of gold nanoparticles via green chemistry approach for biogenic applications. *Materials Today Chemistry, 17*, Article 100327.
Auld, D. S. (2001). Zinc coordination sphere in biochemical zinc sites. *Zinc biochemistry, physiology, and homeostasis: recent insights and current trends*, pp. 85–127.
Bar, H., Bhui, D. K., Sahoo, G. P., Sarkar, P., Pyne, S., & Misra, A. (2009). Green synthesis of silver nanoparticles using seed extract of Jatropha curcas. *Colloids and Surfaces a: Physicochemical and Engineering Aspects, 348*(1–3), 212–216.
Benites-Alfaro, E., Rivera, L. T., Vergara, A. S., Castaneda-Olivera, C. A., & Ccuro, F. M. (2023). Green silver nanoparticles in improvement of physicochemical and microbiological properties of tannery wastewater. *Chemical Engineering Transactions, 101*, 31–36.
Bhagat, D. S., Gurnule, W. B., Pande, S. G., Kolhapure, M. M., & Belsare, A. D. (2020). Biosynthesis of gold nanoparticles for detection of dichlorvos residue from different samples. *Materials Today: Proceedings, 29*, 763–767.
Boomi, P., Poorani, G. P., Selvam, S., Palanisamy, S., Jegatheeswaran, S., Anand, K., Balakumar, C., Premkumar, K., & Prabu, H. G. (2020). Green biosynthesis of gold nanoparticles using Croton sparsiflorus leaves extract and evaluation of UV protection, antibacterial and anticancer applications. *Applied Organometallic Chemistry, 34*(5), Article e5574.
Chandran, S. P., Chaudhary, M., Pasricha, R., Ahmad, A., & Sastry, M. (2006). Synthesis of gold nanotriangles and silver nanoparticles using Aloevera plant extract. *Biotechnology Progress, 22*(2), 577–583.

Chaudhari, P. R., Masurkar, S. A., Shidore, V. B., & Kamble, S. P. (2012). Biosynthesis of silver nanoparticles using Saccharum officinarum and its antimicrobial activity. *Micro and Nano Letters, 7*(7), 646–650.

Cheirmadurai, K., Biswas, S., Murali, R., & Thanikaivelan, P. (2014). Green synthesis of copper nanoparticles and conducting nanobiocomposites using plant and animal sources. *RSC Advances, 4*(37), 19507–19511.

Chen, J., Guo, Y., Zhang, X., Liu, J., Gong, P., Su, Z., Fan, L., & Li, G. (2023). Emerging nanoparticles in food: Sources, application, and safety. *Journal of Agricultural and Food Chemistry, 71*(8), 3564–3582.

Choudhary, S., Kumawat, G., Khandelwal, M., Khangarot, R.K., Saharan, V., Nigam, S., & Harish. (2024). Phyco-synthesis of silver nanoparticles by environmentally safe approach and their applications. *Scientific Reports, 14*(1), 9568.

Das, A., Dey, S., & Das, A. P. (2024). Ecotoxicological impacts of synthetic microfiber pollutants and development of sustainable mitigation strategies. *Environmental Chemistry and Ecotoxicology, 7*, 201–210.

Das, A. P., & John, A. (2024). *Sustainable management of urban plastic waste through circular economic approaches,* 302.

Dubchak, S., Ogar, A., Mietelski, J. W., & Turnau, K. (2010). Influence of silver and titanium nanoparticles on arbuscular mycorrhiza colonization and accumulation of radiocaesium in Helianthus annuus. *Spanish Journal of Agricultural Research, 8*, 103–108.

Faisal, S., Jan, H., Shah, S. A., Shah, S., Khan, A., Akbar, M. T., Rizwan, M., Jan, F., Wajidullah, Akhtar, N., & Khattak, A. (2021). Green synthesis of zinc oxide (ZnO) nanoparticles using aqueous fruit extracts of Myristica fragrans: Their characterizations and biological and environmental applications. *ACS omega, 6*(14), 9709–9722.

Fernando, S. I. D., & Judan Cruz, K. G. (2020). Ethnobotanical biosynthesis of gold nanoparticles and its downregulation of quorum sensing-linked AhyR gene in Aeromonas hydrophila. *SN Applied Sciences, 2*(4), 570.

Gangwar, J., Pratap Singh, A., Marimuthu, N., & Kadanthottu Sebastian, J. (2023). Environmentally sustainable zinc oxide nanoparticles for improved hazardous textile dye removal from water bodies. *AQUA—Water Infrastructure, Ecosystems and Society, 72*(7), 1198–1210.

Ghosh, A., Hegde, R. V., Gholap, S. S., Patil, S. A., & Dateer, R. B. (2021). Green pathways for palladium nanoparticle synthesis: application and future perspectives. *Functionalized nanomaterials for catalytic application*, pp. 303–328.

Goswami, S., Bishnoi, A., Tank, D., Patel, P., Chahar, M., Khaturia, S., Modi, N., Khalid, M., Alam, M. W., Yadav, V. K., & Alreshidi, M. A. (2024) Recent trends in the synthesis, characterization and commercial applications of zinc oxide nanoparticles-a review. *Inorganica Chimica Acta*, 122350.

Goutam, S. P., Saxena, G., Singh, V., Yadav, A. K., Bharagava, R. N., & Thapa, K. B. (2018). Green synthesis of TiO_2 nanoparticles using leaf extract of Jatropha curcas L. for photocatalytic degradation of tannery wastewater. *Chemical Engineering Journal, 336*, 386–396.

Guo, D., Xie, G., & Luo, J. (2013). Mechanical properties of nanoparticles: Basics and applications. *Journal of Physics D: Applied Physics, 47*(1), Article 013001.

Hasan, S. (2015). A review on nanoparticles: Their synthesis and types. *Research Journal of Recent Sciences, 2277*, 2502.

Hazman, Ö., Khamidov, G., Yilmaz, M.A., Bozkurt, M.F., Kargioğlu, M., Tukhtaev, D., & Erol, I. (2024). Environmentally friendly silver nanoparticles synthesized from Verbascum nudatum var. extract and evaluation of its versatile biological properties and dye degradation activity. *Environmental Science and Pollution Research, 31*(23), 33482–33494.

Huang, X., & El-Sayed, M. A. (2010). Gold nanoparticles: Optical properties and implementations in cancer diagnosis and photothermal therapy. *Journal of Advanced Research, 1*(1), 13–28.

Ibrahim, N. H., Taha, G. M., Hagaggi, N. S. A., & Moghazy, M. A. (2024). Green synthesis of silver nanoparticles and its environmental sensor ability to some heavy metals. *BMC Chemistry, 18*(1), 7.

Irshad, M. A., Nawaz, R., Ur Rehman, M. Z., Adrees, M., Rizwan, M., Ali, S., Ahmad, S., & Tasleem, S. (2021). Synthesis, characterization and advanced sustainable applications of titanium dioxide nanoparticles: A review. *Ecotoxicology and Environmental Safety, 212*, Article 111978.

Islam, N. U., Jalil, K., Shahid, M., Rauf, A., Muhammad, N., Khan, A., Shah, M. R., & Khan, M. A. (2019). Green synthesis and biological activities of gold nanoparticles functionalized with Salix alba. *Arabian Journal of Chemistry, 12*(8), 2914–2925.

Islam, F., Shohag, S., Uddin, M. J., Islam, M. R., Nafady, M. H., Akter, A., Mitra, S., Roy, A., Emran, T. B., & Cavalu, S. (2022). Exploring the journey of zinc oxide nanoparticles (ZnO-NPs) toward biomedical applications. *Materials, 15*(6), 2160.

Jadoun, S., Yáñez, J., Aepuru, R., Sathish, M., Jangid, N. K., & Chinnam, S. (2024). Recent advancements in sustainable synthesis of zinc oxide nanoparticles using various plant extracts for environmental remediation. *Environmental Science and Pollution Research, 31*(13), 19123–19147.

Jain, P. K., Lee, K. S., El-Sayed, I. H., & El-Sayed, M. A. (2006). Calculated absorption and scattering properties of gold nanoparticles of different size, shape, and composition: Applications in biological imaging and biomedicine. *The Journal of Physical Chemistry B, 110*(14), 7238–7248.

Jameel, M. S., Aziz, A. A., & Dheyab, M. A. (2020). Green synthesis: Proposed mechanism and factors influencing the synthesis of platinum nanoparticles. *Green Processing and Synthesis, 9*(1), 386–398.

Jeong, S., Choi, S. Y., Park, J., Seo, J. H., Park, J., Cho, K., Joo, S. W., & Lee, S. Y. (2011). Low-toxicity chitosan gold nanoparticles for small hairpin RNA delivery in human lung adenocarcinoma cells. *Journal of Materials Chemistry, 21*(36), 13853–13859.

John, A., & Das, A. P. (2024). Government strategies for achieving circular economy for plastic waste management in Southeast Asia. In *Sustainable management of urban plastic waste through circular economic approaches,* 72–82.

Joudeh, N., & Linke, D. (2022). Nanoparticle classification, physicochemical properties, characterization, and applications: A comprehensive review for biologists. *Journal of Nanobiotechnology, 20*(1), 262.

Kalaiselvi, A., Roopan, S. M., Madhumitha, G., Ramalingam, C., & Elango, G. (2015). Synthesis and characterization of palladium nanoparticles using Catharanthus roseus leaf extract and its application in the photo-catalytic degradation. *Spectrochimica Acta Part a: Molecular and Biomolecular Spectroscopy, 135*, 116–119.

Kamaraj, C., Karthi, S., Reegan, A. D., Balasubramani, G., Ramkumar, G., Kalaivani, K., Zahir, A. A., Deepak, P., Senthil-Nathan, S., Rahman, M. M., & Islam, A. R. M. T. (2022). Green synthesis of gold nanoparticles using Gracilaria crassa leaf extract and their ecotoxicological potential: Issues to be considered. *Environmental Research, 213*, Article 113711.

Karimi, J., & Mohsenzadeh, S. (2015). Rapid, green, and eco-friendly biosynthesis of copper nanoparticles using flower extract of aloe vera. *Synthesis and Reactivity in Inorganic, Metal-Organic, and Nano-Metal Chemistry, 45*(6), 895–898.

Keijok, W. J., Pereira, R. H. A., Alvarez, L. A. C., Prado, A. R., da Silva, A. R., Ribeiro, J., de Oliveira, J. P., & Guimarães, M. C. C. (2019). Controlled biosynthesis of gold nanoparticles with coffea Arabica using factorial design. *Scientific Reports, 9*(1), 16019.

Khalil, M. T., Zhang, P., Han, G., Wu, X., Li, B., & Xiao, M. (2024). Green Synthesis of gold nanoparticles for catalytic reduction of 4-nitrophenol and methylene blue for sustainable development. *Sustainable Polymer and Energy, 2*(1), 10002.

Khan, I., Saeed, K., & Khan, I. (2019). Nanoparticles: Properties, applications and toxicities. *Arabian Journal of Chemistry, 12*(7), 908–931.

Kora, A. J., & Rastogi, L. (2018). Green synthesis of palladium nanoparticles using gum ghatti (Anogeissus latifolia) and its application as an antioxidant and catalyst. *Arabian Journal of Chemistry, 11*(7), 1097–1106.

Kumar, P. V., Kala, S. M. J., & Prakash, K. S. (2019). Green synthesis of gold nanoparticles using Croton Caudatus Geisel leaf extract and their biological studies. *Materials Letters, 236*, 19–22.

Kumar, G. S., Reddy, N. R., Siddiqui, Q. T., Yusuf, K., Pabba, D. P., Kumar, A. S., Kim, J. S., & Joo, S. W. (2024). A facile green synthesis of gold nanoparticles using Canthium parviflorum extract sustainable and energy efficient photocatalytic degradation of organic pollutants for environmental remediation. *Environmental Research, 258*, Article 119471.

Lebaschi, S., Hekmati, M., & Veisi, H. (2017). Green synthesis of palladium nanoparticles mediated by black tea leaves (Camellia sinensis) extract: Catalytic activity in the reduction of 4-nitrophenol and Suzuki-Miyaura coupling reaction under ligand-free conditions. *Journal of Colloid and Interface Science, 485*, 223–231.

Luo, S., Liu, Y., Zhu, Y., Niu, Q., Cheng, M., Ye, S., Yi, H., Shao, B., Shen, M., Wen, X., & Zeng, G. (2021). Perspectives on palladium-based nanomaterials: Green synthesis, ecotoxicity, and risk assessment. *Environmental Science: Nano, 8*(1), 20–36.

Marimuthu, S., Rahuman, A. A., Jayaseelan, C., Kirthi, A. V., Santhoshkumar, T., Velayutham, K., Bagavan, A., Kamaraj, C., Elango, G., Iyappan, M., & Siva, C. (2013). Acaricidal activity of synthesized titanium dioxide nanoparticles using Calotropis gigantea against Rhipicephalus microplus and Haemaphysalis bispinosa. *Asian Pacific Journal of Tropical Medicine, 6*(9), 682–688.

Mekuye, B., & Abera, B. (2023). Nanomaterials: An overview of synthesis, classification, characterization, and applications. *Nano Select, 4*(8), 486–501.

Mohanty, J., Dey, S., Bala, K., & Das, A. P. (2024). Recent advances in microplastic pollution for its sustainable management. In *Sustainable management of urban plastic waste through circular economic approaches,* 27–48.

Mohanty, M., Tripathy, B., Koteswara Reddy, G., & Das, A. P. (2024). Greener technologies for establishing restraint over microplastic pollution. In *Sustainable microbial technology for synthetic and cellulosic microfiber bioremediation,* 239–258.

Nabi, G., Raza, W., & Tahir, M. B. (2020). Green synthesis of TiO_2 nanoparticle using cinnamon powder extract and the study of optical properties. *Journal of Inorganic and Organometallic Polymers and Materials, 30*(4), 1425–1429.

Nadaf, N. Y., & Kanase, S. S. (2019). Biosynthesis of gold nanoparticles by Bacillus marisflavi and its potential in catalytic dye degradation. *Arabian Journal of Chemistry, 12*(8), 4806–4814.

Nagaraja, K., Arunpandian, M., & Oh, T. H. (2024). Green synthesis of platinum nanoparticles using polymer bio-reduction approach and their photocatalytic organic dye degradation. *Journal of Polymers and the Environment, 32*(9), 4538–4549.

Naghdi, S., Sajjadi, M., Nasrollahzadeh, M., Rhee, K. Y., Sajadi, S. M., & Jaleh, B. (2018). Cuscuta reflexa leaf extract mediated green synthesis of the Cu nanoparticles on graphene oxide/manganese dioxide nanocomposite and its catalytic activity toward reduction of nitroarenes and organic dyes. *Journal of the Taiwan Institute of Chemical Engineers, 86*, 158–173.

Narasaiah, P., Mandal, B. K., & Sarada, N. C. (2017). Green synthesis of Pd NPs from Pimpinella tirupatiensis plant extract and their application in photocatalytic activity dye degradation. *IOP Conference Series: Materials Science and Engineering, IOP Publishing, 263*(2), 022013

Narayanan, K. B., & Sakthivel, N. (2008). Coriander leaf mediated biosynthesis of gold nanoparticles. *Materials Letters, 62*(30), 4588–4590.

Nasrollahzadeh, M., Maham, M., Rostami-Vartooni, A., Bagherzadeh, M., & Sajadi, S. M. (2015). Barberry fruit extract assisted in situ green synthesis of Cu nanoparticles supported on a reduced graphene oxide–Fe_3O_4 nanocomposite as a magnetically separable and reusable catalyst for the O-arylation of phenols with aryl halides under ligand-free conditions. *RSC Advances, 5*(79), 64769–64780.

Natrayan, L., Jayakrishna, M., Shanker, K., Muthu, G., Kaliappan, S., & Velmurugan, G. (2024). Green synthesis of silver nanoparticles using lawsonia inermis for enhanced degradation of organic pollutants in wastewater treatment. *Global NEST Journal, 26*(3), 10–30955.

Palani, G., Trilaksana, H., Sujatha, R. M., Kannan, K., Rajendran, S., Korniejenko, K., Nykiel, M., & Uthayakumar, M. (2023). Silver nanoparticles for waste water management. *Molecules, 28*(8), 3520.

Philip, D. (2009). Biosynthesis of Au, Ag and Au–Ag nanoparticles using edible mushroom extract. *Spectrochimica Acta Part a: Molecular and Biomolecular Spectroscopy, 73*(2), 374–381.

Philip, D. (2010). Green synthesis of gold and silver nanoparticles using Hibiscus rosa sinensis. *Physica E: Low-Dimensional Systems and Nanostructures, 42*(5), 1417–1424.

Priyadarsini, S., & Das, A. P. (2025). Lithium bioleaching: A review on microbial-assisted sustainable technology for lithium bio-circularity. *Journal of Water Process Engineering, 69*, Article 106744.

Rahmatullah, M., Sultan, S., Toma, T., Lucky, S., Chowdhury, M., Haque, W., Annay, E., & Jahan, R. (2010). Effect of Cuscuta reflexa stem and Calotropis procera leaf extracts on glucose tolerance in glucose-induced hyperglycemic rats and mice. *African Journal of Traditional, Complementary and Alternative Medicines*, *7*(2).

Rassaei, L., Marken, F., Sillanpää, M., Amiri, M., Cirtiu, C. M., & Sillanpää, M. (2011). Nanoparticles in electrochemical sensors for environmental monitoring. *TrAC Trends in Analytical Chemistry, 30*(11), 1704–1715.

Roopan, S. M., Bharathi, A., Prabhakarn, A., Rahuman, A. A., Velayutham, K., Rajakumar, G., Padmaja, R. D., Lekshmi, M., & Madhumitha, G. (2012). Efficient phyto-synthesis and structural characterization of rutile TiO_2 nanoparticles using Annona squamosa peel extract. *Spectrochimica Acta Part a: Molecular and Biomolecular Spectroscopy, 98*, 86–90.

Sa, N., Alkhayer, K., & Behera, A. (2024). Efficient removal of environmental pollutants by green synthesized metal nanoparticles of Clitoria ternatea. *Heliyon, 10*(9).

Santhosh, P. B., Genova, J., & Chamati, H. (2022). Green synthesis of gold nanoparticles: An eco-friendly approach. *Chemistry, 4*(2), 345–369.

Santhoshkumar, T., Rahuman, A. A., Rajakumar, G., Marimuthu, S., Bagavan, A., Jayaseelan, C., Zahir, A. A., Elango, G., & Kamaraj, C. (2011). Synthesis of silver nanoparticles using Nelumbo nucifera leaf extract and its larvicidal activity against malaria and filariasis vectors. *Parasitology Research, 108*, 693–702.

Sathishkumar, M., Sneha, K., Won, S. W., Cho, C. W., Kim, S., & Yun, Y. S. (2009). Cinnamon zeylanicum bark extract and powder mediated green synthesis of nano-crystalline silver particles and its bactericidal activity. *Colloids and Surfaces B: Biointerfaces, 73*(2), 332–338.

Sathiyavimal, S., Vasantharaj, S., Bharathi, D., Saravanan, M., Manikandan, E., Kumar, S. S., & Pugazhendhi, A. (2018). Biogenesis of copper oxide nanoparticles (CuONPs) using Sida acuta and their incorporation over cotton fabrics to prevent the pathogenicity of Gram negative and gram positive bacteria. *Journal of Photochemistry and Photobiology B: Biology, 188*, 126–134.

Selvarajan, E., & Mohanasrinivasan, V. J. M. L. (2013). Biosynthesis and characterization of ZnO nanoparticles using Lactobacillus plantarum VITES07. *Materials Letters, 112*, 180–182.

Serunting, M. A., Zulfikar, M. A., Setiyanto, H., Setyorini, D. A., & Saraswaty, V. (2023). Green synthesized silver nanoparticles-based sensing for monitoring water pollution: An updated review. *Karbala International Journal of Modern Science, 9*(4), 12.

Sethy, N. K., Arif, Z., Mishra, P. K., & Kumar, P. (2020). Green synthesis of TiO_2 nanoparticles from Syzygium cumini extract for photo-catalytic removal of lead (Pb) in explosive industrial wastewater. *Green Processing and Synthesis, 9*(1), 171–181.

Shaik, M. R., Ali, Z. J. Q., Khan, M., Kuniyil, M., Assal, M. E., Alkhathlan, H. Z., Al-Warthan, A., Siddiqui, M. R. H., Khan, M., & Adil, S. F. (2017). Green synthesis and characterization of palladium nanoparticles using Origanum vulgare L. extract and their catalytic activity. *Molecules, 22*(1), 165.

Shankar, S. S., Ahmad, A., Pasricha, R., & Sastry, M. (2003). Bioreduction of chloroaurate ions by geranium leaves and its endophytic fungus yields gold nanoparticles of different shapes. *Journal of Materials Chemistry, 13*(7), 1822–1826.

Shankar, S. S., Rai, A., Ahmad, A., & Sastry, M. (2004). Rapid synthesis of Au, Ag, and bimetallic Au core–Ag shell nanoparticles using Neem (Azadirachta indica) leaf broth. *Journal of Colloid and Interface Science, 275*(2), 496–502.

Sheny, D. S., Philip, D., & Mathew, J. (2012). Rapid green synthesis of palladium nanoparticles using the dried leaf of Anacardium occidentale. *Spectrochimica Acta Part a: Molecular and Biomolecular Spectroscopy, 91*, 35–38.

Siddiqi, K. S., & Husen, A. (2016). Green synthesis, characterization and uses of palladium/platinum nanoparticles. *Nanoscale Research Letters, 11*, 1–13.

Singh, D., Chauhan, U., Maurya, S., & Gupta, G. (2024). Exploring the potential of grasses in the phytoremediation of various xenobiotic compounds. In *Sustainable management of environmental pollutants through phytoremediation*, 120–134.

Sivaranjani, V., & Philominathan, P. J. W. M. (2016). Synthesize of Titanium dioxide nanoparticles using Moringa oleifera leaves and evaluation of wound healing activity. *Wound Medicine, 12*, 1–5.

Song, J. Y., Jang, H. K., & Kim, B. S. (2009). Biological synthesis of gold nanoparticles using Magnolia kobus and Diopyros kaki leaf extracts. *Process Biochemistry, 44*(10), 1133–1138.

Soundarrajan, C., Sankari, A., Dhandapani, P., Maruthamuthu, S., Ravichandran, S., Sozhan, G., & Palaniswamy, N. (2012). Rapid biological synthesis of platinum nanoparticles using Ocimum sanctum for water electrolysis applications. *Bioprocess and Biosystems Engineering, 35*, 827–833.

Sperling, R. A., Gil, P. R., Zhang, F., Zanella, M., & Parak, W. J. (2008). Biological applications of gold nanoparticles. *Chemical Society Reviews, 37*(9), 1896–1908.

Tolaymat, T. M., El Badawy, A. M., Genaidy, A., Scheckel, K. G., Luxton, T. P., & Suidan, M. (2010). An evidence-based environmental perspective of manufactured silver nanoparticle in syntheses and applications: A systematic review and critical appraisal of peer-reviewed scientific papers. *Science of the Total Environment, 408*(5), 999–1006.

Velayutham, K., Rahuman, A. A., Rajakumar, G., Santhoshkumar, T., Marimuthu, S., Jayaseelan, C., Bagavan, A., Kirthi, A. V., Kamaraj, C., Zahir, A. A., & Elango, G. (2012). Evaluation of Catharanthus roseus leaf extract-mediated biosynthesis of titanium dioxide nanoparticles against Hippobosca maculata and Bovicola ovis. *Parasitology Research, 111*, 2329–2337.

Vijayakumar, S., Vaseeharan, B., Malaikozhundan, B., & Shobiya, M. (2016). Laurus nobilis leaf extract mediated green synthesis of ZnO nanoparticles: Characterization and biomedical applications. *Biomedicine & Pharmacotherapy, 84*, 1213–1222.

Zahra, Z., Habib, Z., Chung, S., & Badshah, M. A. (2020). Exposure route of TiO_2 NPs from industrial applications to wastewater treatment and their impacts on the agro-environment. *Nanomaterials, 10*(8), 1469.

Zhang, Y. X., Zheng, J., Gao, G., Kong, Y. F., Zhi, X., Wang, K., Zhang, X. Q., & Cui, D. X. (2011). Biosynthesis of gold nanoparticles using chloroplasts. *International Journal of Nanomedicine*, 2899–2906.

Nano-Enabled Sensors for Environmental Monitoring

Jolina Rodrigues and Navinchandra Shimpi

Abstract For ecosystems and human health, it is essential to maintain the quality of the air, water, and soil, which calls for environmental monitoring. Because of their unique physicochemical properties and high surface area-to-volume ratio, which offer enhanced sensitivity, selectivity, and real-time monitoring capabilities, nano-enabled sensors have emerged as highly useful instruments in this field. These sensors detect trace levels of contaminants such organic poisons, heavy metals, and greenhouse gasses using nanomaterials like metal–organic frameworks, carbon nanotubes, and graphene derivatives. This chapter focuses on current advancements in nano-enabled sensor technologies, with a focus on various nanomaterial modifications that improve their performance in a variety of environmental conditions. These sensors are particularly helpful for tracking dynamic environmental changes and identifying pollution sources in real time. Integration with wireless networks and the Internet of Things, which enables remote, comprehensive environmental tracking, further increases their importance. The creation of nano-enabled sensors is a significant step toward pollution reduction and sustainable environmental management. By addressing concerns like sensor stability, cost-effectiveness, and environmental impact, nano-enabled sensors have the potential to support data-driven decision-making for ecological preservation and pollution management.

Keywords Sensors · Environmental monitoring · Nanomaterials · Nanotechnology and sensing technology

J. Rodrigues · N. Shimpi (✉)
Laboratory of Materials Science, Department of Chemistry, University of Mumbai, Santa Cruz, Mumbai, India
e-mail: navin_shimpi@rediffmail.com; navinshimpi@chemistry.mu.ac.in

J. Rodrigues
Department of Mechanical and Nuclear Engineering, Khalifa University of Science and Technology, Abu Dhabi, UAE

E. Priyadarshini et al. (eds.), *Advanced Nanotechnology Applications for Environmental Sustainability*, Sustainable Environmental Waste Management Strategies,
https://doi.org/10.1007/978-3-032-12933-8_7

1 Introduction

The escalating concerns over global environmental degradation and its adverse impact on public health have intensified the need for accurate, real-time, and high-resolution environmental monitoring systems. Hazardous pollutants such as nitrogen oxides (NO_x), sulfur dioxide (SO_2), ammonia (NH_3), carbon monoxide (CO), volatile organic compounds (VOCs), and particulate matter (PM) are still released into ecosystems by anthropogenic activities. These pollutants contribute to climate change and both acute and chronic health problems, such as neurological impairments, cardiovascular disorders, and respiratory dysfunction (Kampa & Castanas, 2008; Landrigan et al., 2018). Due to the intricate spatiotemporal dynamics of these contaminants, sensor platforms must be sensitive, selective, scalable, economical, and able to adapt to a variety of environmental circumstances (Kampa & Castanas, 2008; Landrigan et al., 2018).

Over the past ten years, advancements in sensor technologies have drastically changed the methods used to identify, measure, and analyze environmental contaminants. Due to the growth of the Internet of Things (IoT) and smart cities, traditional large instruments have been increasingly replaced or augmented by small sensing devices that can continuously monitor an area in-situ (Wang et al., 2021). The incorporation of nanomaterials, whose quantum-scale characteristics open up new possibilities for sensor functionality, is essential to this evolution. Nanostructures improve charge transport, chemical recognition, and signal amplification mechanisms in sensing platforms because of their high surface-to-volume ratio, customizable surface chemistry, and quantum confinement effects (Liu et al., 2019; Sharma et al., 2019). The combination of sensor engineering and nanotechnology has produced multi-purpose, low-power, and extremely responsive systems that can detect pollutants at parts-per-billion (ppb) levels. For gas, humidity, and aerosol detection, metal oxide nanoparticles (e.g., SnO_2, ZnO), carbon-based nanostructures (graphene, CNTs), and novel materials like metal–organic frameworks (MOFs) and MXenes have shown remarkable potential (Xie et al., 2020; Jiang et al., 2022). In addition to improved sensitivity and selectivity, sensors provided by nanomaterials facilitate integration into wireless networks, flexible substrates, and wearable technology, allowing for decentralized environmental diagnostics that are data-rich and easy to use (Zhao et al., 2021).

Additionally, by precisely adjusting the interaction dynamics between target analytes and sensing surfaces, functionalization techniques such amine modification, surface doping, and heterojunction engineering help to overcome cross-sensitivity and enhance repeatability (Tailoring sensing performance of nanomaterials via surface functionalization: A review. xxxx). These developments not only push the boundaries of environmental sensing but also close the gap between public health and material science, promoting the creation of intelligent, sustainable monitoring ecosystems that are essential for pollution control, early warning systems, and regulatory compliance.

2 Detection Method for Nano-Enabled Sensors

By improving sensor technologies' sensitivity, selectivity, and adaptability to various stimuli, the use of nanomaterials has greatly advanced them. Based on their methods of detection, nano-enabled sensors can be roughly divided into three types: mechanical/optical, magnetic/thermal, and electrochemical. These platforms all take advantage of distinct transduction principles that are impacted by the physicochemical characteristics of the nanostructures.

2.1 Electrochemical Sensors

Changes in electrical signals, such as current, voltage, or impedance, are used by electrochemical sensors to detect chemical reactions. Improved electron transfer and analyte adsorption are made possible by the increased surface area and catalytic activity of nanomaterials such as metal nanoparticles, carbon nanotubes (CNTs), and graphene derivatives. This results in improved detection limits and quicker reaction times (Wang et al., 2020a). Depending on the electrical parameter being monitored, these sensors are divided into subtypes, including conductometric, impedimetric, amperometric, and potentiometric sensors. For example, chemiresistive sensors are based on metal oxide nanostructures (such as ZnO and SnO_2) that change their conductivity when exposed to reducing or oxidizing gasses (Sharma et al., 2019). The selectivity of nanomaterials for certain environmental analytes, such as heavy metals, ammonia, or volatile organic compounds, is further improved by functionalizing them with selective recognition components (e.g., enzymes, aptamers) (Zhang et al., 2021).

2.2 Mechanical and Optical Sensors

Physical deformations in nanostructured membranes or cantilevers brought on by mass loading or tension from analyte adsorption are detected by mechanical sensors, such as piezoresistive or piezoelectric devices. Because of their remarkable mechanical strength and flexibility, nanomaterials such as silicon nanowires, graphene, and MXenes are perfect candidates for pressure and strain sensors utilized in structural health diagnostics and environmental monitoring (Lee et al., 2019). Chemical or physical interactions are converted into quantifiable optical signals, such as absorbance, fluorescence, reflectance, or surface plasmon resonance (SPR), by optical sensors. Because noble metal nanoparticles (such Au and Ag) have localized surface plasmon resonance (LSPR), which is extremely sensitive to variations in the surrounding medium's refractive index, contaminants at trace quantities can be detected with extreme sensitivity (Yuan et al., 2020). Furthermore, fluorescence-based sensing of environmental pollutants and pathogens is being investigated using quantum dots and photonic nanomaterials (Chen et al., 2021a).

2.3 Magnetic and Thermal Sensors

Magnetic nanomaterials (such as FeO_4 and $CoFeO_4$ nanoparticles) that react to external magnetic fields or exhibit magnetoresistive changes when they interact with target analytes are used in magnetic sensors. These sensors can be incorporated into lab-on-a-chip platforms for quick diagnostics and are especially helpful in biosensing applications (Song et al., 2020). Conversely, thermal sensors identify variations in temperature, thermal conductivity, or heat capacity brought on by adsorption or chemical processes. Bismuth telluride (Bi_2Te_3) and carbon-based hybrids are two examples of nanostructured thermoelectric materials that are being employed more and more in industrial settings to detect combustion gasses or exothermic reactions. For high-resolution temperature monitoring, structural engineering can also precisely control heat flow by adjusting the thermal conductivity of nanomaterials (Kumar et al., 2021a).

3 Nanomaterials in Environmental Sensing

Nanomaterials' remarkable surface reactivity, adjustable physicochemical characteristics, and ability to integrate functionally into flexible platforms have made them essential parts of next-generation environmental sensors. Their increased adsorption kinetics, electron transport, and catalytic activity due to their nanoscale size make them extremely effective for detecting contaminants at the trace level. Sensor devices used to monitor gases, heavy metals, organic pollutants, and biological toxins are greatly impacted by the choice and engineering of nanomaterials in terms of sensitivity, selectivity, and stability.

3.1 Metallic Nanoparticles and Metal Oxides

Metallic nanoparticles, such as Au, Ag, Pt, and Pd, are frequently employed in chemical and biosensing applications because of their high conductivity, catalytic properties, and surface plasmon resonance (SPR). These particles reduce overpotentials for analyte detection and improve charge transfer at the sensing interface by acting as electron mediators or electrocatalytic sites (Zhou et al., 2021). To bind heavy metal ions and organophosphates in water, for example, gold nanoparticles (AuNPs) have been functionalized with thiol or amine groups (Cheng et al., 2019). SnO_2, ZnO,

WO_3, and TiO_2 are examples of metal oxide semiconductors (MOS), which are the foundation of many chemiresistive gas sensors. Because of charge transfer processes and redox interactions on their surfaces, these oxides show variations in electrical resistance when gases are adsorbed (Wang et al., 2021). By radically increasing surface-to-volume ratios and active site exposure, nanoscale engineering of these oxides into nanorods, nanospheres, or hollow structures makes it possible to detect NO_2, NH_3, CO, and VOCs at ppb levels (Lee et al., 2020).

3.2 *Carbon-Based Nanomaterials and Quantum Dots*

Graphene, carbon nanotubes (CNTs), and reduced graphene oxide (rGO) are examples of carbon nanomaterials that have remarkable mechanical, electrical, and thermal characteristics. They are ideal substrates for chemical sensors due to their high carrier mobility and π–π interactions with gaseous and organic compounds. Certain environmental pollutants, including benzene, toluene, xylene, and even microbiological infections, can be selectively detected using functionalized graphene, or CNTs, which have been modified with metal oxides or polymers (Novikov et al., 2020). Strong quantum confinement effects and size-dependent photoluminescence are characteristics of quantum dots (QDs), which are usually semiconductor nanocrystals such as CdSe, ZnS, or graphene QDs. These properties, where analyte binding results in fluorescence quenching or amplification, are utilized in optical sensors for the detection of metal ions and pesticides (Chen et al., 2021b). Particularly, carbon quantum dots (CQDs) are attracting interest due to their biocompatible nature, environmentally friendly manufacturing, and adjustable fluorescence in heavy metal ion detection (Liu et al., 2022).

3.3 *Polymeric-Based Composites and Hybrid Materials*

By incorporating nanoparticles into organic matrices, polymeric nanocomposites provide a synergistic blend of processability, flexibility, and functional sensitivity. Conducting polymers with reversible redox reactivity and environmental stability, as PEDOT: PSS, polypyrrole (PPy), and polyaniline (PANI), are commonly employed as sensing layers. These polymers improve charge transport and analyte binding when paired with carbon nanotubes or nanoparticles, allowing for ambient chemiresistive and electrochemical sensing (Wang et al., 2020b). Metal–organic frameworks (MOFs), zeolitic imidazolate frameworks (ZIFs), and polymer-metal oxide composites are examples of hybrid nanomaterials that use a variety of functions to enhance sensing performance. For example, MOFs are appropriate for gas separation and sensing at ultra-low concentrations due to their highly porous structures and selective adsorption sites (Zhang et al., 2022a). Under varying temperature and humidity conditions, MOF integration with graphene or polymers enhances signal transduction even more while preserving structural integrity and selectivity.

4 Nanofabrication and Integration in Sensors

The development of nano-enabled environmental sensors is contingent upon the identification of useful nanomaterials as well as accurate fabrication, patterning, and integration methods that enable scalable, application-specific, and miniature sensing platforms. Controlled assembly and alignment of nanostructures into useful devices is made possible by nanofabrication, while system-level deployment for environmental monitoring, real-time data collection, and signal processing are guaranteed by integration with microelectronics.

For modifying the physicochemical characteristics of nanomaterials for sensing applications, it is essential to fabricate them with certain dimensions, shapes, and crystallinities. For producing homogeneous nanostructures with high purity and tunability, bottom-up synthesis techniques such sol–gel processing, hydrothermal/solvothermal procedures, chemical vapor deposition (CVD), and electrochemical deposition are frequently used (Yin et al., 2021). For instance, well-aligned ZnO nanorods and SnO_2 nanofibers with improved gas-sensing performance have been produced via hydrothermal growth because of their directed growth and larger surface area (Wang et al., 2020c). The production of porous and hierarchical nanostructures appropriate for chemiresistive and optical sensors is also being investigated more and more using template-assisted and self-assembly techniques. While maintaining nanoscale control over porosity, surface roughness, and defect states parameters that significantly affect analyte adsorption and electron transfer kinetics these techniques enable scalable production.

Determining sensing geometries, facilitating multiplexed analyte detection, and attaining repeatability across sensor arrays all depend on micro- and nanopatterning. To create microelectrode patterns, nanogaps, and interdigitated electrode structures, nanomaterials can be precisely positioned on substrates using processes such photolithography, electron beam lithography (EBL), nanoimprint lithography (NIL), and soft lithography (Zhou et al., 2019). New techniques like laser scribing, inkjet printing, and aerosol jet printing provide scalable, maskless, and inexpensive patterning methods that work with flexible substrates. These methods are especially pertinent to the development of printed sensor arrays for wearable and disposable environmental monitoring devices using paper, textiles, or polymeric films (Liu et al., 2021). Nanomaterials need to have surface functionalized in order to give sensors functioning in intricate environmental matrices selectivity, stability, and anti-interference capabilities (Jiang et al., 2020). Functionalization is the process of attaching molecular recognition components, like polymers, silanes, thiols, amines, or aptamers, to nanostructured surfaces in order to facilitate selective interaction with target analytes. MOFs functionalized with amine or hydroxyl and graphene derivatives improve the adsorption of acidic gases such as SO_2 or NO_2 for gas sensors.

Water sensors also use thiol-based self-assembled monolayers (SAMs) to trap heavy metal ions. Additionally, functional coatings can add hydrophobic or hydrophilic qualities to adjust sensor performance in temperature or humidity variations (Zhao et al., 2021).

Nanoscale sensing materials have to be completely integrated with wireless communication modules, complementary metal–oxide–semiconductor (CMOS) platforms, and microelectromechanical systems (MEMS) so as to be translated into useful devices. Compact, low-power, multiplexed sensor devices that are appropriate for in-situ and real-time environmental monitoring are made possible by this integration (Ali et al., 2021). Backend-of-line (BEOL) processing techniques that work with CMOS circuits, including spin coating, inkjet printing, or dielectrophoretic assembly of nanomaterials, are a part of on-chip integration. Additionally, sensor, signal amplification, data storage, and transmission modules are being co-integrated on a single platform through the development of system-on-chip (SoC) and lab-on-chip designs. These initiatives improve the tracking of pollutants in space and time by fostering the growth of environmental monitoring networks enabled by the Internet of Things (IoT) (Nguyen et al., 2022).

5 Mechanism of Detection and Signal Transduction

The fundamental dynamics of interactions between target analytes and nanomaterials provide the basis of environmental sensing. These interactions result in quantifiable changes that are converted into mechanical, optical, thermal, or electrical signals. High surface area, adjustable band topologies, and catalytic activity are just a few of the distinctive physicochemical properties of nanomaterials that allow for quick, sensitive, and selective detection at extremely low analyte concentrations. Depending on the type of sensor, processes like adsorption, charge transfer, surface interaction, redox reactions, and resonance phenomena control the detecting mechanism.

The target molecule's particular or non-specific adsorption onto the surface of the nanomaterial initiates the detection process. The material's surface characteristics and charge distribution are changed by this interaction, which can entail physisorption (van der Waals forces, hydrogen bonding) or chemisorption (covalent or ionic bonding) (Jiang et al., 2020). For instance, oxidizing or reducing gases can react with the surface of metal oxide semiconductors (such as SnO_2 and ZnO), changing their carrier concentration and producing a detectable shift in resistance (Sharma et al., 2019). Functionalization techniques that increase analyte affinity and improve sensor specificity include the addition of selective ligands, enzymes, or receptors. Additionally, size-selective molecule sieving is made possible by nanoscale confinement and porosity in materials like MOFs and carbon aerogels, which improves the detection of gases like CO_2, NH_3, and VOCs in ambient circumstances (Zhang et al., 2022a). Electrochemical and chemiresistive sensors focus heavily on charge transfer mechanisms. The analyte and sensor surface exchange electrons during analyte interaction, changing the electrical characteristics of the nanomaterial. For example, reducing gases have the opposite impact on p-type semiconductors than oxidizing gases, which increase hole concentration and decrease resistance (Wang et al., 2020d). Surface plasmon resonance (SPR) and localized surface plasmon resonance (LSPR) are used in optical sensors to provide ultra-sensitive and label-free detection. At particular wavelengths, noble metal nanoparticles (such as Au and Ag) display collective electron oscillations. The resonance peak shifts when an analyte adsorbs onto its surface, changing the local refractive index. This phenomenon is utilized in the detection of gases, heavy metals, and pathogens (Homola, 2008; Yuan et al., 2020).

Current, voltage, impedance, or capacitance changes following analyte interaction are measured in electrochemical transduction. Redox processes at nanostructured electrodes cause fluctuations in electron flow in amperometric and potentiometric sensors used for heavy metal ion detection (Zhou et al., 2021). Light-matter interactions like absorbance, fluorescence, luminescence, or scattering are the foundation of optical transduction. Real-time monitoring of airborne contaminants or water quality is made possible by the conversion of molecular binding into spectrum shifts or intensity changes by quantum dots, graphene oxide, and SPR-active nanoparticles (Chen et al., 2021b). Thermal transduction depends on how chemical interactions alter the thermal conductivity, heat capacity, or thermopower of nanomaterials. By monitoring temperature-dependent voltage changes during gas adsorption, thermoelectric nanocomposites (such as Bi_2Te_3 and CNT-based hybrids) can identify exothermic or endothermic events (Kumar et al., 2021a). Physical deformation is transformed into an electrical signal by mechanical transduction. Under stress (such as from pressure or analyte mass), nanomaterials such as graphene, MXenes, and piezoelectric nanowires deform, changing resistance or producing voltage. According to Lee et al. (2019), these are helpful in microfluidic environmental platforms and tactile sensors.

6 Applications of Nano-Enabled Sensors in Environmental Monitoring

Environmental monitoring has greatly improved since the use of nanomaterials in sensor systems allowed for the real-time, highly sensitive, and selective detection of contaminants in a variety of ecosystems. The pressing need for ongoing monitoring of soil, water, and air quality as well as more general ecological and climatic dynamics is met by their incorporation into wearable, portable, and distant sensing platforms. Key application domains are described in the sections that follow.

6.1 Air Quality and Atmospheric Monitoring

Air pollution has become a major environmental hazard to human health and climate stability, including cardiovascular disorders, respiratory ailments, and early death.

The World Health Organization reports that more than 90% of people on the planet breathe air that contains more pollutants than recommended levels, including particulate matter (PM), nitrogen oxides (NO_x), ozone (O_3), sulfur dioxide (SO_2), carbon monoxide (CO), and volatile organic compounds (VOCs) (Landrigan et al., 2018). Sensors that can operate in real-time under a variety of environmental conditions must be sensitive, selective, portable, and able to monitor these airborne contaminants effectively. High surface area, improved electron transport characteristics, and the capacity to detect low gas concentrations in the parts per million (ppm) to parts per billion (ppb) range make nano-enabled sensors a significant improvement over traditional air quality monitoring system (Sharma et al., 2019). ZnO, SnO_2, WO_3, and In_2O_3 are examples of metal oxide semiconductors (MOS) that are frequently employed as chemiresistive materials. These sensors work on the basis of the idea that redox interactions between gas molecules and oxygen species adsorbed on the metal oxide surface cause a change in resistance (Wang et al., 2021). For example, n-type semiconductors' resistance increases when exposed to oxidizing gases like NO_2 because they extract electrons, whereas reducing gases like CO or NH_3 reduce resistance by donating electrons. Air monitoring has additionally made use of carbon-based nanomaterials, such as graphene, carbon nanotubes (CNTs), and reduced graphene oxide (rGO), because of their remarkable conductivity, surface reactivity, and ease of functionalization (Novikov et al., 2020). Functionalization with metal nanoparticles or polymers improves their selectivity for particular gases, including ammonia, benzene, and formaldehyde. Parallel to this, label-free, real-time monitoring of gaseous and particulate pollutants has been made possible using optical sensors based on surface plasmon resonance (SPR) and localized surface plasmon resonance (LSPR) phenomena in gold (Au) and silver (Ag) nanoparticles. According to Yuan et al. (2020), these sensors identify quantifiable variations in the resonance wavelength caused by analyte adsorption-induced changes in the refractive index close to the metal surface. Continuous geospatial mapping of air quality is now possible because of recent developments that have also seen the integration of nano-enabled sensors with wireless sensor networks and Internet of Things (IoT) platforms. For the purpose of collecting environmental data and monitoring individual exposure, these systems can be integrated into mobile units, smart city infrastructure, or personal wearables (Zhou et al., 2020).

6.2 *Water Quality and Contaminant Detection*

Food security, sustainable development, and public health all depend on having access to clean water. But there are serious risks to freshwater supplies due to growing contamination from household trash, agricultural runoff, and industrial effluents. Concerning pollutants include pesticides, pharmaceutical residues, microbial pathogens, heavy metal ions (e.g., Pb^{2+}, Hg^{2+}, and Cd^{2+}), and new contaminants such microplastics and endocrine-disrupting substances (Jiang et al., 2020). In addition to being sensitive and specific, analytical techniques that are portable, affordable, and able to be monitored in real time are necessary for the detection of these contaminants at trace levels. Nano-enabled sensors' large surface area, adjustable surface chemistry, and quick analyte interaction dynamics have made them effective instruments for evaluating the quality of water. According to Zhou et al. (2021), these sensors use nanomaterials including metal nanoparticles, carbon nanostructures, quantum dots, and polymeric nanocomposites to convert chemical or biological recognition events into quantifiable electrical or optical signals. Heavy metals and organic contaminants are commonly detected using electrochemical sensors, especially those based on graphene, carbon nanotubes (CNTs), and gold nanoparticles. These nanostructures facilitate the addition of selective recognition components like aptamers, enzymes, or molecularly imprinted polymers (MIPs), increase electron transfer rates, and lower detection limits. For example, using anodic stripping voltammetry, graphene-modified electrodes functionalized with thiol or amine groups have shown selective detection of Hg^{2+} and Pb^{2+} ions at nanomolar concentrations (Chen et al., 2020). Through changes in photoluminescence, fluorescence-based sensors particularly those that use semiconductor quantum dots (such as CdTe or ZnS) or carbon quantum dots (CQDs) offer optical contamination detection. Label-free and real-time monitoring of metal ions, phenols, or nitroaromatic chemicals is made possible by the analyte's interaction with the dot surface, which causes fluorescence to be quenched or enhanced (Liu et al., 2022). Furthermore, CQDs are becoming more well-known because of their green synthesis methods, aqueous solubility, and non-toxicity. Utilizing nanostructured gold or silver substrates, surface-enhanced Raman scattering (SERS) has been used for the ultra-sensitive detection of microbiological pollutants and trace organics. By amplifying the vibrational fingerprints of target molecules, SERS makes it possible to identify contaminants like atrazine, bisphenol A, and E. coli in complex matrices without the need for labels (Wang et al., 2019). Furthermore, biosensors that use enzyme-nanomaterial hybrids have shown promise in detecting wastewater's levels of urea, nitrate, and biochemical oxygen demand (BOD). Nanomaterials aid in immobilization and signal amplification in these systems, which use biocatalytic activity to transform analytes into electroactive species or optical signals (Kumar et al., 2021b).

The integration of these nanosensors into portable and wireless monitoring platforms enables point-of-use applications for decentralized water monitoring in rural, industrial, and disaster-struck regions. Coupled with smartphone-based readouts and IoT frameworks, nano-enabled sensors represent a transformative shift toward smart and sustainable water quality management.

6.3 Soil and Agricultural Monitoring

Environmental factors like moisture content, pH, nutrient levels, salt, and residual agrochemicals have a direct impact on soil quality and agricultural output. Environmental preservation, resource optimization, and precision agriculture all depend on accurate, real-time monitoring of these variables. Conventional soil analysis techniques can require a lot of work and time, and they don't have the spatial or temporal resolution required for contemporary farming methods. By enabling the on-site, quick, and extremely sensitive detection of important soil and crop health indices, nano-enabled sensor technologies present potent substitutes. Nanomaterials have been successfully integrated into soil sensors to monitor chemical, physical, and biological soil parameters because of their multifunctionality, large surface-to-volume ratio, and tunable surface chemistry (Cheng et al., 2020). Soil pH, nitrate, phosphate, and potassium levels all essential nutrients for plant growth can be detected with great sensitivity using electrochemical sensors based on carbon nanotubes (CNTs), graphene, and metal oxide nanostructures (e.g., ZnO, CuO, TiO_2) (Patel et al., 2021).

When compared to conventional probes, soil moisture sensors augmented with nanocomposites (such as CNTs/PDMS or rGO/polyaniline) have better capacitance and conductivity profiles, which result in increased responsiveness and stability. Especially in arid or semi-arid areas, these sensors are critical for water use efficiency, drought control, and irrigation scheduling (Wang et al., 2020e). Moreover, molecularly imprinted polymers (MIPs) on nanostructured platforms have been used to create nanosensors for the detection of pesticides and herbicides. By selectively binding to pesticide residues including atrazine, glyphosate, and paraquat, these sensors enable voltammetric or colorimetric approaches to detect them at trace amounts in soil and runoff water (Singh et al., 2019). Metal oxide nanoparticles and MOF-based hybrids are used as gas sensors to track the release of greenhouse gases (such as CH_4 and N_2O) and volatile organic compounds (VOCs) from soil. These gases are connected to fertilizer application, microbial activity, and the breakdown of organic matter (Zhang et al., 2022b). Sustainable fertilizer management techniques are informed by the monitoring of these emissions, which aids in the assessment of soil health. Additionally, biosensors that incorporate nanomaterials can identify enzymes, plant hormones, and pathogen biomarkers, providing information about soil microbiota and early crop disease detection. For instance, nitrate reductase activity or biological oxygen demand (BOD) can be measured using enzyme-nanoparticle conjugates, which represent nitrogen cycling and microbial respiration (Kumar et al., 2021c). Remote monitoring and data analytics across agricultural landscapes are made possible by the growing integration of nano-enabled sensors into wireless sensor networks (WSNs) and Internet of Things platforms. For the future of intelligent and sustainable agriculture, this capability supports early warning systems, automated fertilization, and site-specific management.

6.4 Climate and Ecosystem Monitoring

One of the most important worldwide issues of the twenty-first century is climate change and ecosystem degradation, which have an impact on human livelihoods, food security, water availability, and biodiversity. Early warning systems, climate modeling, and biodiversity conservation methods all depend on the ability to track minute changes in climatic variables and ecosystem health in real time. Unprecedented sensitivity and spatiotemporal resolution are provided by nano-enabled sensors, which make them effective instruments for tracking vital variables like greenhouse gas emissions, temperature variations, soil respiration, plant physiological reactions, and the health of aquatic ecosystems.

The identification of greenhouse gases (GHGs), such as carbon dioxide (CO_2), methane (CH_4), and nitrous oxide (N_2O), is one of the most straightforward uses of nano-sensing in climate science. The radiative forcing of the Earth is largely caused by these gases. When functionalized with catalytic nanoparticles or dopants, metal oxide semiconductor (MOS)-based sensors (such as SnO_2, ZnO, and In_2O_3) have shown great sensitivity and quick response times for detecting low concentrations of CH_4 and N_2O in the atmosphere (Wang et al., 2021). Through resistive or optical transduction processes, metal–organic frameworks (MOFs) have been utilized to selectively adsorb and detect CO_2 due to their tunable porosity and functional groups (Zhang et al., 2022c). Soil respiration, a crucial measure of microbial and root activity, serves as a crucial stand-in for carbon cycling and soil health in terrestrial ecosystems. Real-time evaluations of soil carbon fluxes and microbial activity in response to environmental stressors including fertilization, temperature fluctuations, and drought are made possible by the ability of gas sensors based on nanomaterials to detect changes in CO_2 content in soil headspace (Kumar et al., 2021b). Nano-enabled optical and electrochemical sensors are used to track temperature, pH, dissolved oxygen, nutrient load (such as phosphates and nitrates), and new contaminants like endocrine disruptors and microplastics in aquatic environments. High accuracy in identifying contaminants and biomarkers of ecosystem stress, such as oxidative enzymes and algae toxins, has been demonstrated by quantum dots, graphene-based composites, and plasmonic nanoparticles (Chen et al., 2021b; Liu et al., 2022). Reactive oxygen species (ROS), ethylene, and abscisic acid are examples of plant stress signals that are capable of being measured by nano-enabled biosensors in vegetation and forest ecosystems. These molecular markers provide early warning signs of disease invasion, salinity, or drought. They offer multi-scale insights into vegetation dynamics and forest health when combined with satellite or drone-based imaging systems (Nguyen et al., 2022). Additionally, the deployment of distributed monitoring systems across sensitive and remote ecological zones is made possible by the integration of these sensors with wireless sensor networks (WSNs) and the Internet of Things (IoT). These platforms make it easier to gather long-term data and analyze the connections between ecosystems and climate, which is essential for making informed environmental decisions.

7 Challenges, Risk and Environmental Impact

The increasing use of nanomaterials into sensor systems raises questions about their toxicity and long-term environmental destiny. The reactivity and bioavailability of engineered nanomaterials (ENMs), including metal oxides (ZnO, TiO_2), carbon-based nanostructures (CNTs, graphene), and quantum dots, influences and persist in ecosystems (Keller et al., 2013). These substances can endanger aquatic life, microbial communities, and higher trophic levels when they are released since they can build up in soil, water, and air. Silver nanoparticles (AgNPs), for example, have potent antibacterial qualities that, although advantageous for sensing applications, may disturb the beneficial microbial ecosystems in soil or wastewater (Bondarenko et al., 2013). Furthermore, there are serious toxicological concerns regarding the oxidative stress and membrane damage brought on by the production of reactive oxygen species (ROS) in cells exposed to specific nanomaterials (such as TiO_2 NPs under UV radiation) (Nel et al., 2006).

Regulatory frameworks have remained disjointed and frequently lag behind innovation, despite the quick advancements in nanotechnology. Although nanoparticles are covered by current laws like the European Union's REACH (Registration, Evaluation, Authorization and Restriction of Chemicals) regulations, precise characterization, risk assessment, and labelling guidelines are still being developed (CENIHR, 2009). Standardizing nanoparticle size, surface characteristics, and aggregation behaviour is a significant difficulty since these factors affect how the particles interact with biological systems and make toxicity evaluation more difficult. Furthermore, the manufacturing of sensors exposes workers to airborne nanoparticles, necessitating strict occupational safety regulations that have not yet been widely implemented (Maynard & Kuempel, 2005). Long-term operating stability, minimal integration costs, and scalable synthesis are critical to the economic feasibility of nano-enabled sensors. Nevertheless, the environmental benefits of certain nanofabrication techniques may be outweighed by their energy-intensive nature (e.g., chemical vapor deposition, atomic layer deposition) and the use of rare or dangerous precursors (Gao & Kanaoka, 2019). Although sustainable alternatives have arisen to lessen the environmental load, such as green synthesis using plant extracts or microbial methods, their industrial scalability and repeatability are still limited. To measure the entire environmental impact of nanosensors, from the extraction of raw materials to the disposal of end-of-life devices, life cycle evaluations, or LCAs, are essential. For example, although effective, carbon nanotube-based sensors have a larger potential for ecotoxicity and cumulative energy consumption than some polymer-based substitutes (Pourzahedi et al., 2017).

8 Future Prospects

Environmental monitoring is being revolutionized by the combination of machine learning (ML) with sensor technologies, which allow for the accurate, adaptive, and real-time analysis of complex sensor datasets. Conventional sensors frequently produce noisy and non-linear outputs, particularly in dynamic environmental conditions. It is possible to train machine learning methods like principal component analysis (PCA), support vector machines (SVMs), and artificial neural networks (ANNs) to identify patterns, reduce noise, and improve sensitivity and selectivity (Xu et al., 2020). For instance, ML helps gas sensors based on nanomaterials (such as metal oxides, or MOFs) differentiate between gas mixes or overlapping volatile organic compounds in urban air. Additionally, automatic anomaly detection and predictive maintenance in sensor networks are increasingly supported by deep learning models, extending the longevity and resilience of systems (Rajab et al., 2018). In order to attain greater sensitivity, durability, and multifunctionality, next-generation environmental sensors are depending more and more on hybrid sensor platforms, which integrate the special qualities of several nanomaterials. For example, customizable electrical responses and enhanced adsorption capacities for a variety of environmental analytes are made possible by integrating carbon-based materials (graphene, CNTs) with metal oxides or conductive polymers (Anjum et al., 2021). Hybrid systems increase the usefulness of a single sensor unit by simultaneously detecting gasses, humidity, temperature, and even biological poisons. Additionally, sensors can operate in dynamic environmental conditions like greenhouses or coastal areas thanks to nanocomposites that exhibit stimuli-responsive behaviour (such as photo-, thermo-, or pH-responsive) (Han et al., 2019).

The requirement for continuous, dispersed monitoring over wide geographic areas is driving up demand for flexible, low-power, autonomous environmental sensors. Self-powered, wearable, or deployable sensors that function without the need for external power or human interaction are becoming possible because to advancements in printed electronics, stretchable substrates, and energy-harvesting technologies (such as triboelectric, piezoelectric, and solar) (Dagdeviren et al., 2017). In situations when conventional wired infrastructure is impractical, such systems are perfect for smart cities, agriculture, and disaster areas. Real-time environmental data transmission and decision-making are further facilitated by integration with wireless modules (such as IoT protocols like LoRaWAN or NB-IoT) (Khan et al., 2016). Although nano-enabled sensors show great promise, there are still several obstacles to commercialization and policy. Commercialization and public acceptance of nanomaterials are restricted by the absence of uniform standards and regulatory frameworks for their classification, safety evaluation, and environmental release (OECD, 2021). To reduce environmental and health risks, transparency in the lifetime, labeling, and disposal of nanomaterials is essential. Additionally, scaling up nanomaterial synthesis, guaranteeing repeatability, and attaining cost-effectiveness present difficulties for both startups and established enterprises. Lab-scale breakthroughs can be quickly transformed into market-ready products with the help of policy assistance through green subsidies, patent facilitation, and public–private partnerships (Bawa & Johnson, 2009).

9 Conclusion

Next-generation environmental monitoring will be accomplished by nano-enabled sensors, which provide revolutionary capabilities because to the special physicochemical characteristics of nanomaterials. Through the ability to detect contaminants in air, water, and soil matrices with extreme sensitivity and in real time, these sensors immediately support resource conservation and environmental preservation. Their incorporation into small, energy-efficient platforms is a perfect fit with ecological stewardship and sustainable development objectives. This chapter has examined the novel production, functionalization, and transduction processes of nanomaterials, showcasing their potential to tackle intricate environmental issues. Intelligent instruments for ongoing environmental diagnostics are being developed using nano-enabled sensors, which range from hybrid nanostructures with multipurpose sensing capabilities to smart, networked systems that use IoT and machine learning. Significantly, the development of biodegradable, flexible, and self-powered sensor systems offers a viable route to widespread implementation with a little environmental impact. Particularly in climate-sensitive and resource-constrained areas, these technologies have the potential to strengthen environmental management systems' resilience, lower energy usage in monitoring operations, and promote circular economy models. However, it is crucial to address the toxicological and environmental effects of using nanomaterials in order to guarantee their long-term viability. Harmonizing innovation and environmental responsibility will require the creation of safe-by-design frameworks, thorough life cycle assessments, and international regulatory standards.

In summary, sensors using nanotechnology are catalysts for long-term change rather than merely being analytical instruments. They enable society to move toward proactive, data-driven environmental governance and a more resilient, sustainable future by bridging the gap between cutting-edge nanoscience and environmental policy.

References

Ali, M. A., et al. (2021). Nanomaterial-CMOS integration: Challenges and solutions for next-generation sensors. *IEEE Sensors Journal, 21*(15), 17089–17101.

Anjum, M. W., et al. (2021). Advanced nanomaterials and hybrid architectures for next-generation environmental sensors. *Sensors and Actuators B: Chemical, 330*, Article 129346.

Bawa, R., & Johnson, T. L. (2009). Emerging issues in nanomedicine and ethics. *Medical Law International, 10*(1), 23–35.

Bondarenko, O., et al. (2013). Toxicity of Ag, CuO and ZnO nanoparticles to selected environmentally relevant test organisms. *Science of the Total Environment, 449*, 456–466.

Chen, L., et al. (2020). Electrochemical sensing of heavy metal ions using functionalized graphene-based nanocomposites: A review. *Sensors and Actuators B: Chemical, 310*, Article 127889.
Chen, L., et al. (2021a). Quantum dot-based sensors for environmental applications. *Sensors and Actuators b: Chemical, 329*, Article 129211.
Chen, L., et al. (2021b). Quantum dot-based fluorescent sensors for environmental pollutant detection. *Environmental Science: Nano, 8*(3), 703–728.
Cheng, M., et al. (2019). Functional gold nanoparticle-based sensors for environmental monitoring. *Sensors, 19*(9), 2046.
Cheng, M., et al. (2020). Smart soil sensors for precision agriculture based on nanomaterials. *Sensors, 20*(7), 1881.
Dagdeviren, C., et al. (2017). Recent progress in flexible and stretchable piezoelectric devices for mechanical energy harvesting, sensing and actuation. *Extreme Mechanics Letters, 9*, 269–281.
Gao, W., & Kanaoka, C. (2019). Environmental impacts of nanomanufacturing processes. *Clean Technologies and Environmental Policy, 21*, 103–115.
Han, T., et al. (2019). Multifunctional hybrid nanomaterials for environmental monitoring: A review. *Environmental Research, 169*, 71–85.
Homola, J. (2008). Surface plasmon resonance sensors for detection of chemical and biological species. *Chemical Reviews, 108*(2), 462–493.
Jiang, Y., et al. (2020). Functionalization strategies for nanomaterial-based sensors targeting environmental analytes. *Chemical Society Reviews, 49*(22), 8197–8231.
Jiang, Y., et al. (2022). Nanostructured materials for environmental monitoring and remediation. *Environmental Nanotechnology, Monitoring & Management*, *12*, 100265.
Kampa, M., & Castanas, E. (2008). Human health effects of air pollution. *Environmental Pollution, 151*(2), 362–367.
Keller, A. A., et al. (2013). Global life cycle releases of engineered nanomaterials. *Journal of Nanoparticle Research, 15*, 1692.
Khan, Y., et al. (2016). Monitoring of vital signs with flexible and wearable medical devices. *Advanced Materials, 28*(22), 4373–4395.
Kumar, A., et al. (2021a). Thermal and thermoelectric properties of nanostructured materials for sensing applications. *Nano Energy, 84*, Article 105915.
Kumar, A., et al. (2021b). Advances in biosensor-based ecosystem monitoring using nanotechnology. *TrAC Trends in Analytical Chemistry, 136*, Article 116193.
Kumar, A., et al. (2021c). Nanotechnology-enabled biosensors for monitoring soil microbiota and crop health. *TrAC Trends in Analytical Chemistry, 136*, Article 116193.
Landrigan, P. J., et al. (2018). The Lancet Commission on pollution and health. *The Lancet, 391*(10119), 462–512.
Lee, C., et al. (2019). Graphene-based piezoresistive strain sensors. *Materials Science and Engineering: R: Reports, 126*, 1–29.
Lee, J., et al. (2020). Hollow nanostructured metal oxides for gas sensing: A review. *ACS Applied Materials & Interfaces, 12*(3), 3109–3130.
Liu, H., et al. (2021). Printed and flexible sensors for environmental monitoring. *Advanced Materials Technologies, 6*(2), 2000917.
Liu, H., et al. (2022). Carbon quantum dots for environmental sensing: Synthesis, applications, and challenges. *Journal of Hazardous Materials, 423*, Article 127150.
Liu, X., et al. (2019). Recent advances in nanomaterials for gas sensors. *Advanced Materials, 33*(3), 2002646.
Maynard, A. D., & Kuempel, E. D. (2005). Airborne nanostructured particles and occupational health. *Journal of Nanoparticle Research, 7*, 587–614.
Nel, A., et al. (2006). Toxic potential of materials at the nanolevel. *Science, 311*(5761), 622–627.
Nguyen, T., et al. (2022). Integration of nanomaterial sensors with IoT platforms for environmental monitoring: A review. *Environmental Science & Technology, 56*(5), 2891–2908.
Novikov, S., et al. (2020). Functionalized carbon nanomaterials for environmental gas sensors. *Carbon, 167*, 1–21.

OECD. (2021). *Nanomaterials in wastewater treatment: Current status and future outlook.* Organisation for Economic Co-operation and Development.

Patel, P., et al. (2021). Nanotechnology-based sensors for soil nutrient detection: A review. *Environmental Nanotechnology, Monitoring & Management, 16*, Article 100517.

Pourzahedi, L., et al. (2017). Life cycle environmental impacts of carbon nanomaterials for sensor applications. *Environmental Science & Technology, 51*(16), 10386–10396.

Rajab, K. Z., et al. (2018). A review on machine learning algorithms for the development of smart environment applications. *Artificial Intelligence Review, 50*, 379–407.

Scientific Committee on Emerging and Newly Identified Health Risks (CENIHR). (2009). *Risk assessment of products of nanotechnologies*. European Commission.

Sharma, B., et al. (2019). Nanostructured materials for environmental monitoring and remediation. *Environmental Nanotechnology, Monitoring & Management, 12*, Article 100265.

Sharma, B., et al. (2019). A survey on gas sensing technology. *Sensors, 19*(2), 376.

Singh, R., et al. (2019). Molecularly imprinted polymer-based nanosensors for the detection of agrochemicals: Recent trends and applications. *Biosensors and Bioelectronics, 142*, Article 111538.

Song, Y., et al. (2020). Magnetic nanoparticles for biosensing and diagnostic applications. *Sensors, 20*(24), 6730.

Tailoring sensing performance of nanomaterials via surface functionalization: A review. *Nano Today, 39*, 101180.

Wang, Y., et al. (2019). Recent advances in surface-enhanced Raman spectroscopy for environmental analysis: Nanostructures and detection strategies. *Environmental Science: Nano, 6*(11), 3200–3223.

Wang, C., et al. (2020a). Recent progress in electrochemical sensors based on carbon nanomaterials for the detection of environmental pollutants. *TrAC Trends in Analytical Chemistry, 132*, Article 116045.

Wang, Y., et al. (2020b). Polymer nanocomposites for electrochemical sensors: Fundamentals and applications. *ACS Sensors, 5*(5), 1360–1377.

Wang, C., et al. (2020c). Synthesis of oriented ZnO nanostructures for enhanced gas sensing performance. *Advanced Functional Materials, 30*(34), 2002175.

Wang, C., et al. (2020d). Metal oxide gas sensors: Sensitivity and selectivity tuning strategies. *Sensors and Actuators b: Chemical, 330*, Article 129304.

Wang, C., et al. (2020e). Advanced materials for smart farming: Sensors for soil and water monitoring. *ACS Nano, 14*(9), 10828–10835.

Wang, C., et al. (2021). Metal oxide gas sensors: Sensitivity and selectivity tuning strategies. *Sensors and Actuators b: Chemical, 330*, Article 129304.

Wang, C., et al. (2021). Trends in low-cost air quality sensor research: A review. *Environmental International, 145*, 106159.

Xie, G., et al. (2020). OF-based sensors for environmental pollutant detection. *Chemical Society Reviews, 51*(1), 145–179

Xu, X., et al. (2020). Machine learning for intelligent sensing and diagnostics: Recent advances and future prospects. *IEEE Transactions on Industrial Informatics, 16*(9), 5626–5635.

Yin, X., et al. (2021). Recent advances in nanomaterial synthesis and applications for gas sensors. *Sensors and Actuators b: Chemical, 346*, Article 130489.

Yuan, K., et al. (2020). Plasmonic nanomaterials for environmental sensing and monitoring. *Advanced Optical Materials, 8*(9), 1902064.

Zhang, Y., et al. (2021). Electrochemical sensors based on nanomaterials for detection of heavy metal ions. *Micromachines, 12*(1), 74.

Zhang, Z., et al. (2022a). Hybrid metal-organic framework nanocomposites for chemiresistive gas sensing. *Advanced Functional Materials, 32*(9), 2108181.

Zhang, Z., et al. (2022b). Metal-organic frameworks in environmental field: Soil gas monitoring and emission reduction. *Journal of Environmental Chemical Engineering, 10*(3), Article 107389.

Zhang, Z., et al. (2022c). Metal-organic frameworks for environmental sensing and greenhouse gas detection. *ACS Applied Materials & Interfaces, 14*(3), 3256–3271.
Zhao, C., et al. (2021). Surface chemistry tuning of nanomaterials for enhanced selectivity in environmental sensing. *Nano Today, 39*, Article 101180.
Zhao, C., et al. (2021). Flexible sensors for environmental monitoring. *Materials Today, 41*, 28–45.
Zhou, Q., et al. (2019). Lithographic techniques in nanosensor fabrication: From microfabrication to nanofeatures. *ACS Applied Nano Materials, 2*(11), 7232–7248.
Zhou, Q., et al. (2020). Trends in low-cost air quality sensor research: A review. *Environment International, 145*, Article 106159.
Zhou, X., et al. (2021). Noble metal nanostructures for chemical and biosensing applications. *Biosensors and Bioelectronics, 174*, Article 112804.

Green Nano-Materials as Photo-Catalyst for Environmental Applications

Supriya Tripathy, Kailash Nath Pandey, and Navinchandra Gopal Shimpi

Abstract Green nanoparticles' remarkable catalytic qualities and environmentally benign manufacturing have made them promising photocatalysts for a range of environmental applications. The creation and application of green nanomaterials in photocatalytic processes for environmental remediation are examined in this chapter. These materials, which are frequently produced using ecological and biological processes, provide efficient means of lowering harmful compounds in soil and air, cleaning water and breaking down contaminants. The processes via which green nanomaterials use light energy to speed up reactions are covered, along with variables like stability, surface area and particle size that impact how effective they are at catalyzing reactions. This chapter emphasizes on recent developments in material design techniques, such as surface modification, doping and the creation of hybrid systems, are specifically highlighted in order to improve photocatalytic activity. The technological and environmental benefits of employing green nanomaterials as photocatalysts are evaluated, along with present issues and possible fixes for real-world application alongside the potential of green nanoparticles in developing sustainable remediation solutions by concentrating on their function in mitigating pollution and promoting cleaner environmental practices.

Keywords Green nanomaterials · Photocatalysis · Sustainable synthesis · Photocatalytic efficiency

S. Tripathy · N. G. Shimpi (✉)
Laboratory of Materials Science, Department of Chemistry, University of Mumbai, Santacruz (East), Mumbai, India
e-mail: navin_shimpi@rediffmail.com; navinshimpi@chemistry.mu.ac.in

K. N. Pandey
Defence Materials and Stores Research and Development Establishment (DRDO) GT Road, Kanpur, India

E. Priyadarshini et al. (eds.), *Advanced Nanotechnology Applications for Environmental Sustainability*, Sustainable Environmental Waste Management Strategies,
https://doi.org/10.1007/978-3-032-12933-8_8

1 Introduction

The manipulation of materials at the atom and molecular level is the focus of the science of nanotechnology (Joshi et al., 2008). Materials up to nm in size are synthesized, characterized, explored and exploited in nanoscience (Ealia et al., 2017). The study of materials at the nanoscale has revolutionized material science. This has made it possible to modify novel materials with exceptional properties (Wang, 2018). The physical and chemical characteristics of the bulk material are changed as it is scaled to the nanoscale. Because of the higher surface to volume ratio, the nanoparticles exhibit notable characteristics that set them apart from the bulk material. Materials at the nanoscale find use in many different sectors (Tripathy et al., 2023). Changing materials at the atom and molecular level is the focus of the science of nanotechnology. Any size smaller than a billionth of a hair or smaller than 100 nm, is referred to as nano (Tripathy et al., 2024a). Over the past few decades, nanotechnology has become a promising trend and the most powerful scientific field spanning a wide range of disciplines (Jeevanandam et al., 2018). Researchers in material science have been motivated by nanoscience to investigate novel approaches and create new techniques for engineering unique structures with noteworthy characteristics and effects on the environment (Roco et al., 2005).

One of the main issues brought on by the quick growth of industrialization and urbanization is environmental pollution. In light of this environmental issue, maintaining a clean environment for people is crucial to sustainability (Chen et al., 2019). The nano-structured opportunities to address the problem of environmental sustainability have been presented by photocatalysts with intriguing physiochemical properties (Tripathy et al., 2024b). The synthesis and use of photocatalyst in environmental remediation have advanced significantly in recent years. One of the "green" and successful approaches to addressing environmental pollution issues in recent years has been shown to be semiconductor photocatalytic technology (Thorat et al., 2024). As a result, environmental photocatalysis—which includes water disinfection, hazardous waste removal, air and water purification, deodorization, antimicrobial and self-cleaning—has drawn increasing interest (Huo et al., 2019).

Catalysis is crucial in chemical processes that aim for increased conversion, selectivity and product yield. Catalysis is widely used in the chemical, petrochemical and petroleum industries, all of which have a direct impact on societal development (Dong et al., 2019). It's also critical to note that experts from the scientific and engineering fields have contributed to the growth of catalysis-based research, which is broad and diversified. Continuous progress in material synthesis, reactor design improvement, the creation of new characterization methods and reaction parameter optimization are all necessary to increase the catalytic system's efficiency (Colmenares et al., 2019). Concerns over the depletion of non-renewable resources have also increased as a result of the world's population growth, improved lifestyle and society's ever growing economic demands. The emergence of environmental problems that have raised doubts about the planet's viability has made matters worse (De Vries et al., 2012). It is obvious that energy and the environment are intertwined, necessitating ongoing assessment of how current energy process's function and their environmental impacts. Although the amount of work done in the environmental and energy domains is different, recent studies on catalysis techniques like photocatalysis and thermo-catalysis have been successfully applied in both. Similar to this, the growth and effectiveness of nanotechnology research have removed the barriers to working in a single field of study and created opportunities for multidisciplinary research (Durgalakshmi et al., 2020).

In addition to its special qualities and environmentally friendly methodology, photocatalysis has become a vital technique for solving a number of environmental problems. Sustainable pollution treatment, flexible pollutant removal, water treatment applications, air purification, practical and financial advantages are some of the main reasons why photocatalysis is important in environmental applications. In line with sustainable development objectives, the combination of photocatalysis and environmental applications presents a viable strategy for tackling pressing pollution issues. Photocatalytic systems are becoming more effective, cost-effective and suitable for a range of environmental conditions as this field of study develops.

2 Types of Green Nanomaterials for Photocatalysis

Several forms of green nanomaterials are used in photocatalytic applications, each with a particular advantage for restoring the environment. A wide variety of ecologically friendly materials designed especially for light-driven catalytic applications are referred to as "green nanomaterials" for photocatalysis. TiO_2 nanoparticles produced using plant extract-mediated sol–gel techniques are widely used because of their high photocatalytic effectiveness and minimal environmental impact, making them a significant category of metal oxide-based nanomaterials (Tahir et al., 2020a, 2020b). Likewise, ZnO nanostructures and Fe_2O_3 nanomaterials made from plant extracts or biological templates have exceptional photocatalytic activity while reducing hazardous synthesis byproducts. Another important category of materials are carbon-based ones, such as graphene oxide made with safe reducing agents, carbon quantum dots made from biomass waste, and g-C_3N_4 produced utilizing green thermal methods. The remarkable electron transport characteristics and visible light absorption capabilities of these carbon compounds improve photocatalytic efficiency. ZnO/carbon quantum dot hybrids and TiO_2/graphene composites are examples of hybrid and composite nanomaterials that use synergistic effects to enhance photocatalytic performance while upholding green synthesis principles. With plant-mediated metal nanoparticles, microbially synthesized nanomaterials, and algae-based nanostructures that use natural biological processes to generate nanoparticles, biogenic nanomaterials are arguably the most sustainable category (Wellia et al., 2017). These environmentally friendly synthesis methods usually use renewable reducing and capping chemicals, water as the main solvent, and ambient conditions. In addition to addressing worries about the environmental impact of traditional nanomaterial synthesis methods, the resultant photocatalysts exhibit remarkable activity for breaking down environmental pollutants, generating hydrogen, and lowering CO_2. This advances sustainable nanotechnology applications for environmental remediation (Nagajyothi et al., 2019).

2.1 Metal Oxide Nanoparticles

Given their remarkable qualities and environmentally friendly uses, metal oxide nanoparticles are essential components in green photocatalysis. The quintessential photocatalyst, TiO_2 nanoparticles provide exceptional photochemical stability, minimal toxicity, and economic viability. Green synthesis of these nanoparticles using plant extracts as reducing agents preserves their high photocatalytic effectiveness while lowering their negative effects on the environment. Another well-known type, ZnO nanostructures, have excellent quantum efficiency and are easily made from biological templates derived from plants or microbes. They can take on a variety of shapes, such as nanorods, nanoflowers, and nanosheets that improve catalytic active sites. Fe_2O_3 nanoparticles can absorb visible light, which overcomes TiO_2's bandwidth constraints. They can also be produced environmentally by using plant polyphenols. Recent developments include CeO_2 nanoparticles, which are made using biological reduction techniques with no environmental impact and take use of oxygen vacancy defects for enhanced photocatalytic activity. By improving charge separation and visible light usage, hybrid systems that combine these metal oxides with carbon-based materials or noble metal nanoparticles have completely changed photocatalytic efficiency. Natural extracts are used as stabilizing and reducing agents in the green production of these hybrid photocatalysts, which does away with the necessity for dangerous chemicals. In accordance with the principles of green chemistry, these metal oxide nanoparticles exhibit exceptional efficacy in breaking down persistent organic pollutants, generating hydrogen via water splitting, and converting CO_2 into useful molecules. Metal oxide nanoparticles are positioned as crucial elements in using sustainable nanotechnology to address environmental concerns due to the ongoing improvement of synthesis techniques and performance optimization (Arularasu et al., 2018).

2.2 Carbon-Based Nanomaterials

With regard to their distinct electrical characteristics, structural adaptability, and potential for ecologically friendly synthesis, carbon-based nanomaterials have become outstanding prospects for green photocatalysis. With its exceptional electron mobility and wide surface area, graphene oxide—which is made using modified Hummer's techniques and natural reducing agents like ascorbic acid or plant extracts—dramatically improves charge separation and catalytic active sites in photocatalytic devices. Using hydrothermal carbonization to create carbon quantum dots from biomass waste, agricultural wastes, or food byproducts is a sustainable way to produce photoluminescent materials with exceptional visible light harvesting capabilities. Because of its suitable band gap and low recombination rates, graphitic carbon nitride (g-C_3N_4), which is produced by thermally condensing nitrogen-rich precursors like urea or melamine, has completely changed visible-light photocatalysis (Ng et al., 2012). The pyrolysis of biomass and agricultural waste yields biochar-derived nanoparticles, which add advantageous functional groups that improve pollutant adsorption and offer affordable substrates for photocatalyst immobilization. These carbon nanomaterials are excellent both as stand-alone photocatalysts and as complementary parts of hybrid systems, where they increase the mobility of charge carriers and broaden the range of light absorption that conventional semiconductor photocatalysts can absorb (Syed et al., 2019). In addition to facilitating energy conversion processes like hydrogen evolution and CO_2 reduction, their uses include environmental remediation, which includes the breakdown of persistent organic pollutants, pathogen disinfection, and heavy metal reduction. In line with green chemistry principles, these carbon-based nanomaterials are produced sustainably using waste materials, renewable resources, and environmentally friendly processes. This provides promising avenues for tackling environmental issues through cutting-edge photocatalytic technologies (Madima et al., 2020).

2.3 Biopolymer-Based Nanomaterials

Nanomaterials based on biopolymers are a cutting-edge development in green photocatalysis, providing remarkable sustainability due to their biodegradability and renewable source. For photocatalytic systems, naturally occurring polymers like cellulose, chitosan, alginate, and starch make great templates, stabilizers, and functional matrices. By improving dispersion and avoiding agglomeration, cellulose nanocrystals and nanofibrils—which are produced from plentiful plant biomass through mechanical or chemical processes—make perfect substrates for metal oxide photocatalysts (Jaleh et al., 2021). Unique amino and hydroxyl functional groups added by chitosan, which is derived from crab shells, aid in the binding of metal ions and the subsequent creation of evenly distributed nanoparticles with enhanced photocatalytic activity. Brown seaweed-derived alginate-based materials allow for regulated gelation procedures for encasing photocatalysts, resulting in recyclable and effective systems. By increasing surface area, improving light scattering, and improving the adsorption of target pollutants, these biopolymers not only act as platforms for green synthesis but also boost photocatalytic performance (Suresh et al., 2021). Synergistic effects are shown by nanocomposites that combine biopolymers with conventional photocatalysts such as ZnO or TiO_2, where the biopolymer matrix enhances charge carrier separation and visible light absorption while preserving biocompatibility. These materials' biodegradability removes worries about nanomaterial accumulation and provides substantial benefits in environmental applications. Under mild conditions, biopolymer-based photocatalysts have proven to be very effective in selective organic transformations, air filtration, and water purification, establishing them as sustainable substitutes for traditional systems. By using renewable feedstocks, ecologically friendly synthesis techniques, and end-of-life degradability while retaining strong photocatalytic performance, these materials perfectly exemplify the concepts of green chemistry and help address urgent environmental issues (Badranova et al., 2016).

2.4 Composite Nanomaterials and Nano-Hybrids

In order to overcome the inherent limitations of individual materials and improve overall photocatalytic performance, composite nanomaterials and nano-hybrids intelligently combine a variety of components to represent a transformational frontier in green photocatalysis. These complex structures usually combine secondary metal oxides, carbon-based materials, or biopolymers with semiconductor metal oxides to produce systems that work well together. This strategy is best demonstrated by metal oxide/graphene composites, in which the remarkable electron mobility of graphene sheets enables quick charge transfer from photoexcited TiO_2 or ZnO, greatly lowering electron–hole recombination and expanding photocatalytic activity into visible light regions. Strong redox potential is maintained while reduction and oxidation reactions can be spatially separated thanks to Z-scheme heterojunctions, which are created by joining semiconductors with complementary band structures (Harun-Ur-Rashid et al., 2024). Environmentally safe matrices that improve dispersion, stability, and recyclability while adding more functional groups for pollutant adsorption are provided by biopolymer-reinforced composites that contain chitosan, cellulose, or alginate. While hierarchical three-dimensional structures enhance accessible surface area and light harvesting, core–shell architectures carefully arrange materials to maximize light absorption, charge separation, and surface reactions. Through surface plasmon resonance effects, plasmonic nano-hybrids containing noble metal nanoparticles activate visible light, enhancing charge carrier production and local electromagnetic fields. Environmentally friendly techniques such as biological templating, hydrothermal/solvothermal methods using non-toxic solvents, and microwave-assisted energy-saving procedures are used in the green synthesis of these complex compounds. These composite nanomaterials exhibit remarkable adaptability in a variety of environmental applications, such as CO_2 conversion, hydrogen production, air treatment, and water purification, while upholding the principles of green chemistry through increased durability, decreased waste production, and increased efficiency (Mishra et al., 2023).

3 Green Synthesis Methods for Photocatalytic Nanomaterials

See Fig. 1.

3.1 *Plant-Based Synthesis*

Plant-based green synthesis techniques for photocatalytic nanomaterials have become viable substitutes for traditional chemical processes, providing notable benefits in terms of resource efficiency and environmental friendliness. With their diverse phytochemical profiles that include proteins, polyphenols, flavonoids, and terpenoids, plant extracts work as multipurpose agents in the creation of nanoparticles, offering both stabilizing and reducing properties. Plant leaf extracts from Aloe vera, Camellia sinensis (green tea), and Azadirachta indica (neem) have shown remarkable effectiveness in creating metal oxide nanoparticles with regulated shape

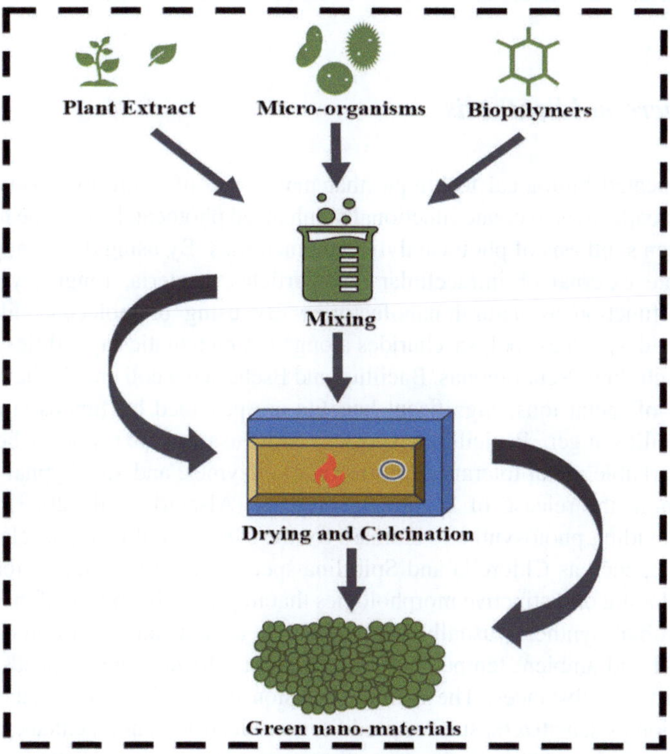

Fig. 1 General synthesis mode of green nanomaterials

and increased photocatalytic activity (Aslam et al., 2021). By using gentle conditions, such as ambient temperature, neutral pH, and an aqueous medium, these plant-mediated synthesis pathways can drastically cut down on energy use and get rid of dangerous compounds. In order to create nanoparticles with distinct surface characteristics that frequently improve photocatalytic activity, the technique typically entails combining plant extract with metal precursor solutions, where phytochemicals enable controlled reduction and stability. Naturally occurring organic acids found in citrus, pomegranate, and grape fruit extracts help regulate particle development and avoid agglomeration (Yazdi et al., 2019). Furthermore, rice husk, maize cob, and sugarcane bagasse extracts are examples of agricultural waste products that beautifully illustrate the concepts of the circular economy by turning what would otherwise be discarded materials into useful photocatalysts. Due to the addition of plant chemicals that make the material more sensitive to visible light or alter its surface characteristics, the resultant plant-synthesized nanomaterials frequently show improved photocatalytic activity for breaking down organic pollutants, generating hydrogen, and lowering CO_2. While creating efficient photocatalysts for environmental applications, this synthesis method complies with several green chemistry principles, including waste prevention, the use of renewable feedstocks, the reduction of derivatives, and the creation of safer compounds (Mirzaeifard et al., 2024).

3.2 Microbial Synthesis

A sophisticated biological technique that makes use of microorganisms' distinct metabolic capacities to create functionally enhanced photocatalysts is the microbial-based green synthesis of photocatalytic nanomaterials. By using different processes to generate external or intracellular nanoparticles, bacteria, fungus, yeasts, and algae all function as natural nanofactories. By using biomolecules like reductases, peptides, and exopolysaccharides along with enzymatic capabilities, bacterial species including Pseudomonas, Bacillus, and Escherichia coli enable the controlled reduction of metal ions. Significant benefits are provided by fungus strains such as Aspergillus niger, Penicillium species, and Fusarium oxysporum because of their remarkable metal tolerance, profusion of enzymes, and vast hyphal networks that promote the release of stabilizing proteins (Alsaiari et al., 2023). In addition to providing photosynthetic biomass for the synthesis of nanoparticles, marine microalgae, such as Chlorella and Spirulina species, often produce materials with intrinsic doping or distinctive morphologies that improve absorption of visible light. The microbial synthesis usually takes place in an aqueous environment with a neutral pH and ambient temperature, which greatly lowers energy needs and gets rid of harmful substances. The natural inclusion of proteins and metabolites that can function as sensitizers, surface modifiers, or electron donors, enhancing charge separation and broadening light absorption ranges, frequently results in improved performance from these biologically produced photocatalysts. Furthermore, by adjusting growing conditions, metal ion concentration, and harvesting time, microbial synthesis provides exceptional control over particle size, shape, and crystallinity. With the removal of dangerous chemicals, significant energy savings, continuous production potential, and the ability to integrate bioremediation with nanomaterial synthesis when metal-contaminated waste streams are utilized as precursors, this method exhibits remarkable sustainability and is a perfect fit with green chemistry principles (Roy et al., 2021).

3.3 Bio-Templates and Bio-Reduction Techniques

Utilizing biological structures and processes, bio-templates and bio-reduction procedures are advanced green synthesis methods for photocatalytic nanomaterials that produce extremely effective and ecologically friendly photocatalysts. The growth and organization of nanomaterials with remarkable control over shape, crystal orientation, and hierarchical assembly are guided by biological templates, which are naturally occurring structures with finely ordered properties. Renewable scaffolds with an abundance of hydroxyl and amino groups are provided by polysaccharide templates such as cellulose, chitosan, and starch. These scaffolds coordinate metal ions and initiate the formation of nanoparticles. The development of high-surface-area photocatalysts with improved light-harvesting capabilities is facilitated by their fibrous or porous architectures. The exact spatial arrangements of functional groups provided by protein templates, such as collagen, silk fibroin, and peptide assemblies, allow for the controlled crystallization and development of nanostructures with distinctive exposed crystal facets that maximize photocatalytic activity. Under mild conditions, bio-reduction processes transform metal ions into nanomaterials by using naturally occurring reducing agents derived from biological sources (Mofokeng et al., 2025). For regulated nanoparticle formation, enzymes like oxidoreductases mediate electron transfer processes, and antioxidants originating from plants, such as polyphenols, flavonoids, and vitamins, aid in metal reduction while also encasing the resultant nanoparticles. Through cellular processes or released biomolecules, microbial reduction uses the metabolic processes of bacteria and fungi to convert metal precursors into nanomaterials.

These environmentally friendly methods produce photocatalysts with special benefits, such as enhanced stability from natural capping agents, enhanced charge separation from discrete defect topologies, and biomolecule-functionalized surfaces that improve visible light absorption. These techniques produce high-performance photocatalytic materials for environmental applications while demonstrating sustainable nanotechnology by removing hazardous chemicals, using renewable resources, lowering energy inputs, and frequently working in aqueous systems (Zhang et al., 2023).

3.4 Waste-Derived Synthesis

Waste-derived synthesis has emerged as a groundbreaking approach in the green fabrication of photocatalytic nanomaterials, ingeniously transforming environmental liabilities into valuable catalytic resources. Agricultural waste—including rice husks, sugarcane bagasse, corn cobs, and fruit peels—provides silica-rich or carbon-rich precursors that can be converted into functional photocatalysts through thermal or chemical treatments (Rodriguez-Padron et al., 2020). The high silica content in rice husks enables direct conversion to silicon dioxide nanostructures, which serve as excellent supports for photocatalytic metals or as templates for mesoporous materials with enhanced surface area. Fruit peels from oranges, bananas, and pomegranates contain natural reducing agents like citric acid, ascorbic acid, and polyphenols that facilitate metal ion reduction while simultaneously providing stabilizing biomolecules that prevent agglomeration (Bhole et al., 2024).

Industrial byproducts including fly ash, red mud, and slag offer mineral-rich sources for synthesizing photocatalytic materials, particularly iron oxides, titanium dioxide, and various mixed metal oxides. These waste streams typically require minimal processing beyond extraction and purification to yield active nanomaterials. Electronic waste provides a valuable urban mining source for precious metals like silver, gold, and platinum that can be recovered and transformed into plasmonic photocatalysts with superior visible light activity. The leaching of metals from e-waste using mild, biodegradable solvents followed by green reduction creates nanomaterials with exceptional performance (Chao et al., 2023). By tackling waste management issues and producing high-performance photocatalysts at the same time, waste-derived synthesis is the pinnacle of green chemistry. These methods produce materials with special qualities that are frequently improved by the waste source's natural composition while drastically lowering manufacturing costs, resource consumption, the removal of hazardous precursors, and energy needs. The resultant photocatalysts show exceptional effectiveness in environmental applications such as air filtration, water purification, and the generation of renewable energy (Chen et al., 2022).

4 Photocatalytic Mechanisms and Principles

Through a number of basic physicochemical mechanisms, photocatalysis transforms light energy into chemical reactivity. Fundamentally, the process starts when photons are absorbed by a semiconductor photocatalyst, which produces electron–hole pairs when the photon energy reaches or surpasses the band gap of the material. By moving electrons from the valence band to the conduction band, this photoexcitation produces positively charged holes and negatively charged electrons, which act as potent oxidizing and reducing agents, respectively. Charge carrier dynamics have a major impact on photocatalysis efficiency. Given that recombination is the main mechanism for energy loss, electrons and holes must move to the catalyst surface after production (Saravanan et al., 2017). These charge carriers engage in redox reactions with adsorbed substances at the surface. While holes oxidize water or hydroxyl ions to create hydroxyl radicals (•OH) or directly oxidize substrates, electrons usually decrease electron acceptors like oxygen to form superoxide radicals (O_2^{-}•). Through non-selective oxidation pathways, these reactive oxygen species (ROS)—mainly hydrogen peroxide, superoxide radicals, and hydroxyl radicals—are the main active species in charge of breaking down organic contaminants. Organic compounds are eventually mineralized to CO_2 and H_2O as a result of the degradation, which usually proceeds through successive oxidation processes (Bei Ran et al., 2023). The important variables that affect photocatalytic efficiency include surface area and active site availability, crystallinity and defect structure, surface adsorption qualities, charge carrier lifespan and mobility, and band gap energy, which determines the light absorption range. In order to improve charge separation, increase the range of light absorption, and speed up surface reactions, advanced photocatalytic systems use techniques like heterojunction formation, doping, and co-catalyst loading. This increases quantum efficiency and broadens the range of practical applications in energy conversion, environmental remediation, and selective organic transformations (Mandada, 2021) (Fig. 2).

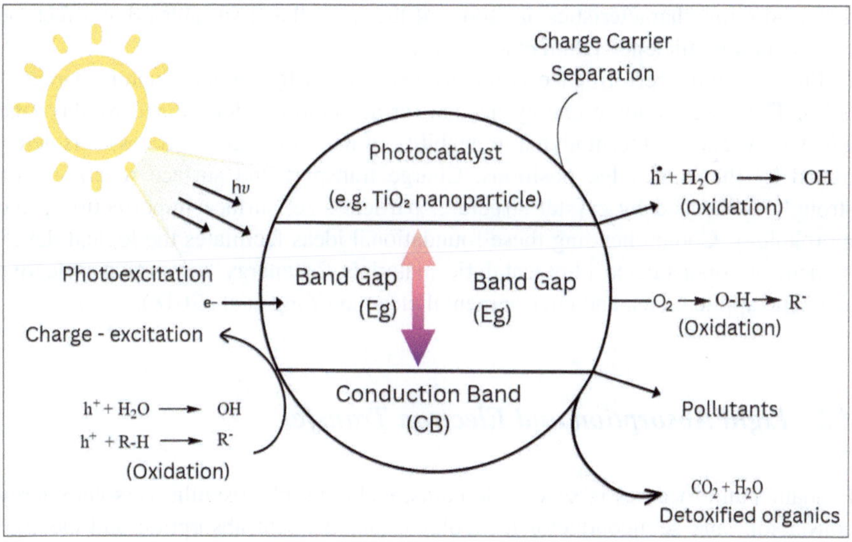

Fig. 2 Photocatalytic mechanisms and principles

4.1 Fundamentals of Photocatalysis

Incorporating a catalyst that has been activated by photon energy, photocatalysis is a light-driven process that speeds up chemical reactions. This phenomenon is made possible by a number of crucial processes that work together to transform light energy into chemical reaction. When photons with energy equal to or greater than the material's band gap reach the catalyst surface, a semiconductor photocatalyst absorbs the light, starting the process. Electron–hole pairs are created when this photoexcitation moves electrons from the valence band to the conduction band (Gurylev, 2021). The reactive species that propel further chemical changes are represented by these charge carriers. Charge carrier dynamics have a major impact on photocatalysis's efficacy. Prior to recombination, a crucial competitive process that reduces photocatalytic efficiency, electrons and holes must move to the catalyst surface after production. Adsorbed species then engage in redox reactions with surface-reaching charge carriers (Tahir et al., 2020a, 2020b). Whereas valence band holes oxidize water or hydroxide ions to create hydroxyl radicals, conduction band electrons usually decrease electron acceptors such as molecular oxygen to create superoxide radicals. The main mechanisms for breaking down target compounds or promoting desired chemical reactions are these reactive oxygen species and direct hole oxidation. Light absorption range, charge carrier lifespan, recombination rate, surface area, and adsorption characteristics are some of the interrelated variables that affect the total quantum efficiency (Rani et al., 2022).

Photocatalytic performance is mostly determined by semiconductor characteristics. The lowest photon energy needed for activation is determined by the band gap energy, and the thermodynamic viability of particular redox processes is determined by the band edge positions. Charge transport and surface reactivity are strongly influenced by crystal structure, particle size, surface imperfections, and morphology. Comprehending these foundational ideas facilitates the logical development of sophisticated photocatalytic materials for energy conversion, selective synthetic applications, and environmental cleanup (Yang et al., 2018).

4.2 Light Absorption and Electron Transfer

Fundamental processes in several domains, including photosynthesis, solar energy conversion, and semiconductor technology, include light absorption and electron transfer.

When photons, or light particles, interact with matter and are taken up by electrons in atoms or molecules, light absorption takes place. In this process, an electron is excited to a higher energy level by the photon's energy, usually to a state known as the "excited state" where the electron has more energy. This phenomenon is essential to activities like photosynthesis in plants, which produce glucose by absorbing sunlight through pigments like chlorophyll. The next important step after light absorption is electron transport (Mohamed et al., 2012). An electron flow is produced during this process as the excited electron travels between molecules or atoms. This transfer takes place inside a number of proteins and other molecules in the chloroplasts' thylakoid membrane during photosynthesis. Energy-rich molecules like ATP (adenosine triphosphate) and NADPH are produced in tandem with the flow of electrons and are subsequently utilized in the Calvin cycle to create glucose.

Similar to this, in solar cells, electrons in semiconductor materials are excited by sunlight, forming electron–hole pairs. An electrical current is produced when these couples are split apart and the electrons move through the substance. Following light absorption, electron transfer is necessary for energy conversion in both artificial and biological systems. It is also a vital field of research for enhancing the energy efficiency of renewable energy sources (Zhang et al., 2012).

4.3 *Reactive Oxygen Species (ROS) Formation*

Free radicals like superoxide anion (O_2^-), hydroxyl radicals (OH•), and non-radical molecules like hydrogen peroxide (H_2O_2) are examples of reactive oxygen species (ROS), which are extremely reactive molecules that include oxygen (Nosaka et al., 2017). Depending on their levels and context, these species, which are naturally created during a variety of biochemical processes, especially in cellular metabolism, can have both positive and negative impacts. Superoxide radicals (ROS) are primarily produced in the mitochondria during cellular respiration, where oxygen molecules undergo partial reduction (Huang et al., 2017). ROS can also be produced by enzymes such as xanthine oxidases and NADPH oxidases during regular cellular processes. Furthermore, smoking, pollution, and UV radiation are examples of environmental variables that might contribute to the creation of ROS. ROS are crucial for immune defence and cellular signalling at low to moderate concentrations. They play a role in immune cell defence against infections, gene expression control, and signalling pathway activation. Oxidative stress, on the other hand, happens when ROS levels surpass the body's antioxidant defence capacity (He et al., 2013).

Excess ROS can cause oxidative damage by harming lipids, proteins, and DNA, among other biological constituents. This is associated with age and a number of illnesses, such as diabetes, cancer, heart disease, and neurological diseases including Parkinson's and Alzheimer's. The body uses both enzymatic (like superoxide dismutase) and non-enzymatic (like vitamins C and E) antioxidants to counteract ROS and preserve cellular integrity. In order to prevent oxidative damage and sustain health, it is essential to maintain a balance between the generation of ROS and antioxidant defences, even though ROS are essential for cellular function (Ma et al., 2019).

5 Applications of Green Photocatalysts in Environmental Remediation

Green photocatalysts, which use renewable solar energy to propel chemical processes that break down contaminants, represent a sustainable frontier in environmental remediation technologies. These eco-friendly substances have become effective instruments for dealing with a range of pollution issues. Green photocatalysts use sophisticated oxidation processes to efficiently degrade medicines, industrial colours, and persistent organic contaminants in water purification (Danish et al., 2021). These catalysts produce reactive oxygen species in response to light, which mineralize pollutants into innocuous byproducts. In wastewater systems, BiOBr, g-C_3N_4, and ZnO nanostructures have shown exceptional effectiveness in breaking down antibiotics and endocrine disruptors. Another crucial use is air purification, where photocatalytic surfaces may break down pollutants that comprise sulfur, nitrogen oxides, and volatile organic compounds (VOCs) (Nunes et al., 2021). Under normal lighting conditions, TiO_2-based materials that have been altered with earth-abundant components have been successfully integrated into building materials to create self-cleaning surfaces that constantly filter the surrounding air. Photocatalytic methods can convert petroleum hydrocarbons and stubborn pesticides into biodegradable substances, which is advantageous for soil cleanup (Bano et al., 2021) (Fig. 3).

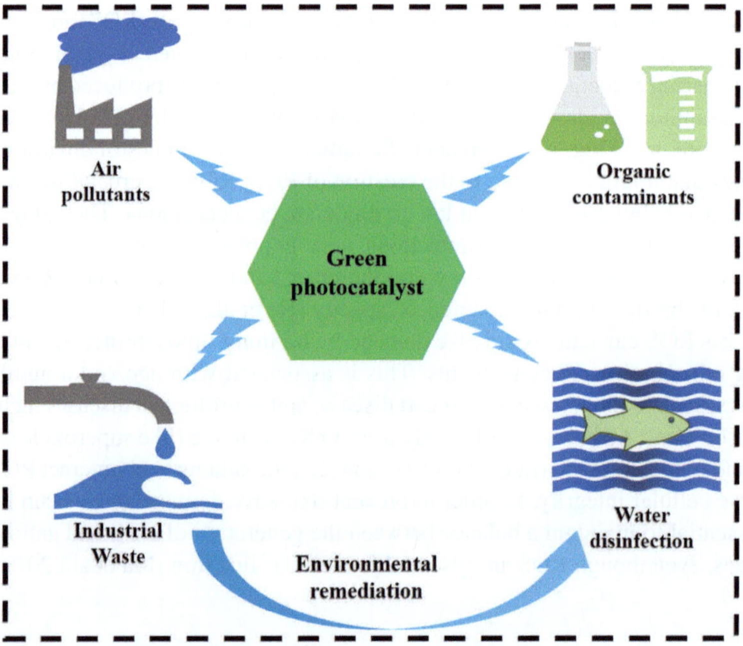

Fig. 3 Application of photocatalysis in environmental remediation

The technology's foundational circular economy ideas are exemplified by green photocatalysts made from waste materials like mining wastes or rice husk ash. In order to increase the range of usable solar energy, recent developments have concentrated on creating visible-light-responsive photocatalysts using methods including doping, heterojunction creation, and biomimetic approaches (Singh et al., 2021). Carbon-based photocatalysts and metal–organic frameworks (MOFs) are examples of promising materials that combine high reactivity with little environmental impact. A new generation of multipurpose environmental remediation technologies that can handle challenging pollution issues while upholding sustainability principles is ushered in by the incorporation of green photocatalysts into membrane systems, 3D-printed structures, and smart materials (Malik et al., 2023).

6 Challenges and Limitations in Green Photocatalysis

Green photocatalysts have a number of important obstacles that prevent them from being widely used, despite their promise environmental uses. These restrictions cover the fields of technology, application, and economics. The limited use of the solar spectrum is a basic problem. UV light, which makes up only 4–5% of the sun spectrum, is the only light that can activate many photocatalysts, including conventional TiO_2 (Khan et al., 2020). Even though band gap engineering has been used to create visible-light-responsive materials, their effectiveness is generally diminished by quick electron–hole recombination or decreased redox potential. Another crucial issue is catalyst stability. During operation, a lot of photocatalysts experience photo-corrosion, which degrades the material and may cause metal ions to leak into treated medium (Nasrollahzadeh et al., 2020). In addition to shortening the catalyst's lifespan, this could result in secondary contamination problems. Scalability continues to be a challenge. Due to restrictions in mass transfer, problems with light penetration in turbid solutions, and difficulties with catalyst recovery, laboratory-scale results frequently do not translate to industrial applications. Implementation costs are further raised by the need for specialized reactor designs. Real-world conditions make performance inconsistent (Zelekew et al., 2023). Variable pH, natural organic matter, and inorganic ion impacts in the water matrix can all considerably reduce photocatalytic performance. Similar to this, variations in humidity and competing contaminants can significantly lower efficiency in air purification applications. As modern photocatalysts are frequently manufactured using energy-intensive techniques or costly precursors, concerns about economic feasibility remain. Currently, the total cost of ownership—which includes reactor design, operation, maintenance, and catalyst synthesis—often surpasses that of traditional treatment technologies. The sector is further complicated by issues with standardization, as varied testing procedures make it impossible to compare performance and impede regulatory approval. Multidisciplinary strategies combining environmental chemistry, engineering, and materials science are needed to address these complex issues (Tripathy et al., 2025).

7 Future Directions and Prospects

Green photocatalysis's future depends on overcoming present constraints using creative methods from several scientific fields. New research directions have a lot of potential to turn this technology from a curious lab experiment into a useful environmental remedy. With artificial intelligence and computational screening speeding up the search for photocatalysts with ideal band structures, rational materials design is a significant breakthrough. By precisely adjusting electrical characteristics, these methods may be able to overcome the problem of solar spectrum utilization while preserving significant oxidative powers. Photocatalysts with more surface area, better light harvesting, and better charge separation are being produced using hierarchical nano structuring. Pathways to effective visible-light-responsive devices that make use of a wider spectrum of solar energy are provided by Z-scheme heterojunctions, plasmonic coupling, and quantum dot sensitization. Advances from biomimetic techniques that are modelled after natural photosynthesis are encouraging. Self-healing properties and improved stability in a variety of environmental circumstances may be made possible by combining light-harvesting complexes with strong inorganic structures. There are intriguing opportunities for integration with developing technology. The efficiency of remediation is increased when photocatalysis is combined with membrane procedures, microfluidics, or electrochemical systems. Furthermore, technologies driven by renewable energy might make it possible for remote areas to run continuously. The environmental impact of producing catalysts is being lessened by advances in green manufacturing that include hydrothermal, solvothermal, and mechanochemical processes. Sustainability qualities are further enhanced by circular economy strategies that use waste products as precursors. In order to verify performance claims and guarantee actual environmental benefits, standardized testing procedures and life cycle assessments are becoming more prevalent. These advancements raise the possibility that green photocatalysis could soon become commercially viable for a wide range of environmental applications, especially when combined with declining production costs and growing regulatory push for sustainable remediation methods. Photocatalysis has the potential to be a key element of next-generation environmental remediation solutions as these research avenues develop.

References

Alsaiari, N. S., et al. (2023). Plant and microbial approaches as green methods for the synthesis of nanomaterials: Synthesis, applications, and future perspectives. *Molecules, 28*(1), 463. https://doi.org/10.3390/molecules28010463

Arularasu, M. V., et al. (2018). An innovative approach for green synthesis of iron oxide nanoparticles: Characterization and its photocatalytic activity. *Polyhedron*. https://doi.org/10.1016/j.poly.2018.09.036

Aslam, M., et al. (2021). Recent development in the green synthesis of titanium dioxide nanoparticles using plant-based biomolecules for environmental and antimicrobial applications. *Journal of Industrial and Engineering Chemistry, 98*, 1–16. https://doi.org/10.1016/j.jiec.2021.04.010

Badranova, G. U., et al. (2016). Biopolymer-based hydrogels for encapsulation of photocatalytic TiO_2 nanoparticles prepared by the freezing/thawing method. *Journal of Molecular Liquids, 223*, 16–20. https://doi.org/10.1016/j.molliq.2016.07.135

Bano, S., et al. (2021). Advanced application of green materials in environmental remediation. In *Applications of advanced green materials* (pp. 481–502). https://doi.org/10.1016/B978-0-12-820484-9.00019-2

Bhole, A., et al. (2024). Treatment of waste using waste-derived materials and free energy: A practical concept of circular-economy. *Sustainable Chemistry and Pharmacy, 37*, Article 101341. https://doi.org/10.1016/j.scp.2023.101341

Chao, J., et al. (2023). Trash to treasure: Green synthesis of novel Ag_2O/Ag_2CO_3 Z-scheme heterojunctions with highly efficient photocatalytic activities derived from waste mussel shells. *Chemical Engineering Journal, 454*, Article 140259. https://doi.org/10.1016/j.cej.2022.140259

Chen, P., et al. (2019). Directional electron delivery and enhanced reactants activation enable efficient photocatalytic air purification on amorphous carbon nitride co-functionalized with O/La. *Applied Catalysis B: Environmental, 242*, 19–30. https://doi.org/10.1016/j.apcatb.2018.09.078

Chen, Z., et al. (2022). Recent advances in waste-derived functional materials for wastewater remediation. *Eco-Environment & Health, 1*(2), 86–104. https://doi.org/10.1016/j.eehl.2022.05.001

Colmenares, J. C., et al. (2019). Thermo-photo-catalysis: Environmental and energy applications. *Chemsuschem*. https://doi.org/10.1002/cssc.201900175

Danish, M. S. S., et al. (2021). Photocatalytic applications of metal oxides for sustainable environmental remediation. *Metals, 11*(1), 80. https://doi.org/10.3390/met11010080

De Vries, J. G., et.al. (2012). Homogeneous and heterogeneous catalysis in industry. *Catalysis Science & Technology, 2*(10). https://doi.org/10.1039/c2cy90039d

Dong, F., et al. (2019). Editorial: Photocatalysis for environmental applications. *Frontiers in Chemistry, 7*, 303. https://doi.org/10.3389/fchem.2019.00303

Durgalakshmi, D., et al. (2020). Principles and mechanisms of green photocatalysis. In M. Naushad, S. Rajendran, & E. Lichtfouse (Eds.), *Green photocatalysts* (Environmental Chemistry for a Sustainable World, vol. 34). Springer. https://doi.org/10.1007/978-3-030-15608-4_1

Ealia, S. A. M., et al. (2017). *Materials Science and Engineering, 263*, Article 032019. https://doi.org/10.1088/1757-899X/263/3/032019

Gurylev, V. (2021). Photocatalysis: Fundamentals. In *Nanostructured photocatalyst via defect engineering*. Springer. https://doi.org/10.1007/978-3-030-81911-8_1

Harun-Ur-Rashid, et al. (2024). Hybrid nanocomposite fabrication of nanocatalyst with enhanced and stable photocatalytic activity. *Topics in Catalysis, 67*, 17–45. https://doi.org/10.1007/s11244-023-01809-4

He, W., et al. (2013). Photogenerated charge carriers and reactive oxygen species in ZnO/Au hybrid nanostructures with enhanced photocatalytic and antibacterial activity. *Journal of the American Chemical Society, 136*(2), 750–757. https://doi.org/10.1021/ja410800y

Huang, S., et al. (2017). Enhancing reactive oxygen species generation and photocatalytic performance via adding oxygen reduction reaction catalysts into the photocatalysts. *Applied Catalysis b: Environmental, 218*, 174–185. https://doi.org/10.1016/j.apcatb.2017.06.030

Huo, W., et al. (2019). Synthesis of Bi_2WO_6 with gradient oxygen vacancies for highly photocatalytic NO oxidation and mechanism study. *Chemical Engineering Journal, 361*, 129–138. https://doi.org/10.1016/j.cej.2018.12.071

Jaleh, B., et al. (2021). Biopolymer-derived (nano)catalysts for hydrogen evolution via hydrolysis of hydrides and electrochemical and photocatalytic techniques: A review. *International Journal of Biological Macromolecules, 182*, 1056–1090. https://doi.org/10.1016/j.ijbiomac.2021.04.087

Jeevanandam, J., et al. (2018). Review on nanoparticles and nanostructured materials: History, sources, toxicity and regulations. *Beilstein Journal of Nanotechnology, 9*, 1050–1074. https://doi.org/10.3762/bjnano.9.98

Joshi, M., et al. (2008). Characterization techniques for nanotechnology applications in textiles. *Indian Journal of Fibre & Textile Research, 33*, 304–317. https://www.researchgate.net/publication/235672359

Khan, K., et al. (2020). Recent progress, challenges, and prospects in two-dimensional photocatalyst materials and environmental remediation. *Nano-Micro Letters, 12*, 167. https://doi.org/10.1007/s40820-020-00504-3

Ma, H.-Y., et al. (2019). Roles of reactive oxygen species (ROS) in the photocatalytic degradation of pentachlorophenol and its main toxic intermediates by TiO_2/UV. *Journal of Hazardous Materials*. https://doi.org/10.1016/j.jhazmat.2019.02.080

Madima, N., et al. (2020). Carbon-based nanomaterials for remediation of organic and inorganic pollutants from wastewater: A review. *Environmental Chemistry Letters, 18*, 1169–1191. https://doi.org/10.1007/s10311-020-01001-0

Malik, A. Q., et al. (2023). A review on the green synthesis of nanoparticles, their biological applications, and photocatalytic efficiency against environmental toxins. *Environmental Science and Pollution Research, 30*, 69796–69823. https://doi.org/10.1007/s11356-023-27437-9

Mandada, P. (2021). Introduction, basic principles, mechanism, and challenges of photocatalysis. In *Handbook of nanomaterials for wastewater treatment* (pp. 137–154). https://doi.org/10.1016/B978-0-12-821496-1.00016-7

Mirzaeifard, Z., et al. (2024). Economical, one-pot, and green synthesis of plant-based carbon quantum dots for efficient visible-light photocatalytic dye degradation and water purification. *Journal of the Taiwan Institute of Chemical Engineers, 163*, Article 105655. https://doi.org/10.1016/j.jtice.2024.105655

Mishra, K., et al. (2023). Hybrid semiconductor photocatalyst nanomaterials for energy and environmental applications: Fundamentals, designing, and prospects. *Advanced Sustainable Systems, 2300095*. https://doi.org/10.1002/adsu.202300095

Mofokeng, L. E., et al. (2025). A comprehensive review on biogenic metallic, metal oxide and metal sulfide nanoparticles for water treatment and sensing applications. *Journal of Industrial and Engineering Chemistry, 144*, 48–76. https://doi.org/10.1016/j.jiec.2024.09.039

Mohamed, H. H., et al. (2012). The role of electron transfer in photocatalysis: Fact and fictions. *Applied Catalysis B: Environmental, 128*, 91–104. https://doi.org/10.1016/j.apcatb.2012.05.045

Nagajyothi, P. C., et al. (2019). Green synthesis: Photocatalytic degradation of textile dyes using metal and metal oxide nanoparticles—Latest trends and advancements. *Critical Reviews in Environmental Science and Technology*, 1–107. https://doi.org/10.1080/10643389.2019.1705103

Nasrollahzadeh, M., et al. (2020). Green-synthesized nanocatalysts and nanomaterials for water treatment: Current challenges and future perspectives. *Journal of Hazardous Materials*, 123401. https://doi.org/10.1016/j.jhazmat.2020.123401

Ng, Y. H., et al. (2012). A perspective on fabricating carbon-based nanomaterials by photocatalysis and their applications. *Energy & Environmental Science, 5*(11), 9307. https://doi.org/10.1039/c2ee22128d

Nosaka, Y., et al. (2017). Generation and detection of reactive oxygen species in photocatalysis. *Chemical Reviews, 117*(17), 11302–11336. https://doi.org/10.1021/acs.chemrev.7b00161

Nunes, D., et al. (2021). Metal oxide-based photocatalytic paper: A green alternative for environmental remediation. *Catalysts, 11*(4), 504. https://doi.org/10.3390/catal11040504

Ran, B., et al. (2023). Photocatalytic antimicrobials: Principles, design strategies, and applications. *Chemical Reviews, 123*(22), 12371–12430. https://doi.org/10.1021/acs.chemrev.3c00326

Rani, B., et al. (2022). Fundamentals principle of photocatalysis. In *Nanostructured materials for visible light photocatalysis* (pp. 1–22). https://doi.org/10.1016/B978-0-12-823018-3.00009-9

Roco, M. C., et al. (2005). Societal implications of nanoscience and nanotechnology: Maximizing human benefit. *Journal of Nanoparticle Research, 7*(1), 1–13. https://doi.org/10.1007/s11051-004-2336-5

Rodríguez-Padrón, D., et al. (2020). Waste-derived materials: Opportunities in photocatalysis. In M. Muñoz-Batista, A. Navarrete Muñoz, & R. Luque (Eds.), *Heterogeneous photocatalysis* (Topics in Current Chemistry Collections). Springer. https://doi.org/10.1007/978-3-030-49492-6_1

Roy, M., et.al. (2021). Bio-inspired synthesis of nanomaterials. In A. K. Tyagi & R. S. Ningthoujam (Eds.), *Handbook on synthesis strategies for advanced materials* (Indian Institute of Metals Series). Springer. https://doi.org/10.1007/978-981-16-1807-9_18

Saravanan, R., et al. (2017). Basic principles, mechanism, and challenges of photocatalysis. In M. Khan, D. Pradhan, & Y. Sohn (Eds.), *Nanocomposites for visible light-induced photocatalysis* (Springer Series on Polymer and Composite Materials). Springer. https://doi.org/10.1007/978-3-319-62446-4_2

Singh, N. B., et al. (2021). Applications of green synthesized nanomaterials in water remediation. *Current Pharmaceutical Biotechnology, 22*(29), 733–761. https://doi.org/10.2174/1389201021666201027160029

Suresh, R., et al. (2021). Recent progress in green and biopolymer based photocatalysts for the abatement of aquatic pollutants. *Environmental Research, 199*, Article 111324. https://doi.org/10.1016/j.envres.2021.111324

Syed, N., et al. (2019). Carbon-based nanomaterials via heterojunction serving as photocatalyst. *Frontiers in Chemistry, 7*. https://doi.org/10.3389/fchem.2019.00713

Tahir, M. B., et al. (2020a). Nanomaterials for photocatalysis. In *Nanotechnology and photocatalysis for environmental applications* (pp. 65–76). https://doi.org/10.1016/b978-0-12-821192-2.00005-x

Tahir, M. B., et al. (2020b). Photocatalysis fundamentals. In *Nanotechnology and Photocatalysis for Environmental Applications* (pp. 57–64). https://doi.org/10.1016/B978-0-12-821192-2.00004-8

Thorat, D., et.al. (2024). Graphene: A wonder material of the twenty-first century. In *Graphene-based photocatalysts* (pp. 3–23). Springer. https://doi.org/10.1007/978-3-031-66260-7_1

Tripathy, S., et.al. (2023). Nano-biomaterials: Perspectives for medical applications in the diagnosis and treatment of diseases. *Materials Research Forum LLC*, *145*, 92–130. https://doi.org/10.21741/9781644902370-4

Tripathy, S., et al. (2024a). Fabrication of nanofillers: The future research. In S. Thomas, N. G. Shimpi, & P. Mishra (Eds.), *Handbook of nanofillers* (pp. 1–29). Springer Nature. https://doi.org/10.1007/978-981-99-3516-1_22-2

Tripathy, S., et.al. (2024b). Silks. In *Reference Module in Materials Science and Materials Engineering*. Elsevier. https://doi.org/10.1016/B978-0-323-95486-0.00121-6

Tripathy, S., et.al. (2025). Hydrothermal-ultrasonication synthesis of $NiFe_2O_4$ nanospheres for temperature-controlled xylene gas sensor. *ChemistrySelect*, *e202500146*. https://doi.org/10.1002/slct.202500146

Wang, G. (2018). Nanotechnology: The new features. arXiv:1812.04939. [cs.ET]. arXiv. https://arxiv.org/pdf/1812.04939.pdf

Wellia, D. V., et al. (2017). Introduction of nanomaterials for photocatalysis. In M. Khan, D. Pradhan, & Y. Sohn (Eds.), *Nanocomposites for visible light-induced photocatalysis* (Springer Series on Polymer and Composite Materials). Springer. https://doi.org/10.1007/978-3-319-62446-4_1

Yang, X., et al. (2018). Photocatalysis: From fundamental principles to materials and applications. *ACS Applied Energy Materials, 1*(12), 6657–6693. https://doi.org/10.1021/acsaem.8b01345

Yazdi, M. E. T., et al. (2019). Eco-friendly and plant-based synthesis of silver nanoparticles using *Allium giganteum* and investigation of its bactericidal, cytotoxicity, and photocatalytic effects. *Materials Technology*, 1–8. https://doi.org/10.1080/10667857.2019.1583408

Zelekew, O. A., et al. (2023). Recent progress on plant extract-mediated biosynthesis of ZnO-based nanocatalysts for environmental remediation: Challenges and future outlooks. *Advances in Colloid and Interface Science, 317*, Article 102931. https://doi.org/10.1016/j.cis.2023.102931

Zhang, L., et al. (2012). Kinetics and mechanisms of charge transfer processes in photocatalytic systems: A review. *Journal of Photochemistry and Photobiology c: Photochemistry Reviews, 13*(4), 263–276. https://doi.org/10.1016/j.jphotochemrev.2012.07.002
Zhang, D., et al. (2023). Biosynthesis of metal nanoparticles: Bioreduction and biomineralization. *Nanotechnology Reviews, 12*(1), 20230170. https://doi.org/10.1515/ntrev-2023-0170

Nanocellulose-Based Photocatalysts for Environmental Remediation

Twinkle Gupta and Shilpa Sharma

Abstract Nanocellulose, a promising bionanomaterial, is gradually being recognized for its potential in environmental remediation due to its abundance, cost-effectiveness, eco-friendliness, high surface area, non-toxicity, and exceptional mechanical properties. This chapter provides a comprehensive overview of various types of nanocellulose, their properties, and their application as a photocatalyst in environmental remediation. It also discusses recent advancements in nanocellulose-based photocatalysts for pollutant degradation, highlighting their effectiveness in breaking down organic and inorganic contaminants under light irradiation. The potential integration of nanocellulose with other materials to enhance its photocatalytic performance is also explored. Finally, challenges and future directions in this field are addressed, including the need to scale up production and improve photocatalytic performance for practical applications in wastewater treatment. This chapter provides insights into the state-of-the-art developments and prospects of nanocellulose-based photocatalysts for sustainable and efficient wastewater treatment solutions.

Keywords Nanocellulose · Photocatalyst · Bionanomaterial · Wastewater treatment · Heavy metals

1 Introduction

The rapid emergence of industrialization, population growth, and unplanned rapid urbanization has led to severe water and land pollution. The primary sources of freshwater pollution are directly associated with continuous discharge of untreated or inadequately treated industrial waste and effluents (Thomas et al., 2021). Several types of industrial effluents, organic as well as inorganic pollutants such as pharmaceutical waste, dyestuff, finishing chemicals, heavy metal salts, plastics, fibers,

T. Gupta · S. Sharma (✉)
Department of Biological Sciences and Engineering, Netaji Subhas University of Technology, Dwarka, New Delhi, India
e-mail: shilpa.sharma@nsut.ac.in

E. Priyadarshini et al. (eds.), *Advanced Nanotechnology Applications for Environmental Sustainability*, Sustainable Environmental Waste Management Strategies,
https://doi.org/10.1007/978-3-032-12933-8_9

volatile organic compounds, oil emulsions, and bleaching agents have been observed in wastewater (Mehta et al., 2021). For instance, during the dyeing process, approximately 1–20% of global dye production is lost and discharged into textile effluents (Bhuin et al., 2024). Also, significant water pollution can arise from pharmaceutical waste and heavy metal ions entering the environment through a variety of channels, such as wastewater discharge, agricultural runoff, chemical manufacturing, mining, electroplating, and applying fertilizers and pesticides. Toxic heavy metals such as Cd(II), Cr(VI), Hg(II), Cu(II), and Pb(II) are non-biodegradable, persistent, and hazardous to both human health and ecosystems. The genotoxic, mutagenic, and carcinogenic properties of these pollutants make them hazardous, with evidence linking them to increased cancer incidence and reproductive imbalances, particularly among workers in industries such as textiles, dye manufacturing, pharmaceuticals, and petrochemicals (Muniasamy et al., 2024). Additionally, untreated wastewater used for irrigation can have detrimental effects on crops, further underscoring the necessity of proper wastewater treatment before environmental release (Akpan & Hameed, 2009).

To address this issue, researchers are exploring advanced wastewater treatment technologies, which include chemical decomposition, enzymatic degradation, electrochemical processes, advanced oxidation processes, membrane separation, adsorption, etc. (Tee et al., 2022). Unfortunately, these approaches have certain disadvantages, like high operation costs and failure to remove organic contaminants from wastewater completely. In this context, photocatalysis has emerged as a green technology for water treatment and is considered an effective method for utilizing solar energy to degrade pollutants into non-toxic substances (Ravelli et al., 2009). The photocatalysis technique primarily requires a semiconductor (photocatalyst) such as ZnO, TiO_2, CdS, Fe_2O_4, SnO_2, etc., and exposure to a light source, such as UV, visible, or near-infrared (UV–Vis–NIR) light. The photocatalysts catalyze chemical reactions by absorbing light (photon energy) to create a transient state, generating electron–hole pairs that produce free radicals, enabling degradation of pollutants through photocatalytic reactions (Chawla et al., 2021). However, the effectiveness of semiconductor photocatalysts remains a key concern for researchers due to challenges related to light absorption, bandgap optimization, and charge carrier lifespan, which need to be addressed. In this context, nanomaterials as support systems for photocatalysts are highly desirable, as they can significantly enhance the efficiency of photocatalysts in removing pollutants. Nanomaterials, including inorganic, polymer-based, and nanocomposites, exhibit unique physical, chemical, optical, and electrical properties and high surface area, making them valuable in various applications like environmental remediation, photocatalysis, battery production, etc. However, their use in environmental applications faces challenges, primarily due to toxicity risks to humans and ecosystems. The toxicity of these nanomaterials varies based on factors such as composition, size, crystalline structure, morphology, and concentration (Bradford et al., 2022). Therefore, before deploying these nanomaterials in living systems and the environment, it is crucial to assess their long-term toxicity. An ideal nanomaterial photocatalyst should possess non-toxicity, affordability, stability, and exceptional photoactivity. In this context, innovative nanocellulose-based materials have demonstrated immense potential for wastewater purification and pollution mitigation, offering cost-effective, non-toxic, and scalable eco-friendly solutions, with significant potential in industrial scalability (Reshmy et al., 2022). Nanocellulose-based photocatalytic materials, known for their extensive surface area and exceptional mechanical strength, are highly suitable as substrates (Malhotra et al., 2023; Sudhaik et al., 2023). Studies have shown that nanocellulose acts as an interfacial bridge between polymers and inorganic nanomaterials, enhancing interfacial chemistry in composites. The oxygen-containing functional groups of nanocellulose serve as anchoring sites for various nanoscale photocatalysts, thereby stabilizing them and enhancing their photodegradation performance for removal of aquatic pollutants under light irradiation. In addition, nanocellulose improves electron distribution and facilitates their transfer to the photocatalyst, thereby increasing catalytic efficiency. Besides, hybrid nanocellulose/metal oxide or metal nanoparticle composites offer improved mechanical strength, UV resistance, thermal stability, and low toxicity. Overall, nanocellulose has emerged as a renewable and environmentally friendly template for the development of highly efficient photocatalysts (Oun et al., 2020).

The word cloud plot representing bibliometric research on the application of nanocellulose for waste-water remediation highlights frequently occurring terms such as “nanocellulose”, “water”, “photocatalyst”, “dye”, and “nanocomposite” (Fig. 1). The prevalence of these terms confirms that the literature is focused on utilizing advanced materials, particularly biopolymer nanocomposites, for water treatment and pollutant remediation applications.

This chapter consolidates and shares valuable insights into the development of nanocellulose-based photocatalysts to remove hazardous contaminants from water through photocatalysis. It also highlights nanocellulose sources, types, extraction methods, and mechanisms of photocatalytic degradation. An overview of recent developments of nanocellulose-based photocatalysts for wastewater treatment and insights into future research directions in the field are also detailed in the chapter.

2 Mechanism of Photocatalysis:

Photocatalysis is a light-driven chemical reaction that involves heat activation, where light and heat interact synergistically to initiate and enhance the chemical reactions. When light irradiates a semiconductor photocatalyst, an electron transitions from the valence band (VB) to the conduction band (CB), creating a hole in the VB. The negative CB potential and positive VB potential enable strong reduction and oxidation capabilities of excited electrons and holes. Photoexcited electrons (e^-) reduce O_2 in water to form superoxide radicals ($\bullet O_2^-$), while photo-generated holes (h^+) oxidize water molecules, producing hydroxyl radicals ($\bullet OH$)(B. Liu et al., 2020). These high numbers of photo-generated e^- and h^+ could significantly generate more active radicals for enhanced photocatalytic performance. Interestingly, these reactive species oxidize any organic pollutant into simple carbon dioxide, water, and other degraded products (mineral acids), leaving no harmful byproducts (Fawzi Suleiman

Fig. 1 Word cloud from bibliographic research about pollutant removal using nanocellulose-based photocatalysts

Khasawneh & Palaniandy, 2021). Figure 2 illustrates the generalised photocatalytic mechanism for pollutant degradation. The photocatalytic reactions for pollutant degradation are shown below:

$$\text{Photocatalyst} + hv \rightarrow e^-_{cb} + hv^+_{vb} \quad (1)$$

$$O_2 + e^- \rightarrow O_2^- \quad (2)$$

$$h^+ + H_2O \rightarrow OH \quad (3)$$

$$OH^- + h^+ \rightarrow H^+ + OH \quad (4)$$

$$\text{R - H} + \text{OH} \rightarrow \text{RCOO} \rightarrow \text{CO} + \text{H}_2\text{O} + \text{inorganic ion} \quad (5)$$

$$\text{Pollutants} \xrightarrow{\text{photocatalyst} + hv} \text{intermediates} \rightarrow \text{CO}_2 + \text{H}_2\text{O} \quad (6)$$

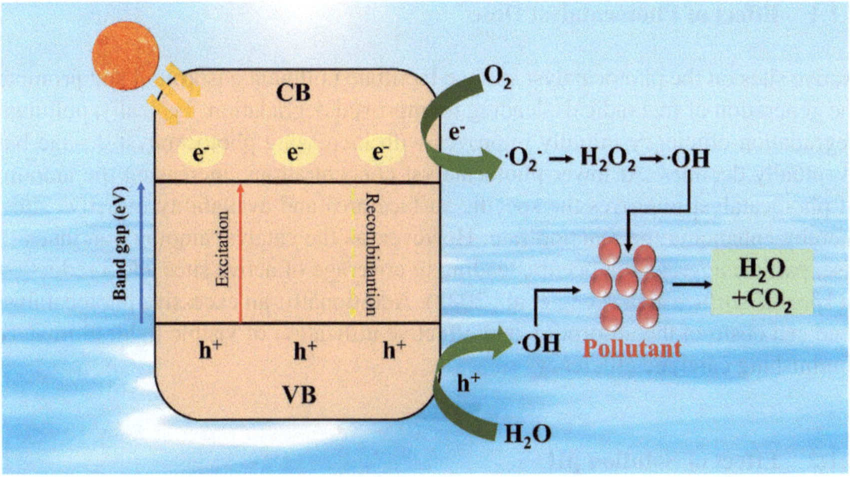

Fig. 2 Generalised photocatalytic mechanism of pollutant degradation

2.1 Factors Affecting Photocatalytic Degradation

The efficiency of photocatalytic pollutant degradation depends on various parameters, such as photocatalyst dose, pollutant concentration, solution pH, light intensity, photocatalyst morphology, and temperature. Figure 3 illustrates different factors impacting the photocatalytic degradation of pollutants.

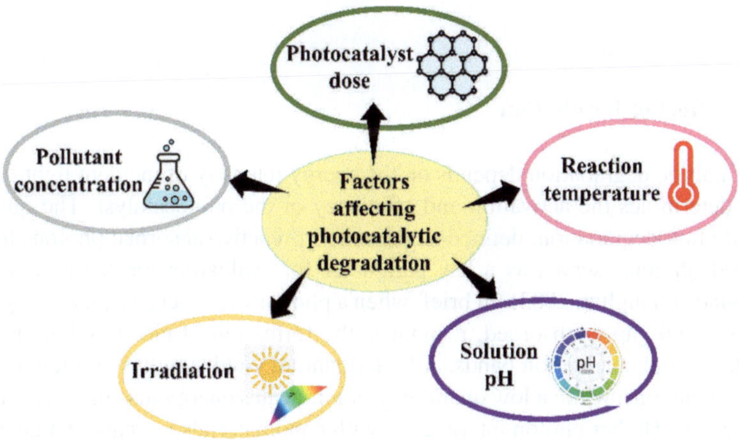

Fig. 3 Factors affecting photocatalytic degradation process

2.1.1 Effect of Photocatalyst Dose

Active sites on the photocatalyst surface facilitate pollutant adsorption and promote the generation of free radicals, leading to improved degradation. Typically, pollutant degradation efficiency initially improves with increasing photocatalyst dosage but eventually declines. At lower photocatalyst concentrations, increasing the amount of photocatalyst improves the specific surface area and availability of active sites, thereby enhancing the reaction rate. However, as the catalyst amount continues to rise, particle aggregation occurs, leading to coverage of active sites and a reduction in specific surface area (Feng et al., 2022). Additionally, an excessive photocatalyst load can obstruct the absorption and effective utilization of visible light, ultimately diminishing catalytic efficiency.

2.1.2 Effect of Solution pH

The pH of the solution may depend on the surface charge, energy band position, and nature and concentration of active species on the photocatalyst. Extreme alkali and acidic conditions can negatively impact the degradation performance (Matsumoto et al., 1989). At highly acidic environments, excess H^+ ions react with generated $\cdot O_2^-$, reducing the availability of active oxygen species. Conversely, in highly alkaline conditions, excessive OH^- ions interact with holes and consume H^+, leading to a decrease in photocatalytic degradation efficiency. However, the optimal pH for photocatalysis varies depending on the specific pollutant (Zhang et al., 2021). At the point of zero charge (pH_{pzc}), the photocatalyst surface carries no net charge. Below pH_{pzc}, the surface becomes positively charged, while it turns negative at higher pH levels.

2.1.3 Effect of Irradiation

Photocatalytic degradation depends on the energy intensity of incident light quanta, which determines the activation and efficiency of the photocatalyst. The quantum yield of photodegradation, defined as the ratio of reactive absorbed photons to total absorbed photons, serves as a key parameter for evaluating the light stability of compounds (Gauglitz, 2003). In brief, when a photon with energy equal to or greater than the band gap is absorbed, it induces the formation of electron–hole pairs in the valence and conduction bands. A high quantum yield signifies efficient photon-induced reactions, while a low quantum yield indicates energy dissipation as heat or fluorescence. Higher photon interactions with nanomaterials increase charge carrier generation, enhancing the degradation rate.

2.1.4 Effect of Pollutant Concentration

The pollutant (dye, heavy metal ions, etc.) dose significantly impacts photocatalytic degradation efficiency, as lower concentrations allow better interaction with reactive species, while higher concentrations lead to active site saturation, light shielding, and intermediate accumulation. An optimal pollutant concentration exists where degradation is most effective (Jamal et al., 2020; Panthi et al., 2015).

2.1.5 Effect of Temperature

The light-mediated degradation reactions are typically monitored under ambient conditions, such as standard room temperature and pressure. As the photocatalytic process depends on light, temperature has minimal influence within the 20–80 °C range. However, below 0 °C or above 80 °C, photocatalytic activity is limited by the slow desorption of formed product, which occurs at a lower rate than surface degradation and reactant adsorption (Shabna et al., 2024). Thus, photocatalytic degradation reactions are generally unaffected by temperature variations. A slight enhancement in activity is observed at higher temperatures, likely due to faster diffusion of OH^- ions from the catalyst surface towards the pollutant. However, high temperatures deplete dissolved O_2 concentration from the reaction mixture, promoting electron–hole recombination on the photocatalyst surface (Hassan et al., 2015).

3 Nanocellulose: Types, Sources, and Preparation

Cellulose, the most abundant biopolymer, is composed of straight chain of D-glucose molecules (ranging from 15 to 15,000 ring units) connected by β-1,4-glycosidic bonds, which allows the molecule to twist and bend, contributing to its structural flexibility. Cellulose serves as the primary constituent of plant cell walls, accompanied by hemicellulose, lignin, and minor quantities of extractives (Zhao et al., 2012). It is estimated that approximately 10^{10} to 5×10^{10} tons of dry lignocellulosic materials are generated globally each year. Lignocellulosic materials serve as excellent raw resources due to their abundance, renewability, sustainability, and potential for surface modification (Zhao et al., 2012). However, in their untreated form, they generally lack functional groups necessary for effective adsorption, and their adsorption capacity varies depending on the specific biomass source. The nanoscale confinement of nanocellulose has significantly enhanced its effectiveness as a sustainable nanomaterial. It has a high surface area, exceptional mechanical properties, remarkable hydrophilicity, low density, and a chemically modifiable surface that enhances adsorption capabilities (Hitam & Jalil, 2023). Additionally, the abundance of -OH groups present in nanocellulose provides an effective means to anchor metal cations for surface modifications. Moreover, 100% biodegradability and non-toxic nanocellulose pose no harm to humans or the environment. This makes it particularly noteworthy in addressing safety concerns, with diverse applications, which has accelerated their use as an advanced generation of biobased adsorbents, nanocatalysts, etc. These characteristics make nanocellulose a versatile material, widely utilized as template support, an independent functional material, a reinforcing component in hybrid materials, and a key element in the development of sustainable green composites for diverse applications.

3.1 Sources and Types of Nanocellulose

Cellulose is naturally found in wood, hemp, cotton, sisal, ramie, jute, fruit waste, leaf fibers, and marine biomass, where it serves as the primary structural component in plant cells. However, timber remains the primary source of cellulose fibers. The growing concern about environmental sustainability and deforestation driven by rising timber demand has sparked interest in utilizing non-wood cellulose materials (Oun & Rhim, 2016).

Other possible sources, like agricultural waste, namely sugarcane bagasse, rice straw, banana peel, corn stalk, and other byproducts, have also been explored for the extraction of cellulose and nanocellulose. Apart from plant sources, marine invertebrates, algae, fungi, and bacteria are also main sources of cellulose extraction (Gupta & Sharma, 2024). Bacterial cellulose differs from plant-derived cellulose as it lacks wax, lignin, pectin, and hemicellulose. Additionally, its polymer chain length and molecular weight can be precisely regulated during fermentation (Reiniati et al., 2017).

Nanocellulose is classified into three primary types, as distinguished by the synthesis conditions that influence their size, composition, and characteristics: cellulose nanofibers (CNFs), cellulose nanocrystals (CNCs), and bacterial nanocellulose (BNC). Cellulose-based fibrous structures are called CNFs, with diameters ranging from 5 to 30 nm and lengths extending to several microns. Despite being tightly twisted submicron-scale fibers, they are widely recognized as nanofibers. CNFs are composed of highly crystalline nanofibrils with unique properties and dimensions. On the other hand, CNCs are charged nanoparticles derived from cellulose, characterized by a high aspect ratio. CNCs typically exhibit rod-like, ribbon-like, or whisker-like shapes, with lengths varying from tens of nanometers to several microns and widths between 3 and 50 nm. BNC, nanofiber material, is a chemically pure biopolymer synthesized through fermentation by various bacteria, including *Acetobacter*, *Sarcina ventriculi*, *Azotobacter*, and *Pseudomonas* in different media. During the fermentation process, microbes either remain suspended or attach to cellulose fibers, leading to the formation of a swollen gel structure. This requires a subsequent purification step, which involves elimination of microorganisms and removal of cell debris and residual media from the cellulose matrix (De Amorim et al., 2020). Unlike

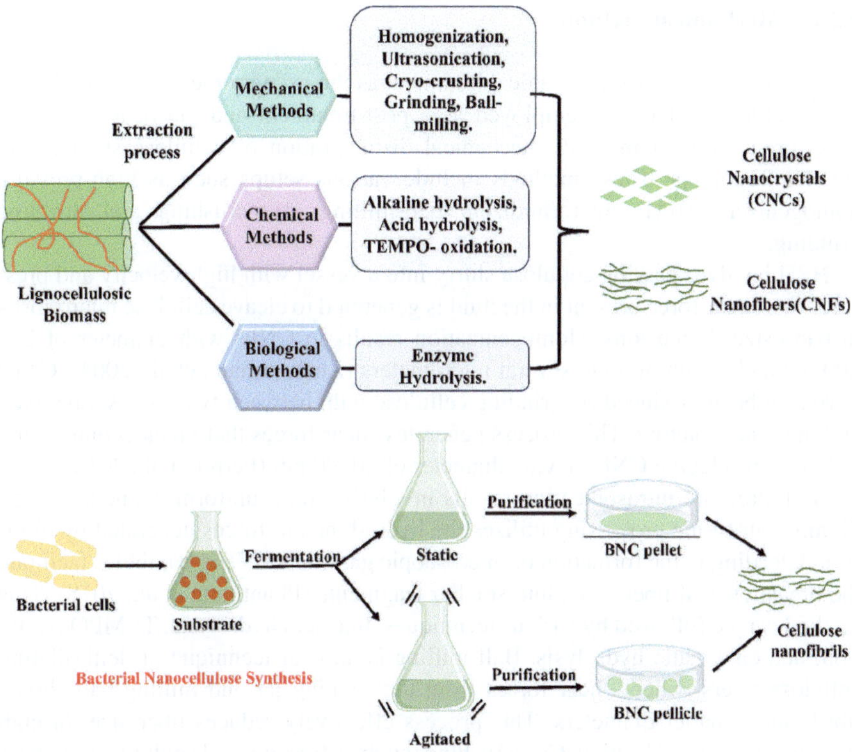

Fig. 4 Extraction route of various types of nanocellulose

synthetic processes, BNC production does not involve the polymerization or aggregation of smaller particles (Fig. 4). Instead, its characteristic structure arises from the assembly of glucose chains, forming a network of cellulose nanofibrils (R, Philip, et al., 2021). The elemental structure of BNC shows a high transparency level in comparison to plant origins. However, BNC production is more challenging because of the fermentation process, which remains relatively costly at the industrial level.

3.2 Extraction of Nanocellulose

Various techniques have been employed to isolate nanocellulose from lignocellulosic materials, such as mechanical methods, acid hydrolysis, chemical or TEMPO-mediated oxidation, and biological methods (Fig. 4). The properties of the isolated nanocellulose are largely influenced by the cellulose source (i.e., chemical structure) and the specific isolation method used.

3.2.1 Mechanical Methods

Most studies emphasize mechanical treatment as the primary method for producing CNFs, although it is also employed as a post-treatment and purification step in CNC production. It involves mechanical disintegration of cellulose strands into smaller fragments. These methods include various setups such as high-pressure homogenizers (HPH), microfluidizers, ball-milling, cryo-crushing, and ultra-fine grinding.

HPH involves passing cellulose slurry into a vessel with high velocity and pressure. The shear force present in the fluid is generated to cleave cellulose microfibrils in nano-size dimensions. Homogenization results in CNFs with diameter of 20–100 nm and length of up to several micrometers (Zimmermann et al., 2004). CNFs have also been produced by grinding cellulose pulp between two stones, one fixed and the other rotating. This process generates shear forces that break cellulose and H-bonds, producing CNFs having diameter of 20–90 nm (Ferrer et al., 2012).

High-pressure ultrasonication results in CNFs with a uniform diameter of 15–20 nm. Ultrasonic processing utilizes the hydrodynamic forces generated by ultrasound, leading to the formation of microscopic gas bubbles. These bubbles facilitate the breakdown of fiber walls into smaller fragments (Phanthong et al., 2018). This method can be followed by various techniques, like acid hydrolysis, TEMPO oxidation, and enzymatic hydrolysis. Ball milling is another technique of defibrillating cellulose fibers, where shear forces from the rotating jar and milling balls break fibrils into smaller diameters. This process effectively reduces fiber size through repeated impact and friction. Cryocrushing involves freezing and applying high shear forces to break the cell wall, while steam explosion uses steam pressure to rupture cellulose structure, facilitating hemicellulose extraction and lignin depolymerization (Panthapulakkal & Sain, 2012).

3.2.2 Chemical Methods

Chemical approaches stand out as energy-efficient production methods for large-scale production of nanocellulose. Studies have reported that chemical agents play a crucial role in defibrillation of cellulose during nanocellulose production. Acid/ alkaline hydrolysis and TEMPO-oxidation are the most common chemical pretreatment methods for nanocellulose production.

Acid hydrolysis is widely used method for isolating CNCs from cellulose. Unlike mechanical breakdown, this process targets and eliminates amorphous region of the microfibrils while preserving the crystalline regions. The CNCs obtained using this process are mostly 4–70 nm in width and 100–6000 nm in length (Pawcenis et al., 2022). Alkaline hydrolysis, using an alkali solution as a swelling agent, increases the internal surface area while reducing cellulose crystallinity and polymerization. Both methods dissolve lignin and hemicellulose by breaking glycosidic ether bonds, but their high water demand for neutralization lowers thermal stability and crystallinity (Thipchai et al., 2024). Alternatively, TEMPO oxidation selectively oxidizes primary -OH groups at C_6 position of cellulose into $-COO^-$ groups, crucial for stabilizing nanocellulose suspensions (Shen et al., 2017). In this process, the crystallinity of nanocellulose remains largely unaffected. However, there are some drawbacks, including use of hazardous chemicals, challenges in recycling, inefficient removal of impurities, and limited oxidation sites. Therefore, researchers have integrated different pretreatment techniques to enhance the removal of impurities from cellulose.

3.2.3 Biological Methods

Identifying an efficient alternative to conventional chemical and physical synthesis methods that is both cost-effective and environmentally sustainable is essential. In this regard, the biological approach, known as "green synthesis," is gaining global attention for its scalability and eco-friendliness. Enzymatic hydrolysis is an effective technique for producing nanocellulose from lignocellulosic biomass, consisting of multiple stages. The process begins with pretreatment to isolate pure cellulose by removing lignin, hemicellulose, and extractives, followed by enzymatic hydrolysis using various combinations of cellulases like endoglucanases, exoglucanases, β-glucosidases, and xylanases. (Karim et al., 2017). Since cellulose is composed of a complex structure with diverse chemical components, no single enzyme can effectively break it down. This approach minimizes chemical waste and lowers energy requirements. Enzymatic hydrolysis typically precedes mechanical disintegration in nanocellulose production (Gabric et al., 2016). Nanofibers generated through enzymatic treatment with endoglucanase exhibit a superior structure, featuring higher molar mass and aspect ratio compared to those obtained through acid hydrolysis (Henriksson et al., 2007). Interestingly, enzyme hydrolysis can be employed for the extraction of BNC, CNFs, and CNCs.

3.3 Nanocellulose-Based Photocatalysts

An ideal photocatalyst should be eco-friendly, stable, highly photoactive, and cost-effective. The shift from petrochemical-based feedstocks to environmentally friendly biomaterials, such as cellulose-based materials, is paving the way for a cleaner and more sustainable future (Yang et al., 2011). Among them, nanocellulose has been recognized as the most abundant renewable organic resource, possessing unique properties, including chirality, hydrophilicity, and wide chemical versatility, demonstrating significant potential as an environmentally friendly and biocompatible support material. Moreover, nanocellulose is often regarded as an excellent host material for nanoparticles, as it enhances their stability, controls their growth, and preserves their unique morphology (Zeng et al., 2010). This control over morphological features leads to variations in specific surface area and pore structure, which in turn influence the overall photocatalytic performance. Nanocellulose itself lacks intrinsic photocatalytic activity under visible and UV light. Incorporating conventional metal oxide photocatalysts such as titanium dioxide, zinc oxide, graphene oxide, carbon nanotubes, etc., significantly enhances its photocatalytic performance. Table 1 provides a summary of the applications of nanocellulose-based photocatalysts for the enhanced photocatalytic degradation of various organic pollutants.

Recently, a membrane developed by incorporating HKUST-1 and modified carbon nitride onto nano-fibrillated cellulose (NFC) was demonstrated to remove formaldehyde with 86.9% efficiency and photocatalytically degrade it with 48.45% efficiency under xenon lamp irradiation. The synergy between HKUST-1's adsorption capacity and CNx's photocatalytic activity on NFC membrane enhanced performance, with excellent stability over ten cycles (Chen et al., 2024). In another very recent study, CNFC@Bi_2O_3 composites were synthesized using CNFs as a template via a one-step calcination method. The resulting Z-type heterojunction exhibited excellent visible-light response, achieving a high malachite green dye degradation rate of 758.64 mg g^{-1} in 5 h compared to BC@Bi_2O_3, which used bamboo cellulose as a template (425.76 mg g^{-1}). Additionally, CNFC@Bi_2O_3 outperformed other variants, including CNF/0.3Bi (473.04 mg g^{-1}) and CNF/1.7Bi (545.52 mg g^{-1}), demonstrating its superior photocatalytic activity. The material's enhanced performance was attributed to abundant surface functional groups and efficient charge separation (D. Yang et al., 2025). Similarly, nanocellulose-supported S-scheme heterojunction developed by anchoring Ag_2WO_4, NiO, and MoO_3 nanoparticles onto nanocellulose matrix exhibited exceptional photocatalytic performance, achieving 99.6% norfloxacin degradation via reactive oxygen species generation within 30 min of visible light irradiation (Sambyal et al., 2025) In another study, porous Ag–AgCl@TiO_2@CNFs composites were synthesized using physical/chemical crosslinking, photoreduction, and freeze-drying techniques. The CNFs matrix provided strong mechanical stability and enhanced adsorption capacity due to the presence of abundant polar functional groups, while the heterostructure facilitated efficient photocatalytic degradation (99.54% efficiency) and retained 99.07% efficiency after five reuse cycles (Yu et al., 2025).

TiO_2 is the most widely studied compound with over 40,000 publications in the past decade. (Athawale et al., 2020). However, the application of TiO_2 nanoparticles as a standalone photocatalyst becomes challenging due to low specific surface area and insufficient active sites due to the tendency to agglomerate (Hitam & Jalil, 2022). In a recent study, CNCs were used as support material for dispersing TiO_2 nanoparticles, and the resulting nanocomposite was demonstrated to photodegrade methylene blue dye with 98.5% efficiency in 40 min. The CNCs in the nanocomposite enhanced the dispersion of TiO_2, increased the surface area and adsorption capacity for the dye (Morshed et al., 2020). In another study, in-situ hybridization of TiO_2 with hydroxypropyl methyl cellulose enhanced the nanocomposite's surface area and extended its light absorption into the visible region. The resulting nanocomposite exhibited significantly higher photocatalytic activity than pure TiO_2 and demonstrated substantial photostability after five experimental cycles (Nsib et al., 2014). Nanocellulose-templated CeO_2/TiO_2 was prepared through a hydrolysis precipitation reaction and exhibited enhanced photocatalytic performance than CeO_2/TiO_2 prepared without

Table 1 Nanocellulose-based photocatalysts for the enhanced photodegradation of the pollutants

Catalyst	Model pollutant	Light Source	Degradation efficiency (%)	References
Cellulose nanocrystal-titananium dioxide nanorods	RhB	Ozone-free 450 W xenon short-arc lamp	High photocatalytic activity with enhanced light absorption	(Nair et al., 2020)
Nanocellulose–TiO_2/polyester nonwoven fabric (NC-TiO_2/PET) composite	MB Acid red (AR)	6 W UV lamp ($\lambda \approx 280$ nm)	90.02% 91.14%	(Tan et al., 2023)
Graphene oxide/TiO_2/bacterial cellulose nanocomposite	(MO)	UV lamp (main 365 nm, 175 w)	100%	(L.-P. Liu et al., 2017)
Immobilized TiO_2 nanoparticles and laccase on BNC membrane	Reactive red X-3B	UV irradiation	95%	(Li et al., 2017)
NC/ZnO	RhB CF	Visible light (250W) Sunlight Sunlight	35% 76% 75%	(Tavker & Sharma, 2018)
Nanocellulose templated CeO2/TiO_2	RhB MO Cr(VI)	Xenon lamp (350 W, 400 nm)	Complete removal	(Sun et al., 2021)
NC10-TiO_2 NC20-TiO_2 NC80-TiO_2	MO	UV light 39 W lamp ($\lambda =$ 365 nm	99% 72% 23%	(Toro et al., 2022)
Fe-doped ZnO/nanocellulose nanocomposite	MB	Visible irradiation	98.84%	(Vasheghani Farahani et al., 2022)

(continued)

Table 1 (continued)

Catalyst	Model pollutant	Light Source	Degradation efficiency (%)	References
Ag_3PO_4/nanocellulose composite	MO	Sunlight	90% in DI and 70% in wastewater	(Lebogang et al., 2019)
CdS/MoS_2/Montmorillonite (CMMt) and TEMPO-oxidized cellulose nanofiber (TOCN)-mediated polyacrylamide (PAM)	Tetracyline (TC)	Visible light	90.03%	(Yue et al., 2022)
Polyvinylidene fluoride (PVDF) nanocomposite membranes enhanced with nanocellulose (NC) and titanium dioxide (TiO_2)	MB	100 W UV light irradiation	98%	(Rahman et al., 2024)
CNCs/zinc oxide	MB	UV light irradiation Visible light	96.11% 85.60%	(Cumba et al., 2022)
Carbonized cellulose nanofibrils/$Ti_3C_2T_x$ MXene/g-C_3N_4	MB RhB TC	Visible light	96.5% 95.4% 86.5%	(Zu et al., 2022)
Cu-doped ZnO nanoparticles immobilized onto cellulose nanofibrous membranes	MB	Xenon light	99.8%	(Nguyen et al., 2024)
TiO_2—CNC hybrid	MB	UVC lamps ((λ = 254 nm)	98.5%	(Morshed et al., 2020)
Nanocellulose/TiO_2 nanocomposite	Mefenamic acid	89% 450 W high pressure mercury lamp	89%	(Rathod et al., 2018)
Nanocellulose/MoS_2	RhB	Tungsten lamp (200W)	96%	(Tavker & Sharma, 2020)

(continued)

Table 1 (continued)

Catalyst	Model pollutant	Light Source	Degradation efficiency (%)	References
3D g-C3N4 based nanocellulose aerogel	MO MB	500w Xe lamp (($\lambda > 400$ nm)	99% 99%	(Ma et al., 2021)
Ag–AgCl@TiO_2@cellulose nanofiber composites	NF	Visible light	99.6%	(Sambyal et al., 2025)

nanocellulose as a template. The nanocellulose-templated CeO_2/TiO_2 nanocomposite degraded methyl orange (MO) and rhodamine B dyes (250 mL, 50 ppm) and reduced Cr(VI) solution (250 mL, 50 ppm) within 70, 50, and 60 min, respectively, under visible light. The nanocellulose template created a uniform network in solution, facilitating nucleation via surface -OH groups, in addition to preventing aggregation and growth of CeO_2/TiO_2 during preparation, resulting in smaller grain size, high dispersion, and increased specific surface area of CeO_2/TiO_2. Furthermore, this nanocomposite reduced the recombination of holes and electrons, which improved photocatalytic degradation (Sun et al., 2021). In another study, a nanocomposite was fabricated through a hydrothermal process applied to a titania gel film coated onto CNFs integrated within a bulk cellulosic substrate (filter paper). This treatment converted the titania gel into anatase nanoparticles, which were uniformly distributed across the surface of cellulose nanofibers while preserving the hierarchical network structure of the original filter paper and exhibiting significant photocatalytic efficiency in degrading aqueous solutions of methylene blue (MB) and MO when exposed to UV light (Luo et al., 2014).

Zinc oxide is increasingly being recognized as a comparable alternative to the widely used TiO_2, primarily due to its unique properties, including low cost, reduced toxicity, and a wide band gap (E_g) of 3.37 eV, which facilitates strong redox potential (Hitam et al., 2020). However, it also tends to agglomerate, significantly limiting its practical applications. Additionally, its photocatalytic performance is hindered by rapid electron–hole recombination and weak absorption in the visible light range. To overcome these challenges, ZnO nanoparticles supported on CNCs (ZnO/CNCs) were synthesized and utilized for MB dye photodegradation. The CNCs were derived from the *Nypa fruticans* trunk in Vietnam through hydrolysis using a citric/hydrochloric acid mixture, facilitating ZnO deposition through electrostatic precipitation. The ZnO nanocrystals were evenly distributed on the CNCs due to strong interactions between CNC carboxyl groups and Zn^{2+} ions. This synergy enhanced photodegradation efficiency (95%), with CNCs providing adsorption sites for MB and ZnO preventing agglomeration while generating reactive OH• and O_2^- radicals under UVC light (Nang An et al., 2020).

Attempts have also been made to shift the energy band gap of ZnO to the visible light range by incorporating ZnO into nanocellulose. For instance, ZnO nanoparticles green synthesized from *Tinospora Cordifolia* plant stem extract were incorporated into nanocellulose prepared from sugarcane bagasse. The synthesized nanocomposite exhibited photodegradation of enrofloxacin (EF) antibiotic with 97% efficiency under visible light. This may be attributed to the electrostatic interaction between the lone pair of electrons on the EF surface and the surface reactive species (Nahi et al., 2021).

Bacterial cellulose, also serves as an excellent support for metal oxide nanoparticles. Its ultrafine 3D network of well-separated nano and microfibrils provides high specific surface area, while –OH binding sites and the fibrous network enhance metal oxide adsorption (Li et al., 2017). A study reported synthesis of bacterial cellulose/ZnO nanocomposite by immersing bacterial cellulose in zinc salt solution. On treatment with sodium hydroxide, ZnO nanoparticles were formed and distributed homogenously on the bacterial cellulose fiber surface. The synthesized bacterial cellulose/ZnO film exhibited high photocatalytic efficiency (91% photodegradation of MO dye) under UV-irradiation within 2 h as well as antibacterial activity against tested Gram-positive and Gram-negative bacterial strains (Wahid et al., 2019).

The examples above highlight nanocellulose's role as an emerging support material, significantly boosting the photocatalytic activity of semiconductor photocatalysts.

4 Challenges and Future Directions:

Nanocellulose holds immense potential as a material for environmental remediation; however, several challenges and limitations hinder its practical applications. To achieve widespread commercialization and market adoption in this field, nanocellulose applications must be scalable, cost-effective, and capable of generating valuable by-products. Despite its promising potential, most research on nanocellulose production remains confined to laboratory-scale studies (Yayayürük & Yayayürük, 2021). Managing key factors such as porosity, permeability, and specific adsorption remains a major hurdle. Native nanocellulose membranes exhibit moderate adsorption and photocatalytic degradation performance, but further advancements are needed to develop composite materials with enhanced pollutant removal efficiency. To improve material properties, alternative processing methods should prioritize achieving reproducible single nanofiber geometries while minimizing fiber agglomeration within a web-like structure. A one-step process that ensures controlled porosity and surface properties can address these challenges. Reducing the number of processing steps by incorporating controllable charges on the nanofiber surfaces, avoiding traditional wet chemistry techniques, and utilizing environmentally friendly solvents is essential for creating sustainable and efficient solutions.

A key challenge in commercializing nanocellulose composites is achieving uniform dispersion of cellulose and inorganic nano-adsorbents. The tendency of the composite components to aggregate can adversely impact the material's morphology, effectiveness, and overall performance. To tackle this issue, researchers have investigated various strategies, such as employing cellulose solvents (Salama et al., 2014), utilizing soluble cellulose (Salama et al., 2021), and implementing methods to control inorganic mineralization (Salama, 2017).

A key challenge in using modified nanocellulose for environmental remediation is the lack of clear toxicology, posing economic and regulatory barriers. However, ecotoxicology studies remain limited despite their importance. (Lin & Dufresne, 2014). Smaller nanoparticles pose higher risks when inhaled, as their size facilitates cellular and systemic absorption. While most nanocellulose materials are less toxic, some studies have reported adverse effects on cell viability and proliferation, particularly at high concentrations (Lin & Dufresne, 2014).

Additionally, a significant challenge occurs due to gradual deactivation of photocatalysts, often caused by surface fouling or inactive species formation. This deactivation limits the durability and effectiveness of nanocellulose-based photocatalysts in real-world applications, emphasizing the need for innovative approaches to enhance their longevity and maintain their performance over time (Jaffar et al., 2022).

5 Conclusion

Cellulose, the Earth's most abundant and eco-friendly biopolymer, is readily available in significant amounts from plants, crops, and forest residues. The discovery of cellulose nanomaterials marks a milestone in materials science, gaining global attention for applications like bio-based packaging, optoelectronics, and tissue regeneration. Nanocellulose's large surface area, eco-friendliness, and easy functionalization make it promising for environmental uses, such as water treatment and carbon sequestration. This chapter systematically summarizes recent advancements in nanocellulose-based materials for environmental health remediation. It examines various nanocellulose synthesis methods and their role in improving the photocatalytic efficiency of non-metal and metal photocatalysts. Photocatalytic remediation stands out as an eco-friendly and efficient method to degrade pollutants into harmless products. Nanocellulose, being biodegradable and non-toxic, provides excellent support for photocatalysts, offering large surface areas, active sites, and improved pollutant adsorption. By optimizing the choice of dopants and nanocellulose type, hybrid photocatalysts with enhanced visible light harvesting, inhibited charge recombination, and efficient charge separation can be designed.

Despite significant progress in the past two decades, there is still potential for innovation. Utilizing agricultural residues like rice straw, corn husks, and sugarcane bagasse as bio-based sources for nanocellulose not only addresses waste management but also creates economic and environmental benefits. This approach promotes sustainable and greener solutions for photocatalytic environmental remediation and other photo-driven applications. However, challenges include high energy consumption, low yield during preparation can be explored for better enhancement. Modifications enhance its functionality for pollutant removal, and integrated methods show synergistic benefits. Future research should focus on cost-effective extraction, stability tests, and improving its performance in real-world wastewater treatment.

References

Akpan, U. G., & Hameed, B. H. (2009). Parameters affecting the photocatalytic degradation of dyes using TiO_2-based photocatalysts: A review. *Journal of Hazardous Materials, 170*(2–3), 520–529. https://doi.org/10.1016/j.jhazmat.2009.05.039

Athawale, A., Bokare, A., Singh, H., Nguyen, V.-H., Vo, D.-V. N., Kumar, D., & Sharma, A. (2020). Synthesis of Ag_2O coated TiO_2 nanoparticles by sonochemically activated methods for enhanced photocatalytic activities. *Topics in Catalysis, 63*(11–14), 1056–1065. https://doi.org/10.1007/s11244-020-01374-0

Bhuin, A., Udayakumar, S., Gopalarethinam, J., Mukherjee, D., Girigoswami, K., Ponraj, C., & Sarkar, S. (2024). Photocatalytic degradation of antibiotics and antimicrobial and anticancer activities of two-dimensional ZnO nanosheets. *Scientific Reports, 14*(1), 10406. https://doi.org/10.1038/s41598-024-59842-6

Bradford, S. A., Shen, C., Kim, H., Letcher, R. J., Rinklebe, J., Ok, Y. S., & Ma, L. (2022). Environmental applications and risks of nanomaterials: An introduction to CREST publications during 2018–2021. *Critical Reviews in Environmental Science and Technology, 52*(21), 3753–3762. https://doi.org/10.1080/10643389.2021.2020425

Chawla, H., Chandra, A., Ingole, P. P., & Garg, S. (2021). Recent advancements in enhancement of photocatalytic activity using bismuth-based metal oxides Bi2MO6 (M = W, Mo, Cr) for environmental remediation and clean energy production. *Journal of Industrial and Engineering Chemistry, 95*, 1–15. https://doi.org/10.1016/j.jiec.2020.12.028

Chen, Z., Zhang, X., Xu, K., He, X., Li, J., Zhang, L., & Wang, G. (2024). Facile fabrication of nanocellulose-supported membrane composited with modified carbon nitride and HKUST-1 for efficient photocatalytic degradation of formaldehyde. *International Journal of Biological Macromolecules, 268*, Article 131937. https://doi.org/10.1016/j.ijbiomac.2024.131937

Cumba, R. M. T., Ligalig, C. B., Tingson, J. M. D., Molina, M. P., Alguno, A. C., Deocaris, C. C., Latayada, F., Primadona, I., & Capangpangan, R. Y. (2022). Photocatalytic activity of cellulose nanocrystals/Zinc oxide nanocomposite against thiazine dye under UV and visible light irradiation. *ASEAN Journal of Chemical Engineering, 22*(1), 168. https://doi.org/10.22146/ajche.72331

De Amorim, J. D. P., De Souza, K. C., Duarte, C. R., Da Silva Duarte, I., De Assis Sales Ribeiro, F., Silva, G. S., De Farias, P. M. A., Stingl, A., Costa, A. F. S., Vinhas, G. M., & Sarubbo, L. A. (2020). Plant and bacterial nanocellulose: Production, properties and applications in medicine, food, cosmetics, electronics and engineering. A review. *Environmental Chemistry Letters, 18*(3), 851–869. https://doi.org/10.1007/s10311-020-00989-9

Fawzi Suleiman Khasawneh, O., & Palaniandy, P. (2021). Removal of organic pollutants from water by Fe_2O_3/TiO_2 based photocatalytic degradation: A review. *Environmental Technology & Innovation, 21*, 101230. https://doi.org/10.1016/j.eti.2020.101230

Feng, H., Zhang, Y., & Cui, F. (2022). Enhanced photocatalytic activity of Cu_2O for visible light-driven dye degradation by carbon quantum dots. *Environmental Science and Pollution Research, 29*(6), 8613–8622. https://doi.org/10.1007/s11356-021-16337-5

Ferrer, A., Filpponen, I., Rodríguez, A., Laine, J., & Rojas, O. J. (2012). Valorization of residual Empty Palm Fruit Bunch Fibers (EPFBF) by microfluidization: Production of nanofibrillated cellulose and EPFBF nanopaper. *Bioresource Technology, 125*, 249–255. https://doi.org/10.1016/j.biortech.2012.08.108

Gabric, M. V., Boncina, T., Humar, M., & Pohleven, F. (2016). Laccase treatment of norway spruce wood surface improves resistance and copper fixation of treated wood. *Drewno. Prace Naukowe, Doniesienia, Komunikaty = Wood. Research Papers, Reports, Announcements, 59*(198), 19–33. https://doi.org/10.12841/wood.1644-3985.179.10

Gauglitz, G. (2003). Photophysical, photochemical and photokinetic properties of photochromic systems. In *Photochromism* (pp. 15–63). Elsevier. https://doi.org/10.1016/B978-044451322-9/50006-3

Gupta, T., & Sharma, S. (2024). Nanocellulose scaffolds for skin tissue engineering. In V. Gautam, R. Kumar, K. Das Manandhar, & S. C. Kamble (Eds.), *Nanomedicine* (pp. 273–301). Springer Nature Switzerland. https://doi.org/10.1007/978-3-031-72467-1_11

Hassan, S. S. M., El Azab, W. I. M., Ali, H. R., & Mansour, M. S. M. (2015). Green synthesis and characterization of ZnO nanoparticles for photocatalytic degradation of anthracene. *Advances in Natural Sciences: Nanoscience and Nanotechnology, 6*(4), Article 045012. https://doi.org/10.1088/2043-6262/6/4/045012

Henriksson, M., Henriksson, G., Berglund, L. A., & Lindström, T. (2007). An environmentally friendly method for enzyme-assisted preparation of microfibrillated cellulose (MFC) nanofibers. *European Polymer Journal, 43*(8), 3434–3441. https://doi.org/10.1016/j.eurpolymj.2007.05.038

Hitam, C. N. C., & Jalil, A. A. (2022). Recent advances on nanocellulose biomaterials for environmental health photoremediation: An overview. *Environmental Research, 204*, Article 111964. https://doi.org/10.1016/j.envres.2021.111964

Hitam, C. N. C., & Jalil, A. A. (2023). A review on biohydrogen production through photofermentation of lignocellulosic biomass. *Biomass Conversion and Biorefinery, 13*(10), 8465–8483. https://doi.org/10.1007/s13399-020-01140-y

Hitam, C. N. C., Jalil, A. A., Izan, S. M., Azami, M. S., Hassim, M. H., & Chanlek, N. (2020). The unforeseen relationship of Fe_2O_3 and ZnO on fibrous silica KCC-1 catalyst for fabricated Z-scheme extractive-photooxidative desulphurization. *Powder Technology, 375*, 397–408. https://doi.org/10.1016/j.powtec.2020.07.114

Jaffar, S. S., Saallah, S., Misson, M., Siddiquee, S., Roslan, J., Saalah, S., & Lenggoro, W. (2022). Recent development and environmental applications of nanocellulose-based membranes. *Membranes, 12*(3), 287. https://doi.org/10.3390/membranes12030287

Jamal, N., Radhakrishnan, A., Raghavan, R., & Bhaskaran, B. (2020). Efficient photocatalytic degradation of organic dye from aqueous solutions over zinc oxide incorporated nanocellulose under visible light irradiation. *Main Group Metal Chemistry, 43*(1), 84–91. https://doi.org/10.1515/mgmc-2020-0009

Karim, Z., Afrin, S., Husain, Q., & Danish, R. (2017). Necessity of enzymatic hydrolysis for production and functionalization of nanocelluloses. *Critical Reviews in Biotechnology, 37*(3), 355–370. https://doi.org/10.3109/07388551.2016.1163322

Lebogang, L., Bosigo, R., Lefatshe, K., & Muiva, C. (2019). Ag_3PO_4/nanocellulose composite for effective sunlight driven photodegradation of organic dyes in wastewater. *Materials Chemistry and Physics, 236*, Article 121756. https://doi.org/10.1016/j.matchemphys.2019.121756

Li, G., Nandgaonkar, A. G., Wang, Q., Zhang, J., Krause, W. E., Wei, Q., & Lucia, L. A. (2017). Laccase-immobilized bacterial cellulose/TiO_2 functionalized composite membranes: Evaluation for photo- and bio-catalytic dye degradation. *Journal of Membrane Science, 525*, 89–98. https://doi.org/10.1016/j.memsci.2016.10.033

Lin, N., & Dufresne, A. (2014). Nanocellulose in biomedicine: Current status and future prospect. *European Polymer Journal, 59*, 302–325. https://doi.org/10.1016/j.eurpolymj.2014.07.025

Liu, B., Wu, H., & Parkin, I. P. (2020). New Insights into the fundamental principle of semiconductor photocatalysis. *ACS Omega, 5*(24), 14847–14856. https://doi.org/10.1021/acsomega.0c02145

Liu, L.-P., Yang, X.-N., Ye, L., Xue, D.-D., Liu, M., Jia, S.-R., Hou, Y., Chu, L.-Q., & Zhong, C. (2017). Preparation and characterization of a photocatalytic antibacterial material: Graphene oxide/TiO_2/bacterial cellulose nanocomposite. *Carbohydrate Polymers, 174*, 1078–1086. https://doi.org/10.1016/j.carbpol.2017.07.042

Luo, Y., Xu, J., & Huang, J. (2014). Hierarchical nanofibrous anatase-titania–cellulose composite and its photocatalytic property. *CrystEngComm, 16*(3), 464–471. https://doi.org/10.1039/C3CE41906A

Ma, Z., Zhou, P., Zhang, L., Zhong, Y., Sui, X., Wang, B., Ma, Y., Feng, X., Xu, H., & Mao, Z. (2021). A recyclable 3D g-C_3N_4 based nanocellulose aerogel composite for photodegradation of organic pollutants. *Cellulose, 28*(6), 3531–3547. https://doi.org/10.1007/s10570-021-03748-3

Malhotra, M., Sudhaik, A., Sonu, Raizada, P., Ahamad, T., Nguyen, V.-H., Van Le, Q., Selvasembian, R., Mishra, A. K., & Singh, P. (2023). An overview on cellulose-supported photocatalytic materials for the efficient removal of toxic dyes. *Industrial Crops and Products, 202*, 117000. https://doi.org/10.1016/j.indcrop.2023.117000

Matsumoto, Y., Yoshikawa, T., & Sato, E. (1989). Dependence of the band bending of the oxide semiconductors on pH. *Journal of the Electrochemical Society, 136*(5), 1389–1391. https://doi.org/10.1149/1.2096927

Mehta, M., Sharma, M., Pathania, K., Jena, P. K., & Bhushan, I. (2021). Degradation of synthetic dyes using nanoparticles: A mini-review. *Environmental Science and Pollution Research, 28*(36), 49434–49446. https://doi.org/10.1007/s11356-021-15470-5

Morshed, M. N., Al Azad, S., Deb, H., Shaun, B. B., & Shen, X. L. (2020). Titania-loaded cellulose-based functional hybrid nanomaterial for photocatalytic degradation of toxic aromatic dye in water. *Journal of Water Process Engineering, 33*, Article 101062. https://doi.org/10.1016/j.jwpe.2019.101062

Muniasamy, S. K., Gopala Krishna, Gvt., Murali, V., & Ravindiran, G. (2024). The adsorption behaviour of biochar derived from Prosopis juliflora with surface modified by KOH against Allura red dye. *Desalination and Water Treatment, 318*, 100337. https://doi.org/10.1016/j.dwt.2024.100337

Nahi, J., Radhakrishnan, A., & Beena, B. (2021). Green synthesis of zinc oxide incorporated nanocellulose with visible light photocatalytic activity and application for the removal of antibiotic enrofloxacin from aqueousmedia. *Materials Today: Proceedings, 41*, 583–589. https://doi.org/10.1016/j.matpr.2020.05.253

Nair, S. S., Chen, J., Slabon, A., & Mathew, A. P. (2020). Converting cellulose nanocrystals into photocatalysts by functionalisation with titanium dioxide nanorods and gold nanocrystals. *RSC Advances, 10*(61), 37374–37381. https://doi.org/10.1039/D0RA05961G

Nang An, V., Van, T. T. T., Nhan, H. T. C., & Van Hieu, L. (2020). Investigating methylene blue adsorption and photocatalytic activity of ZnO/CNC nanohybrids. *Journal of Nanomaterials, 2020*, 1–10. https://doi.org/10.1155/2020/6185976

Nguyen, P., Tran, T. N., Le, M. T., Ullah, A., Phan, D., & Kim, I. (2024). Enhanced dye removal and antibacterial efficacy of copper-doped ZnO nanoparticles on cellulose nanofibers. *Advanced Materials Interfaces, 11*(35), 2400468. https://doi.org/10.1002/admi.202400468

Nsib, M. F., Hajji, F., Mayoufi, A., Moussa, N., Rayes, A., & Houas, A. (2014). In situ synthesis and characterization of TiO_2/HPM cellulose hybrid material for the photocatalytic degradation of 4-NP under visible light. *Comptes Rendus. Chimie, 17*(7–8), 839–848. https://doi.org/10.1016/j.crci.2014.01.010

Oun, A. A., & Rhim, J.-W. (2016). Characterization of nanocelluloses isolated from Ushar (*Calotropis procera*) seed fiber: Effect of isolation method. *Materials Letters, 168*, 146–150. https://doi.org/10.1016/j.matlet.2016.01.052

Oun, A. A., Shankar, S., & Rhim, J.-W. (2020). Multifunctional nanocellulose/metal and metal oxide nanoparticle hybrid nanomaterials. *Critical Reviews in Food Science and Nutrition, 60*(3), 435–460. https://doi.org/10.1080/10408398.2018.1536966

Panthapulakkal, S., & Sain, M. (2012). Preparation and characterization of cellulose nanofibril films from wood fibre and their thermoplastic polycarbonate composites. *International Journal of Polymer Science, 2012*, 1–6. https://doi.org/10.1155/2012/381342

Panthi, G., Park, M., Kim, H.-Y., Lee, S.-Y., & Park, S.-J. (2015). Electrospun ZnO hybrid nanofibers for photodegradation of wastewater containing organic dyes: A review. *Journal of Industrial and Engineering Chemistry, 21*, 26–35. https://doi.org/10.1016/j.jiec.2014.03.044

Pawcenis, D., Leśniak, M., Szumera, M., Sitarz, M., & Profic-Paczkowska, J. (2022). Effect of hydrolysis time, pH and surfactant type on stability of hydrochloric acid hydrolyzed nanocellulose. *International Journal of Biological Macromolecules, 222*, 1996–2005. https://doi.org/10.1016/j.ijbiomac.2022.09.289

Phanthong, P., Reubroycharoen, P., Hao, X., Xu, G., Abudula, A., & Guan, G. (2018). Nanocellulose: Extraction and application. *Carbon Resources Conversion, 1*(1), 32–43. https://doi.org/10.1016/j.crcon.2018.05.004

Rahman, M. R., James, A., Mohamed Said, K. A., Namakka, M., Khandaker, M. U., Jiunn, W. H., Al-Humaidi, J. Y., Althomali, R. H., & Rahman, M. M. (2024). A TiO_2 grafted bamboo derivative nanocellulose polyvinylidene fluoride (PVDF) nanocomposite membrane for wastewater treatment by a photocatalytic process. *Materials Advances, 5*(19), 7617–7636. https://doi.org/10.1039/D4MA00716F

Rathod, M., Moradeeya, P. G., Haldar, S., & Basha, S. (2018). Nanocellulose/TiO_2 composites: Preparation, characterization and application in the photocatalytic degradation of a potential endocrine disruptor, mefenamic acid, in aqueous media. *Photochemical & Photobiological Sciences, 17*(10), 1301–1309. https://doi.org/10.1039/c8pp00156a

Ravelli, D., Dondi, D., Fagnoni, M., & Albini, A. (2009). Photocatalysis. A multi-faceted concept for green chemistry. *Chemical Society Reviews, 38*(7), 1999. https://doi.org/10.1039/b714786b

Reiniati, I., Hrymak, A. N., & Margaritis, A. (2017). Recent developments in the production and applications of bacterial cellulose fibers and nanocrystals. *Critical Reviews in Biotechnology, 37*(4), 510–524. https://doi.org/10.1080/07388551.2016.1189871

Reshmy, R., Philip, E., Thomas, D., Madhavan, A., Sindhu, R., Binod, P., Varjani, S., Awasthi, M. K., & Pandey, A. (2021a). Bacterial nanocellulose: Engineering, production, and applications. *Bioengineered, 12*(2), 11463–11483. https://doi.org/10.1080/21655979.2021.2009753

Reshmy, R., Thomas, D., Philip, E., Paul, S. A., Madhavan, A., Sindhu, R., Binod, P., Pugazhendhi, A., Sirohi, R., Tarafdar, A., & Pandey, A. (2021b). Potential of nanocellulose for wastewater treatment. *Chemosphere, 281*, 130738. https://doi.org/10.1016/j.chemosphere.2021.130738

Reshmy, R., Philip, E., Madhavan, A., Pugazhendhi, A., Sindhu, R., Sirohi, R., Awasthi, M. K., Pandey, A., & Binod, P. (2022). Nanocellulose as green material for remediation of hazardous heavy metal contaminants. *Journal of Hazardous Materials, 424*, Article 127516. https://doi.org/10.1016/j.jhazmat.2021.127516

Salama, A. (2017). Dicarboxylic cellulose decorated with silver nanoparticles as sustainable antibacterial nanocomposite material. *Environmental Nanotechnology, Monitoring & Management, 8*, 228–232. https://doi.org/10.1016/j.enmm.2017.08.003

Salama, A., Abouzeid, R., Leong, W. S., Jeevanandam, J., Samyn, P., Dufresne, A., Bechelany, M., & Barhoum, A. (2021). Nanocellulose-based materials for water treatment: adsorption, photocatalytic degradation, disinfection, antifouling, and nanofiltration. *Nanomaterials, 11*(11), 3008. https://doi.org/10.3390/nano11113008

Salama, A., Neumann, M., Günter, C., & Taubert, A. (2014). Ionic liquid-assisted formation of cellulose/calcium phosphate hybrid materials. *Beilstein Journal of Nanotechnology, 5*, 1553–1568. https://doi.org/10.3762/bjnano.5.167

Sambyal, S., Chauhan, V., Shandilya, P., & Priye, A. (2025). Photocatalytic norfloxacin degradation enabled by a dual S-scheme nanocellulose-based Ag_2 WO_4 /NiO/MoO_3 tertiary heterojunction. *Catalysis Science & Technology, 15*(6), 1865–1881. https://doi.org/10.1039/D4CY01243G

Shabna, S., Singh, C. J. C., Dhas, S. D. S. J., Jeyakumar, S. C., & Biju, C. S. (2024). An overview of prominent factors influencing the photocatalytic degradation of cationic crystal violet dye employing diverse nanostructured materials. *Journal of Chemical Technology & Biotechnology, 99*(5), 1027–1055. https://doi.org/10.1002/jctb.7593

Shen, X.-J., Huang, P.-L., Chen, J.-H., Wu, Y.-Y., Liu, Q.-Y., & Sun, R.-C. (2017). Comparison of acid-hydrolyzed and TEMPO-oxidized nanocellulose for reinforcing alginate fibers. *BioResources, 12*(4), 8180–8198. https://doi.org/10.15376/biores.12.4.8180-8198

Sudhaik, A., Raizada, P., Ahamad, T., Alshehri, S. M., Nguyen, V.-H., Van Le, Q., Thakur, S., Thakur, V. K., Selvasembian, R., & Singh, P. (2023). Recent advances in cellulose supported photocatalysis for pollutant mitigation: A review. *International Journal of Biological Macromolecules, 226*, 1284–1308. https://doi.org/10.1016/j.ijbiomac.2022.11.241

Sun, H., Guo, Y., Zelekew, O. A., Abdeta, A. B., Kuo, D.-H., Wu, Q., Zhang, J., Yuan, Z., Lin, J., & Chen, X. (2021). Biological renewable nanocellulose templated CeO2/TiO2 synthesis and its photocatalytic removal efficiency of pollutants. *Journal of Molecular Liquids, 336*, Article 116873. https://doi.org/10.1016/j.molliq.2021.116873

Tan, J., Deng, H., Lu, F., Chen, W., Su, X., & Wang, H. (2023). Antibacterial nanocellulose-TiO_2/polyester fabric for the recyclable photocatalytic degradation of dyes. *Polymers, 15*(22), 4376. https://doi.org/10.3390/polym15224376

Tavker, N., & Sharma, M. (2018). Enhanced photocatalytic activity of nanocellulose supported zinc oxide composite for RhB dye as well as ciprofloxacin drug under sunlight/visible light. In *AIP conference proceedings* (p. 030013). https://doi.org/10.1063/1.5035215

Tavker, N., & Sharma, M. (2020). Designing of waste fruit peels extracted cellulose supported molybdenum sulfide nanostructures for photocatalytic degradation of RhB dye and industrial effluent. *Journal of Environmental Management, 255*, Article 109906. https://doi.org/10.1016/j.jenvman.2019.109906

Tee, G. T., Gok, X. Y., & Yong, W. F. (2022). Adsorption of pollutants in wastewater via biosorbents, nanoparticles and magnetic biosorbents: A review. *Environmental Research, 212*, Article 113248. https://doi.org/10.1016/j.envres.2022.113248

Thipchai, P., Sringarm, K., Punyodom, W., Jantanasakulwong, K., Thanakkasaranee, S., Panyathip, R., Arjin, C., & Rachtanapun, P. (2024). Production of nanocellulose from sugarcane bagasse and development of nanocellulose conjugated with polylysine for fumonisin B1 toxicity absorption. *Polymers, 16*(13), 1881. https://doi.org/10.3390/polym16131881

Toro, R. G., Adel, A. M., De Caro, T., Brunetti, B., Al-Shemy, M. T., & Caschera, D. (2022). A facile one-pot approach to the fabrication of nanocellulose-titanium dioxide nanocomposites with promising photocatalytic and antimicrobial activity. *Materials, 15*(16), 5789. https://doi.org/10.3390/ma15165789

Vasheghani Farahani, M. S., Nikzad, M., & Ghorbani, M. (2022). Fabrication of Fe-doped ZnO/nanocellulose nanocomposite as an efficient photocatalyst for degradation of methylene blue under visible light. *Cellulose, 29*(13), 7277–7299. https://doi.org/10.1007/s10570-022-04735-y

Wahid, F., Duan, Y.-X., Hu, X.-H., Chu, L.-Q., Jia, S.-R., Cui, J.-D., & Zhong, C. (2019). A facile construction of bacterial cellulose/ZnO nanocomposite films and their photocatalytic and antibacterial properties. *International Journal of Biological Macromolecules, 132*, 692–700. https://doi.org/10.1016/j.ijbiomac.2019.03.240

Yang, D., Li, L., Xu, W., Jia, Z., Lei, J., & Liu, J. (2025). Bamboo nanocellulose derived carbon loaded bismuth oxide: An effective nanocomposite for the solar photocatalytic degradation of malachite green. *Colloids and Surfaces A: Physicochemical and Engineering Aspects, 711*, Article 136315. https://doi.org/10.1016/j.colsurfa.2025.136315

Yang, J., Yu, J., Fan, J., Sun, D., Tang, W., & Yang, X. (2011). Biotemplated preparation of CdS nanoparticles/bacterial cellulose hybrid nanofibers for photocatalysis application. *Journal of Hazardous Materials, 189*(1–2), 377–383. https://doi.org/10.1016/j.jhazmat.2011.02.048

Yayayürük, A. E., & Yayayürük, O. (2021). Recent advances in environmental analysis towards green nanomaterials. *Current Analytical Chemistry, 17*(4), 449–460. https://doi.org/10.2174/1573411016999200719154826

Yu, W., Li, J., Li, C., Liu, W., & Zhang, S. (2025). Innovative ag-AgCl@TiO2@cellulose nanofiber porous composites with Z-scheme heterojunction for enhanced adsorption and photocatalytic degradation of formaldehyde. *Carbohydrate Polymers, 357*, Article 123505. https://doi.org/10.1016/j.carbpol.2025.123505

Yue, Y., Shen, S., Cheng, W., Han, G., Wu, Q., & Jiang, J. (2022). Construction of mechanically robust and recyclable photocatalytic hydrogel based on nanocellulose-supported CdS/MoS2/Montmorillonite hybrid for antibiotic degradation. *Colloids and Surfaces A: Physicochemical and Engineering Aspects, 636*, Article 128035. https://doi.org/10.1016/j.colsurfa.2021.128035

Zeng, J., Liu, S., Cai, J., & Zhang, L. (2010). TiO_2 immobilized in cellulose matrix for photocatalytic degradation of phenol under weak UV light irradiation. *The Journal of Physical Chemistry C, 114*(17), 7806–7811. https://doi.org/10.1021/jp1005617

Zhang, J., Si, M., Jiang, L., Yuan, X., Yu, H., Wu, Z., Li, Y., & Guo, J. (2021). Core-shell Ag@nitrogen-doped carbon quantum dots modified $BiVO_4$ nanosheets with enhanced photocatalytic performance under Vis-NIR light: Synergism of molecular oxygen activation and surface plasmon resonance. *Chemical Engineering Journal, 410*, Article 128336. https://doi.org/10.1016/j.cej.2020.128336

Zhao, X., Zhang, L., & Liu, D. (2012). Biomass recalcitrance. Part II: Fundamentals of different pre-treatments to increase the enzymatic digestibility of lignocellulose. *Biofuels, Bioproducts and Biorefining, 6*(5), 561–579. https://doi.org/10.1002/bbb.1350

Zimmermann, T., Pöhler, E., & Geiger, T. (2004). Cellulose fibrils for polymer reinforcement. *Advanced Engineering Materials, 6*(9), 754–761. https://doi.org/10.1002/adem.200400097

Zu, W., Jiang, C., Liu, W., Hou, Q., & Ji, X. (2022). Fabrication of a carbonized cellulose nanofibrils/$Ti_3C_2T_x$ MXene/g-C_3N_4 heterojunction for visible-light-driven photocatalysis. *Langmuir, 38*(43), 13109–13120. https://doi.org/10.1021/acs.langmuir.2c01818

Harnessing Nanomaterials for the Mitigation of Micro-nanoplastics

Saurabh Dubey, Srijita De, Musaddique Mahfuz Ahmed, Aman Paul, Rahul Deka, Debashreeta Singha, Omkar Suresh Deshmukh, and Dipankar Bandyopadhyay

Abstract The broad environmental spread of micro-nanoplastics (MNPs) has set off the need for innovative solutions for their management. Recently, nanomaterials have emerged as efficient plastic degrading agents because of their high specific surface area, ease of tunability, and excellent plasmonic properties. This chapter examines various nanomaterials that transform and reduce (MNPs) in the environment. We begin by analyzing widely used nanomaterials such as metal, metal oxide, carbon, and organic nanocomposites. We explain their degradation mechanism, including chemical oxidation, photothermal breakdown, and catalytic transformation. The chapter discusses how the presence of nanomaterials supports rapid degradation of (MNPs). The underlined phenomena such as Fenton-like reactions, photocatalysis, and surface plasmon resonance help break down (MNPs) into smaller, less harmful, mineralized end-products. In the end, the chapter discusses the ways of integrating some of such nanomaterials into filtration and capture systems for efficient environmental remediation.

Keywords Microplastics · Nanoplastics · Degradation · Nanomaterials · Environmental remediation

---

S. Dubey · S. De · D. Bandyopadhyay (✉)
Centre for Nanotechnology, Indian Institute of Technology Guwahati, Assam, India
e-mail: dipban@iitg.ac.in

M. M. Ahmed · A. Paul · D. Singha · O. S. Deshmukh · D. Bandyopadhyay
Department of Chemical Engineering, Indian Institute of Technology Guwahati, Assam, India

R. Deka · D. Bandyopadhyay
Jyoti and Bhupat Mehta School of Health Sciences and Technology, Indian Institute of Technology Guwahati, Assam, India

E. Priyadarshini et al. (eds.), *Advanced Nanotechnology Applications for Environmental Sustainability*, Sustainable Environmental Waste Management Strategies,
https://doi.org/10.1007/978-3-032-12933-8_10

1 Introduction

Microplastics are objects with one of the dimensions between 1 μm and 5 mm, while nanoplastics are objects smaller than 1 μm. The spread of micro-nanoplastics (MNPs) has negatively influenced ecosystems, be it soil, water, and air. These tiny objects originate from the products containing them as an ingredient known as primary MNPs. Sometimes, these tiny objects form because of the fragmentation of larger plastic debris known as secondary MNPs. Traditional MNP mitigation methods, such as mechanical filtration or chemical treatments, negatively influence the environment. Nanomaterials have at least one of their dimensions (length, width, or height) within the nanoscale, typically between 1 and 100 nm. To visualize a nanometer, it is one-millionth of a millimeter, about 100,000 times smaller than the diameter of a human hair. They are so tiny that they are invisible to the naked eye or with a conventional microscope. Scientists have engineered nanomaterials through atomic and molecular methods to generate properties that do not exist at bigger measurement scales.

Figure 1 pictorially outlines the content of the chapter that includes an array of nanomaterials that can mitigate environmental MNPs. The chapter starts with the common nanoparticles like metal, metal oxide, carbon, and organic nanocomposites before describing their degradation mechanisms, such as catalytic transformation, chemical oxidation, and photothermal breakdown (Priyadarsini & Das, 2025; Reddy Koteswara et al., 2025). Further, it elaborates the special cases wherein some nanoparticles promoting the quick breakdown of MNPs, Fenton-like reactions, photocatalysis, and surface plasmon resonance in breaking MNPs into smaller, less hazardous, mineralized end products (Vanarse et al., 2025). The chapter also discusses the potential of some of these nanomaterials into capture and filtration systems for effective environmental restoration.

2 Sources

Plastics are the second most cosmopolitan entity after humans, found almost everywhere. Primary microplastics are an essential class of plastic particles smaller than 5 mm in size (Hale et al., 2020). The utility of microplastics comes in handy in controlling the viscosity, stability, and morphological appearance of the respective product and how well it can respond to wear and tear (Guan et al., 2021). It can be traced in many fields, such as cosmetic items, formulation of cleaning items, and the fishery industry for preparing nets and turf (Giechaskiel et al., 2024; Patil et al., 2024). With the omnipresent nature of microplastics, there is evidence of leakage of its remnants in the environment (Das et al., 2024a, 2024b, 2024c, 2024d; Mohanty et al., 2024; Priyadarshini et al., 2024). Typically, such environmental discharge comes from (i) textile industries (e.g., synthetic fiber clothing materials), (ii) automobile industries (e.g., wear and tear of tires along the road) (Giechaskiel et al.,

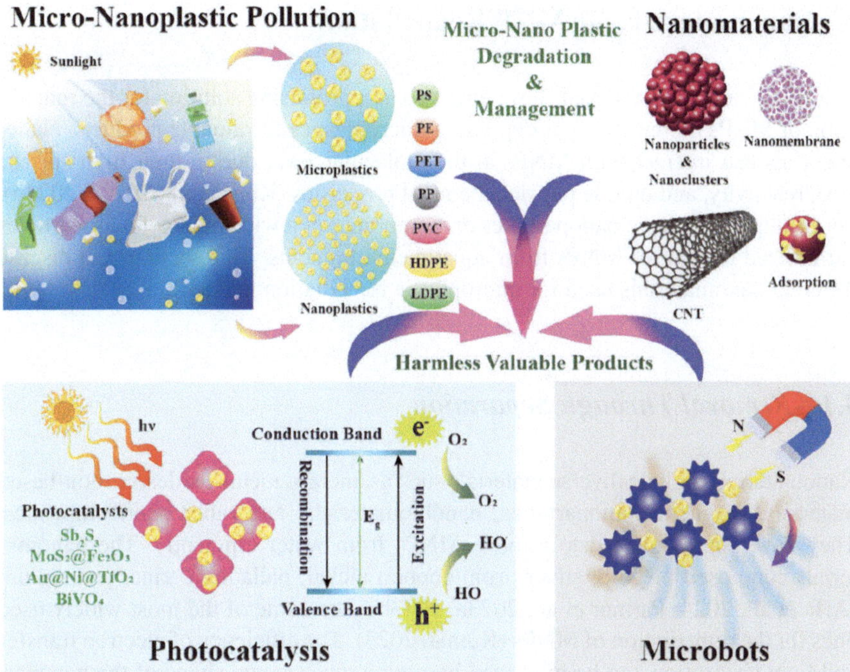

Fig. 1 The illustration shows the source of MNPs and nanomaterials used for the degradation and management

2024), (iii) paint industries (e.g., degradation of paint marking on walls/roads alike) (Fang et al., 2024).

As the name suggests, secondary microplastics arise from the action of specific extrinsic forces towards large-sized plastics suspended in the environment. For example, UV radiation exposure, thermal gradients, freezing temperatures, and mechanochemical friction-based degradation from oceanic waves reduce the number of macroplastics that come their way into microplastics (Dubey et al., 2022; Song et al., 2017). In the case of oceanic microplastic pollution, sources such as carrier ships, oceanic tourist activities, and crude oil drilling activities can become detrimental (Huber et al., 2022; Jaafarzadeh et al., 2024). Other indirect causes of microplastic generation can be through fishing activities wherein nylon-based nets are used, eventually contributing to microplastic pollution (Kane & Clare, 2019; Ziani et al., 2023). This naturally raises the question of the fate of MNPs generated in massive tonnes globally. The fate of MNPs can be governed by degradation processes such as UV photodegradation, mechanical wear and tear due to the action of strong river currents, microorganisms-mediated biodegradation, and chemical abrasion. This can lead to the leaching of MNPs into various strata of environment niches and affect both flora and fauna (Alimi et al., 2023; Auta et al., 2017; Kurniawan et al., 2023; Yang et al., 2021).

3 Nanomaterials for MNP Remediation

In recent years, nanomaterials have emerged as a promising solution for the remediation of MNPs by removing or degrading microplastics to valuable products. Nanomaterials can interact with MNPs at the molecular level due to their high surface area, reactivity, and unique physicochemical properties (Khan et al., 2024a, 2024b). For example, magnetic nanoparticles or membranes of a wide range of nanoparticles are utilized to remove MNPs from aquatic environments. This section will discuss different nanomaterials used for microplastic remediation in detail.

3.1 Removal Through Separation

Nanomaterials include diverse materials such as, metals, metal-oxides, carbon-based materials, magnetic nanoparticles, nanobiomaterials, and other nanocomposites. They have been explored to remove MNPs from water efficiently. The nanomaterials composed of gold, silver, iron, copper, nickel, palladium, zinc, or titanium (Arfa et al., 2024; Parmar et al., 2023a, 2023b)., are some of the most widely used ones for the remediation of MNPs (Kumar, 2023). The efficiency of electron transfer and light absorption can be improved by tuning the size and shape of the nanoparticles. π–π interaction facilitates the binding of polyethylene (PE), polypropylene (PP), and polystyrene (PS) MP/NP with silver nanoparticles (Li et al., 2020). This enhances the adsorption efficiency of MP/NP onto the surface of the particles, eventually improving the efficiency of MP removal or degradation. Similarly, MP/NP of PET, HDPE, and PS binds with the Fe_3O_4 because of hydrogen bonding, leading to higher adsorption efficiency (Zandieh & Liu, 2022).

The magnetic property of iron nanoparticles like Fe_3O_4 enhances the separation of MP/NP from aqueous medium and allows convenient particle recovery aided by an external magnetic field. Some other examples of magnetic nanomaterials used to remove microplastics are Fe–C–NH_2 composite nanoparticles (Zhou et al., 2022), iron oxide modified with a hydrophobic coating, or magnetic ZIF-8 nanocomposites, etc. (Yu et al., 2024). On the other hand, the graphene-based nanomaterials facilitate the efficient adsorption of MPs and NPs because of their distinctive properties like high surface area and tunable surface chemistry (Das et al., 2024a, 2024b, 2024c, 2024d). Further, the single-layered graphene, graphene oxide, reduced graphene oxides, and MXenes are also the widely used 2D nanomaterials in the membrane-based separation of MP/NP (Das et al., 2024a, 2024b, 2024c, 2024d). In addition, transition metal dichalcogenides, boron nitride, and molybdenum disulfides also find their application as a membrane for removing MP/NP (Goh et al., 2022).

3.2 *Catalytic Degradation*

Catalytic capabilities of the nanomaterials can break complex polymer structures into valuable chemicals and environmentally friendly molecules. In this regard, the transition metal oxide-based nanomaterials are the most widely explored catalyst for the degradation of MP/NP. In particular, TiO_2 is considered the most suitable catalyst because of its higher catalytic activity, thermal stability, and economic viability (Du et al., 2022). The heterojunction catalyst, Bi_2O_3@N–TiO_2, is also examined as a catalyst in the catalytic degradation of PET (Du et al., 2022). Similarly, photocatalysis draws significant attention from the scientific community as the process utilizes light to degrade MP/NP (Kumar, 2023). Photocatalytic processes influence light-activated nanomaterials to generate reactive oxygen species (ROS) (Sen et al., 2023), which initiate oxidative reactions that degrade microplastic polymers (Kumar, 2023). Further, anatase TiO_2 shows excellent photocatalytic activity with a 3–3.2 eV band gap among the three crystallographic phases (Du et al., 2022; Kumar, 2023). TiO_2 nanoparticles degrade LDPE, PP, PS, PE, PVC, etc., photo-catalytically irradiating UV light. ZnO, with a bandgap of 3.37 eV, which can be further tuned, acts as another excellent photocatalyst for the same purpose. A wide range of morphology, such as nanorods, nanotubes, nanoflowers, nano/microspheres, nanowires, nano prisms, nanobelts, nanosheets, etc., of ZnO, is maneuvered for photocatalytic degradation of MP/NP (Kumar, 2023). Some other examples of photocatalysts for MP/NP degradation are NiO, Cu_2O, CuO, CdS, and composite nanomaterials TiO_2-rGO, TiO_2–Fe$(St)_3$, TiO_2/nano-graphite, composites, Fe–ZnO, etc. (Indhur et al., 2023; Khan et al., 2024a, 2024b; Kumar, 2023; Yu et al., 2024; Zhou et al., 2022). The MOFs are crystalline metal clusters with high porosity and are composed of metal-oxide clusters and organic linkers. They are suitable for MP adsorption because of their vast surface areas per unit volume. During the synthesis, the change of metal oxides and organic linkers can control MOFs' pore size, volume, and functionality. Different studies have explored using Zr-based UiO-66-X MOFs, MIL-100(Fe), and ED-MIL 101(Cr) UF membranes to remove MP from water and wastewater (Indhur et al., 2023).

4 Degradation Mechanisms

Nanomaterials employ distinct mechanisms to degrade MNPs: photocatalytic degradation, Fenton-like reaction, adsorption-driven process, and thermal and photothermal effect, as shown in Fig. 2.

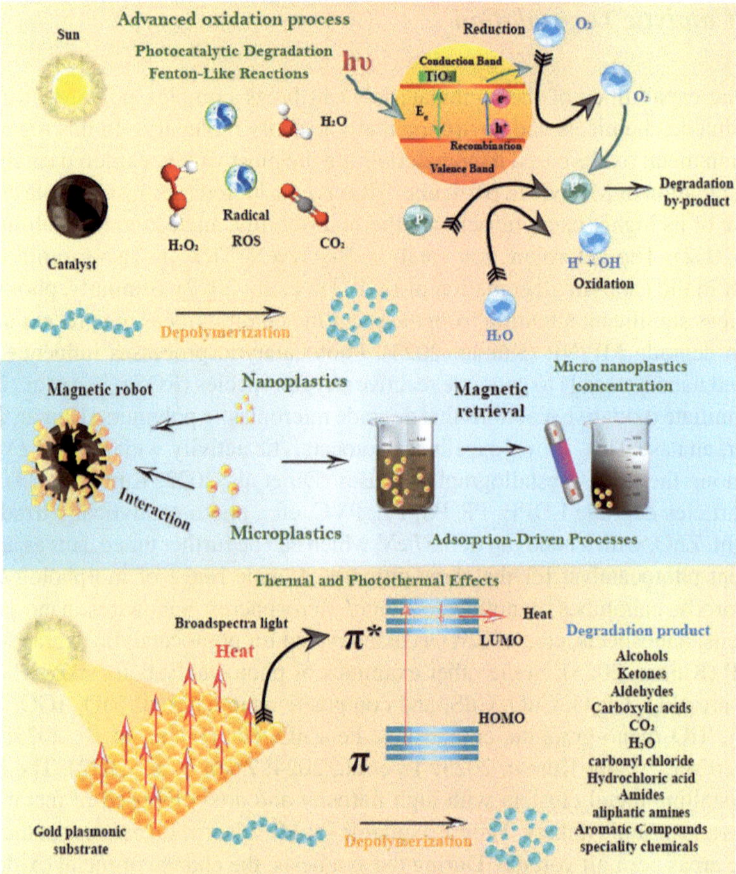

Fig. 2 The figure shows various mechanisms for the degradation of MNPs using advanced oxidation processes

4.1 *Photocatalytic Degradation*

This degradation mechanism cleaves MNPs into smaller, less harmful molecules. This can be achieved using photocatalysts under light irradiation (Gusain et al., 2019; Khan et al., 2024a, 2024b; Koe et al., 2019) the following steps are involved in the photocatalytic degradation process.

- Light irradiation (UV/visible) excites electrons in the photocatalyst.
- Generation of the electron–hole pair.
- Formation of reactive oxygen species (ROS) like hydroxyl radicals (·OH) and superoxide radicals $\left(O_2^{\cdot-}\right)$.

- Attack of the ROS on the entangled polymer chain, leading to chain scission and oxidation into various specialty chemicals such as organic molecules, CO_2, or H_2O.

Some examples include the decomposition of polyethylene and polypropylene into carbon dioxide and water. Various factors affect the efficiency of these reactions, such as microplastic type, light source, property of the catalyst used, and environmental conditions (Ganesan et al., 2024; Koe et al., 2019).

4.2 Fenton-Like Reactions

These are advanced oxidation processes promising for microplastic degradation due to their strong oxidative power (Ortiz et al., 2022). In this process, reactive oxygen species (ROS) generation occurs (Meyerstein, 2021; Walling et al., 2021; Xiao et al., 2024). MNPs and other organic pollutants undergo their breakdown under the action of ROS. Catalytic decomposition of hydrogen peroxide (H_2O_2) in the presence of transition metals such as Fe, Cu, Mn, or Co. Fe-based nanoparticles catalyze the production of hydroxyl radicals from hydrogen peroxide, attacking plastic polymers. Various factors affect the efficiency of these reactions, such as pH, temperature, type of metal catalyst, peroxide concentration, and dissolved oxygen. Due to incomplete degradation, some toxic byproducts may also be produced from different source polymers, as listed in the table.

4.3 Adsorption-Driven Processes

Overcoming microplastic removal challenges exists through the application of Carbon nanotubes together with different nanomaterials as adsorptive agents (Ganesan et al., 2024). While highly effective due to their large surface area and intense interactions with plastic particles. The interaction between microplastic particles and nanomaterial surfaces occurs through intermolecular forces during these processes (Chazovachii et al., 2021). Nanomaterials create both hydrophobic forces and van der Waals interactions with microplastics, which causes their removal through aggregation (Zhou et al., 2022).

The interaction between microplastic particles and nanomaterial surfaces occurs through intermolecular forces during these processes (Chazovachii et al., 2021). Nanomaterials generate both hydrophobic interactions and van der Waals interactions with microplastics, and it causes them to be removed from the solution through aggregation (Zhou et al., 2022). Most MNPs exhibit hydrophobic characteristics due to their water-repellent properties and their belonging to polystyrene groups such as polyethylene and polypropylene. For example, also in this regard, carbon nanotubes and graphene oxide possess strong hydrophobic properties which allow them to remove microplastics from water solutions efficiently (Wang et al., 2023). In effect, when hydrophobic nanomaterials meet microplastics, they are less desirable to remain in the water and more desirable to attach to the hydrophobic nanomaterial.

Weak intermolecular forces named van der Waals Forces develop from the molecular interaction of induced dipole charges between molecules. Nanomaterials bond microplastic particles using the weaker van der Waals forces despite their lower strength than covalent and ionic bonds (Gusain et al., 2019). Carbon nanomaterials enhance these forces by increasing their surface area and utilizing π–π interactions, which leads to microplastic aggregation that results in removal from water systems like CNTs do in water systems.

4.4 Thermal and Photothermal Effects

Thermal and photothermal processes utilize the heat generated by nanoparticles (NPs) upon light absorption to accelerate the degradation of MNPs. This technique is particularly effective for breaking down persistent plastic polymers into smaller, less harmful fragments. Nanoparticles as Photothermal Agents, such as carbon nanotubes (CNTs), graphene-based materials, and plasmonic nanoparticles (e.g., gold (Au) and silver (Ag)), exhibit strong light absorption properties. When exposed to infrared (IR), visible, or ultraviolet (UV) radiation, these nanoparticles efficiently convert absorbed photons into thermal energy (Overchuk et al., 2023). Localized Heating Effect Unlike traditional heating, which requires large-scale energy input, photothermal degradation allows for targeted heating at the nanoscale. This localized heat generation enhances the thermal decomposition of MNPs while minimizing unnecessary energy dissipation (Cui et al., 2023).

The photothermal effect induces thermo-oxidative degradation, where the elevated temperature disrupts polymeric bonds. Long-chain hydrocarbons in plastics undergo chain scission reactions, breaking into shorter chains or monomers. The process can also generate radicals that further degrade the plastic structure. Carbon nanotubes (CNTs) are widely used for photothermal applications due to their high optical absorption and thermal conductivity (Overchuk et al., 2023). Table 1 classifies the probable end products after the degradation of different plastic debris.

When CNTs are exposed to near-infrared (NIR) radiation, they absorb the energy and rapidly convert it into heat. This localized heating effect can raise the surrounding temperature to several hundred degrees Celsius, melting and degrading MNPs nearby. Studies have shown that CNT-assisted photothermal degradation of polyethylene (PE) and polystyrene (PS) under IR radiation leads to significant polymer fragmentation (Overchuk et al., 2023). The main advantages of photothermal degradation are non-contact degradation, controlled degradation, and scalability. Photothermal degradation using nanomaterials like CNTs represents a highly efficient, selective, and energy-saving approach for microplastic breakdown (Cui et al., 2023).

Table 1 Classification of polymer type, initial product, intermediate product, and final product after undergoing polymer degradation via advanced oxidation processes

Photocatalytic degradation

Polymer type	Initial product	Intermedia oxidation products	Final products	Toxic byproducts
Polyethylene (PE) and polypropylene (PP)	Alkane and alkene fragments (C_2H_4, C_3H_6, C_4H_8)	Alchols, ketones, aldehydes (e.g., formaldehyde, acetone, propanal)	Carboxylic acids (formic acid, acetic acid), CO_2, H_2O	Formaldehyde and acetaldehyde
Polystyrene (PS)	Styrene monomers, dimers, and oligomers	Benzaldehyde, benzoic acid, phenol, styrene oxide	CO_2, H_2O	Aromatic compounds
Polypropylene terephthalate (PET)	Terephthalic acid, ethylene glycol	Hydroxyterephthalic acid, benzoic acid, benzophenone	CO_2, H_2O	Aromatic compounds
Polyvinyl chloride (PVC)	Vinyl chloride monomers, chlorinated hydrocarbons (e.g., chloroethane, dichloroethane)	Hydrochloric acid (HCl), carbonyl chloride (phosgene), aldehydes	CO_2, H_2O	Hydrochloric acid (HCl), dioxins, and furans
Polyamide (Nylon)	Caprolactam (for Nylon-6), hexamethylenediamine, adipicacid (for nylon-6,6)	Amides, carboxylic acids, aliphatic amines	CO_2, H_2O	Carbon monoxide, NH_3, HCN, and pyridines

Fenton-like reactions

Polymer type	Final products
Polyethylene (PE)/polypropylene (PP)	Alkane, alkenes, ketones, aldehydes, and carboxylic (CO_2 and H_2O)
Polystyrene (PS)	Styrene monomers, benzaldehyde, benzoic acid, and phenol
Polyethylene terephthalate (PET)	Terephthalic acid, benzoic acid, and smaller organic acids
Polyvinyl chloride (PVC)	Chlorinated hydrocarbons and HCl

4.5 *Nanomaterials for Detection and Removal of Microplastics*

Nanoplastics are ubiquitous in everything from water bodies to living organisms. Therefore, detecting nanoplastics becomes crucial for curbing the threats of deadly diseases they pose to living organisms. Sophisticated high-end microscopic techniques like field emission scanning electron microscopy (FESEM) and atomic force microscopy (AFM) are essential for viewing MNPs surface morphology and topographical features. In order to determine their intricate chemical structure and the bond types involved, it is further characterized by Raman spectroscopy and FTIR. Surface-enhanced Raman Spectroscopy (SERS) is a powerful analytical tool that enhances the Raman signals of MNPs. Recently, a plasmonic substrate fabricated using gold and is characterized by its dense structure of inverted pyramidal cavities. This method detects microplastics as tiny as 0.36 μm in air and 0.45 μm in water with a 100-fold signal enhancement (Xu et al., 2020).

Another group of researchers utilized silver colloid particles to develop a SERS method to identify polystyrene (PS), polyethylene (PE), and polypropylene (PP) in both pure water and seawater. The test conditions were optimized to enhance detection upon aggregation of colloid particles with NaCl. 100 and 500 nm plastic particles were detected with an optimal SERS enhancement factor of 4×10^6 (Lv et al., 2020). Recently, 2D nanomaterials have played a crucial role in curbing microplastic pollution. Nanomaterials like graphene, transition metal dichalcogenides (TMDCs), boron nitride, and molybdenum disulfide (MoS_2) produce superior 2D membranes for highly efficient filtration processes. These materials are insoluble in water. They are dispersed in organic solvents, and their pore size is in nanometres, allowing them to trap MNPs in their membranes. Graphene has atomic thinness with a surface area that allows for efficient filtration. The presence of lamellar structures of TMDCs enhances microplastic intercalation. It is highly stable and has a significant amount of mechanical strength. The hexagonal lattice structure of boron nitride is used in the adsorption of microplastic, and it is chemically inert. MoS_2 portrays good stability and is very selective in microplastic removal as it has sizeable interlayer spacing, aiding in better adsorption efficiency (Das et al., 2024a, 2024b, 2024c, 2024d).

The coral-inspired nano-material-based adsorbents are one of the effective ways to detect MNPs in water bodies. For instance, mesoporous magnetic Fe_3O_4 nanoparticles were coated with adhesive polymerized dopamine (PDA), yielding Fe_3O_4@PDA absorbents. The absorbent formed had increased the efficiency of microplastic removal by 30% compared to pristine Fe_3O_4 nanoparticles. The increase in efficiency of the composite is attributed to the formation of hydrogen bonding, p–p stacking, and hydrophobic interactions between the microplastic and the composite. Adding the PDA to the nanoparticle resulted in a fast and excellent adsorption process and better environmental storage of Fe_3O_4 nanoparticles (Li et al., 2024). The emergence of a quantum dot (QD) based fluorescent probe for detecting MNPs in complex matrices offers a superior alternative to traditional detection techniques such as FTIR and Raman spectroscopy. The FTIR struggles with sensitivity when the particle size is smaller and for high organic content, while in Raman spectroscopy, the signals are masked by fluorescence interference in biological samples. Therefore, the QDs are semiconductors with extraordinary optical characteristics, strong photoluminescence, and size-dependent emission, which turn them into efficient fluorescent probes. Their nanoscale size and diverse functionality for selective binding to microplastic allow signal amplification, catering to precise detection in biological and environmental settings. The signal-to-noise ratio is enhanced by 50% for hydrophilic

QDs, enabling the detection of microplastics as tiny as 0.2 μm in biological and 0.5 μm in environmental samples (Adebiyi & Ajana, 2024).

4.6 Risks of Nanomaterials

The nanomaterials impart unique features due to their sizes and are revolutionary in plastics mitigation. However, they have been reported to have potentially harmful effects on humans and the environment. Nanoparticles can permeate living organisms' physiological barriers through different routes, such as inhalation, ingestion, or skin. Exposure of nanoparticles to the human body initiates unfavorable biological reactions and toxic effects. Effects include building up oxidative stress, leading to inflammation, cytotoxicity, genotoxicity, and cell death (Kumah et al., 2023). The excessive formation of reactive oxygen species increases cell membrane penetration, releasing the cytochrome protein. This causes protein activity imbalance and events, resulting in cell apoptosis. The toxicity level of nanoparticles varies with their size, composition, crystallinity, hydrophobicity, and aggregation. For example, the hazardous effect of carbon nanofibers and carbon nanoparticles is higher than that of carbon nanotubes (Magrez et al., 2006).

The nanoparticles such as Fe_3O_4, ZnO, and Ag have tremendous potential in plastic degradation. Despite their notable contribution, different studies have been reported on the toxic effects produced by these nanoparticles. For example, when introduced to Ag nanoparticles, epithelial cells, and macrophages display changes in antioxidant defense and homeostasis balance (Zhang et al., 2015). The effect of nanoparticles is also evident in human dermal and cervical cell lines. They cause a decrease in glutathione levels and increase cell membrane damage (Mukherjee et al., 2012). Nanoparticles such as TiO_2, SiO_2, and Ag induce neurotoxicity, membrane rupture, and autophagy dysfunction in human neuroblastoma (Kumah et al., 2023). Carbon-based nanoparticles, such as carbon nanotubes employed in drug delivery, should be scrutinized before application. Single and multi-walled carbon nanotubes develop changes in cell morphology, lessen cell viability, disrupt cell-to-cell interactions in human lung epithelial cells, and induce fibrosis (Magrez et al., 2006).

The environment has been another vulnerable host to the disposal of nanoparticles directly or indirectly. Nanoparticles are emitted to the environment either during the fabrication of nano-enabled products, their usage, or while discarding them. Findings have been reported on nanoparticles' impact on aquatic and soil environments. ZnO nanoparticles display hazardous effects in plants, such as altering oxidant-antioxidant metabolites, affecting carbon/nitrogen metabolism, and inhibiting mitosis (Kang et al., 2024). The nanoparticles reach the plant organs via absorption and accumulate in the tissues, resulting in plant nutrient imbalance and photosynthesis perturbation. This generates abiotic stress, triggering ROS overproduction by perturbing cellular redox homeostasis. ZnO affects the overall growth and development of plants by reducing seed germination, inhibiting rhizome elongation, decreasing biomass, and phenotypic structural damage. Terrestrial and aquatic microorganisms like bacteria, fungi, and others. They are used to maintain nutrient cycles and purification in soil and water, respectively, and are significantly affected when introduced to TiO_2 nanoparticles (Fiordaliso et al., 2022). In marine microalgae, TiO_2 nanoparticles cause DNA damage and physiological inhibition, leading to imbalanced aquatic ecosystems. These nanoparticles are also reported to adversely affect organisms like zebrafish, carp, fathead minnow, and rainbow trout. Liquid peroxides form on cell membranes, eventually leading to membrane and tissue damage. Studies on nematodes showed that TiO_2 exposure altered reproduction, development, and growth (Fiordaliso et al., 2022).

4.7 Mitigation Strategies

Multiple mitigation strategies are available for removing MNPs, including screening, sedimentation, filtration, membrane bioreactors, reverse osmosis, electrocoagulation, and microbots, to name a few. An antimony sulfide (Sb_2S_3)-based microbot is synthesized by functionalizing magnetite nanoparticles for microplastic degradation and removal from the water environment. The microbots are operated via magnetic field and light irradiation, working orthogonally independently. The rotating magnetic field enables the microbots to capture poly(3-hydroxybutyrate) (PHB) and poly(lactic acid) microplastics. The photocatalytic activity of Sb_2S_3 nanoparticles allows the bots to degrade microplastics under UV irradiation without the need for any external fuel source (Jancik-Prochazkova et al., 2023). Another magnetic core–shell $MoS_2@Fe_2O_3$ micromotors are developed to eliminate polystyrene efficiently (PS) microplastics.

Micromotors exhibit photophoretic motion under solar light and degrade microplastics due to the effect of reactive oxygen species (ROS) production and localized heat generation (de la Asunción-Nadal et al., 2024). Figure 3a shows the photocatalytic $Au@Ni@TiO_2$ micromotors are actuated by UV light using active–passive particle interaction to remove microplastics in H_2O and H_2O_2 solutions. While individual micromotors remove microplastics by phoretic motion, assemblies of micromotors depend on shoveling and pushing interactions for microplastic elimination, stimulated by an external magnetic field (Wang et al., 2019). A similar microbot is prepared with another material, bismuth vanadate ($BiVO_4$) (Beladi-Mousavi et al., 2021). Using an adsorption–desorption mechanism, a sponge is prepared from chitin and graphene oxide to remove neutrally, positively, and negatively charged PS microplastics from wastewater.

The key reasons for PS adsorption are hydrogen bond interactions, electrostatic interactions, π–π interactions, and intra-particle diffusion (Sun et al., 2020a, 2020b). Xie and his research team report that a "hedgehog" shaped sunflower pollen grains-based microsubmarines are functionalized with a nickel layer for magnetic actuation in the application of oil and microplastic removal. These microsubmarines have

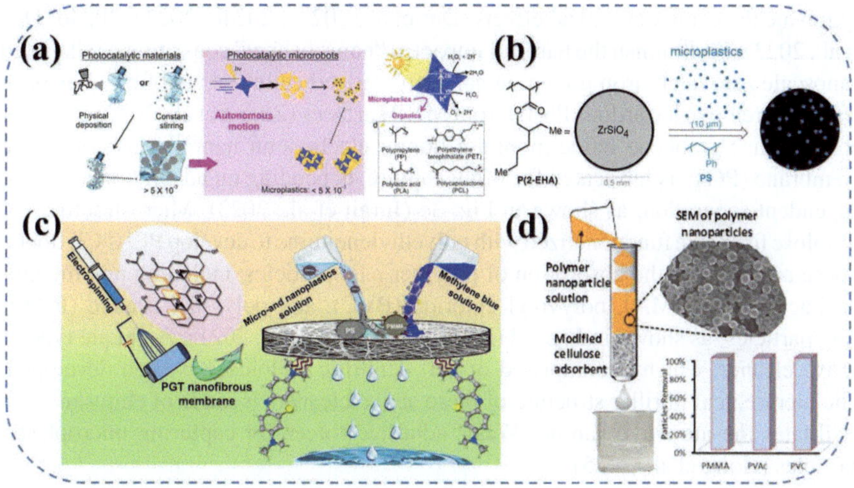

Fig. 3 Figure **a** shows photocatalytic mechanisms for the degradation (Wang et al., 2019), **b** shows adhesive microbot (Chazovachii et al., 2021), **c** illustrates the electrospun fabricated membrane (Juraij et al., 2023), and **d** depicts the modified cellulose adsorbent for the capture of micro-nano plastics (Batool & Valiyaveettil, 2021)

highly ordered spikes, nanopores, and superoleophilicity, making them more affinitive toward oil and plastic absorption (Sun et al., 2020a, 2020b). Different types of coagulants, including magnesium hydroxide, iron (III) chloride, and aluminum chloride combined with anionic polyacrylamide (PAM), are utilized for the absorption of microplastics, especially polyethylene (PE) from water (Li et al., 2022a, 2022b; Ma et al., 2019). Figure 3b shows the Zirconium silicate beads coated with poly(2-ethylhexyl acrylate) are used as adhesives for removing 99% PS microplastics in aqueous suspensions (Chazovachii et al., 2021). Researchers have developed a unique multilayered MXene-derived γ-Fe_2O_3/Pt/TiO_2 microrobots to capture nanoplastics. The microbot is synthesized by thermal annealing of $Ti_3C_2T_x$ MXene microparticles, followed by the deposition of a Pt layer and the surface decoration with magnetic γ-Fe_2O_3 nanoparticles. Due to these microbots' negative photogravitaxis and programmable zeta potential, they can capture nanoplastics in 3D space under UV light irradiation (Urso et al., 2022).

A self-driven magneto robot made of ion-exchange resin microspheres functionalized with superparamagnetic Fe_3O_4 nanoparticles. They have been utilized to remove and separate MNPs. Driven by self-induced ionic gradient and phoretic force (Li et al., 2022a, 2022b). On a sustainable note, scientists have designed a bioCap based on the surface functionalization of wood sawdust with polyphenols for the capture of various plastics, including PMMA, PE, PVC, polypropylene (PP), and polyethylene terephthalate (PET) MNPs. Here, the driving force includes strong and multiple molecular interactions (Wang et al., 2023). A plant-based hygroscopic nanocellulose network has been established to quantify colloidal MNPs, facilitated by the influence of pH and ionic strength (Leppänen et al., 2022). MNPs act as vectors for heavy metals, which aggravate the ecotoxicological effects (Das et al., 2024a, 2024b, 2024c, 2024d; Dey et al., 2023). To diminish the harmful impact, a "nano-fishnet" consisting of alkylated nanoscale zerovalent iron grown on naturally derived cellulose nanofibers has been crafted to remove hazardous Pb(II) from MNP surfaces (Zhang et al., 2023). Another polyurethane/graphene oxide-montmorillonite electrospun nanofibrous composite membrane (PGT) is fabricated for MNP capture, depending on adsorption and size-dependent separation, as shown in Fig. 3c (Juraij et al., 2023). Microstructures of cellulose fibers are functionalized with polyethyleneimine to develop PEI@CE fibers. These are used for the absorption of polymer nanoparticles, including polymethyl-methacrylate (PMMA), polyvinyl chloride (PVC), and polyvinyl acetate (PVAc) nanoparticles, as shown in Fig. 3d (Batool & Valiyaveettil, 2021). A unique type of active cleaner, which is comprised of soft dendritic colloids, has been developed. The hierarchical fibrillar structure of these active cleaners is made of chitosan. This facilitates the attractive van der Waals adhesive forces for capturing microplastic particles (Hong et al., 2025).

4.8 Future Directions and Challenges

Photocatalytic degradation presents a promising method for breaking down microplastics, but challenges remain in improving efficiency, selectivity, and environmental safety. Future research should focus on optimizing photocatalyst design, reaction conditions, and real-world applications to address plastic pollution effectively. Additionally, improving catalyst properties such as surface area, bandgap energy, and doping remains challenging for maximizing degradation efficiency. Complete mineralization of microplastics ideally leads to CO_2 and H_2O, but partial degradation often generates toxic intermediates that can persist in the environment. Research should push to optimize photocatalysts and reaction conditions to minimize harmful byproducts.

Understanding kinetics and degradation mechanisms is needed as different microplastic types (e.g., PE, PS, PP, PET) degrade at varying rates, necessitating tailored degradation strategies. The role of light sources requires further exploration, particularly in enhancing the efficiency of visible-light-responsive catalysts as energy-efficient alternatives to UV degradation. Environmental conditions, including pH, temperature, dissolved oxygen levels, and natural organic matter, introduce variability, requiring deeper investigation into their combined effects. Incomplete mineralization remains a concern, as MPs often fragment into smaller particles rather than fully degrading. The formation of toxic byproducts, including oxidation-derived intermediates such as aldehydes, ketones, and oligomers, necessitates further investigation to minimize environmental risks. Scaling up degradation strategies from controlled laboratory settings to real-world environmental systems requires optimization to enhance efficiency and feasibility. Developing non-toxic, eco-friendly, and recyclable photocatalysts is essential for ensuring sustainable and practical degradation solutions.

Efficient recovery of CNTs from water after capturing microplastics remains challenging, as it must be achieved without causing secondary contamination. The potential ecotoxicity of CNTs is another concern, as their nanoscale size and bioaccumulation risks could threaten aquatic organisms if released into the environment. Efficient recovery and separation of CNTs or metal nanoparticles after degradation are essential to prevent environmental accumulation. Future research should address the formation of toxic byproducts during oxidation, as it can generate harmful intermediates such as formaldehyde, benzene, and chlorinated compounds. The focus should be on improving nanoparticle stability and reusability, as prolonged exposure to high temperatures can lead to degradation. Incomplete degradation remains a concern, potentially resulting in micro-sized fragments or hazardous intermediates forming. Utilization of hybrid photothermal systems by integrating CNTs with plasmonic nanoparticles (e.g., Au, Ag) or semiconductors (e.g., TiO_2, ZnO) to enhance degradation efficiency.

5 Conclusions

There are rising concerns related to micro nanoplastics (MNPs) in almost all sorts of places, be it water, soil, or air. These MNPs have even penetrated human internal organs, making it challenging to address them in situ. This chapter revolves around the problems associated with MNPs and their adverse health impacts. Here, more emphasis is given to nanomaterial-based solutions to MNPs mitigation and degradation problems. The nanomaterials-driven approach efficiently addresses this crisis through innovative and sustainable environmental strategies. The unique properties of nanomaterials include high specific surface area, increased reactivity, ease of functionalization, plasmonic properties, and electronic properties. These properties support the wide use of these materials for solving MNPs degradation problems. Nanomaterials such as metal (e.g., Fe), metal oxide (e.g., TiO_2), non-metal, nonmetal oxide, semiconductors, carbon-based nanomaterials (e.g., graphene oxide, carbon nanotubes), nanocomposites, nanomaterials as nanostructures, and some organic nanocomposites have been discussed in detail while some of the mechanisms of degradation have also been explained thoroughly.

The possible mechanistic pathways, such as photothermal breakdown, chemical oxidation, catalytic transformations, e.g., Fenton, Fenton-like reactions, and photocatalytic degradation, have been systematically discussed. The surface plasmon resonance induced heat generation, leading to the molecular disintegration of MNPs into less harmful products or by-products, e.g., organic compounds, mineralized end products, and gasses. The in-depth discussion about how incorporating nanomaterials accelerates the chemical, photothermal, photocatalytic, and catalytic breakdown processes, changing harmful MNPs into less harmful by-products. Finally, with nanomaterials, it is possible to achieve efficient degradation, precise detection, and effective removal of MNPs.

References

Adebiyi, V. A., & Ajana, M. O. (2024). Quantum dot fluorescence for high-sensitivity microplastic detection in biological and environmental samples: A comprehensive review. https://doi.org/10.2139/SSRN.5025210

Alimi, O. S., Claveau-Mallet, D., Lapointe, M., Biu, T., Liu, L., Hernandez, L. M., Bayen, S., & Tufenkji, N. (2023). Effects of weathering on the properties and fate of secondary microplastics from a polystyrene single-use cup. *Journal of Hazardous Materials, 459*. https://doi.org/10.1016/j.jhazmat.2023.131855

Arfa, N., Parmar, P. R., & Bandyopadhyay, D. (2024). Self-organized synthesis of nanoserpent clusters on templated starch for sensitive detection of pancreatic α-amylase. *ACS Sustainable Chemistry and Engineering, 12*(13), 5075–5085. https://doi.org/10.1021/ACSSUSCHEMENG.3C06403

Auta, H. S., Emenike, C. U., & Fauziah, S. H. (2017). Distribution and importance of microplastics in the marine environmentA review of the sources, fate, effects, and potential solutions. *Environment International, 102*. https://doi.org/10.1016/j.envint.2017.02.013

Batool, A., & Valiyaveettil, S. (2021). Surface functionalized cellulose fibers—A renewable adsorbent for removal of plastic nanoparticles from water. *Journal of Hazardous Materials, 413*, Article 125301. https://doi.org/10.1016/J.JHAZMAT.2021.125301

Beladi-Mousavi, S. M., Hermanová, S., Ying, Y., Plutnar, J., & Pumera, M. (2021). A maze in plastic wastes: Autonomous motile photocatalytic microrobots against microplastics. *ACS Applied Materials and Interfaces, 13*(21), 25102–25110. https://doi.org/10.1021/ACSAMI.1C04559/SUPPL_FILE/AM1C04559_SI_005.AVI

Chazovachii, P. T., Rieland, J. M., Sheffey, V. V., Jugovic, T. M. E., Zimmerman, P. M., Eniola-Adefeso, O., Love, B. J., & McNeil, A. J. (2021). Using adhesives to capture microplastics from water. *ACS ES and T Engineering, 1*(12), 1698–1704. https://doi.org/10.1021/ACSESTENGG.1C00272/ASSET/IMAGES/LARGE/EE1C00272_0004.JPEG

Cui, X., Ruan, Q., Zhuo, X., Xia, X., Hu, J., Fu, R., Li, Y., Wang, J., & Xu, H. (2023). Photothermal nanomaterials: A powerful light-to-heat converter. *Chemical Reviews, 123*(11), 6891–6952. https://doi.org/10.1021/ACS.CHEMREV.3C00159/ASSET/IMAGES/LARGE/CR3C00159_0029.JPEG

Das, A., Dey, S., Bala, K., Gnanasekaran, R., & Das, A. P. (2024a). Microbial bioremediation technology for sustainable treatment and management of synthetic microfiber waste. *Sustainable Microbial Technology for Synthetic and Cellulosic Microfiber Bioremediation, 221–238*. https://doi.org/10.1007/978-3-031-62600-5_12

Das, A., Dey, S., & Das, A. P. (2024b). Ecotoxicological impacts of synthetic microfiber pollutants and development of sustainable mitigation strategies. *Environmental Chemistry and Ecotoxicology*. https://doi.org/10.1016/J.ENCECO.2024.12.003

Das, A., Ray, S. K., Mohanty, M., Mohanty, J., Dey, S., & Das, A. P. (2024c). Ecotoxicity of microplastic wastes and their sustainable management: A review. *Environmental Chemistry and Ecotoxicology, 6*, 144–152. https://doi.org/10.1016/J.ENCECO.2024.05.003

Das, T. K., Basak, S., & Ganguly, S. (2024d). 2D nanomaterial for microplastic removal: A critical review. *Chemical Engineering Journal, 492*, Article 152451. https://doi.org/10.1016/J.CEJ.2024.152451

de la Asunción-Nadal, V., Solano, E., Jurado-Sánchez, B., & Escarpa, A. (2024). Photophoretic MoS_2–Fe_2O_3 Piranha micromotors for collective dynamic microplastics removal. *ACS Applied Materials and Interfaces, 16*(36), 47396–47405. https://doi.org/10.1021/ACSAMI.4C06672/ASSET/IMAGES/LARGE/AM4C06672_0005.JPEG

Dey, S., Tripathy, B., Kumar, M. S., & Das, A. P. (2023). Ecotoxicological consequences of manganese mining pollutants and their biological remediation. *Environmental Chemistry and Ecotoxicology*. https://doi.org/10.1016/J.ENCECO.2023.01.001

Du, H., Wang, Q., Chen, G., & wang, J. (2022). Photo/electro-catalytic degradation of micro- and nano-plastics by nanomaterials and corresponding degradation mechanism. *TrAC Trends in Analytical Chemistry, 157*, 116815. https://doi.org/10.1016/J.TRAC.2022.116815

Dubey, S., Parmar, P. R., Vanarse, V. B., & Bandyopadhyay, D. (2022). Instability facilitated Micro Nano-plastic disintegration under thermal influence. In *Proceedings of CUChE Alumni Symposium 2022 on "Circular Economy on Sustainable Basis: The Role of Chemical Engineers."*

Fang, C., Zhou, W., Hu, J., Wu, C., Niu, J., & Naidu, R. (2024). Paint has the potential to release microplastics, nanoplastics, inorganic nanoparticles, and hybrid materials. *Environmental Sciences Europe, 36*(1). https://doi.org/10.1186/s12302-024-00844-6

Fiordaliso, F., Bigini, P., Salmona, M., & Diomede, L. (2022). Toxicological impact of titanium dioxide nanoparticles and food-grade titanium dioxide (E171) on human and environmental health. *Environmental Science: Nano, 9*(4), 1199–1211. https://doi.org/10.1039/D1EN00833A

Ganesan, S., Kokulnathan, T., Sumathi, S., & Palaniappan, A. (2024). Efficient photocatalytic degradation of textile dye pollutants using thermally exfoliated graphitic carbon nitride (TE–g–C_3N_4). *Scientific Reports, 14*(1), 1–13. https://doi.org/10.1038/s41598-024-52688-y

Giechaskiel, B., Grigoratos, T., Mathissen, M., Quik, J., Tromp, P., Gustafsson, M., Franco, V., & Dilara, P. (2024). Contribution of road vehicle tyre wear to microplastics and ambient air pollution. *Sustainability (Switzerland), 16*(2). https://doi.org/10.3390/su16020522

Goh, P. S., Kang, H. S., Ismail, A. F., Khor, W. H., Quen, L. K., & Higgins, D. (2022). Nanomaterials for microplastic remediation from aquatic environment: Why nano matters? *Chemosphere, 299*, Article 134418. https://doi.org/10.1016/J.CHEMOSPHERE.2022.134418

Guan, Q. F., Yang, H. Bin, Zhao, Y. X., Han, Z. M., Ling, Z. C., Yang, K. P., Yin, C. H., & Yu, S. H. (2021). Microplastics release from victuals packaging materials during daily usage. *EcoMat, 3*(3). https://doi.org/10.1002/eom2.12107

Gusain, R., Gupta, K., Joshi, P., & Khatri, O. P. (2019). Adsorptive removal and photocatalytic degradation of organic pollutants using metal oxides and their composites: A comprehensive review. *Advances in Colloid and Interface Science, 272*, Article 102009. https://doi.org/10.1016/J.CIS.2019.102009

Hale, R. C., Seeley, M. E., La Guardia, M. J., Mai, L., & Zeng, E. Y. (2020). A global perspective on microplastics. *Journal of Geophysical Research: Oceans, 125*(1). https://doi.org/10.1029/2018JC014719

Hong, H., Bang, R. S., Verster, L., & Velev, O. D. (2025). Designing of self-dispersing soft dendritic microcleaners for microplastics capture and recovery. *Advanced Functional Materials, 2423494*. https://doi.org/10.1002/ADFM.202423494

Huber, M., Archodoulaki, V. M., Pomakhina, E., Pukánszky, B., Zinöcker, E., & Gahleitner, M. (2022). Environmental degradation and formation of secondary microplastics from packaging material: A polypropylene film case study. *Polymer Degradation and Stability, 195*. https://doi.org/10.1016/j.polymdegradstab.2021.109794

Indhur, R., Amoah, I., Bux, F., & Kumari, S. (2023). Nanomaterials for microplastic removal from wastewater: Current state of the art nanomaterials and future prospects. *ACS ES and T Water, 3*(12). https://doi.org/10.1021/acsestwater.3c00203

Jaafarzadeh, N., Reshadatian, N., Feyzi Kamareh, T., Sabaghan, M., Feizi, R., & Jorfi, S. (2024). Study of the litter in the urban environment as primary and secondary microplastics sources. *Scientific Reports, 14*(1), 1–9. https://doi.org/10.1038/s41598-024-80611-y

Jancik-Prochazkova, A., Jašek, V., Figalla, S., Pumera, M., Jancik-Prochazkova, A., Pumera, M., Jašek, V., & Figalla, S. (2023). Photocatalytic microplastics "On-The-fly" degradation via motile quantum materials-based microrobots. *Advanced Optical Materials, 11*(22), 2300782. https://doi.org/10.1002/ADOM.202300782

Juraij, K., Ammed, S. P., Chingakham, C., Ramasubramanian, B., Ramakrishna, S., Vasudevan, S., & Sujith, A. (2023). Electrospun polyurethane nanofiber membranes for microplastic and nanoplastic separation. *ACS Applied Nano Materials, 6*(6), 4636–4650. https://doi.org/10.1021/ACSANM.3C00112/SUPPL_FILE/AN3C00112_SI_004.MP4

Kane, I. A., & Clare, M. A. (2019). Dispersion, accumulation, and the ultimate fate of microplastics in deep-marine environments: A review and future directions. *Frontiers in Earth Science, 7*. https://doi.org/10.3389/feart.2019.00080

Kang, M., Liu, Y., Weng, Y., Wang, H., & Bai, X. (2024). A critical review on the toxicity regulation and ecological risks of zinc oxide nanoparticles to plants. *Environmental Science: Nano, 11*(1), 14–35. https://doi.org/10.1039/D3EN00630A

Khan, M. S., Asif, M. I., Asif, M., Khan, M. R., Mustafa, G., & Adeel, M. (2024a). Nanomaterials for the catalytic degradation and detection of microplastics: A review. *Topics in Catalysis, 2024*, 1–18. https://doi.org/10.1007/S11244-024-01927-7

Khan, S., Noor, T., Iqbal, N., & Yaqoob, L. (2024b). Photocatalytic dye degradation from textile wastewater: A review. *ACS Omega, 9*(20), 21751–21767. https://doi.org/10.1021/ACSOMEGA.4C00887/ASSET/IMAGES/LARGE/AO4C00887_0007.JPEG

Koe, W. S., Lee, J. W., Chong, W. C., Pang, Y. L., & Sim, L. C. (2019). An overview of photocatalytic degradation: Photocatalysts, mechanisms, and development of photocatalytic membrane. *Environmental Science and Pollution Research, 27*(3), 2522–2565. https://doi.org/10.1007/S11356-019-07193-5

Kumah, E. A., Fopa, R. D., Harati, S., Boadu, P., Zohoori, F. V., & Pak, T. (2023a). Human and environmental impacts of nanoparticles: A scoping review of the current literature. *BMC Public Health, 23*(1), 1–28. https://doi.org/10.1186/S12889-023-15958-4

Kumar, R. (2023). Metal oxides-based nano/microstructures for photodegradation of microplastics. *Advanced Sustainable Systems, 7*(6). https://doi.org/10.1002/adsu.202300033

Kurniawan, T. A., Haider, A., Ahmad, H. M., Mohyuddin, A., Umer Aslam, H. M., Nadeem, S., Javed, M., Othman, M. H. D., Goh, H. H., & Chew, K. W. (2023). Source, occurrence, distribution, fate, and implications of microplastic pollutants in freshwater on environment: A critical review and way forward. *Chemosphere, 325*. https://doi.org/10.1016/j.chemosphere.2023.138367

Leppänen, I., Lappalainen, T., Lohtander, T., Jonkergouw, C., Arola, S., & Tammelin, T. (2022). Capturing colloidal nano- and microplastics with plant-based nanocellulose networks. *Nature Communications, 13*(1), 1–12. https://doi.org/10.1038/s41467-022-29446-7

Li, P., Zou, X., Wang, X., Su, M., Chen, C., Sun, X., & Zhang, H. (2020). A preliminary study of the interactions between microplastics and citrate-coated silver nanoparticles in aquatic environments. *Journal of Hazardous Materials, 385*. https://doi.org/10.1016/J.JHAZMAT.2019.121601

Li, B., Zhao, J., Ge, W., Li, W., & Yuan, H. (2022a). Coagulation-flocculation performance and floc properties for microplastics removal by magnesium hydroxide and PAM. *Journal of Environmental Chemical Engineering, 10*(2), Article 107263. https://doi.org/10.1016/J.JECE.2022.107263

Li, W., Wu, C., Xiong, Z., Liang, C., Li, Z., Liu, B., Cao, Q., Wang, J., Tang, J., & Li, D. (2022b). Self-driven magnetorobots for recyclable and scalable micro/nanoplastic removal from nonmarine waters. *Science Advances, 8*(45), 1731. https://doi.org/10.1126/SCIADV.ADE1731/SUPPL_FILE/SCIADV.ADE1731_DATA_S1.ZIP

Li, Y., Chen, H., Li, S., Feng, L., Wang, Z., Wang, D., Wang, Q., & Wang, H. (2024). Corals-inspired magnetic absorbents for fast and efficient removal of microplastics in various water sources. *RSC Advances, 14*(17), 11908–11913. https://doi.org/10.1039/D4RA02521K

Lv, L., He, L., Jiang, S., Chen, J., Zhou, C., Qu, J., Lu, Y., Hong, P., Sun, S., & Li, C. (2020). In situ surface-enhanced Raman spectroscopy for detecting microplastics and nanoplastics in aquatic environments. *Science of the Total Environment, 728*. https://doi.org/10.1016/j.scitotenv.2020.138449

Ma, B., Xue, W., Hu, C., Liu, H., Qu, J., & Li, L. (2019). Characteristics of microplastic removal via coagulation and ultrafiltration during drinking water treatment. *Chemical Engineering Journal, 359*, 159–167. https://doi.org/10.1016/J.CEJ.2018.11.155

Magrez, A., Kasas, S., Salicio, V., Pasquier, N., Seo, J. W., Celio, M., Catsicas, S., Schwaller, B., & Forró, L. (2006). Cellular toxicity of carbon-based nanomaterials. *Nano*

Letters, 6(6), 1121–1125. https://doi.org/10.1021/NL060162E/SUPPL_FILE/NL060162ESI20060410_072509.PDF

Meyerstein, D. (2021). Re-examining Fenton and Fenton-like reactions. *Nature Reviews Chemistry, 5*(9), 595–597. https://doi.org/10.1038/s41570-021-00310-4

Mohanty, M., Tripathy, B., Koteswara Reddy, G., & Das, A. P. (2024). Greener technologies for establishing restraint over microplastic pollution. In *Sustainable microbial technology for synthetic and cellulosic microfiber bioremediation* (pp. 239–258). https://doi.org/10.1007/978-3-031-62600-5_13

Mukherjee, S. G., O'Claonadh, N., Casey, A., & Chambers, G. (2012). Comparative in vitro cytotoxicity study of silver nanoparticle on two mammalian cell lines. *Toxicology in Vitro, 26*(2), 238–251. https://doi.org/10.1016/J.TIV.2011.12.004

Ortiz, D., Munoz, M., Nieto-Sandoval, J., Romera-Castillo, C., de Pedro, Z. M., & Casas, J. A. (2022). Insights into the degradation of microplastics by Fenton oxidation: From surface modification to mineralization. *Chemosphere, 309*, Article 136809. https://doi.org/10.1016/J.CHEMOSPHERE.2022.136809

Overchuk, M., Weersink, R. A., Wilson, B. C., & Zheng, G. (2023). Photodynamic and photothermal therapies: Synergy opportunities for nanomedicine. *ACS Nano, 17*(9), 7979–8003. https://doi.org/10.1021/ACSNANO.3C00891/ASSET/IMAGES/LARGE/NN3C00891_0007.JPEG

Parmar, P. R., Mahanta, J., Dubey, S., Mandal, T. K., & Bandyopadhyay, D. (2023a). Real-time monitoring of glycemic starch using biosynthesized gold nanoparticle-amylase composites. *ACS Sustainable Chemistry and Engineering, 11*(28), 10509–10519. https://doi.org/10.1021/ACSSUSCHEMENG.3C02292

Parmar, P. R., Yadav, N., Dubey, S., Arfa, N., Ghoshal, T., Mahanta, J., Kakati, N., Vanarse, V. B., & Bandyopadhyay, D. (2023b). Microdroplet sensor for point-of-care-testing of glycemic index using gold-amylase nanocomposite. *IEEE Journal on Flexible Electronics, 2*(6), 457–463. https://doi.org/10.1109/JFLEX.2023.3298312

Patil, P. M., Mahamuni-Badiger, P., Ingavale, R. R., Patel, P. R., & Dhanavade, M. J. (2024). Usage of microplastic beads in pharmaceuticals and cosmetics industry: A review. *Microplastic Pollution, 51–72.* https://doi.org/10.1007/978-981-99-8357-5_4

Priyadarshini, S., Jagatee, S., & Das, A. P. (2024). Synthetic fabrics and microfiber pollution–An assessment of their global impact. *Environmental Science and Engineering, Part F2333*, 137–157. https://doi.org/10.1007/978-3-031-51792-1_8

Priyadarsini, S., & Das, A. P. (2025). Lithium bioleaching: A review on microbial-assisted sustainable technology for lithium bio-circularity. *Journal of Water Process Engineering, 69*, Article 106744. https://doi.org/10.1016/j.jwpe.2024.106744

Reddy Koteswara, G., Bhargavi, S., Yamuna, D., Anjani Devi, M., Jeevana Lakshmi Kumar, G., Manas, T., Adarsh, S., Bhavya Sai, V., Nath, H., Alreshidi, M. A., Togun, H., Ram, D. K., & Das, A. P. (2025). Recent advancements in the remediation of marine hydrocarbon and plastic pollutants. *Environmental Science and Engineering, Part F218*, 127–142. https://doi.org/10.1007/978-3-031-81615-4_7

RENEWABLE ENERGY GENERATION AND VALUE ADDITION FROM ENVIRONMENTAL. (2024).

Sen, P., Bhattacharya, P., Mukherjee, G., Ganguly, J., Marik, B., Thapliyal, D., Verma, S., Verros, G. D., Chauhan, M. S., & Arya, R. K. (2023). Advancements in doping strategies for enhanced photocatalysts and adsorbents in environmental remediation. *Technologies, 11*(5), 144. https://doi.org/10.3390/TECHNOLOGIES11050144

Song, Y. K., Hong, S. H., Jang, M., Han, G. M., Jung, S. W., & Shim, W. J. (2017). Combined effects of UV exposure duration and mechanical abrasion on microplastic fragmentation by polymer type. *Environmental Science and Technology, 51*(8), 4368–4376. https://doi.org/10.1021/ACS.EST.6B06155/ASSET/IMAGES/LARGE/ES-2016-06155H_0005.JPEG

Sun, C., Wang, Z., Chen, L., & Li, F. (2020a). Fabrication of robust and compressive chitin and graphene oxide sponges for removal of microplastics with different functional groups. *Chemical Engineering Journal, 393*, Article 124796. https://doi.org/10.1016/J.CEJ.2020.124796

Sun, M., Chen, W., Fan, X., Tian, C., Sun, L., & Xie, H. (2020b). Cooperative recyclable magnetic microsubmarines for oil and microplastics removal from water. *Applied Materials Today, 20*, Article 100682. https://doi.org/10.1016/J.APMT.2020.100682

Urso, M., Ussia, M., Novotný, F., & Pumera, M. (2022). Trapping and detecting nanoplastics by MXene-derived oxide microrobots. *Nature Communications, 13*(1), 1–14. https://doi.org/10.1038/s41467-022-31161-2

Vanarse, V. B., Ahmed, M. M., Singha, D., Paul, A., & Bandyopadhyay, D. (2025). Depolymerization: A catalyst for environmental sustainability. *ACS Symposium Series, 103–129.* https://doi.org/10.1021/BK-2025-1498.CH006

Walling, S. A., Um, W., Corkhill, C. L., & Hyatt, N. C. (2021). Fenton and Fenton-like wet oxidation for degradation and destruction of organic radioactive wastes. *NPJ Materials Degradation, 5*(1), 1–20. https://doi.org/10.1038/s41529-021-00192-3

Wang, L., Kaeppler, A., Fischer, D., & Simmchen, J. (2019). Photocatalytic TiO_2 micromotors for removal of microplastics and suspended matter. *ACS Applied Materials and Interfaces, 11*(36), 32937–32944. https://doi.org/10.1021/ACSAMI.9B06128/SUPPL_FILE/AM9B06128_SI_010.AVI

Wang, Y., Wang, M., Wang, Q., Wang, T., Zhou, Z., Mehling, M., Guo, T., Zou, H., Xiao, X., He, Y., Wang, X., Rojas, O. J., & Guo, J. (2023). Flowthrough capture of microplastics through polyphenol-mediated interfacial interactions on wood sawdust. *Advanced Materials, 35*(36), 2301531. https://doi.org/10.1002/ADMA.202301531

Xiao, J., Guo, S., Wang, D., & An, Q. (2024). Fenton-like reaction: Recent advances and new trends. *Chemistry—A European Journal, 30*(24), e202304337. https://doi.org/10.1002/CHEM.202304337

Xu, G., Cheng, H., Jones, R., Feng, Y., Gong, K., Li, K., Fang, X., Tahir, M. A., Valev, V. K., & Zhang, L. (2020). Surface-enhanced raman spectroscopy facilitates the detection of microplastics <1 μm in the environment. *Environmental Science and Technology, 54*(24). https://doi.org/10.1021/acs.est.0c02317

Yang, H., Chen, G., & Wang, J. (2021). Microplastics in the marine environment: Sources, fates, impacts and microbial degradation. *Toxics, 9*(2), 1–19. https://doi.org/10.3390/TOXICS9020041

Yu, T., Huang, X., Zhang, X. F., Li, K., Liu, S. P., Dai, N., Zhang, K., Zhang, Y. X., & Li, H. (2024). A review of nanomaterials with excellent purification potential for the removal of micro- and nanoplastics from liquid. *DeCarbon, 5*, Article 100064. https://doi.org/10.1016/J.DECARB.2024.100064

Zandieh, M., & Liu, J. (2022). Removal and degradation of microplastics using the magnetic and nanozyme activities of bare iron oxide nanoaggregates. *Angewandte Chemie International Edition, 61*(47), Article e202212013. https://doi.org/10.1002/ANIE.202212013

Zhang, S., Zhang, X., Liu, H., Qu, W., Guan, Z., Zeng, Q., Jiang, C., Gao, H., Zhang, C., Lei, R., Xia, T., Wang, Z., Yang, L., Chen, Y., Wu, X., Cui, Y., Yu, L., & Wang, A. (2015). Modifying effect of COMT gene polymorphism and a predictive role for proteomics analysis in children's intelligence in endemic fluorosis area in Tianjin, China. *Toxicological Sciences, 144*(2), 238–245. https://doi.org/10.1093/TOXSCI/KFU311

Zhang, Y., Dong, Z., Deng, Z., Shi, S., Tang, C., & Hu, X. (2023). An iron "nano-fishnet" for the rapid removal and surface clean-up of micro/nanoplastics from seawater. *Environmental Science: Nano, 10*(9), 2566–2577. https://doi.org/10.1039/D3EN00344B

Zhou, G., Huang, X., Xu, H., Wang, Q., Wang, M., Wang, Y., Li, Q., Zhang, Y., Ye, Q., & Zhang, J. (2022). Removal of polystyrene nanoplastics from water by CuNi carbon material: The role of adsorption. *Science of the Total Environment, 820*, Article 153190. https://doi.org/10.1016/J.SCITOTENV.2022.153190

Ziani, K., Ioniță-Mîndrican, C. B., Mititelu, M., Neacșu, S. M., Negrei, C., Moroșan, E., Drăgănescu, D., & Preda, O. T. (2023). Microplastics: A real global threat for environment and food safety: A state of the art review. *Nutrients, 15*(3). https://doi.org/10.3390/NU15030617

Microplastic Degradation and Management Via Nanomaterials

Esin Balci and Aysun Sofuoglu

Abstract Microplastics (MPs) are tiny particles that result from the breakdown of plastics, and are commonly present in various environmental media with high stability and resistance. Through mechanisms such as bioaccumulation and environmental contamination, MPs serve as carriers for various pollutants, enabling the transfer of these toxic substances from invertebrates to high trophic levels in food chain. Consequently, their environmental presence has become a matter of significant concern to the public and the scientific community. Therefore, their elimination from the environment or reducing their spread is vital for a cleaner and more sustainable environment and the existence of healthy living organisms. Although conventional methods can partially eliminate MPs, their limitations indicate that more effective and sustainable solutions need to be developed. This requires both advanced technologies and efficient materials that can eliminate MPs from the environmental media. This chapter addresses the urgent need for effective remediation methods by providing systematic information focusing on the important function of nanomaterials (NMs) in the removal of MPs. This study elaborates on the usage for NMs in the elimination of distinct types of MPs, examining their oxidative, photocatalytic, and enzymatic degradation mechanisms. It presents important outcomes from existing research and also reveals shortcomings in the current research. It provides different perspectives on the applicability of NMs in laboratory research and the environment, which may lead to significant advances in environmental remediation applications. Thus, a contribution is made to the advancement of sustainable solutions to reduce MP pollution by elucidating new approaches and mechanisms.

E. Balci
Faculty of Engineering, Department of Environmental Engineering, Izmir Institute of Technology, Urla/İzmir, Türkiye
e-mail: esinbalci@iyte.edu.tr

A. Sofuoglu (✉)
Faculty of Engineering, Department of Chemical Engineering, Izmir Institute of Technology, Urla/İzmir, Türkiye
e-mail: aysunsofuoglu@iyte.edu.tr

E. Priyadarshini et al. (eds.), *Advanced Nanotechnology Applications for Environmental Sustainability*, Sustainable Environmental Waste Management Strategies,
https://doi.org/10.1007/978-3-032-12933-8_11

Keywords Microplastics · Nanomaterials · Microplastic pollution · Environmental remediation

1 Introduction

Microplastics (MPs) are emerging pollutants detected in nearly every environment on Earth, from oceans and rivers to groundwater, soil, air, and even remote polar regions (Duis & Coors, 2016; Goh et al., 2022; Khan et al., 2024; Wang et al., 2020). They are broadly described as residues of plastics, including polyethylene (PE), polyester, polyvinyl chloride (PVC), polystyrene (PS) in various forms like microbeads, fibers, fragments, and granules (Bayo et al., 2020). These particles have irregular or regular shapes, not exceeding 5 mm in size and they can be divided into two types depending on their origin: primary and secondary MPs (Thompson et al., 2004). Primary MPs are micro-sized plastics originating from cleaning products, fibers in clothing, urban dust, worn tires, ships, road paints, etc. (Chellasamy et al., 2023; Monira et al., 2022), while secondary MPs are formed as a result of physical (wave action, air abrasion, etc.), chemical (ultraviolet (UV) radiation) and biological degradation of large-scale plastic fragments (Goh et al., 2022; Xiao et al., 2024). The major cause of the problem associated with MPs is global plastic pollution, and there are an estimated 51 trillion plastic fragments in surface waters worldwide (Edo et al., 2020). Due to their negative impact on all forms of life and the environment either through their presence in ecosystems or interactions with heavy metals, persistent organic pollutants, and other emerging industrial contaminants, they are regarded as extremely harmful (Gao et al., 2021). Significant amounts of MPs enter rivers, shorelines, surface water, and beaches owing to the large volumes of wastewater released from treatment plants (Bellasi et al., 2020; Waller et al., 2017). As a result of their small size and buoyancy, MPs get into the food of aquatic creatures (Das et al., 2024b; Mohanty et al., 2024; Ray et al., 2024). They pose a significant risk to marine life because aquatic creatures such as fish, mussels, and oysters can ingest or absorb MPs due to their resemblance to their prey, which can result in physical harm, problems in their internal organs, and even fatal effects (Alimba & Faggio, 2019; Chatterjee & Sharma, 2019; Prata et al., 2019). In soil, the sources of MPs include airborne deposition, sewage sludge applications, compost and plastic mulch operation, and tires abrasion (He et al., 2018; Petroody et al., 2021). On an annual basis, approximately 44,000–300,000 tons of MPs in North America and 63,000–430,000 tons in Europe reach agricultural soils through sewage sludge (Gionfra, 2018). Microplastics may alter soil ecosystems and affect fertility in terrestrial environments (Li et al., 2020). Furthermore, long-distance transport, meteorological conditions, and emission sources can impact on the concentration, spatial distribution, and MPs' removal (Allen et al., 2019). They pose a danger to many living organisms, especially humans, when they are taken into the body directly or indirectly via bioaccumulation. In contrast to large-scale plastic fragments, MPs in the environmental media are not appropriately collected for recycling and are difficult to manage (Ahmed et al., 2024). The sorption of organic pollutants such as polychlorinated biphenyls, polycyclic aromatic hydrocarbons, polybrominated diphenyl ethers, per- and polyfluoroalkyl substances on MPs or MPs' interaction with heavy metals, pharmaceutical materials, and pesticides have been reported (Bakir et al., 2012; Rochman et al., 2013). Therefore, MPs not only serve as sorption sites for pollutants but also as point sources of pollution in aquatic ecosystems (Dey et al., 2023a, 2023b; Khatun et al., 2023; Mallick et al., 2024; Mishra et al., 2019). In other words, they act like a carrier for hydrophobic organic compounds, thereby exacerbating ecological risks and undermining environmental sustainability (Gallo et al., 2018). Effective strategies and techniques need to be developed to eliminate or minimize pollution caused by MPs, which is a global-scale issue (Priyadarshini & Das, 2024; Satapathy et al., 2024). As concerns about MPs increase, many EU nations have joined together to declare the prohibitions on the usage of MPs in personal care items (UNEP, 2015). Similarly, prohibitions on MPs in cosmetic products have been incorporated into U.S. federal law through the Microbead-Free Water Act (2015). Recently, nanomaterials (NMs) have begun to be used as advanced tools to aid in the elimination of numerous pollutants (Chellasamy et al., 2022; Indhur et al., 2023). NMs within the size range of 1–100 nm demonstrate outstanding mechanical strength, a large surface area, the ability to be recycled, enhanced chemical reactivity and functionality, as well as requiring less energy and cost for production, and being easy to fabricate. These properties make NMs suitable for the MPs' efficacious elimination (Chellasamy et al., 2022; Gong et al., 2018; Kokkinos et al., 2020). Many different NMs such as nanoscale zeolites, metal oxides, carbon nanotubes (CNTs), halloysite nanotubes, and several noble metals have been used in environmental applications and different fields including ceramics, microelectronics, bioengineering, chemical engineering, medicine, etc. (Balci et al., 2022, 2023). So far, many studies have explored innovative combinations of various methods/materials and NMs for MP removal; hence, this chapter elaborates on the utilization of NMs for the MPs' removal from the environment and offers different insights into the fate of MPs in the presence of NMs. In this context, we delve into the NMs that can potentially remove MPs from the environment, the action mechanisms of NMs, and the properties that facilitate their interactions with MPs.

2 MPs' Remediation Methods

2.1 Traditional Methods

To reduce the detrimental effects of MPs, several removal methods have been developed using microbial techniques and engineering strategies like filtration and adsorption, as well as thermal treatment and biodegradation (Goh et al., 2022). Adsorption has been particularly effective and widely applied for the elimination of MPs from wastewater through materials such as CNTs and biochar particles, which have the capability to adsorb MPs. On the other hand, while these materials can effectively remove MPs, they also facilitate the entry of new pollutants into the environment. Sand filtration is an easy and effective method; however, it may not fully eliminate small-size MPs (Hidayaturrahman & Lee, 2019). Membrane filtration with 99.4% efficiency in the removal of MPs has major drawbacks related to membrane fouling, which can cause irreversible damage and reduce the effectiveness of the technologies in long-term applications (Lares et al., 2018). These methods are generally inadequate and require integration with other approaches. The advanced oxidation processes (AOPs), including photochemical and electrochemical oxidation have proven to be successful in reducing MP pollution (Ali et al., 2021). Nevertheless, AOPs have several constraints such as high energy consumption and operating costs, and the necessity for advanced catalytic systems (Mahbub & Duket, 2023). While chemical treatment processes are effective in the degradation of MPs, their implementation is often hindered by economic and technical challenges (Das et al., 2024a, 2024b; Dey et al., 2023a, 2023b, 2024). Similarly, biodegradation also represents a more environmentally sustainable, non-invasive, and cost-effective approach, but its efficiency is constrained by the structural properties of MPs (melting point, crystallinity, elasticity, transition-temperature, molecular weight), and surface properties (surface area, hydrophilicity, and hydrophobicity) (Bacha et al., 2021). Enzymes are crucial for the biodegradation process; however, their activities can be significantly influenced by fluctuations in various factors (temperature and pH), leading to structural changes and eventually denatured enzymes. Beyond that, this method is time-consuming and has variable efficiency, limited applicability and challenges to scale (Peng et al., 2021; Balci et al., 2025).

2.2 *Nanotechnology-Based Methods*

Traditional MP removal techniques, including chemical, biological, and physical methods, are not wholly effective (Rizvi et al., 2024). Nanotechnology-based approaches for MP treatment have gained considerable attention in recent research. Nanomaterials (NMs) have a great potential to remove MPs from the environment by significantly improving conventional processes (Ouda et al., 2023). Currently, ferritic, metallic, bimetallic, zeolitic, and magnetic NMs have been used to eliminate MPs from the environment (Devi et al., 2022). NMs with high surface-to-volume ratio have catalytically active surfaces (Zandieh & Liu, 2022) and owing to NMs' nanoscale size, catalysts can more easily reach restricted environments, including cells used in biomedical research and environmental systems.

2.2.1 Nanomaterial-Based Adsorbents

Adsorption is the physical or chemical attachment of a molecule from a liquid or gas phase to a solid surface. It includes various mechanisms such as hydrophobic interactions, hydrogen bonding, van der Waals forces, electrostatic attractions, π–π

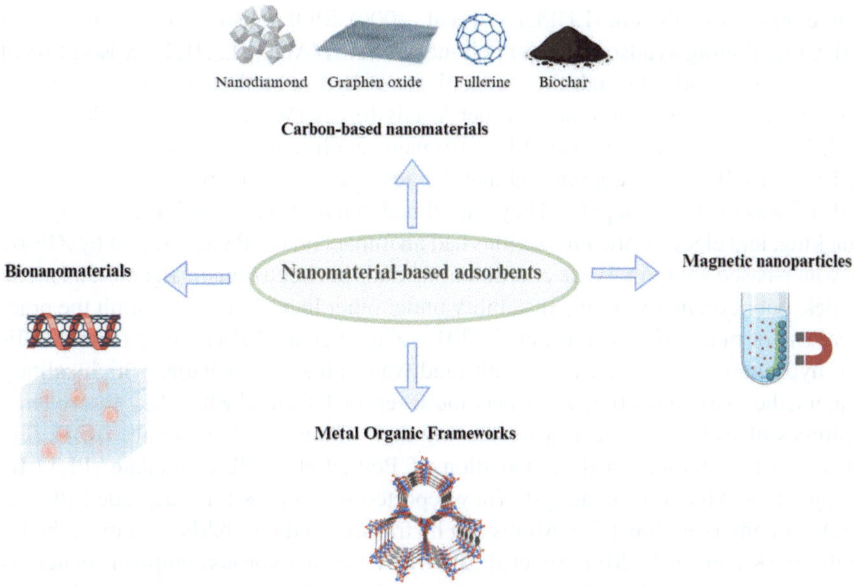

Fig. 1 Types of NM-based adsorbents

interactions, pore filling, and intraparticle diffusion. A wide variety of nanomaterial-based adsorbents such as biochar, metal oxides, metal organic skeletons, carbon-based NMs, and magnetic NMs have been widely employed in the MPs' remediation (Fig. 1). Having a high surface area and porosity, and a further modification with various functional groups to target specific pollutants make them desirable. Additionally, they exhibit high reactivity, strong adsorption capacity, and favorable cost-effectiveness.

Nanomaterials Derived from Metal Organic Frameworks

Metal–organic frameworks (MOFs) have become a highly effective material in separation technology owing to their physico-chemical properties, including exceptionally high surface areas, porous structures, rapid functionalization, and the capability to tune their structure by selecting suitable metal ions and organic ligands. They have porous lattice-like structures containing linear or branched organic ligands and metal nodes attached to secondary bond units. They have considerable potential and high efficacy as both adsorbents and catalysts. They have been utilized in the applications of adsorption, electrocatalytic, photocatalytic and advanced oxidative degradation of MPs (Liu et al., 2022). By altering their pore size and composition, they can be utilized for certain purposes to selectively capture MPs of various shapes and sizes (Honarmandrad et al., 2023). They act as homogeneous catalysts that can cleave carbon–carbon bonds in polypropylene (PP), high-density polyethylene (HDPE) and low-density polyethylene (LDPE) (Lal et al., 2006). MOFs adsorb onto the surfaces of MPs, functioning as adsorbents for polymeric chains (Ali et al., 2021). Adsorption of MPs occurs mostly through the mechanisms of hydrophobic interaction, electrical attraction, π–π interaction and van der Waals forces (Wu et al., 2024). Wan et al. (2022) focused on the removal of PS MPs using zeolitic imidazolate frameworks-67 (ZIF-67) MOFs. They determined that the adsorption efficiency of PS (5 mg/L) by ZIF-67 was over 90% at pH 8. They underlined that hydrogen bond interactions, π–π stacking, and electrostatic interactions had an impact on the PS adsorption by ZIF-67 in water-based solution. Weak coordination bonds between organic ligands and metal nodes, poor conductivity, and instability under other harsh conditions limit the practical use of many MOFs (Raza et al., 2019; Zhang et al., 2014). Since many MOFs are hydrophilic and they cannot withstand water attacks, particularly high salinity waters, their structures frequently become severely deformed when they absorb large volumes of water, significantly decreasing their efficiency (Peng et al., 2024). Suo et al. (2017) studied on decomposition of Polyethylene Terephthalate (PET) by using ZIF-8 MOFs as a catalyst. They reported that ZIF-8 fully degraded PET to PET monomers within 1.5 h. MOFs can be transformed into NMs with much better stability (Kaneti et al., 2017; Shi et al., 2020). These nanoporous composite materials transformed from MOFs demonstrated synergistic effects by merging metallic and carbonaceous structures. Pasanen et al. (2023) produced magnetic ZIF-8 (Nano-Fe-ZIF-8) nanocomposite and then utilized it as an adsorbent to remove PS (1.1 μm) in water. They underlined that Nano-Fe-ZIF-8 removed almost all of PS within a short time. Haris et al. (2023) developed nanopillared structures made of two-dimensional (2D) MOF with carbon encapsulated iron oxide (C@FeO) nanopillars. While C@FeO adsorbent could remove 72% of MPs (100 mg/L) in water within 60 min, 2D MOF-C-FeO was able to remove 100% of MPs for the same duration. You et al. (2022) prepared ZIF-8@Aerogel growing in situ ZIF-8 on wood-based aerogel to eliminate PS and poly(1,1-difluoroethylene) (PVDF) in aqueous media. They determined that the 91% of PVDF and 86% of PS in the presence of ZIF-8@Aerogel was successfully adsorbed. Compared to aerogel, they underlined that ZIF-8@Aerogel enhanced the removal efficiency of PVDF (60–110 nm) and PS (90–140 nm) by 34.5% and 43.3%, respectively. Chen et al. (2023) created a superhydrophobic MOF-based coated sponge. These sponges could rapidly adsorb 100% of MPs (40 mg/L) containing PS (~150 μm), PE (~6.5 μm), PVC (~6.5 μm), and PP (6.5 and 100 μm). In conclusion, NMs with MOFs were proven to be effective and stable adsorbents or catalysts.

Carbon-Based Nanomaterials

Carbon-based NMs such as graphene oxide (GO), activated carbon (AC), metal-modified carbon, biochar (BC), and CNTs represent the most frequently employed materials for the MPs' removal due to their tunable physicochemical and structural properties. The key factor influencing adsorption is the interaction between MPs and carbon-based NMs, which is determined by the surface structure of the carbon, the ionic composition of the water, and the characteristics of the adsorbate. Many MPs have significant hydrophobicity (Zheng et al., 2024). In this regard, CNTs can bond with MPs via hydrophobicity or π-π interactions. Tang et al. (2021) fabricated magnetic CNTs (M-CNTs) to eliminate MPs containing PE, PET, polyamide (PA) via magnetic forces. They reported that M-CNT adsorbent (5 mg/L) fully removed MPs (5 mg/L) within 5 h. Wang et al. (2021) carried out the PS adsorption (1 μm, 100 mg/mL) utilizing magnetic biochar (MBC), and modified MBC with $Mg(OH)_2$ and ZnO. They found that MBC, Mg-MBC, and Zn-MBC removed 94.8%, 98.8% and 99.5% of PS, respectively. They suggested that this adsorption mechanism was related to electrostatic interaction and chemical bonding interactions between MPs and MBC. Yuan et al. (2020) conducted adsorption of PS in water using three-dimensional reduced graphene oxide (3D-RGO). They determined that almost 90% of 100 mg/L PS was removed employing the 3D-RGO. The adsorption mechanism between 3D-RGO and PS was related to the strong π-π interactions. Overall, carbon-based NMs with different surface characteristics and porous structures, as mentioned above, are quite effective in removing MPs. Holistic research on the interactions between carbon-based NMs and MPs is needed since it may be important to study their effectiveness for the mitigation of MP pollution, especially in industrial processes.

Magnetic Nanomaterials

Magnetic NMs are environmentally friendly materials with low cost and low toxicity (Fig. 2). They can attach to the MPs' hydrophobic surfaces through electrostatic interaction or specific binding groups under the external magnetic field (Yeap et al., 2012). Magnetic extraction method with NMs is particularly effective for small-sized MPs (<20 μm). Grbic et al. (2019) created functionalized Fe-NMs with silane to facilitate MPs' adsorption. While functionalized Fe-NMs with silane eliminated 93% of PE, PET, PS, PVC, and PP (size <1 mm), they were able to remove 92% of PE and PS between 10 μm and 20 μm from seawater using magnetic extraction. Kang et al. (2019) synthesized a magnetic carbon nanosprings (Manganese N-doped carbid nanoparticles; Mn@NCNTs) to eliminate MPs in water. Following 8 h of experiment, the MPs' weight loss was found to be over 40% in the presence of Mn@NCNTs. Compared to other materials (GO (28%), NCNTs (20%), and MnO_2 (<10%)), it exhibited superior efficiency in the weight loss of MPs at same time. Elmacı (2020) tailored a high porous C@Fe_3O_4 magnetic nanocomposites by applying microwave synthesis method and then nanocomposites were used to eliminate PS from drinking water. Within 2 min, PS (20 mg) was removed from drinking water using nanocomposites (5 mg) via external magnet. Shi et al. (2022), nano Fe_3O_4 was utilized to remove PE, PP, PS, and PET between 200 and 900 nm via magnetic adsorption. Nano-Fe_3O_4 (1.3 g/L) eliminated 87% of PE, 85% of PP, 86% of PS, and 63% of PET in the water within 150 min. This variation in removal efficiency was attributed to the dimensional difference of MPs and the density of nano-Fe_3O_4 adsorbed onto the MPs' surfaces. In various studies, bare magnetic Fe_3O_4 was used as adsorbent to eliminate MPs (Heo et al., 2022; Yan et al., 2023; Zandieh & Liu, 2022). Zhang et al.

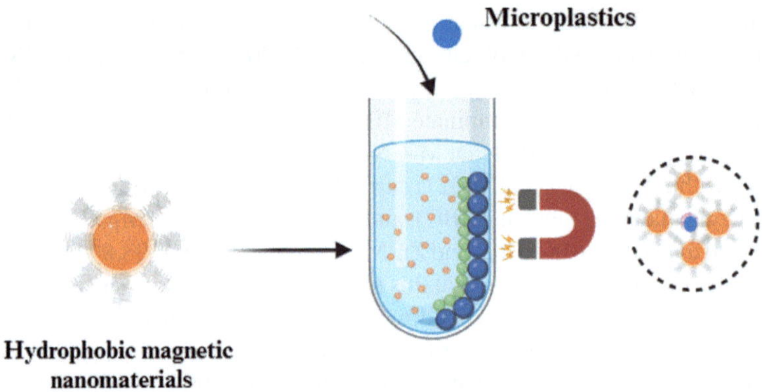

Fig. 2 Magnetic extraction of MPs via NMs

(2023) used a magnetic nanomaterial $CuFe_2O_4$ (0.2 g/L) to eliminate PS (0.04 g/L) MPs from water. $CuFe_2O_4$ NMs exhibited a high removal efficiency (<90%) owing to its magnetic nature. Sacko et al. (2024) synthesized tannic acid-coated magnetite NMs (TA-Fe_3O_4) to remove PS and PET at a concentration of 1.25 g/L. They reported that TA-Fe_3O_4 (2.3 mg/L) adsorbed 83% of PS and 98% of PET within 5 h. Ren et al. (2024) employed functionalized magnetic cobalt ferrite NMs for the removal of MPs. The combination of hydrophobic interactions and van der Waals forces contributed to their efficiency in eliminating MPs. The removal efficiency was nearly 100%, and even after ten cycles of use, the efficiency remained above 98%. Aragon et al. (2025) used bare Fe_3O_4 and Fe_3O_4 magnetic NMs functionalized with sodium alginate to capture PEs. They determined an increase of the removal efficiency when using functionalized Fe_3O_4 compared to Fe_3O_4. The results of all these studies clearly demonstrate that magnetic nanotechnologies are a highly effective approach in the MPs' removal from the environment.

Bionanomaterials

The use of certain adsorbents and catalysts for MPs' treatment may result in secondary pollution. It is essential to develop non-toxic and biodegradable materials that are capable of removing small-sized MPs while minimizing environmental impact. Bionanomaterials are also NMs obtained from biological sources or integrated with organic compounds (Fig. 3). Jalvo et al. (2021) created two distinct polysaccharide nanocrystals (cellulose (CNC) and chitin nanocrystals (ChNC)) using a cast coating process. This coating enhanced their tensile strength, hydrophilicity, and elasticity. While CNC was able to remove MPs with a diameter of 500 nm, ChNCs exhibited the capability to eliminate MPs as small as 2 μm in diameter. In another study, lysozyme amyloid fibril was used as a bionanomaterial to remove PS, and it could remove 98.2% of PS (Peydayesh et al., 2021). Zhuang et al. (2022) tested cellulose nanofibers (CNFs) for the MPs' treatment in water. The cyclic adsorption

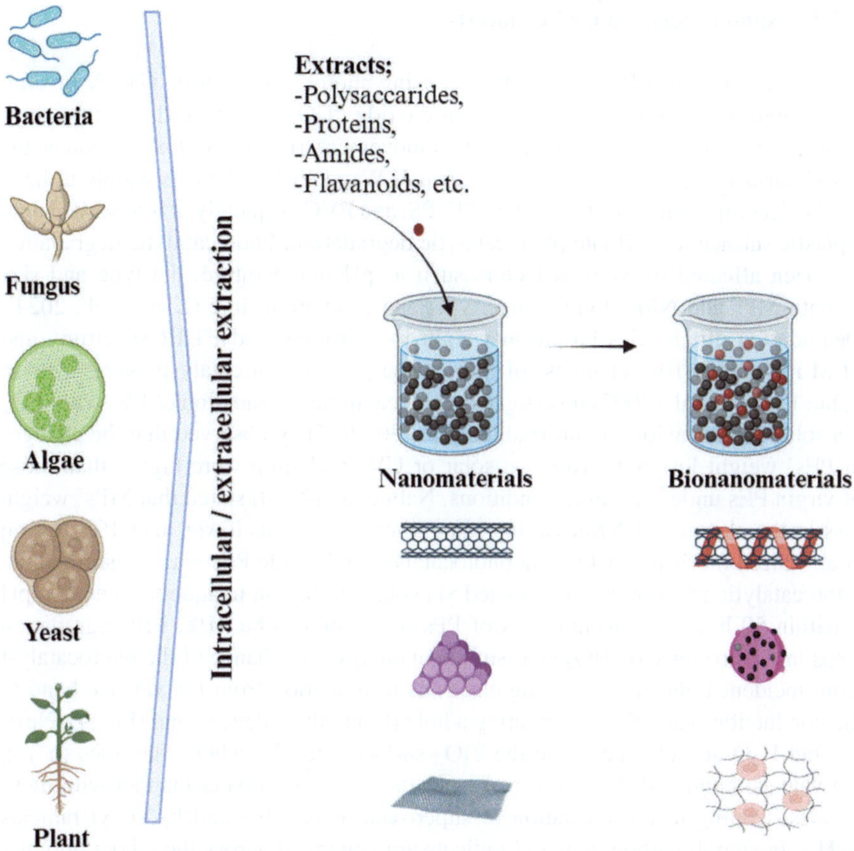

Fig. 3 Bionanomaterials

efficiency was found to be almost 100% in the 1st cycle, while the adsorption efficiency in the 2nd, 3rd, and 4th cycles were 83.1%, 64.7%, and 59.4%, respectively. Bionanomaterials obtained by combining biological extracts with NMs having photocatalytic activity have also been employed on the MPs' removal. Raghavendra et al. (2022) formed ZnO bionanomaterials employing areca nut (*Areca catechu*) extract to remove LDPEs from water. Experiments were carried out in ambient air for 300 h by exposing LDPEs (1 mm) to solar radiation in the presence of ZnO bionanomaterials. The weight loss of LDPEs was found to be %67 while this loss in the absence of bionanomaterials under only solar irradiation at the same time was 1–2%. Vas et al. (2024) used fresh extract from *P. longum* (PL) leaves as an agent for the fabrication of ZnO nanoparticles. PL-ZnO (100 μL) were employed in MP (50–90 nm) degradation studies via UV irradiation. The study underlined that PL-ZnO have a high potential for the degradation of MPs. The findings of all these studies clearly show that bionanomaterials are highly effective in MPs' degradation.

2.2.2 Nanomaterial-Based Catalysts

The degradation of MPs can be achieved using photocatalytic NMs via UV or solar irradiation. Irradiation (e.g. on Titanium dioxide (TiO_2)) leads to the formation of photoexcited electron–hole pairs, and their movement triggers a redox reaction on the NMs' surface (Schultz & Yoon, 2014; Zhu & Wang, 2017). TiO_2 is widely utilized for the decomposition of MPs such as PE, PS, and PVC. Typically, it is adsorbed onto a plastic surface to facilitate photocatalytic degradation. Photocatalytic degradation has been affected by various factors such as pH, temperature, the type and size of both MPs and NMs, light source type, and irradiation time (Xiao et al., 2024). Degradation efficiencies based on weight loss, Raman and FT-IR spectrum, and SEM imaging (>10% changes) of MPs in the presence of catalysts were given in Table 1. Zhao et al. (2007) investigated photocatalytic degradation of PEs with TiO_2 via solar and ultraviolet light irradiation for 300 h. They observed that the changes in PEs' weight loss with TiO_2 via solar or UV irradiation were higher than those of virgin PEs under the same conditions. Nabi et al. (2020) stated that MPs' weight loss in the absence of NMs via the same light source was lower than 1%. Jeyaraj et al. (2023) fabricated TiO_2 as a photocatalyst to degrade PPs via sol–gel method. Photocatalytic reaction was conducted via solar irradiation in aqueous media at pH 3 within 50 h and the weight loss of PPs was found to be 51%. MP degradation steps in the presence of TiO_2 start sufficient energy absorbance of the photocatalyst from incident light (hv), allowing electrons to transition from the valence band to the conduction band, thereby creating a hole (h^+) in the valence band (Fig. 4). Next, O_2 and H_2O are adsorbed onto the TiO_2 surface (step 2), where the holes (h^+) in the valence band and the electrons (e^-) in the conduction band interact with these species, leading to the formation of superoxide ions (O_2^-) and hydroxyl radicals (OH·). In step 4, carbon-centered radicals are generated across the whole polymer structure by OH·. Radicals centered on carbon then interact with O_2, causing peroxide radicals to form. This leads to the formation of metabolites, such as carbonyl and ketone compounds. Eventually, these metabolites are photocatalytically oxidized to CO_2 and H_2O (Dinoop et al., 2019; Tien et al., 2020). Like TiO_2, ZnO, MnO_2, Cu_2O, and CdS also exhibit photocatalytic activity. ZnO nanorods were tested to photocatalytically degrade LDPE under a visible light source for 175 h (Tofa et al., 2019). There was an increase of 30% in the carbonyl index of LDPE residues and the formation of numerous cracks and hollow structures on the LDPEs' surface. The photocatalytic activity for MPs' degradation can be increased by integration of numerous NMs. Uogintė et al. (2023) tested the ability of graphene oxide-based metal oxide NMs (GO-Cu_2O, GO-Mn_2O, and GO-TiO_2) to degrade PEs in aqueous solutions under UV irradiation. At the end of 480 min, the weight loss percentages of PEs in the presence of GO-TiO_2, GO-MnO_2, GO-Cu_2O, and GO were 50.5%, 48%, 39.5%, and 35.6%, respectively. It was clear that the combined photocatalysts showed a higher catalytic efficiency than pure photocatalysts. It can be due to the various adsorption properties of the materials. Cao et al. (2022) fabricated MXene/$Zn_{0.6}Cd_{0.4}S$ nanocomposite to degrade PET. They found that decomposition of PET to organic micromolecules such as glycolate, acetate, ethanol, etc. occurred in the presence of the nanocomposites. Detailed information about the MPs' degradation using NM catalysts was given in Table 1.

2.2.3 Nanozymes

Nanozymes are catalytic NMs with enzyme-like activities (Zandieh & Liu, 2022). Fe_3O_4 is one of the most utilized nanozymes with peroxidase-like properties. Kang et al. (2019) employed a magnetic nanozyme exhibiting peroxymonosulfate activity to degrade cosmetic MPs, and they observed 54% weight loss at 160 °C. Zandieh and Liu (2022) performed the degradation study of MPs such as HDPE, PVC, PS, and PET in water using the Fe_3O_4. They stated that the hydrophilic Fe_3O_4 triggered catalytic degradation of MPs at temperatures around their melting point, resulting in approximately 100% weight loss. Melting MPs make it easier to access the polymer chains, while enzymes can denature at high temperatures. Compared to enzymes, nanozymes have high stability and activity against extreme conditions (Li & Liu, 2021). Some nanozymes are created by immobilizing enzymes or microorganisms on NMs through either physical adsorption or covalent bonding. Immobilization facilitates catalyst recovery, selective depolymerization, and enzyme activity. To enhance the decomposition of MPs, NM complexes were formed with enzymes through immobilization. Schaminger et al. (2021) carried out an immobilization of PETase enzyme on supermagnetic iron oxide NMs to degrade PET. They found that due to the crystallinity of PET, MP particles decomposed through multiple cycles in the presence of immobilized enzymes. Krakor et al. (2021) performed the degradation experiments of polycaprolactone (PCL), using immobilized lipase and cutinase on silica-coated magnetic Fe_3O_4. They reported effective PCL degradation by underlining weight loss and enzyme stability. Jia et al. (2021) immobilized PETase enzyme onto $Co_3(PO_4)_2$ via His-tag and subsequently it was used in PET degradation. The immobilized enzyme maintained 70.2% of initial enzyme activity following 10 degradation cycles with high degradation efficiency. Li et al. (2022) studied solar-driven degradation of PET using immobilized DuraPETase enzyme on Fe_3O_4. They highlighted that PET degradation efficiency in the presence of immobilized enzyme was six-fold higher compared to that in the absence of Fe_3O_4. In another study, Zhou et al. (2021) studied immobilization of lipase on polydopamine-coated Fe_3O_4 to improve enzyme activity and stability in degradation experiments of polycaprolactone. After 24 h of incubation with immobilized enzymes, morphological changes in the MPs' structures were determined. Kotnis et al. (2025) used PETase enzyme immobilized on magnetic NMs in PET degradation. They underlined that PETase's activity was enhanced when it was immobilized on NMs. In addition to enzymes, various microorganisms can also be functionalized with NMs. Ye et al. (2022) studied integration of carbon dot-functionalized polymeric carbon nitrides (CDPCN) with *Methanosarcina barkeri* to form *b*-CDPCN biohybrids to convert MPs into CH_4 via both photoreduction and photooxidation pathway. They achieved 90.2% mineralization of polylactic acid as well as sustained high-grade CH_4 production (>99.5%) in

Table 1 Photocatalytic degradation efficiencies of various MPs in the presence of NMs

MPs	NMs	Light source	Irradiation time (h)	Degradation efficiency (%)	References
PE					
• 10 μm	TiO_2 nanofilm	UV light (256 W) at 254 nm	36	100%	Nabi et al. (2020)
• 15–20 μm	TiO_2 nanocomposite (1 wt%)	UV light (32 W) at 254 nm	300	90%	Zhao et al. (2007)
• 15–20 μm	TiO_2 nanocomposite (1 wt%)	Solar	300	42%	Zhao et al. (2007)
• 300 μm–1 mm	GO-TiO_2 (1:1 v/v)	UV light (72 W) at 350 nm	480	50.5%	Uogintė et al. (2023)
• 300 μm–1 mm	GO-MnO_2 (1:1 v/v)	UV light (72 W) at 350 nm	480	48%	Uogintė et al. (2023)
• 300 μm–1 mm	GO-Cu_2O (1:1 v/v)	UV light (72 W) at 350 nm	480	39.5%	Uogintė et al. (2023)
• 300 μm–1 mm	GO	UV light (72 W) at 350 nm	480	35.6%	Uogintė et al. (2023)
PS					
• 400 nm	TiO_2 nanofilm	UV light (256 W) at 254 nm	12	98.4%	Nabi et al. (2020)
• 700 nm	TiO_2 nanofilm	UV light (256 W) at 365 nm	24	95.3%	Nabi et al. (2020)

(continued)

Table 1 (continued)

MPs	NMs	Light source	Irradiation time (h)	Degradation efficiency (%)	References
• 1 μm	TiO_2 nanofilm	UV light (256 W) at 365 nm	24	93.5%	Nabi et al. (2020)
• 5 μm	TiO_2 nanofilm	UV light (256 W) at 254 nm	24	99.9%	Nabi et al. (2020)
PE with low density					
• 15 nm	TiO_2 nanocomposite (0.1 wt%)	UV light (36 W) at 315 nm	200	68%	Thomas and Sandhyarani (2013)
• 24 μm	ZnO nanorods	Solar	360	78%	Ali et al. (2016)
• 50 μm	TiO_2 nanocomposite (10 wt%)	Visible light (50 W)	175	–	Tofa et al. (2019)
• 80 μm	Polyacrylamide grafted TiO_2 (1 wt%)	UV light (200 W) at 340 nm	520	40%	Liang et al. (2013)
• 80 μm	TiO_2 nanocomposite (1 wt%)	UV light (200 W) at 340 nm	520	30.5%	Liang et al. (2013)
PVC					
• 100 μm	ZnO composite	UV light (22 W) at 365 nm	2	20%	Chakrabarti and Dutta (2008)
• 25–30 μm	TiO_2 nanocomposite (1.5 wt%)	UV light (200 W) at 350 nm	300	27%	Cho and Choi (2001)
• 70 nm	Graphen oxide-nitrogen doped TiO_2 (1:3)	Visible light	3	99%	Kaur and Kaur (2025)

(continued)

Table 1 (continued)

MPs	NMs	Light source	Irradiation time (h)	Degradation efficiency (%)	References
• 100 μm	Fe-Doped ZnO composite	UV light (4W) at 210 nm	3.5	11%	Das et al. (2017)
PP					
• 10–50 μm	TiO_2	Solar	150	51%	Jeyaraj et al. (2023)
• 140 μm	Carbon-coated TiO_2 nanocomposite(2wt%)	UV light (75 W) at 254 nm	500	36%	Kamrannejad et al. (2014)
• 155 μm	ZnO	Visible light	456	> 65%	Uheida et al. (2021)

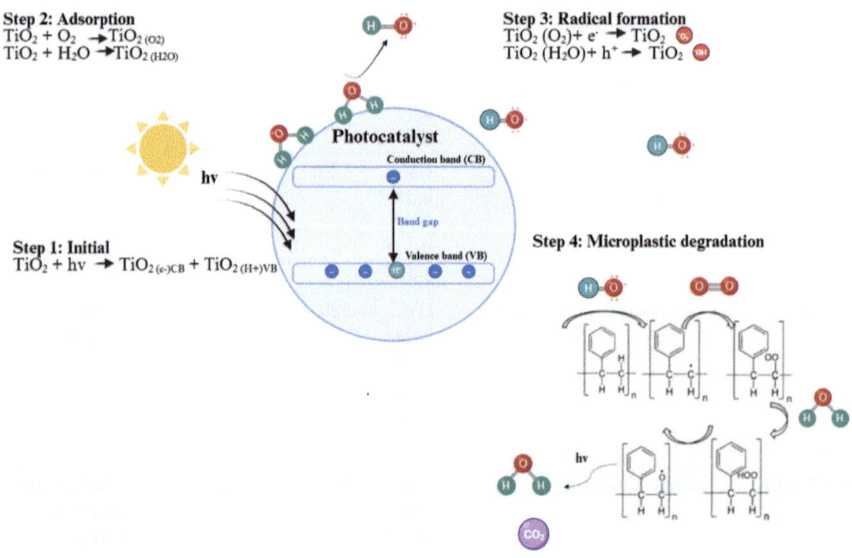

Fig. 4 MP degradation steps in the presence of TiO_2 via photocatalysis

the presence of b-CDPCN. Considering the results of the above-mentioned studies, nanozymes are effective nano-catalysts for MP degradation in the environment.

2.2.4 Nanomaterial-Based Membranes

NMs-based membranes are considered as one of the most successful approaches to eliminate MPs. In recent research on the removal of MPs, thin film based green materials have been used. Gnanasekaran et al. (2021) fabricated a membrane by incorporating MOF (MIL-100 (Fe)) into polysulfane (PSF). By adding 0.5 wt% MIL-100 (Fe) to the PSF matrix (M0.5), 10.3 times more pure water flux was achieved into the PSF membrane. The integration of hydrophilic MIL-100(Fe) with the PSF matrix increased the removal efficiency of MPs from water. Yang et al. (2022) prepared Holey Ti_3C_2 (h-$Ti_3C_2T_x$) nanosheet membranes to remove MPs in aqueous environment. h-$Ti_3C_2T_x$ membranes removed 99.3% of MPs with a large water flux of 196.7 $L\ h^{-1}\ m^{-2}\ k\ Pa^{-1}$. Despite the excellent performance of nanomembranes, blockages due to the interaction of organic compounds in water with hydrophobic membranes is a significant challenge for nanomembrane technology. This fouling causes a reduction in water flow, thus necessitating the usage of chemical or mechanical methods to clean the membrane (Yang et al., 2015). To overcome these difficulties, membranes can be coated with hydrophilic polymers like polyvinyl alcohol and chitosan (Yu et al., 2024).

3 Current Challenges for the Removal of MPs by Nanomaterials

There are some difficulties and limitations in the removal of MPs, which cause environmental pollution, by NMs. The main findings are summarized below.

(1) Contaminated aquatic environments are intricate systems comprising various compounds (dissolved organic matter, ammonia, nitrogen, and phosphorus). In studies conducted using synthetic water, it is essential to take into account the presence of these pollutants together with other environmental factors.
(2) MPs are vulnerable to decomposition in the environment. Their characteristics and toxic effects on the environment may change with time. Surface chemistry studies are very important to show the distinctions in the adsorption behavior of MPs at various ages.
(3) The metabolites and end-products of MP types formed during the degradation process should be further investigated.
(4) Concerns about the ecotoxicity of NMs arise particularly from their usage in the degradation of MPs in aquatic and terrestrial ecosystems. It is important to focus on the separation and recycling of NMs before reaching environmental compartments.
(5) Addition of large amounts of chemicals to remove MPs causes the formation of secondary pollutants.

(6) Iron oxide nanoparticles are highly regarded as a potential option for MPs' adsorption. Nevertheless, detailed research is needed to understand the compatibility between laboratory models and environmental samples and to characterize inter-particle interactions.
(7) Several ways such as weight loss, carbonyl index, turbidity, etc. have been utilized for the photocatalytic degradation of MPs. It might be essential to define quantitative or qualitative parameters to evaluate the effectiveness of MPs' removal.
(8) Although some NMs are recyclable, their effectiveness decreases over time, and they eventually become useless. Cost-effectiveness should be increased by focusing on the long-term reusability of NMs.

4 Conclusion and Future Directions

Microplastics (MPs) in the environment present a major threat to ecosystems and all living organisms. Developing effective strategies is crucial to reduce the presence of MPs in the environment. This chapter represents recent developments in the use of nanomaterials (NMs) for the MPs' removal. NMs such as metal oxides, clay-based nanocomposites, etc. exhibit high MP removal efficiency but they are highly toxic and expensive. These types of NMs can be integrated with different materials to minimize their toxicity and provide them with greater surface area. It is very important to focus on the development of eco-friendly, self-regenerative, and non-toxic NMs. NMs should be fabricated from cost-effective raw materials and should be scalable. Although NMs have the potential to address the constraints of conventional materials during purification, they have challenges such as scalability and reproducibility. Besides, there is a lack of practical application in processing and managing them at large scales. More convincing data needs to be collected at this stage, especially from field tests. Collaborative efforts from academia and industries are essential to accelerate adoption. Additionally, the general focus of most studies examined in this chapter is on the pollution caused by MPs in the water body. However, the usage of NMs for the MPs' removal should not be limited to water, and their use in other environmental media should be extensively investigated. Finally, addressing plastic pollution requires consideration of technological advancements, cultural initiatives, and political actions; otherwise, it may take decades to resolve this environmental issue.

References

Ahmed, S. F., Islam, N., Tasannum, N., Mehjabin, A., Momtahin, A., Chowdhury, A. A., Almomani, F., & Mofijur, M. (2024). Microplastic removal and management strategies for wastewater treatment plants. *Chemosphere, 347*, Article 140648.

Ali, S. S., Qazi, I. A., Arshad, M., Khan, Z., Voice, T. C., & Mehmood, C. T. (2016). Photocatalytic degradation of low density polyethylene (LDPE) films using Titania nanotubes. *Environmental Nanotechnology Monitoring & Management, 5*, 44–53. https://doi.org/10.1016/j.enmm.2016.01.001

Ali, S. S., Elsamahy, T., Koutra, E., Kornaros, M., El-Sheekh, M., Abdelkarim, E. A., Zhu, D., & Sun, J. (2021). Degradation of conventional plastic wastes in the environment: A review on current status of knowledge and future perspectives of disposal. *Science of the Total Environment, 771*, Article 144719.

Alimba, C. G., & Faggio, C. (2019). Microplastic in the marine environment: Current trends in environmental pollution and mechanism of toxicological profile. *Environmental Toxicology and Pharmacology, 68*, 61–74. https://doi.org/10.1016/j.etap.2019.03.001

Allen, S., Allen, D., Phoenix, V. R., Le Roux, G., Duranteza, P., Simonneau, A., Stephane, B., & Galop, D. (2019). Atmospheric transport and deposition of microplastics: In a remote mountain catchment. *Nature Geoscience, 12*, 339–344. https://doi.org/10.1038/s41561-019-0335-5

Aragon, D., Garcia-Merino, B., Barquin, C., Bringas, E., Rivero, M. J., & Ortiz, I. (2025). Advanced green capture of microplastics from different water matrices by surface-modified magnetic nanoparticles. *Separation and Purification Technology, 354*(4), Article 128813. https://doi.org/10.1021/acs.langmuir.4c02420

Bacha, A. U. R., Nabi, I., & Zhang, L. (2021). Mechanisms and the engineering approaches for the degradation of microplastics. *ACS ES&T Engineering, 1*(11), 1481–1501. https://doi.org/10.1021/acsestengg.1c00216

Bakir, A., Rowland, S. J., & Thompson, R. C. (2012). Competitive sorption of persistent organic pollutants onto microplastics in the marine environment. *Marine Pollution Bulletin, 64*, 2782–2789.

Balci, E., Rosales, E., Pazos, M., Sofuoglu, A., & Sanroman, M. A. (2022). Continuous treatment of diethyl hexyl and dibutyl phthalates by fixed-bed reactor: Comparison of two esterase bionanocomposites. *Bioresource Technology, 363*, 127990. https://doi.org/10.1016/j.biortech.2022.127990

Balci, E., Rosales, E., Pazos, M., Sofuoglu, A., & Sanromán, M. A. (2023). Immobilization of esterase from *Bacillus subtilis* on Halloysite nanotubes and applications on dibutyl phthalate degradation. *Environmental Technology & Innovation, 30*, 103113. https://doi.org/10.1080/09593330.2025.2548636

Balci, E., Rosales, E., Pazos Curras, M., Sofuoglu, A., & Sanromán, M. A. (2025). Enhanced catalytic performance of *Rhizomucor miehei*lipase on di-n-butyl and diethylhexyl phthalates: Insights into substrate specificity and immobilization strategy. *Environmental Technology, 46*(27), 5566–5582. https://doi.org/10.1080/09593330.2025.2548636

Bayo, J., Olmos, S., & Lopez-Castellanos, J. (2020). Microplastics in an urban wastewater treatment plant: The influence of physicochemical parameters and environmental factors. *Chemosphere, 238*, Article 124593.

Bellasi, A., Binda, G., Pozzi, A., Galafassi, S., Volta, P., & Bettinetti, R. (2020). Microplastic contamination in freshwater environments: A review, focusing on interactions with sediments and benthic organisms. *Environments, 7*(4), 30.

Cao, B., Wan, S., Wang, Y., Guo, H., Ou, M., & Zhong, Q. (2022). Highly-efficient visible-light-driven photocatalytic H2 evolution integrated with microplastic degradation over MXene/ZnxCd1-xS photocatalyst. *Journal of Colloid and Interface Science, 605*, 311–319. https://doi.org/10.1016/j.jcis.2021.07.113

Chakrabarti, S., & Dutta, B. K. (2008). Dye-senstised photocatalysis degradation of PVC-ZnO composite film. *Int. J. Environmental Technology and Management, 9*(1), 34–36.

Chatterjee, S., & Sharma, S. (2019). Microplastics in our oceans and marine health. *Field Actions Science Reports, 19*, 54–61.

Chellasamy, G., Kiriyanthan, R. M., Maharajan, T., Radha, A., & Yun, K. (2022). Remediation of microplastics using bionanomaterials: A review. *Environmental Research, 208*, Article 112724. https://doi.org/10.1016/j.envres.2022.112724

Chellasamy, G., Ramasundaram, S., Veerapandian, M., Chandran, M., Dhanasekaran, B., Oh, T. H., Govindaraju, S., & Yun, K. (2023). Systematic review on fate and behavior of microplastics towards the environment. *TrAC, Trends in Analytical Chemistry, 169*, Article 117390. https://doi.org/10.1016/S1010-6030(01)00499-3

Chen, X., Ma, H., Ji, X., Han, R., Pang, K., Yang, Z., Liu, Z., & Peng, S. (2023). Engineering green MOF-based superhydrophobic sponge for efficiently synchronous removal of microplastics and pesticides from high-salinity water. *Water Research, 243*, Article 120314. https://doi.org/10.1016/j.watres.2023.120314

Cho, S., & Choi, W. (2001). Solid-phase photocatalytic degradation of PVC–TiO2 polymer composites. *Journal of Photochemistry and Photobiology A: Chemistry, 143*(2–3), 221–228. https://doi.org/10.1016/S1010-6030(01)00499-3

Das, P., Roy, A., & Chakrabarti, S. (2017). Photocatalytic degradation of the nanocomposite film comprising polyvinyl chloride (PVC) and sonochemically synthesized iron-doped zinc oxide: A comparative study of performances between sunlight and UV radiation. *Journal of Polymers and the Environment, 25*, 1231–1241. https://doi.org/10.1007/s10924-016-0894-0

Das, A., Dey, S., Bala, K., Gnanasekaran, R., & Das, A. P. (2024a). Microbial bioremediation technology for sustainable treatment and management of synthetic microfiber waste. In *Sustainable microbial technology for synthetic and cellulosic microfiber bioremediation* (pp. 221–238). https://doi.org/10.1007/978-3-031-62600-5_12

Das, A., Dey, S., & Das, A. P. (2024b). Ecotoxicological impacts of synthetic microfiber pollutants and development of sustainable mitigation strategies. *Environmental Chemistry and Ecotoxicology*. https://doi.org/10.1016/J.ENCECO.2024.12.003

Devi, M. K., Karmegam, N., Manikandan, S., Subbaiya, R., Song, H., Kwon, E. E., Sarkar, B., Bolan, N., Kim, W., Rinklebe, J., & Govarthanan, M. (2022). Removal of nanoplastics in water treatment processes: A review. *Science of the Total Environment, 845*, Article 157168.

Dey, S., Das, A., Mallick, K., Sahu, A., & Das, A. P. (2023a). Environmental petroleum waste: Pollution. *Toxicity, Sustainable Remediation, 159–175.* https://doi.org/10.1007/978-3-031-48220-5_7

Dey, S., Tripathy, B., Kumar, M. S., & Das, A. P. (2023b). Ecotoxicological consequences of manganese mining pollutants and their biological remediation. *Environmental Chemistry and Ecotoxicology*. https://doi.org/10.1016/J.ENCECO.2023.01.001

Dey, S., Bissoyi, A., Koteswara Reddy, G., & Das, A. P. (2024). Synthetic microfibers: An urban emerging pollutant demanding for sustainable environmental solutions. In *Sustainable microbial technology for synthetic and cellulosic microfiber bioremediation* (pp. 107–121). https://doi.org/10.1007/978-3-031-62600-5_7

Dinoop, L. S., Sunil, J. T., & Rajesh, C. (2019). Solid-phase photodegradation of polystyrene by nano TiO_2 under ultraviolet radiation. *Environmental Nanotechnology, Monitoring & Management,* 12, 100229. https://doi.org/10.1016/j.enmm.2019.100229

Duis, K., & Coors, A. (2016). Microplastics in the aquatic and terrestrial environment: Sources (with a specific focus on personal care products), fate and effect. *Environmental Sciences Europe, 28*(1), 2. https://doi.org/10.1186/s12302-015-0069-y

Edo, C., Gonzalez-Pleiter, M., Leganes, F., Fernande-Pinas, F., & Rosal, R. (2020). Fate of microplastics in wastewater treatment plants and their environmental dispersion with effluent and sludge. *Environmental Pollution, 259*, Article 113837. https://doi.org/10.1016/j.envpol.2019.113837

Elmacı, G. (2020). Microwave-assisted rapid synthesis of C@Fe_3O_4 composite for removal of microplastics from drinking water. *Adıyaman University Journal of Science,* 10(1), 207–217. https://doi.org/10.37094/adyujsci.739599

Gallo, F., Fossi, C., Weber, R., Santillo, D., Sousa, J., Ingram, I., Nadal, A., & Romano, D. (2018). Marine litter plastics and microplastics and their toxic chemicals components: The need for urgent preventive measures. *Environmental Science Europe, 30*, 1–14.

Gao, L., Fu, D., Zhao, J., Wu, W., Wang, Z., Su, Y., & Peng, L. (2021). Microplastics aged in various environmental media exhibited strong sorption to heavy metals in seawater. *Marine Pollution Bulletin., 169*, Article 112480. https://doi.org/10.1016/j.marpolbul.2021.112480

Gionfra, S. (2018). Plastic pollution in soil. *Institute of European Environmental Policy,* 1–6.

Gnanasekaran, G., Arthanareeswaran, G., & Mok, Y. S. (2021). A high-flux metal-organic framework membrane (PSF/MIL-100 (Fe)) for the removal of microplastics adsorbing dye contaminants from textile wastewater. *Separation and Purification Technology, 277*, 119655. https://doi.org/10.1016/j.seppur.2021.119655

Goh, P. S., Kang, H. S., Ismail, A. F., Khar, W. H., Quen, L. K., & Higgins, D. (2022). Nanomaterials for microplastics remediation from aquatic environment: Why nano matters? *Chemosphere, 299*, Article 134418. https://doi.org/10.1016/j.chemosphere.2022.134418

Gong, X., Huang, D., Liu, Y., Peng, Z., Zeng, G., Xu, P., Cheng, M., Wang, R., & Wan, J. (2018). Remediation of contaminated soils by biotechnology with nanomaterials: Bio-behavior, applications, and perspectives. *Critical Reviews in Biotechnology, 38*(3), 455–468. https://doi.org/10.1080/07388551.2017.1368446

Grbic, J., Ngvyen, B., Guo, E., You, J. B., Sinton, D., & Rochman, C. M. (2019). Magnetic extraction of microplastics from environmental samples. *Environmental Science & Technology Letters, 6*(2), 68–72. https://doi.org/10.1021/acs.estlett.8b00671

Haris, M., Khan, M. W., Zavabeti, A., Mahmood, N., & Eshtiaghi, N. (2023). Self-assembly of C@FeO nanopillars on 2D-MOF for simultaneous removal of microplastic and dissolved contaminants from water. *Chemical Engineering Journal, 455*, Article 140390. https://doi.org/10.1016/j.cej.2022.140390

He, D., Luo, Y., Lu, S., Liu, M., Song, Y., & Lei, L. (2018). Microplastics in soils: Analytical methods, pollution characteristics, and ecological risks. *Trends in Analytical Chemistry, 109*, 163–172.

Heo, Y., Lee, E.-H., & Lee, S.-W. (2022). Adsorptive removal of micron-sized polystyrene particles using magnetic iron oxide nanoparticles. *Chemosphere, 307*, Article 135672.

Hidayaturrahman, H., & Lee, T.-G. (2019). A study on characteristics of microplastic in wastewater of South Korea: Identification, quantification, and fate of microplastics during treatment process. *Marine Pollution Bulletin, 146*, 696–702. https://doi.org/10.1016/j.marpolbul.2019.06.071

Honarmandrad, Z., Kaykhaii, M., & Gebicki, J. (2023). Microplastic removal from aqueous environment by metal organic frameworks. *BMC Chemistry, 17*, 122. https://doi.org/10.1186/s13065-023-01032-y

Indhur, R., Amoah, I., Bux, F., & Kumari, S. (2023). Nanomaterials for microplastic removal from wastewater: Current state of the art nanomaterials and future prospects. *ACS EST Water, 3*, 3741–3754.

Jalvo, B., Aguilar-Sanchez, A., Ruiz-Caldas, M. X., & Mathew, A. P. (2021). Water filtration membranes based on non-woven cellulose fabrics: Effect of nanopolysaccharide coatings on selective particle rejection, antifouling, and antibacterial properties. *Nanomaterials, 11*(7), 1752.

Jeyaraj, J., Baskaralingam, V., Stalin, T., & Muthuvel, I. (2023). Mechanistic vision on polypropylene microplastics degradation by solar radiation using TiO_2 nanoparticle as photocatalyst. *Environmental Research, 233*, Article 116366. https://doi.org/10.1016/j.envres.2023.116366

Jia, Y., Samak, N. A., Hao, X., Chen, Z., Yang, G., Zhao, X., Mu, T., Yang, M., & Xing, J. (2021). Nano-immobilized of PETase enzyme for enhance polyethylene terephthalate biodegradation. *Biochemical Engineering Journal, 176*, Article 108205. https://doi.org/10.1016/j.bej.2021.108205

Kamrannejad, M. M., Hasanzadeh, A., Nosoudi, N., Mai, L., & Babaluo, A. A. (2014). Photocatalytic degradation of polypropylene/TiO_2 Nano-composites. *Material Research, 17*(4), 1039–1046. https://doi.org/10.1590/1516-1439.267214

Kaneti, Y. V., Tang, J., Salunkhe, R. R., Jiang, X., Yu, A., Wu, K. C. W., & Yamauchi, Y. (2017). Nanoarchitectured design of porous materials and nanocomposites from metal-organic frameworks. *Advanced Materials, 29*(12), 1604898.

Kang, J., Zhou, L., Duan, X., Sun, H., Ao, Z., & Wang, S. (2019). Degradation of cosmetic microplastics via functionalized carbon nanosprings. *Matter, 1*(3), 745–758. https://doi.org/10.1016/j.matt.2019.06.004

Kaur, K., & Kaur, H. (2025). Visible light induced photocatalytic degradation of polyvinyl chloride from water using graphene oxide/nitrogen-doped TiO_2 nanocomposite. *Journal of Environmental Chemical Engineering, 13*(2), Article 115664. https://doi.org/10.1016/j.jece.2025.115664

Khan, M. S., Asif, M. I., Asif, M., Khan, M. R., Mustafa, G., & Adeel, M. (2024). Nanomaterials for the catalytic degradation and detection of microplastics: A review. *Topics in Catalysis, 169*, Article 112480. https://doi.org/10.1007/s11244-024-01927-7

Khatun, R., Singh, S., Dubey, N. K., & Prasad Das, A. (2023). A review on marine-based phytochemicals and their application in biomedical research. In *Recent frontiers of phytochemicals: Applications in food, pharmacy, cosmetics, and biotechnology* (pp. 383–395). Elsevier. https://doi.org/10.1016/B978-0-443-19143-5.00013-X

Kokkinos, P., Mantzavinos, D., & Venieri, D. (2020). Current trends in the application of nanomaterials for the removal of emerging micropollutants and pathogens from water. *Molecules, 25*(9), 2016. https://doi.org/10.3390/molecules25092016

Kotnis, S., Gulati, S., & Sun, Q. (2025). High-efficiency PET degradation with a duo-enzyme system immobilized on magnetic nanoparticles. *Biotechnology and Bioengineering*. https://doi.org/10.1002/bit.28963

Krakor, E., Gessner, I., Wilhelm, M., Brune, V., Hohnsen, J., Frenzen, L., & Mathur, S. (2021). Selective degradation of synthetic polymers through enzymes immobilized on nanocarriers. *MRS Communications, 11*, 363–371.

Lal, S., Anisia, K. S., & Kumar, A. (2006). Depolymerization of HDPR to wax in the presence of a catalysts formed by homonuclear macrocyclic zirconium complex chemically bonded to alumina support. *Applied Catalysis A: General, 303*(1), 9–17. https://doi.org/10.1016/j.apcata.2006.01.016

Lares, M., Ncibi, M. C., Sillanpaa, M., & Sillanpaa, M. (2018). Occurrence, identification and removal of microplastic particles and fibers in conventional activated sludge process and advanced MBR technology. *Water Research, 133*, 236–246.

Li, J., Song, Y., & Cai, Y. (2020). Focus topics on microplastics in soil: Analytical methods, occurrence, transport, and ecological risks. *Environmental Pollution, 257*, Article 113570. https://doi.org/10.1016/j.envpol.2019.113570

Li, Y., & Liu, J. (2021). Nanozyme's catching up: Activity, specificity, reaction conditions and reaction types. *Materials Horizons, 8*, 336–350. https://doi.org/10.1039/D0MH01393E

Li, Z., Chen, K., Yu, L., Shi, Q., & Sun, Y. (2022). Fe_3O_4 nanoparticles-mediated solar-driven enzymatic PET degradation with PET hydrolase. *Biochemical Engineering Journal, 180*, Article 108344. https://doi.org/10.1016/j.bej.2022.108344

Liang, W., Luo, Y., Song, S., Dong, X., & Yu, X. (2013). High photocatalytic degradation activity of polyethylene containing polyacrylamide grafted TiO_2. *Polymer Degradation and Stability, 98*, 1754–1761.

Liu, D., Gu, W., Zhang, L., Wang, L., Zhang, J., Liu, Y., & Lei, J. (2022). Recent advances in MOF-derived carbon-based nanomaterials for environmental applications in adsorption and catalytic degradation. *Chemical Engineering Journal, 427*, Article 131503. https://doi.org/10.1016/j.cej.2021.131503

Mahbub, P., & Duke, M. (2023). Scallability of advanced oxidation processes (AOPs) in industrial applications: A review. *Journal of Environmental Management, 345*, Article 118861. https://doi.org/10.1016/j.jenvman.2023.118861

Mallick, K., Sahu, S., Sahu, A., Dey, S., & Das, A. P. (2024). A critical review of marine microfiber pollution routes, toxicity, and its sustainable remediation. *Environmental Science and Engineering, Part, F2333*, 189–211. https://doi.org/10.1007/978-3-031-51792-1_10

Microbead-Free Waters Act of 2015, Pub. L. No. 114–114 (2015)

Mishra, S., Rath, C. C., & Das, A. P. (2019). Marine microfiber pollution: A review on present status and future challenges. *Marine Pollution Bulletin, 140*, 188–197. https://doi.org/10.1016/J.MARPOLBUL.2019.01.039

Mohanty, J., Dey, S., Bala, K., & Das, A. P. (2024). Recent advances in microplastic pollution for its sustainable management. In *Sustainable management of urban plastic waste through circular economic approaches* (pp. 27–48). https://doi.org/10.1201/9781003465447-3

Monira, S., Roychand, R., Bhuiyan, M. A., Hai, F. I., & Pramanik, B. K. (2022). Identification, classification, and quantification of microplastics in road dust and stormwater. *Chemosphere, 299*, Article 134389.

Nabi, I., Bacha, A. U. R., Li, K., Cheng, H., Wang, T., Liu, Y., Ajmal, S., Yang, Y., Feng, Y., & Zhang, L. (2020). Complete photocataltic mineralization of microplastics on TiO_2 nanoparticle film. *Science, 23*(7), Article 101326. https://doi.org/10.1016/j.isci.2020.101326

Ouda, M., Banat, F., Hasan, S. W., & Karanikolos, G. N. (2023). Recent advances on nanotechnology-driven strategies for remediation of microplastics and nanoplastics from aqueous environments. *Journal of Water Process Engineering, 52*, Article 103543. https://doi.org/10.1016/j.jwpe.2023.103543

Pasanen, F., Fuller, R. O., & Maya, F. (2023). Fast and simultaneous removal of microplasticss and microplastics-derived endocrine disruptors using a magnetic ZIF-8 nanocomposite. *Chemical Engineering Journal, 455*, Article 140405. https://doi.org/10.1016/j.cej.2022.140405

Peng, B., Chen, Z., Chen, J., Zhou, X., Wu, W.-M., & Zhang, Y. (2021). Biodegradation of polylactic acid by yellow mealworms (larvae of Tenebria molitor) via resorce recovery: A sustainable approach for waste management. *Journal of Hazardous Materials, 416*, Article 125803. https://doi.org/10.1016/j.jhazmat.2021.125803

Peng, S., Ji, X., Dong, H., Guo, Z., Ma, H., Han, R., Pang, K., Chen, X., & Wang, X. (2024). Designing MOFs/CuCo-LDHs hybrid-based superhydrophobic sponge for speedy eliminating microplastics, dyes, and oils from high-salinity water. *Colloids and Surfaces a: Physicochemical and Engineering Aspects, 692*, Article 133996. https://doi.org/10.1016/j.colsurfa.2024.133996

Petroody, S. S. A., Hashemi, S. H., & van Gestel, C. A. (2021). Transport and accumulation of microplastics through wastewater treatment sludge processes. *Chemopshere, 278*, Article 130471.

Peydayesh, M., Suta, T., Usuelli, M., Handschin, S., Canelli, G., Bagnani, M., & Mezzenga, R. (2021). Sustainable removal of microplastics and natural organic matter from water by coagulation–flocculation with protein amyloid fibrils. *Environmental Science & Technology, 55*(13), 8848–8858.

Prata, J. C., Da Costa, J. P., Duarte, A. C., & Rocha-Santos, T. (2019). Methods for sampling and detection of microplastics in water and sediment: A critical review. *TrAC, Trends in Analytical Chemistry, 110*, 150–159. https://doi.org/10.1016/j.trac.2018.10.029

Priyadarshini, B., & Das, A. P. (2024). The adverse impact of synthetic microplastics released from disposable face masks pollutants on aquatic microbiota, aquatic species and humans: A review. *Environmental Science and Engineering, Part, F2827*, 67–81. https://doi.org/10.1007/978-3-031-50840-0_5

Raghavendra, V. B., Shankar, S., Govindappa, M., Pugazhendhi, A., Sharma, M., & Nayaka, S. C. (2022). Green synthesis of zinc oxide nanoparticles (ZnO NPs) for effective degradation of dye, polyethylene and antibacterial performance in waste water treatment. *Journal of Inorganic and Organometallic Polymers and Materials, 32*, 614–630. https://doi.org/10.1007/s10904-021-021 42-7

Ray, S. K., Dey, S., Das, A. P., & John, A. (2024). Microplastic waste management through greener technologies: A review. In *Sustainable management of urban plastic waste through circular economic approaches* (pp. 237–261). https://doi.org/10.1201/9781003465447-14

Raza, W., Kukkar, D., Saulat, H., Raza, N., Azam, M., Mehmood, A., & Kim, K. K. H. (2019). Metal-organic frameworks as an emerging tool for sensing various targets in aqueous and biological media. *Trends in Analytical Chemistry, 120*, Article 115654.

Ren, A., Ruis- Ayra, O., Kang, M., & Llorca-Iserm, N. (2024). Durably superhydrophobic magnetic cobalt territes for highly efficient oil-water separation and fast microplastic removal. *Langmuir, 40*(41), 21533–21546. https://doi.org/10.1021/acs.langmuir.4c02420

Rizvi, N. B., Sarwar, A., Waheed, S., Iqbal, Z. F., Imran, M., Javaid, A., Kim, T. H., & Khan, M. S. (2024). Nano-based remediation strategies for micro and nanoplastic pollution. *Journal of Contaminant Hydrology., 265*, Article 104380. https://doi.org/10.1016/j.jconhyd.2024.104380

Rochman, C. M., Hoh, E., Hentschel, B. T., & Kaye, S. (2013). Long-term field measurement of sorption of organic contaminants to five types of plastic pellets: Implications for Plastic Marine Debris. *Environmental Science and Technology, 47*(3), 1646–1654.

Sacko, A., Nure, J. F., Nyoni, H., Mamba, B., Nkambule, B., & Msagati, T. A. M. (2024). The application of Tannic Acid-coated magnetite nanoparticles for recovery of microplastics from the water system. *Water Conservation Science and Engineering, 9*, 41. https://doi.org/10.1007/s41101-024-00275-7

Satapathy, S., Kar, D., Priyadarsini, S., Das, A. P., & Kuanar, A. (2024). Application of microorganisms in emerging waste management strategies and remediation of synthetic and cellulosic microfibers. In *Sustainable microbial technology for synthetic and cellulosic microfiber bioremediation* (pp. 183–206). https://doi.org/10.1007/978-3-031-62600-5_10

Schultz, D. M., & Yoon, T. P. (2014). Solar synthesis: Prospects in visible light photocataysis. *Science, 343*(6174), Article 123976.

Schwaminger, S. P., Fehn, S., Steegmüller, T., Rauwolf, S., Löwe, H., Pflüger-Grau, K., & Berensmeier, S. (2021). Immobilization of PETase enzymes on magnetic iron oxide nanoparticles for the decomposition of microplastic PET. *Nanoscale Advances, 3*, 4395–4399. https://doi.org/10.1039/D1NA00243K

Shi, L., Benetti, D., Li, F., Wei, Q., & Rosei, F. (2020). Phase-junction design of MOF-derived TiO_2 photoanodes sensitized with quantum dots for efficient hydrogen generation. *Applied Catalysis B: Environmental., 263*, Article 118317.

Shi, X., Zhang, X., Gao, W., Zhang, Y., & He, D. (2022). Removal of microplastics from water by magnetic nano-Fe_3O_4. *Science of the Total Environment, 802*, Article 149838.

Suo, Q., Zi, J., Bai, Z., & Qi, S. (2017). The glycolysis of poly(ethylene terephthalate) promoted by metal organic framework (MOF) catalysis. *Catalysis Letters, 147*, 240–252. https://doi.org/10.1007/s10562-016-1897-0

Tang, Y., Zhang, S., Su, Y., Wu, D., Zhao, Y., & Xie, B. (2021). Removal of microplastics from aqueous solutions by magnetic carbon nanotubes. *Chemical Engineering Journal, 406*, Article 126804. https://doi.org/10.1016/j.cej.2020.126804

Thomas, R. T., & Sandhyarani, N. (2013). Enhancement in the photocatalytic degradation of low density polyethylene-TiO_2 nanocomposite films under solar irradiation. *RSC Advances, 33*(3), 14080–14087.

Thompson, R. C., Olsen, Y., Mitchell, R. P., Davis, A., Rowland, S. J., John, A. W., McGonigle, D., & Russel, A. E. (2004). Last at sea: Where is all the plastics? *Science, 304*(5672), 838.

Tien, C.-J., Wang, Z.-X., & Chen, C. S. (2020). Microplastics in water, sediment, and fish from the Fengshan River system: Relationship to aquatic factors and accumulation of polycyclic aromatic hydrocarbons by fish. *Environmental Pollution, 265*, Article 114962.

Tofa, T. S., Kunjali, K. L., Paul, S., & Dutta, J. (2019). Visible light photocatalytic degradation of microplastic residues with zinc oxide nanorods. *Environmental Chemistry Letters, 17*, 1341–1346. https://doi.org/10.1007/s10311-019-00859-z

Uheida, A., Mejia, H. G., Abdel-Rehim, M., Hamd, W., & Dutta, J. (2021). Visible light photocatalysis degradation of polypropylene microplastics in a continuous water flow system. *Journal of Hazardous Materials, 406*, Article 124299. https://doi.org/10.1016/j.jhazmat.2020.124299

Uogintė, I., Pleskytė, S., Skapas, M., Stanionytė, S., & Lujanienė, G. (2023). Degradation and optimization of microplastic in aqueous solutions with graphene oxide-based nanomaterials. *International Journal of Environmental Science and Technology, 20*(9), 9693–9706. https://doi.org/10.1007/s13762-022-04657-z

UNEP. (2015). Plastic in cosmetics: are we polluting the environment through our personal care? Plastic ingredients that contribute to marine microplastic litter. UN.

Vas, N. V., Sundari, S. K., & Jeyachandran, S. (2024). Degradation of microplastics by zinc oxide nanoparticles synthesized using piper longum leaf extract. *Cureus, 16*(9).

Waller, C. L., Griffiths, H. J., Waluda, C. M., Thorpe, S. E., Loaiza, I., Moreno, B., & Hughes, K. A. (2017). Microplastic in the Antarctic marine system: An emerging area of research. *Science of the Total Environment, 598*, 220–227. https://doi.org/10.1016/j.scitotenv.2017.03.283

Wan, H., Wang, j., Sheng, X., Yan, J., Zhang, W., & Xu, Y. (2022). Removal of polystyrene microplastics from aqueous solution using the metal-organic framework material of ZIF-67. *Toxics, 10*(2), 70. https://doi.org/10.3390/toxics10020070

Wang, L., Liu, B., Zhang, W., Li, Q., Lin, B., & Wei, C. (2020). An unrecognized entry pathway of submicrometer plastics into crop root: The split of hole in protective layer. *Journal of Hazardous Materials, 457*, Article 131732.

Wang, J., Sun, C., Huang, Q.-X., Chi, Y., & Yan, J.-H. (2021). Adsorption and thermal degradation of microplastics from aqueous solutions by Mg/Zn modified magnetic biochars. *Journal of Hazardous Materials., 419*, Article 126486. https://doi.org/10.1016/j.jhazmat.2021.126486

Wu, P., Guo, M., Zhang, R.-W., Huang, Q., Wang, G., & Lan, Y.-Q. (2024). When microplastics/plastics meet metal-organic frameworks: Turning threats into opportunities. *Chemical Science, 15*, 17781.

Xiao, Y., Tian, Y., Xu, W., & Zhu, J. (2024). Photodegradation of Microplastics through nanomaterials: Insights into photocatalysts modification and detailed mechanism. *Materials, 17*(11), 2755. https://doi.org/10.3390/ma17112755

Yan, R., Lin, S., Jiang, W., Yu, X., Zhang, L., Zhao, W., & Sui, Q. (2023). Effect of aggregation behavior on microplastics removal by magnetic Fe_3O_4 nanoparticles. *Science of the Total Environments, 898*, Article 165431.

Yang, B., Geng, P., & Chen, G. (2015). One-dimensional structured IrO_2 nanorods modified membrane for electrochemical anti-fouling infiltration of oily wastewater. *Separation and Purification Technology, 156*(3), 931–941. https://doi.org/10.1016/j.seppur.2015.10.040

Yang, L., Cao, X., Cui, J., Wang, Y., Zhu, Z., Sun, H., Liang, W., Li, J., & Li, A. (2022). Holey Ti_3C_2 nanosheets based membranes for efficient separation and removal of microplastics from water. *Journal of Colloid and Interface Science, 617*, 673–682. https://doi.org/10.1016/j.jcis.2022.03.055

Ye, J., Chen, Y., Gao, C., Wang, C., Hu, A., Dong, G., Chen, Z., Zhou, S., & Xiong, Y. (2022). Sustainable conversion of microplastics to methane with ultrahigh selectivity by a biotic-abiotic hybrid photocatalytic system. *Angewandte Chemie, 62*(52), Article e202213244. https://doi.org/10.1002/anie.202213244

Yeap, S. P., Ahmad, A. L., Ooi, B. S., & Lim, J. (2012). Electrostatic stabilization and its role in cooperative magnetphoresis of colloidal magnetic nanoparticles. *Langmuir, 28*, 14878–14891.

You, D., Zhao, Y., Yang, W., Pan, Q., & Li, J. (2022). Metal-organic framework-based wood aerogel for effective removal of micro/nano plastics. *Chemical Research in Chinese Universities, 38*, 186–191. https://doi.org/10.1007/s40242-021-1317-x

Yu, T., Huang, X., Zhang, X. F., Li, K., Liu, S. P., Dai, N., Zhang, K., Zhang, Y. X., & Li, H. (2024). A review of nanomaterials with excellent purification potential for the removal of micro-and nanoplastic from liquid. *DeCarbon, 100064*. https://doi.org/10.1016/j.decarb.2024.100064

Yuan, F., Yue, L., Zhao, H., & Wu, H. (2020). Study on the adsorption of polystyrene microplastics by three-dimensional reduced graphene oxide. *Water Science and Technology, 81*(10), 2163–2175. https://doi.org/10.2166/wst.2020.269

Zandieh, M., & Liu, J. (2022). Removal and degradation of MPs using the magnetic and nanozyme activities of bare iron oxide nanoagregates. *61*(47), e202212013. https://doi.org/10.1002/anie.202212013

Zhang, V., Hu, Y., Ge, J., Jiang, H.-L., & Yu, S.-H. (2014). A facile and general coating approach to moisture/water-resistant metal-organic frameworks with intact porosity. *Journal of American Chemical Society, 136*(49), 16978–16981. https://doi.org/10.1021/ja509960n

Zhang, X., Zhai, Y., Wang, Z., Liu, X., Liu, X., Zhou, Y., Liu, G., & Xu, M. (2023). The removal characteristics and mechanisms of polystyrene microplastics with various induced photoaging degrees by $CuFe_2O_4$. *Separation and Purification Technology, 322*, Article 124245. https://doi.org/10.1016/j.seppur.2023.124245

Zhao, X., Li, Z., Chen, Y., Shi, L., & Zhu, Y. (2007). Solid-phase photocatalytic degradation of polyethylene plastic under UV and solar light irradiation. *Journal of Molecular Catalysis A: Chemical, 268*(1–2), 101–106. https://doi.org/10.1016/j.molcata.2006.12.012

Zheng, H., Chen, Q., & Chen, Z. (2024). Carbon-based adsorbents for micro/nano-plastics removal: Current advances and perspectives. *Water Energy Contaminants & Nanoplastics, 3*, 11. https://doi.org/10.20517/wecn.2023.74

Zhou, H., Mayora-Martinez, C. C., & Pumera, M. (2021). Microplastic removal and degradation by mussel-inspired adhesive magnetic/enzymatic microbots. *Small Methods, 5*(9), 2100230. https://doi.org/10.1002/smtd.202100230

Zhu, S., & Wang, D. (2017). Photocatalysis: Basics principles, diverse forms of implementations and emerging, Scientific Opportunities. *Advanced Energy Materials, 7*(23), 1700841. https://doi.org/10.1002/aenm.201700841

Zhuang, J., Rong, N., Wang, X., Chen, C., & Xu, Z. (2022). Adsorption of small size microplastics based on cellulose nanofiber aerogel modified by quaternary ammonium salt in water. *Separation and Purification Technology, 293*, Article 121133. https://doi.org/10.1016/j.seppur.2022.121133

Advanced Nanomaterials for Microplastic Detection and Degradation: A Sustainable Approach to Environmental Remediation

Jagriti Gupta, Rachita Chauhan, and Jaydeep Bhattacharya

Abstract The widespread occurrence of microplastics (MPs) in ecosystems presents serious health and environmental issues, requiring innovative and effective management strategies. This chapter investigates how nanomaterials might be used to address MPs degradation and regulation. Many advances have been achieved in decomposing MPs into less hazardous components by analysing the unique physicochemical properties of nanomaterials, such as enhanced reactivity, huge surface area, and catalytic potential. The chapter examines a variety of nanomaterials, such as metal oxides, carbon-based nanostructures, and bio-supported nanoparticles, emphasizing their mechanisms of action to identify and degrade MPs. It also discusses their incorporation into remediation technologies, such as photocatalytic degradation, oxidative degradation, enzyme-mimicking properties, and nano-catalysts, as well as adsorption and magnetic extraction. The challenges of scaling up these nanotechnological interventions, potential environmental risks, and regulatory considerations are critically analyzed. This chapter underscores the role of interdisciplinary research in enhancing the safety and efficacy of nanomaterial-based solutions, ultimately contributing to sustainable environmental management practices, showcasing nanotechnology as a pivotal tool in quantifying and mitigating the MPs crisis.

Keywords Microplastics pollution · Detection · Degradation · Environmental remediation · Sustainability

1 Introduction

Microplastic (MPs) pollution has emerged as a critical global environmental issue due to rising plastic production and inadequate waste management. In 2023, global plastic production reached approximately 413.8 million metric tons (Mt), nearly

J. Gupta · R. Chauhan · J. Bhattacharya (✉)
Nanobiotechnology Laboratory, School of Biotechnology, Jawaharlal Nehru University, New Delhi 110067, India
e-mail: jaydeep@jnu.ac.in

E. Priyadarshini et al. (eds.), *Advanced Nanotechnology Applications for Environmental Sustainability*, Sustainable Environmental Waste Management Strategies,
https://doi.org/10.1007/978-3-032-12933-8_12

doubling from 213 Mt in 2000 (Statista, 2025a). This increase has led to a significant surge in plastic waste, estimated at 389.23 Mt in 2023 (Statista, 2025b). However, only 9% of plastic waste is recycled, 19% is incinerated, and nearly 50% is disposed of in sanitary landfills, with the remaining 22% discarded into the environment or burned in open pits in 2019 (OECD, 2022). Environmental degradation transforms this plastic waste into microscale particles, classified as macroplastics (>25 mm), mesoplastics (5–25 mm), MPs (<5000 μm), and nanoplastics (NPs) (<1 μm). A substantial proportion of these contaminants infiltrate aquatic and terrestrial ecosystems, negatively impacting biodiversity (Gupta & Rajamani, 2024; Walker & Fequet, 2023). In 2019, an estimated 22 Mt of plastics leaked into the environment, with MPs accounting for 88% (OECD, 2022). Without policy interventions, cumulative global plastic production could reach 12,000 Mt by 2050, exacerbating MPs pollution (Geyer et al., 2017).

Approximately 6.1 Mt of plastic waste leaks into aquatic systems annually (OECD, 2022), with 1–2 Mt entering oceans (Ritchie et al., 2023). MPs and NPs are also transported across continents and oceans through air currents, sea spray, and fog, with an estimated 0.013–25 Mt transported annually (Allen et al., 2022). The increasing concentration of MPs in marine environments poses severe risks to aquatic organisms (Acarer Arat, 2024). Exposure to 1–3 μm MPs reduced population growth, fertilization rates, and reproductive gene expression in *Brachionus plicatilis*, while inducing oxidative stress (Seong et al., 2024). Additionally, MPs inhibited growth and chlorophyll production in *Scenedesmus obliquus* (Besseling et al., 2014), reduced body size and reproduction in *Daphnia magna* (Besseling et al., 2014), and produced pseudofeces formation in *Mytilus edulis*, which resulted in hunger and mortality (Wegner et al., 2012). Additionally, *Litopenaeus vannamei* survival rates were decreased by polystyrene (PS) NPs, which caused immunological suppression, oxidative stress, and histopathological damage (Hsieh et al., 2023).

Through trophic transmission and bioaccumulation, MPs contamination also impacted terrestrial ecosystems and food supplies (Fig. 1) (O'Connor et al., 2022; Rose et al., 2023). MPs have changed the properties of soil, and impacted the plant growth and microbial populations (Hasan & Tarannum, 2025). They hindered root development and lowered plant productivity by interfering with soil aeration, water retention, and nutrient cycling (Chen et al., 2022a; Zhou et al., 2021b). For example, Ren et al. (2021) reported that PS beads altered soil microbial metabolism by impacting chlorophyll a more than chlorophyll b, which in turn reduced plant development and nutrient availability.

Long-term health impacts of MPs have been concerning since they have been found in human blood, lungs, liver, joints, brain, and placenta (Leslie et al., 2022; Sharma et al., 2024). Research indicates that an average person consumes or inhales 78,000–211,000 MPs particles annually (Cox et al., 2019). This exposure results from environmental contamination, where food is grown, and the omnipresence of MPs in air and water (Cox et al., 2019; Saeedi, 2024). MPs also carry chemical additives and pollutants, potentially causing adverse health effects upon ingestion (Ayejoto et al., 2024; Gupta and Rajamani, 2024). Sun et al. (2022) reported a 31% increase in bioaccumulation and an 18% rise in toxicity due to MPs, influenced by particle

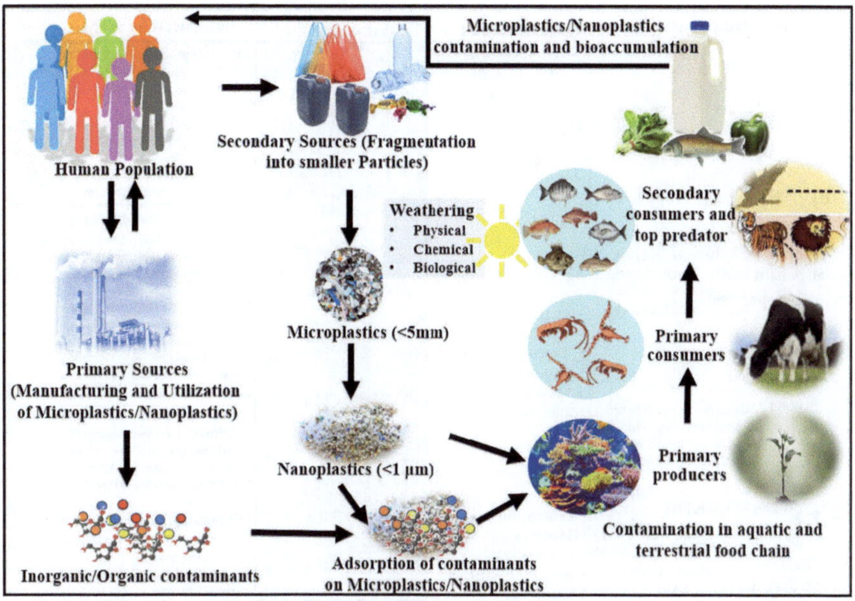

Fig. 1 **a** Sources and fate of microplastics and nanoplastics

size and exposure duration. MPs and NPs persistence disrupts ecosystems, requiring reduced plastic production, advanced waste management, and public awareness for effective mitigation (Gupta & Rajamani, 2024).

2 Challenges in Microplastic Detection and Degradation

2.1 Difficulty in Detecting Microplastics at the Nanoscale

Detecting MPs at the nanoscale is challenging due to their diverse compositions, morphologies, and complex properties (Adhikari et al., 2022). Sample preparation is challenging, requiring contamination control, matrix mitigation, and efficient separation for precise identification (Fig. 2) (Nene et al., 2025). While optical microscopy is a simple, cost-effective, and non-destructive tool for detecting large MPs (>100 μm) in real-time, but is limited by its resolution ($\leq$200 nm) (Song et al., 2015) and inability to determine chemical composition (Forte et al., 2016). It further requires highly sensitive techniques such as Raman spectroscopy, Fourier Transform Infrared Spectroscopy (FTIR), scanning electron microscopy (SEM) coupled with energy-dispersive X-ray spectroscopy (EDX), and Atomic Force Microscopy (AFM), etc. (Fig. 2) (Ivleva et al., 2017). However, environmental aging changes MPs through

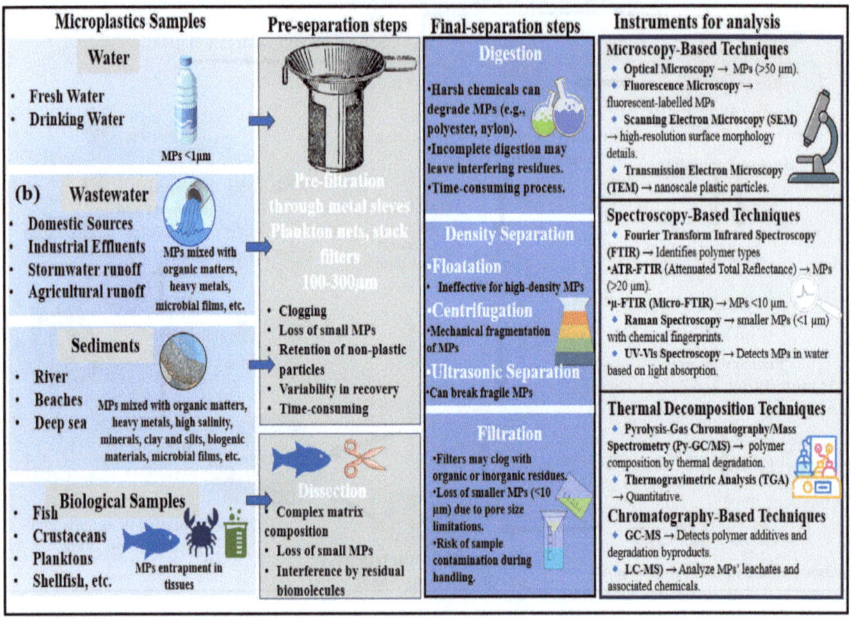

Fig. 2 Analysing microplastics and nanoplastics in simple and complex matrices, sample preparation and identification using different instrumentation

oxidation, UV exposure, and biofilm formation (Luo et al., 2022), making identification more difficult (Nene et al., 2025). These techniques are also expensive, time-consuming, and non-standardized. Additionally, the extraction of MPs from natural organic matter and other pollutants complicates detection and necessitates the use of sophisticated methods like Pyrolysis Gas Chromatography-Mass Spectrometry (Py-GC-MS) (Picó & Barceló, 2020).

2.2 *Limited Efficiency of Conventional Treatment Methods*

Traditional wastewater treatment plants devoid the necessary equipment to remove MPs fully, and large amounts of MPs are released into aquatic habitats. According to reports, up to 95% of MPs might be captured by WWTPs; leaking remaining into natural water bodies, causing plastic pollution (Talvitie et al., 2017). The effectiveness of conventional filtration techniques, like granular activated carbon and membrane bioreactors, in capturing MPs (<10 μm) is limited (Cherniak et al., 2022). The removal of MPs has been investigated using chemical degradation techniques, such as advanced oxidation processes (AOPs) like hydroxyl-radical-based AOPs, Fenton-based AOPs, sulfate-radical-based AOPs, UV/Chlorine advanced oxidation, etc. (Bule Možar et al., 2023; Yang et al., 2024); however, they frequently produce secondary pollutants and require high input of energy (Chen et al., 2021). Biological degradation of MPs using microbial enzymes such as PET hydrolase and PCL cutinase has also shown promise for degradation of PET and PCL, yet their efficiency relies on polymer type and environmental conditions, which leads to incomplete degradation and accumulation in ecosystems (Cai et al., 2023).

2.3 Environmental Concerns Related to Remediation Methods

While existing MPs remediation technologies aim to reduce plastic pollution, many approaches raise environmental concerns. For instance, chemical treatments such as Fenton oxidation and photocatalysis involve the use of reactive radicals, which can generate toxic byproducts, further impacting aquatic life (Chen et al., 2021). Physical filtration techniques, including micro- and nanofiber filters, often require frequent maintenance and disposal due to membrane fouling, leading to secondary waste management issues (Shen et al., 2023). Furthermore, combusting plastic wastes increase air pollution and carbon emissions, affecting the global climate (Shen et al., 2020). Eco-friendly nanocatalysts and biopolymers should be in focus of sustainable solutions to ensure effectiveness without any negative consequences.

3 The Role of Nanomaterials in Microplastic Remediation

Nanomaterials, typically (1–100 nm), have unique physicochemical properties such as a high surface area-to-volume ratio, tunable surface chemistry, enhanced reactivity, and quantum effects, enabling them to effectively address environmental issues, especially in MPs remediation (Yu et al., 2024). They play important role, including detection, quantification, removal, and degradation of MPs (Fig. 3). Fluorescent nanoprobes such as carbon dots (CDs) and quantum dots (QDs) improve MPs visualization (Zhao et al., 2023), metal nanoparticles such as gold (Au) and silver (Ag) enhance Raman signals in Surface-Enhanced Raman Scattering (SERS), increasing sensitivity (Xu et al., 2022). Graphene and metal oxide nanostructures offer high conductivity electrochemical sensing due to their high surface area, conductivity, and functional groups (Baumgarten et al., 2023). Magnetic iron oxide (Fe_3O_4) nanoparticles functionalized with different ligands can effectively separate MPs from complex matrices by heteroaggregation and adsorption (Yan et al., 2023; Zhang et al., 2024). Carbon-based nanomaterials like graphene oxide (GO) and activated carbon adsorb MPs by hydrophobic, π–π, and Van der Waals interactions, improving the efficiency of nanocomposite membranes (Sun et al., 2020a). Semiconductor nanomaterials such as titanium dioxide (TiO_2), zinc oxide (ZnO), and graphitic carbon nitride (g-C_3N_4)

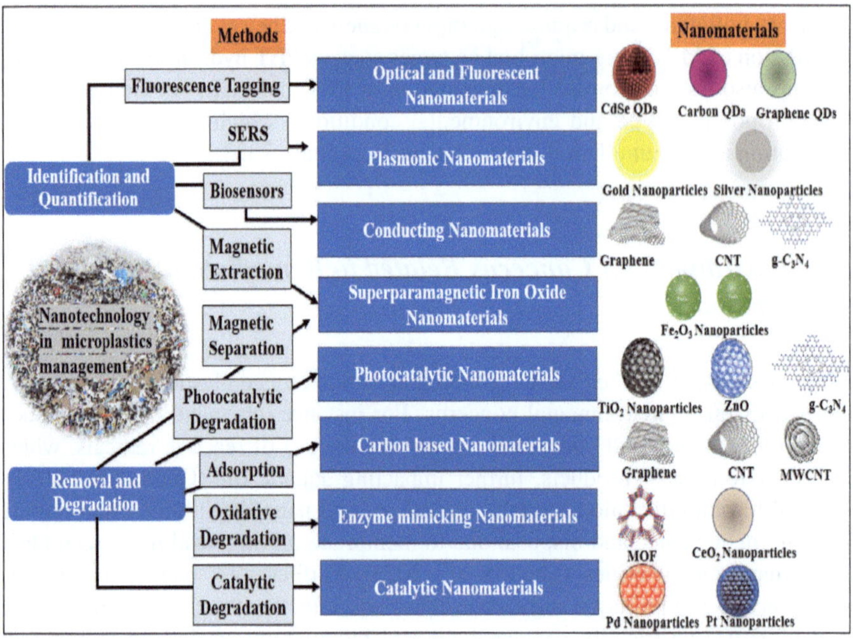

Fig. 3 Overview of different nanomaterials and processes used in microplastics quantification and removal

are reported to break down the MPs into smaller or non-toxic byproducts by photocatalytic degradation by generating reactive oxygen species (ROS) under UV or visible light irradiation (Khan et al., 2024). Additionally, enzyme-mimicking nanomaterials (nanozymes) can also promote MPs degradation by oxidation (Zandieh & Liu, 2022), and plasmonic nanoparticles such as Au and Ag can enhance catalytic degradation through plasmon-assisted photocatalysis (Amirjani et al., 2023).

4 Nanomaterials for Microplastic Detection

4.1 Optical and Fluorescent Nanoprobes

Fluorescent nanoparticles, such as QDs and CDs, can emit detectable fluorescence signals after binding to the surface of MPs. QDs, composed of semiconductor materials like CdSe and CdTe, are reported to detect MPs in complex environmental matrices due to their tunable fluorescence and high photostability. For example, Gianoncelli et al. (2024) used CdSe QDs to identify and track Poly vinyl chloride (PVC) NPs in human ovarian cancer cell line (SK-OV-3) using X-ray fluorescence (XRF) microscopy, and observed distinct fluorescence signals differentiating between NPs and cellular components, which ensured precise localization of MPs without any leaching of QDs, demonstrated its potential for MPs detection in biological systems. However, concerns over their toxicity prompted the exploration of other nanomaterials, such as CDs, due to their biocompatibility. Zhao et al. (2023) detected PS NPs of three different sizes by using Rayleigh scattering and fluorescence mechanism, using nitrogen and chlorine co-doped carbon QDs (N, Cl co-doped CQDs), and found limit of detection (LOD) of 0.4 mg/L. Li et al., (2023a) reported sulfur-doped carbon nanoparticles (S-CNPs) based rapid fluorescence monitoring of polyethylene (PE) MPs within ten minutes. Different MPs exhibited distinct fluorescence signals, which allowed effective differentiation. Additionally, a paper-based fluorescence analysis also demonstrated a linear correlation between fluorescence area and PE mass (5–17 mg), underscoring its potential as a promising probe for MPs detection.

4.2 Plasmonic Nanoparticles for Surface-Enhanced Raman Spectroscopy (SERS)

Plasmonic nanomaterials, such as silver (AgNP) and gold nanoparticles (AuNP), are frequently used in SERS to detect MPs at trace levels due to the enhancement of Raman scattering signals. For example, Xu et al. (2022) fabricated a SERS-based sensor to detect polyethylene terephthalate (PET) MPs using AuNP-doped filter paper as a flexible SERS substrate, and achieved LOD of 0.1 g/L and a maximum enhancement factor of 360.5. The Raman signals were also improved by electromagnetic hot spots (Xu et al., 2022). Caldwell et al. (2021) detected PS of sizes 33 nm or 161 nm and PET of average 62 nm size using 14 and 46 nm AuNP through SERS, and obtained LOD of 10 mg/L with enhancement factors of 446. Kihara et al. (2022) also used AuNP to detect PS NPs (20 nm) using a filter paper-based stable SERS matrix and achieved a LOD of 10 mg/L. Wu et al. (2023) developed a three-dimensional Au@Ag nanostar-based SERS substrate, and reported a LOD of 2.5–25 mg/L for PS MPs. They further applied this probe to successfully detect MPs in tap water, seawater, and soil water samples. Whereas, Wang et al. (2023) reported a Co_3O_4/Co_3S_4/AgNPs-based SERS sensor, and achieved a LOD of 25 mg/L for PS nanospheres (800–1000 nm). The sensor demonstrated detection of PS in real water samples with a LOD of 14 mg/L. Li et al. (2024b) reported a novel honeycomb-like AgNPs@TiO_2 SERS sensor to achieve LOD of 50 mg/L for PS MPs in various water sources with high sensitivity, reproducibility (RSD 9.69%), and recovery rates between 97.6 and 109.7%. Several other studies based on localized surface plasmon resonance (LSPR) or propagating surface plasmon resonance (SPR) of plasmonic nanoparticles to detect refractive index changes at the nanomaterials surface in the presence of MPs and NPs have also been reported. Oh et al. (2021) developed an AuNP-based LSPR sensor functionalized with bio-mimicked peptide probes for selective detection of PS NPs, and achieved 60% enhancement in detection efficiency. Seggio et al. (2024) reported a nano-plasmonic biosensor using an Estrogen Receptor (ER)-grafted gold nanograting (GNG) platform to detect NPs in seawater without any pre-treatment. This sensor enabled rapid detection within three minutes and a linear range of concentration from 1 to 100 ng/mL, and successful quantification of NPs (~30 ng/mL) in real seawater samples.

4.3 *Biosensors and Nano-Enabled Sensor Platforms*

Nanomaterials can integrate into biosensing platforms to enhance MPs/NPs detection and quantification via various mechanisms.

4.3.1 Electrochemical Sensors

Conducting nanomaterials, such as graphene and carbon nanotubes (CNTs), can be incorporated into sensor platforms to improve signal transduction and detection sensitivity. Surucu (2025) reported an electrochemical method to detect polyvinyl chloride (PVC) MPs using a mixed suspension of AgNPs, GO, and multi-walled CNTs deposited on Au electrodes. They observed linearity in the range of concentration 1.00–5.00 mg/mL and a LOD of 0.79 mg/mL. Du et al. (2023) used electrochemical impedance spectroscopy (EIS) to quantify PS MPs of various sizes by using graphene-based electrode, and observed linearity from 0.01 and 25 mg/L, correlation coefficient ($R^2 = 0.9914$) for particle sizes of 1 μm, and $R^2 = 0.9877$ for particle sizes of 0.08–20 μm. Additionally, electrochemical detection of hexamethylenetetramine (HMT) MPs using cyclic voltammetry (CV) and differential pulse voltammetry (DPV) by modifying an indium tin oxide (ITO) electrode with a chitosan-magnesium oxide nanosheet (CHIT-MgO NS) nanocomposite, with a limit of quantitation (LOQ) of 0.10 μM and a LOD of 0.03 μM was reported by Noumani et al. (2024). The sensor showed linearity for concentrations from 0.5 to 4.0 μM and a high sensitivity of 12.908 $\mu A (\mu M)^{-1} cm^{-2}$.

4.3.2 Molecularly Imprinted Polymer (MIP)-Based Sensors

MIPs can mimic natural molecular interactions to provide highly selective recognition sites to MPs. Applications requiring continuous monitoring are enabled by the improved selectivity and reusability of these polymeric sensors. For example, El-Sharif et al. (2022) designed electropolymerized molecularly imprinted polymers (E-MIPs) to detect dioctyl phthalate (DOP), a plasticizer used to increase the flexibility and durability of PVC products. By using low-cost screen-printed electrodes synthesized via CV with 4-aminophenol as the functional monomer, the proposed E-MIPs exhibited an 80% binding efficiency and an imprinting factor of 3 and successfully detected 250 μg (12.8 μM) of DOP within 5 min, with a LOD of 177.1 μg and a LOQ of 536.6 μg, making it a promising tool for freshwater analysis with pre-concentration.

5 Nanomaterial-Based Strategies for Microplastic Mitigation

5.1 Degradation Using Nanomaterials

5.1.1 Photocatalytic Degradation

Various strategies have been reported to mitigate MPs and NPs (Fig. 4a). Photocatalytic degradation of MPs and NPs using semiconductor nanomaterials is an emerging strategy by generating reactive oxygen species (ROS) under irradiation of light. These nanomaterials absorb light and excite electrons to the conduction band (CB) from valence band (VB), to create holes that drive oxidative degradation. The resulting ROS, such as hydroxyl and superoxide radicals, degrade polymer chains into CO_2, H_2O, and other biodegradable byproducts (Fig. 4b) (Ariza-Tarazona et al., 2020). Due to TiO_2's excellent oxidative potential, stability, and non-toxicity, it is widely used as photocatalysts. For example, Llorente-García et al. (2020) reported the degradation of high-density polyethylene (HDPE) and low-density polyethylene (LDPE) MPs in water using visible-light photocatalysis with mesoporous N–TiO_2. They observed that, film-shaped MPs had low degradation due to limited light exposure and oxygenation, compared to smaller MPs. Similarly, Domínguez-Jaimes et al. (2021) used anodized TiO_2 and reported photocatalytic degradation of PS NPs, achieving 23.5% reduction in concentration after 50 h of irradiation. Ullattil and Pumera (2023) developed TiO_2 based superstructured microrobots for phototrapping and photofragmentation of MPs. Several other studies have also reported using TiO_2 and its composites with other nanomaterials for the degradation of different MPs (Table 1).

ZnO nanoparticles also reported to have photocatalytic activity and ability to absorb both UV and visible light to degrade MPs/NPs (Table 1). Furthermore, metal-doped catalysts such Ag–TiO_2 and Fe–ZnO improved charge separation by decreasing recombination rates and increasing degradation efficiency (Piazza et al., 2022; Yuwendi et al., 2022). Carbon-based nanomaterials such as graphene oxide (GO) and carbon nanotubes (CNTs) enhanced photocatalytic activity and improved electron transport by acting as co-catalysts (An et al., 2014; Uoginte et al., 2023). According to recent study, TiO_2/ZnO composites have completely degraded PE and polyester (PES) when exposed to simulated sunlight and an electron scavenger (He et al., 2023). The development of visible-light-active photocatalysts, such as Cu_2O, g-C_3N_4 and $BiVO_4$, further expanded the applicability of photocatalytic degradation, offering sustainable solutions for MPs remediation in wastewater as well as natural

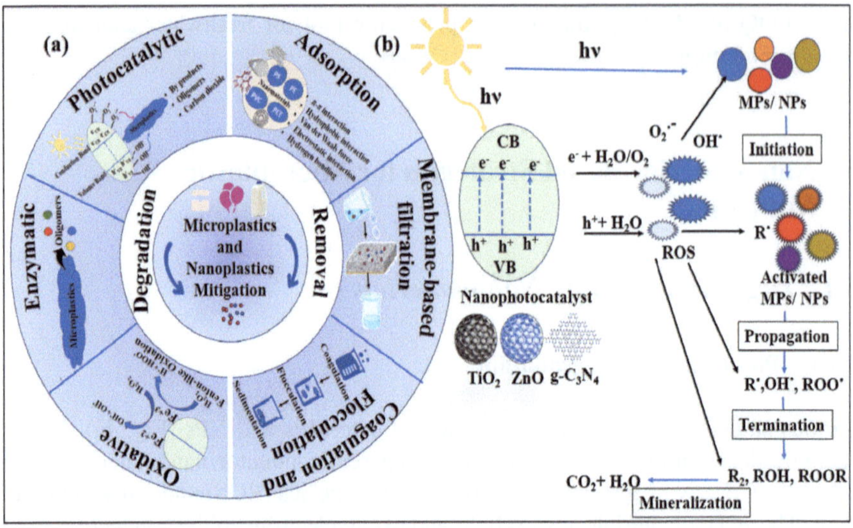

Fig. 4 Various adopted strategies for nanomaterials-based mitigation of microplastics and nanoplastics, **b** schematic of photocatalytic degradation of microplastics and nanoplastics by nanomaterials based photocatalysts

ecosystems (Acuña-Bedoya et al., 2021; Chai et al., 2023; Chu et al., 2025) (Table 1).

5.1.2 Oxidative Degradation

Oxidative degradation of MPs using nanomaterials is a promising strategy that uses AOPs such as photocatalysis, Fenton, and photo-Fenton reactions. These processes generate highly reactive hydroxyl radicals (·OH) that break down MPs into smaller, fragments. The classical homogeneous Fenton reaction involves Fe^{2+} reacting with hydrogen peroxide (H_2O_2) to produce ·OH (Fig. 4a), but its limitations, such as low working pH and iron sludge generation, have led to the development of heterogeneous Fenton processes. Heterogeneous catalysts like zero-valent iron (Fe^0) nanoparticles and zinc oxide (ZnO) nanorods coated with tin oxide (SnO_x) enhance electron transfer, promoting Fe^{3+}/Fe^{2+} recycling under visible light and accelerating MP degradation (Piazza et al., 2022). Similarly, Xue et al. (2024) reported the photocatalysis and photothermal Fenton-like reaction for MPs removal using an α-Fe_2O_3 nanoflower on TiO_2 film with a hierarchical structure of inverse opal-like layer/nanotube arrays (α-Fe_2O_3/TiO_2 HNTAs). Without external H_2O_2, nearly 100% degradation of 310 nm PS spheres is achieved in 4 h at 75 °C under visible-light irradiation. The integration of photocatalysis with Fenton chemistry increases electron production, enhancing charge separation and radical generation and faster degradation (Piazza et al., 2022; Xue et al., 2024).

Table 1 Examples of nanomaterials and their composites for microplastic and nanoplastic mitigation

MPs/NPs	Nanomaterials/Hybrid	Degradation efficiency/removal efficiency/adsorption capacity	Technique/mechanism	References
PE	TiO_2–MWCNTs nanocomposites	35%	Photocatalytic degradation under mercury lamp irradiation Reaction time = 180 h	An et al. (2014)
PET Fiber-MPs	Bi_2O_3@N–TiO_2	10.23 ± 1.9%	Photocatalytic degradation at pH = 9	Zhou et al. (2022)
PET	C, N–TiO_2/SiO_2	9.35–16.22%	Photocatalytic degradation under visible-light at pH = 6 or 8	Ariza-Tarazona et al. (2023)
HDPE	C, N–TiO_2	70%	Photocatalytic degradation under visible light at low pH = 3 and Temp = 0 °C	Ariza-Tarazona et al. (2020
HDPE	Protein-based porous N–TiO_2	2.86%	Photocatalytic degradation under visible light at Reaction Time = 16 h	Ariza-Tarazona et al. (2019)

(continued)

Table 1 (continued)

MPs/NPs	Nanomaterials/Hybrid	Degradation efficiency/removal efficiency/adsorption capacity	Technique/mechanism	References
PE, PES microfibres	TiO_x/ZnO tetrapod	100%	Photocatalytic degradation under UV-A illumination and the addition of an electron scavenger $Na_2S_2O_8$ Reaction time = 480 h and 624 h respectively	He et al. (2023)
PE	Fe- and Ag-modified TiO_2 nanotubes	18%	Photocatalytic degradation under UV-C, Reaction time = 90 min	Yuwendi et al. (2022)
PE and PVC	ZnO nanorods coated with SnOx x < 2/ layer and decorated with Fe^0 NPs	95%	Degradation by photo-Fenton process under visible light and H_2O_2	Piazza et al. (2022)
LDPE	Pt/ZnO Nanorods	–	Photocatalytic degradation under visible light	Tajkia Syeed Tofa et al. (2019)
Polypropylene (PP)	ZnO nanoparticles	–	Photocatalytic degradation under artificial sunlight exposure	Sun and Wang (2023)

(continued)

Table 1 (continued)

MPs/NPs	Nanomaterials/Hybrid	Degradation efficiency/removal efficiency/adsorption capacity	Technique/mechanism	References
PS NPs	Cu_XO nanoparticles	23%	Photocatalytic degradation in visible light, Reaction time = 50 h	Acuña-Bedoya et al. (2021)
PS, PE	CuMgAlTi-LDH	54.2, 33.7%	Photocatalytic degradation under visible light, Reaction time = 300 h	Jiang et al. (2023)
Poly(3-hydroxybutyrate) (PHB) and poly(lactic acid) (PLA)	Sb_2S_3 decorated with Fe_3O_4 nanoparticles microbots	31%	Photocatalytic degradation under UV light and 2.5 wt% of H_2O_2, Reaction time = 5 h	Jancik-Prochazkova et al. (2023)
PVC, PP	Cu-POM nanocluster-infused IPN hydrogel	95%, 93%	Adsorption and photocatalytic degradation under UV light at pH = 6.5	Dutta et al. (2024)
PE	$BiOI/BiVO_4$	75%	Photocatalytic oxidative depolymerization of PE into valuable aliphatic dicarboxylic acids	Chu et al. (2025)
PS	α-Fe_2O_3/g-C_3N_4	9.94%	Photocatalytic degradation under visible light	Chai et al. (2023)

(continued)

Table 1 (continued)

MPs/NPs	Nanomaterials/Hybrid	Degradation efficiency/removal efficiency/adsorption capacity	Technique/mechanism	References
PE	GO GO-MnO_2 GO-Cu_2O GO-TiO_2	35.66% 39.54% 48.06% 50.46%	Adsorption and photocatalytic degradation, Reaction time 480 min	Uoginte et al. (2023)
PS NPs	Cuttlefish bone (CFB)-supported $CoFe_2O_4$ Nanoparticles	99.3%	Adsorption/degradation by Fenton-like process	Qaretapeh et al. (2024)
PS	Graphene-like carbon assembled layered double oxide material (G@LDO)	≥80% at pH 3–11, 60% at pH 1 and 3	Chemisorption	Peng et al. (2022)
PS	Ni/rGO nanocomposite	1.250 g/g, adsorption capacity	Adsorption due to hydrophobic interaction	Karunattu Sajan et al. (2024)
PS-NH_2 NPs, PS-COOH NPs	Ball-milling biochar with Fe_3O_4 at different temp 300, 500, 700 °C	43.67, 82.73 and 57.02%,	Adsorption by hydrophobicity, electrostatic attraction, H-bonding formation and π-π conjunction	Shi et al. (2023a)
PE, PET, PA	Magnetic CNT	100%	Adsorption	Tang et al. (2021)
NH_2-PS NPs (50 nm) NH_2-PS MPs (1.5 μm)	Algae decorated Fe_3O_4 microbots	92% (motion) 71% (static) 70% (motion) 41% (static)	Adsorption	Peng et al. (2024)
PS MPs	Bioinspired magnetic microsubmarines	70%	Adsorption	Sun et al. (2020b)
PS MPs (1 μm)	Fe_3O_4@ polymerized dopamine (PDA) nanoadsorbent	98.5%	Adsorption	Li et al. (2024a)

(continued)

Table 1 (continued)

MPs/NPs	Nanomaterials/Hybrid	Degradation efficiency/removal efficiency/adsorption capacity	Technique/mechanism	References
PE MPs	Polyethylene glycol (PEG)-Fe_3O_4	2.203 g/g adsorption capacity	Adsorption	Zhang et al. (2025)
PS	Spent coffee ground (SCG)-derived magnetic microrobots	64%	Adsorption	Singh et al. (2023)
Metal-doped PS NPs (PAN-Pd@NPs)	Hydrophobic functionalized Magnetic Nanoparticles (HDTMS-FeNPs)	85%	Adsorption	Surette et al. (2023)
PS microspheres	Nano-Fe@ZIF-8	98%	Adsorption	Pasanen et al. (2023)
MPs in potable and seawater	Fe_3O_4-PWA/nOct nanotricomposites	83–99%	Adsorption and magnetic separation	Bhore and Kamble (2022)
MPs, NPs	Nano-melanin with magnetic property (ANM)	89.3%	Adsorption and magnetic separation	Chen et al. (2022b)
PS, PE	$Fe_3O_4@SiO_2$	92.08, 60.67%	Adsorption and magnetic separation	Li et al. (2023b)
PS	Fe_3O_4 core and a carbon shell	0.992 g/g adsorption capacity	Adsorption and magnetic separation	Zhang et al. (2024)
PE, PET, PA	Magnetic CNT	100%	Adsorption	Tang et al. (2021)
PE, PP, PS and PET	Fe_3O_4 Nps	86.87, 85.05%, 86.11%, and 62.83	Adsorption	Shi et al. (2022)
PS, PMMA, PA, PVDF, PVC	Self-driven magnetorobot	>90%	Adsorption and magnetic separation	Li et al. (2022)
PVC NPs PMMA NPs	Zinc- terephthalic acid (TPA) MOF	97% 95%	Adsorption by electrostatic interaction	Chinglenthoiba et al. (2024)

(continued)

Table 1 (continued)

MPs/NPs	Nanomaterials/Hybrid	Degradation efficiency/removal efficiency/adsorption capacity	Technique/mechanism	References
PS NPs	Mesoporous UiO-66 (−OH, $-NH_2$) MOF	100%	Adsorption by electrostatic interaction	Pedrero et al. (2024)
MPs	TiO_2/ Ni-MOF-based superhydrophobic sponge	100%	Adsorption by capillary force, grasping force of ODSOSS cage, lipophilicity, hydrogen bonding, electrostatic, π–π and Van der Waal interaction	Chen et al. (2023)
PS NPs	Cr-MOF	96%	Adsorption by electrostatic, π–π, and acid–base interactions	Modak et al. (2023)
MPs	C@FeO nanopillars on 2D-MOF	0.725 g/g, adsorption capacity, 100% removal efficiency	Adsorption and magnetic separation	Haris et al. (2023)

5.1.3 Enzymatic Degradation

Nanomaterials enhance enzymatic degradation of MPs by improving enzyme stability, activity, and substrate interaction. Functionalized nanomaterials like metal–organic frameworks (MOFs), carbon-based nanomaterials, and magnetic nanoparticles can act as carriers for enzymes such as cutinase, PETase, and laccase, improving the breakdown of MPs (Qamar et al., 2022). These nanohybrids provide a large surface area for enzyme attachment, enhanced stability, and allow reuse of enzyme. For instance, Schwaminger et al. (2021) used His-tagged PETase immobilized on SPIONS, which improved enzyme stability and enabled easy magnetic separation. Zhou et al. (2021a) developed polydopamine-coated Fe_3O_4 magrobots loaded with lipase to degrade polycaprolactone (PCL). Hu et al. (2024) engineered a heterogeneous pepsin catalyst, PEP@M-MIL100(Fe), by attaching carboxylated ferrite MOFs to pepsin, and achieved 95.88% degradation of diisononyl phthalate (DINP) MPs within 150 min. Similarly, Hou et al. (2014) reported the biodegradation of Bisphenol A (BPA), a common plastic additive, using laccase immobilized on TiO_2-decorated PVDF membranes.

Moreover, nanomaterials can act like enzymes. Zandieh and Liu (2022) found that bare Fe_3O4 nanoaggregates exhibited peroxidase-like activity, enabling the complete degradation of high-density PE, PP, PVC, PS, and PET with 100% efficiency. These strategies may offer a sustainable and efficient solution for MPs removal while minimizing secondary pollutants.

5.2 *Removal Using Nanomaterials*

5.2.1 Adsorption

Nanomaterials with high surface area and functionalized surfaces, such as carbon-based materials (GO, CNTs), metal oxides, and MOFs, efficiently adsorb MPs from water (Bodzek, 2023). Their functional groups enable strong electrostatic and hydrophobic interactions with plastic particles, enhancing removal efficiency (Fig. 4a). For instance, Sun et al. (2020a) developed chitin-GO sponges that adsorbed PS, PS-COOH, and PS-NH_2 with efficiencies of 89.8%, 72.4%, and 88.9%, respectively, via electrostatic interactions, hydrogen bonding, and π–π interactions. Similarly, Viswanathan et al. (2024) designed hexadecyltrimethoxysilane-functionalized graphitic carbon nitride (h-GCN) hydrophobic sponges, achieving near-complete removal of poly (methyl methacrylate) (PMMA). Additional examples of carbon-based nanomaterials, MOF and composites are summarized in Table 1, including nanocomposites with magnetic properties for enhanced separation efficiency.

5.2.2 Filtration and Membrane Technologies

Nanomaterial-based membrane filtration is an effective method for removing MPs and NPs from water. Advanced membranes, such as holey Ti_3C_2 nanosheets and holey reduced graphene oxide (rGO)-based membranes, have demonstrated exceptional removal efficiencies of 99.3% and 99.9%, respectively, for PS microspheres and beads (Yang et al., 2022a, 2022b). Similarly, polyurethane-GO-montmorillonite (GO-Mt)-loaded nanofibrous composite membranes efficiently filter various MPs, including acrylonitrile butadiene styrene (ABS), PS, and poly (methyl methacrylate) (PMMA) (Juraij et al., 2023). Additionally, PVDF membranes embedded with carbon nano-onions (CNOs) effectively trap MPs at a concentration of 140 MPs/L (Nkosi et al., 2022). For NPs, sodium alginate (SA)/GO/chitosan (CS) composite membranes show size-dependent removal efficiencies of 99.87% (500 nm) and 97.10% (50 nm) (Li et al., 2023c). These advanced filtration technologies leverage nanomaterials' high surface area and tunable porosity to achieve superior MP and NP removal.

5.2.3 Flocculation and Magnetic Extraction

Nanomaterial-based flocculation, coagulation and magnetic extraction have emerged as effective strategies for removing MPs/NPs from water. Functionalized nanomaterials enhance aggregation, facilitating sedimentation and removal. Shi et al. (2023b) demonstrated that CTAB-modified magnetic biochar effectively aggregated PS and carboxylate-PS, achieving adsorption capacities of 163.4 mg/g and 159.6 mg/g, respectively. Bakhteeva et al. (2023) reported near-complete removal (100% for PE and 88% for PET) using Fe–C–NH_2 nanoparticles via magnetic sedimentation and heteroaggregates formation. Similarly, Yan et al. (2023) achieved 83.1%–92.9% removal of fluorescent PS MPs using Fe_3O_4 nanoparticles through heteroaggregation. Groppe et al. (2022) evaluated the significance of pH, particle concentration, and the magnetic characteristics of superparamagnetic iron oxide nanoparticles (SPIONs) using magnetic particle spectroscopy (MPS) to investigate the aggregation and removal of positively charged PS by negatively charged SPIONS in situ in real-time. Notably, SPION stability improved the magnetic removal and inhibited the formation of any large agglomerates, whereas imbalances in concentration resulted in residual nanoparticles or NPs in dispersion. Functionalized SPIONs with various surface coatings increased capture efficiency by improving affinity towards various MPs. For example, Gaß et al. (2024) functionalized SPIONs (γ-Fe_2O_3, 9.6 nm) with different surface groups such as PAC18, PAC12NC18, and PAC12Imida named as SPION A, B, and C, respectively. These SPIONs interacted with PS, polymethylmethacrylate (PMMA), and melamine resin (MR) across a size range from 100 to 100 μm. At pH 7, SPION A exhibited a negative ζ-potential, while SPION B and C were positively charged, enabled selective electrostatic interactions with oppositely charged MPs and NPs. This study demonstrated efficient magnetic separation, particularly for the smaller plastic fractions (100–300 nm), by achieving removal rates of up to 10^{13} mg^{-1} SPION, with the highest mass removal for particles between 2.5 and 5 μm. These studies highlighted the potential of magnetics nanomaterials in enhancing flocculation, and magnetic extraction-based remediation of plastic pollutants.

6 Future Perspectives and Conclusion

Integrating nanomaterials into MPs detection and degradation offers significant potential for sustainable environmental remediation. Nanotechnology-based sensors, such as surface-enhanced Raman scattering (SERS) using plasmonic nanoparticles, have already demonstrated detection limits as low as 1 ng/L for MPs in water. However, further research is needed to enhance the selectivity and efficiency of these nano-enabled sensors, ensuring precise differentiation of MPs types in complex matrices. Additionally, nanomaterial-driven degradation processes, such as photocatalysis using TiO_2 and ZnO nanoparticles, have shown high efficiency in breaking down MPs under UV light. The incorporation of nanomaterials with photocatalysis and photothermal Fenton-like reaction has further improved the degradation efficiency of MPs by almost 100% under UV and visible light. Apart from that high surface area, functionalization and magnetic properties further improved the removal processes through adsorption, flocculation and filtration. Despite these promising advancements, challenges remain in scaling up cost-effective nanomaterial synthesis and ensuring the long-term environmental safety of engineered nanoparticles. Addressing these concerns by green synthesis methods, utilizing plant extracts, biopolymers, or natural precursors, reduces toxicity and enhances biodegradability, minimizing hazardous byproducts. Additionally, designing recyclable and reusable nanomaterials, such as magnetic nanoparticles, allows for multiple retrieval and reuse cycles, reducing material waste and operational costs. Integrating nanotechnology into existing wastewater treatment systems can further enhance MPs removal through hybrid approaches like adsorption, photocatalysis, and biofiltration, ensuring compatibility with current infrastructure.

Acknowledgements Jagriti gratefully acknowledges the Department of Science and Technology, India, for the National Post Doctoral Fellowship in Nanoscience and Technology (Award Letter Number: JNC/AO/A.1707(28)/25-OW-182).

Conflicts of Interest The authors declare no conflict of interest.

Data Availability Not applicable.

References

Acarer Arat, S. (2024). An overview of microplastic in marine waters: Sources, abundance, characteristics and negative effects on various marine organisms. *Desalination and Water Treatment, 317*, Article 100138. https://doi.org/10.1016/j.dwt.2024.100138

Acuña-Bedoya, J. D., Luévano-Hipólito, E., Cedillo-González, E. I., Domínguez-Jaimes, L. P., Hurtado, A. M., & Hernández-López, J. M. (2021). Boosting visible-light photocatalytic degradation of polystyrene nanoplastics with immobilized CuxO obtained by anodization. *Journal of Environmental Chemical Engineering, 9*(5), Article 106208. https://doi.org/10.1016/j.jece.2021.106208

Adhikari, S., Kelkar, V., Kumar, R., & Halden, R. U. (2022). Methods and challenges in the detection of microplastics and nanoplastics: A mini-review. *Polymer International, 71*(5), 543–551. https://doi.org/10.1002/pi.6348

Allen, D., Allen, S., Abbasi, S., Baker, A., Bergmann, M., Brahney, J., Butler, T., Duce, R. A., Eckhardt, S., Evangeliou, N., Jickells, T., Kanakidou, M., Kershaw, P., Laj, P., Levermore, J., Li, D., Liss, P., Liu, K., Mahowald, N., et al. (2022). Microplastics and nanoplastics in the marine-atmosphere environment. *Nature Reviews Earth & Environment, 3*(6), 393–405. https://doi.org/10.1038/s43017-022-00292-x

Amirjani, A., Amlashi, N. B., & Ahmadiani, Z. S. (2023). Plasmon-enhanced photocatalysis based on plasmonic nanoparticles for energy and environmental solutions: A review. *ACS Applied Nano Materials, 6*(11), 9085–9123. https://doi.org/10.1021/acsanm.3c01671

An, Y., Hou, J., Liu, Z., & Peng, B. (2014). Enhanced solid-phase photocatalytic degradation of polyethylene by TiO_2–MWCNTs nanocomposites. *Materials Chemistry and Physics, 148*(1–2), 387–394. https://doi.org/10.1016/j.matchemphys.2014.08.001

Ariza-Tarazona, M. C., Siligardi, C., Carreón-López, H. A., Valdéz-Cerda, J. E., Pozzi, P., Kaushik, G., Villarreal-Chiu, J. F., & Cedillo-González, E. I. (2023). Low environmental impact remediation of microplastics: Visible-light photocatalytic degradation of PET microplastics using bio-inspired C,N-TiO_2/SiO_2 photocatalysts. *Marine Pollution Bulletin, 193*, Article 115206. https://doi.org/10.1016/j.marpolbul.2023.115206

Ariza-Tarazona, M. C., Villarreal-Chiu, J. F., Barbieri, V., Siligardi, C., & Cedillo-González, E. I. (2019). New strategy for microplastic degradation: Green photocatalysis using a protein-based porous N–TiO_2 semiconductor. *Ceramics International, 45*(7), 9618–9624. https://doi.org/10.1016/j.ceramint.2018.10.208

Ariza-Tarazona, M. C., Villarreal-Chiu, J. F., Hernández-López, J. M., Rivera De La Rosa, J., Barbieri, V., Siligardi, C., & Cedillo-González, E. I. (2020). Microplastic pollution reduction by a carbon and nitrogen-doped TiO2: Effect of pH and temperature in the photocatalytic degradation process. *Journal of Hazardous Materials, 395*, Article 122632. https://doi.org/10.1016/j.jhazmat.2020.122632

Ayejoto, D. A., Egbueri, J. C., Onuba, L. N., Agbasi, J. C., & Nweke, N. D. (2024). toxicological effects of ingested microplastics on human health. In J. C. Egbueri, J. O. Ighalo, & C. B. Pande (Eds.), *Microplastics in African and Asian environments* (pp. 427–461). Springer Nature. https://doi.org/10.1007/978-3-031-64253-1_21

Bakhteeva, I. A., Medvedeva, I. V., Filinkova, M. S., Byzov, I. V., Minin, A. S., Zhakov, S. V., Uimin, M. A., Patrakov, E. I., Novikov, S. I., Suntsov, A. Y., & Demin, A. M. (2023). Removal of microplastics from water by using magnetic sedimentation. *International Journal of Environmental Science and Technology, 20*(11), 11837–11850. https://doi.org/10.1007/s13762-023-04776-1

Baumgarten, L. G., Freitas, A. A., Santana, E. R., Winiarski, J. P., Dreyer, J. P., & Vieira, I. C. (2023). Graphene and gold nanoparticle-based bionanocomposite for the voltammetric determination of bisphenol A in (micro)plastics. *Chemosphere, 334*, Article 139016. https://doi.org/10.1016/j.chemosphere.2023.139016

Besseling, E., Wang, B., Lürling, M., & Koelmans, A. A. (2014). Nanoplastic affects growth of *S. obliquus* and reproduction of *D. magna*. *Environmental Science & Technology, 48*(20), 12336–12343. https://doi.org/10.1021/es503001d

Bhore, R. K., & Kamble, S. B. (2022). Nano adsorptive extraction of diverse microplastics from the potable and seawater using organo-polyoxometalate magnetic nanotricomposites. *Journal of Environmental Chemical Engineering, 10*(6). https://doi.org/10.1016/j.jece.2022.108720

Bodzek, M. (2023). Remediation of nano- and microplastics in water environment using nanomaterials. *Desalination and Water Treatment, 316*, 557–573. https://doi.org/10.5004/dwt.2023.30172

Bule Možar, K., Miloloža, M., Martinjak, V., Cvetnić, M., Kušić, H., Bolanča, T., Kučić Grgić, D., & Ukić, Š. (2023). Potential of advanced oxidation as pretreatment for microplastics biodegradation. *Separations, 10*(2), 132. https://doi.org/10.3390/separations10020132

Cai, Z., Li, M., Zhu, Z., Wang, X., Huang, Y., Li, T., Gong, H., & Yan, M. (2023). Biological degradation of plastics and microplastics: A recent perspective on associated mechanisms and influencing factors. *Microorganisms, 11*(7), 1661. https://doi.org/10.3390/microorganisms11071661

Caldwell, J., Taladriz-Blanco, P., Rothen-Rutishauser, B., & Petri-Fink, A. (2021). Detection of sub-micro- and nanoplastic particles on gold nanoparticle-based substrates through surface-enhanced Raman scattering (SERS) spectroscopy. *Nanomaterials, 11*(5), 1149. https://doi.org/10.3390/nano11051149

Chai, C., Liang, H., Yao, R., Wang, F., Song, N., Wu, J., & Li, Y. (2023). Photocatalytic degradation of polyethylene and polystyrene microplastics by α-Fe_2O_3/g-C_3N_4. *Environmental Science and Pollution Research, 30*(58), 121702–121712. https://doi.org/10.1007/s11356-023-31000-x

Chen, Y., Miller, C. J., & Waite, T. D. (2021). Heterogeneous fenton chemistry revisited: Mechanistic insights from ferrihydrite-mediated oxidation of formate and oxalate. *Environmental Science & Technology, 55*(21), 14414–14425. https://doi.org/10.1021/acs.est.1c00284

Chen, G., Li, Y., Liu, S., Junaid, M., & Wang, J. (2022a). Effects of micro(nano)plastics on higher plants and the rhizosphere environment. *Science of the Total Environment, 807*, Article 150841. https://doi.org/10.1016/j.scitotenv.2021.150841

Chen, Y., Ouyang, L., Liu, N., Li, F., Li, P., Sun, M., Qin, H., Li, Y., Xiang, X., & Wu, L. (2022b). PH-responsive magnetic artificial melanin with tunable aggregation-induced stronger magnetism for rapid remediation of plastic fragments. *Journal of Hazardous Materials, 435*. https://doi.org/10.1016/j.jhazmat.2022.128962

Chen, X., Ma, H., Ji, X., Han, R., Pang, K., Yang, Z., Liu, Z., & Peng, S. (2023). Engineering green MOF-based superhydrophobic sponge for efficiently synchronous removal of microplastics and pesticides from high-salinity water. *Water Research, 243*, Article 120314. https://doi.org/10.1016/j.watres.2023.120314

Cherniak, S. L., Almuhtaram, H., McKie, M. J., Hermabessiere, L., Yuan, C., Rochman, C. M., & Andrews, R. C. (2022). Conventional and biological treatment for the removal of microplastics from drinking water. *Chemosphere, 288*, Article 132587. https://doi.org/10.1016/j.chemosphere.2021.132587

Chinglenthoiba, C., Mahadevan, G., Zuo, J., Prathyumnan, T., & Valiyaveettil, S. (2024). Conversion of PET bottle waste into a terephthalic acid-based metal-organic framework for removing plastic nanoparticles from water. *Nanomaterials (Basel, Switzerland), 14*(3). https://doi.org/10.3390/nano14030257

Chu, D., Weng, J., Liu, X., Wang, H., Cui, Y., Nie, L., & Li, Z. (2025). Photocatalytic oxidation of polyethylene to dicarboxylic acid over BiOI/$BiVO_4$ p-n heterojunction under visible light. *CCS Chemistry,* 1–19. https://doi.org/10.31635/ccschem.025.202405274

Cox, K. D., Covernton, G. A., Davies, H. L., Dower, J. F., Juanes, F., & Dudas, S. E. (2019). Human consumption of microplastics. *Environmental Science & Technology, 53*(12), 7068–7074. https://doi.org/10.1021/acs.est.9b01517

Domínguez-Jaimes, L. P., Cedillo-González, E. I., Luévano-Hipólito, E., Acuña-Bedoya, J. D., & Hernández-López, J. M. (2021). Degradation of primary nanoplastics by photocatalysis using different anodized TiO_2 structures. *Journal of Hazardous Materials, 413*, Article 125452. https://doi.org/10.1016/j.jhazmat.2021.125452

Du, H., Chen, G., & Wang, J. (2023). Highly selective electrochemical impedance spectroscopy-based graphene electrode for rapid detection of microplastics. *Science of the Total Environment, 862*, Article 160873. https://doi.org/10.1016/j.scitotenv.2022.160873

Dutta, S., Misra, A., & Bose, S. (2024). Polyoxometalate nanocluster-infused triple IPN hydrogels for excellent microplastic removal from contaminated water: Detection, photodegradation, and upcycling. *Nanoscale, 16*(10), 5188–5205. https://doi.org/10.1039/D3NR06115A

El-Sharif, H. F., Patel, S., Ndunda, E. N., & Reddy, S. M. (2022). Electrochemical detection of dioctyl phthalate using molecularly imprinted polymer modified screen-printed electrodes. *Analytica Chimica Acta, 1196*, Article 339547. https://doi.org/10.1016/j.aca.2022.339547

Forte, M., Iachetta, G., Tussellino, M., Carotenuto, R., Prisco, M., De Falco, M., Laforgia, V., & Valiante, S. (2016). Polystyrene nanoparticles internalization in human gastric adenocarcinoma cells. *Toxicology in Vitro, 31*, 126–136. https://doi.org/10.1016/j.tiv.2015.11.006

Gaß, H., Kloos, T. M., Höfling, A., Müller, L., Rockmann, L., Schubert, D. W., & Halik, M. (2024). Magnetic removal of micro- and nanoplastics from water—From 100 nm to 100 μm debris size. *Small (Weinheim an der Bergstrasse, Germany), 20*(10), 2305467. https://doi.org/10.1002/smll.202305467

Geyer, R., Jambeck, J. R., & Law, K. L. (2017). Production, use, and fate of all plastics ever made. *Science Advances, 3*(7), Article e1700782. https://doi.org/10.1126/sciadv.1700782

Gianoncelli, A., Tufoni, C., Žižić, M., Zingaro, F., Ceccone, G., Bonanni, V., Salomé, M., Villalobos-Portillo, E., Spina, R. L., Cassano, D., Ricci, G., & Pascolo, L. (2024). Ovarian cell accumulation of model PVC nanoplastics labelled with CdSe—QDs investigated by x-ray fluorescence microscopy. *X-Ray Spectrometry,* xrs.3465. https://doi.org/10.1002/xrs.3465

Groppe, P., Wintzheimer, S., Eigen, A., Gaß, H., Halik, M., & Mandel, K. (2022). Real-time monitoring of magnetic nanoparticle-assisted nanoplastic agglomeration and separation from water. *Environmental Science: Nano, 9*(7), 2427–2439. https://doi.org/10.1039/D2EN00131D

Gupta, J., & Rajamani, P. (2024). Beyond microplastics: Concern of nanoplastic pollution on human and environmental health. *Plastic Pollution, 28*(4), 2–8.

Haris, M., Khan, M. W., Zavabeti, A., Mahmood, N., & Eshtiaghi, N. (2023). Self-assembly of C@FeO nanopillars on 2D-MOF for simultaneous removal of microplastic and dissolved contaminants from water. *Chemical Engineering Journal, 455*, Article 140390. https://doi.org/10.1016/j.cej.2022.140390

Hasan, M. M., & Tarannum, M. N. (2025). Adverse impacts of microplastics on soil physicochemical properties and crop health in agricultural systems. *Journal of Hazardous Materials Advances, 17*, Article 100528. https://doi.org/10.1016/j.hazadv.2024.100528

He, Y., Rehman, A. U., Xu, M., Not, C. A., Ng, A. M. C., & Djurišić, A. B. (2023). Photocatalytic degradation of different types of microplastics by TiOx/ZnO tetrapod photocatalysts. *Heliyon, 9*(11), Article e22562. https://doi.org/10.1016/j.heliyon.2023.e22562

Hou, J., Dong, G., Ye, Y., & Chen, V. (2014). Enzymatic degradation of bisphenol-A with immobilized laccase on TiO_2 sol–gel coated PVDF membrane. *Journal of Membrane Science, 469*, 19–30. https://doi.org/10.1016/j.memsci.2014.06.027

Hsieh, S.-L., Hsieh, S., Xu, R.-Q., Chen, Y.-T., Chen, C.-W., Singhania, R. R., Chen, Y.-C., Tsai, T.-H., & Dong, C.-D. (2023). Toxicological effects of polystyrene nanoplastics on marine organisms. *Environmental Technology & Innovation, 30*, Article 103073. https://doi.org/10.1016/j.eti.2023.103073

Hu, S., Kong, H., Sun, Y., Wu, R., Xu, J., & Guo, M. (2024). Construction of metal-organic framework-based heterogeneous pepsin and its degradation performance and mechanism for phthalic acid esters. *ACS Applied Materials & Interfaces, 16*(30), 39241–39250. https://doi.org/10.1021/acsami.4c04799

Ivleva, N. P., Wiesheu, A. C., & Niessner, R. (2017). Microplastic in aquatic ecosystems. *Angewandte Chemie International Edition, 56*(7), 1720–1739. https://doi.org/10.1002/anie.201606957

Jancik-Prochazkova, A., Jasek, V., Figalla, S., & Pumera, M. (2023). Photocatalytic microplastics "On-The-fly" degradation via motile quantum materials-based microrobots. *Advanced Optical Materials, 11*(22). https://doi.org/10.1002/adom.202300782

Jiang, S., Yin, M., Ren, H., Qin, Y., Wang, W., Wang, Q., & Li, X. (2023). Novel CuMgAlTi-LDH photocatalyst for efficient degradation of microplastics under visible light irradiation. *Polymers, 15*(10), 2347. https://doi.org/10.3390/polym15102347

Juraij, K., Ammed, S. P., Chingakham, C., Ramasubramanian, B., Ramakrishna, S., Vasudevan, S., & Sujith, A. (2023). Electrospun polyurethane nanofiber membranes for microplastic and nanoplastic separation. *ACS Applied Nano Materials, 6*(6), 4636–4650. https://doi.org/10.1021/acsanm.3c00112

Karunattu Sajan, M., Kirubalan, M. R., Rajendran, A. S., & Natesan, A. L. F. (2024). Exploring the effective adsorption of polystyrene microplastics from aqueous solution with magnetically separable nickel/reduced graphene oxide (Ni/rGO) nanocomposite. *Environmental Science and Pollution Research International*. https://doi.org/10.1007/s11356-024-33726-8

Khan, M. S., Asif, M. I., Asif, M., Khan, M. R., Mustafa, G., & Adeel, M. (2024). Nanomaterials for the catalytic degradation and detection of microplastics: A review. *Topics in Catalysis*. https://doi.org/10.1007/s11244-024-01927-7

Kihara, S., Chan, A., In, E., Taleb, N., Tollemache, C., Yick, S., & McGillivray, D. J. (2022). Detecting polystyrene nanoplastics using filter paper-based surface-enhanced Raman spectroscopy. *RSC Advances, 12*(32), 20519–20522. https://doi.org/10.1039/D2RA03395J

Leslie, H. A., Van Velzen, M. J. M., Brandsma, S. H., Vethaak, A. D., Garcia-Vallejo, J. J., & Lamoree, M. H. (2022). Discovery and quantification of plastic particle pollution in human blood. *Environment International, 163*, Article 107199. https://doi.org/10.1016/j.envint.2022.107199

Li, W., Wu, C., Xiong, Z., Liang, C., Li, Z., Liu, B., Cao, Q., Wang, J., Tang, J., & Li, D. (2022). Self-driven magnetorobots for recyclable and scalable micro/nanoplastic removal from nonmarine waters. *Science Advances, 8*(45), eade1731. https://doi.org/10.1126/sciadv.ade1731

Li, H., Li, Y., Liu, F., Xu, Y., & Li, H. (2023a). Identification and quantification of polyethylene microplastics with sulfur-doped carbon nanoparticles in fluorescence mode and construction of paper-based analytical method. *Sensors and Actuators B: Chemical, 394*, Article 134340. https://doi.org/10.1016/j.snb.2023.134340

Li, W., Liu, S., Huang, K., Qin, S., Liang, B., & Wang, J. (2023b). Preparation of magnetic Janus microparticles for the rapid removal of microplastics from water. *Science of the Total Environment, 903*, Article 166627. https://doi.org/10.1016/j.scitotenv.2023.166627

Li, Z., Xie, W., Zhang, Z., Wei, S., Chen, J., & Li, Z. (2023c). Multifunctional sodium alginate/chitosan-modified graphene oxide reinforced membrane for simultaneous removal of nanoplastics, emulsified oil, and dyes in water. *International Journal of Biological Macromolecules, 245*, Article 125524. https://doi.org/10.1016/j.ijbiomac.2023.125524

Li, Y., Chen, H., Li, S., Feng, L., Wang, Z., Wang, D., Wang, Q., & Wang, H. (2024a). Corals-inspired magnetic absorbents for fast and efficient removal of microplastics in various water sources. *RSC Advances, 14*(17), 11908–11913. https://doi.org/10.1039/D4RA02521K

Li, Z., Han, K., Zhang, A., Wang, T., Yan, Z., Ding, Z., Shen, Y., Zhang, M., & Zhang, W. (2024b). Honeycomb-like AgNPs@TiO_2 array SERS sensor for the quantification of micro/nanoplastics in the environmental water samples. *Talanta, 266*, Article 125070. https://doi.org/10.1016/j.talanta.2023.125070

Llorente-García, B. E., Hernández-López, J. M., Zaldívar-Cadena, A. A., Siligardi, C., & Cedillo-González, E. I. (2020). First insights into photocatalytic degradation of HDPE and LDPE microplastics by a mesoporous N–TiO_2 coating: Effect of size and shape of microplastics. *Coatings, 10*(7), 658. https://doi.org/10.3390/coatings10070658

Luo, H., Liu, C., He, D., Xu, J., Sun, J., Li, J., & Pan, X. (2022). Environmental behaviors of microplastics in aquatic systems: A systematic review on degradation, adsorption, toxicity and biofilm under aging conditions. *Journal of Hazardous Materials, 423*, Article 126915. https://doi.org/10.1016/j.jhazmat.2021.126915

Modak, S., Kasula, M., & Esfahani, M. R. (2023). Nanoplastics removal from water using metal-organic framework: Investigation of adsorption mechanisms, kinetics, and effective environmental parameters. *ACS Applied Engineering Materials, 1*(2), 744–755. https://doi.org/10.1021/acsaenm.2c00174

Nene, A., Sadeghzade, S., Viaroli, S., Yang, W., Uchenna, U. P., Kandwal, A., Liu, X., Somani, P., & Galluzzi, M. (2025). Recent advances and future technologies in nano-microplastics detection. *Environmental Sciences Europe, 37*(1), 7. https://doi.org/10.1186/s12302-024-01044-y

Nkosi, S. D., Malinga, S. P., & Mabuba, N. (2022). Microplastics and heavy metals removal from fresh water and wastewater systems using a membrane. *Separations, 9*(7), 166. https://doi.org/10.3390/separations9070166

Noumani, A., Verma, D., Kaushik, A., Khosla, A., & Solanki, P. R. (2024). Electrochemically microplastic detection using chitosan-magnesium oxide nanosheet. *Environmental Research, 252*(Pt 3), Article 118894. https://doi.org/10.1016/j.envres.2024.118894

O'Connor, J. D., Lally, H. T., Koelmans, A. A., Mahon, A. M., O'Connor, I., Nash, R., O'Sullivan, J. J., Bruen, M., Heerey, L., & Murphy, S. (2022). Modelling the transfer and accumulation of microplastics in a riverine freshwater food web. *Environmental Advances, 8*, Article 100192. https://doi.org/10.1016/j.envadv.2022.100192

OECD. (2022). *Global plastics outlook: Economic drivers*. OECD. https://doi.org/10.1787/de747aef-en

Oh, S., Hur, H., Kim, Y., Shin, S., Woo, H., Choi, J., & Lee, H. H. (2021). Peptide specific nanoplastic detection based on sandwich typed localized surface plasmon resonance. *Nanomaterials, 11*(11), 2887. https://doi.org/10.3390/nano11112887

Pasanen, F., Fuller, R. O., & Maya, F. (2023). Fast and simultaneous removal of microplastics and plastic-derived endocrine disruptors using a magnetic ZIF-8 nanocomposite. *Chemical Engineering Journal, 455*. https://doi.org/10.1016/j.cej.2022.140405

Pedrero, D., Edo, C., Fernández-Piñas, F., Rosal, R., & Aguado, S. (2024). Efficient removal of nanoplastics from water using mesoporous metal organic frameworks. *Separation and Purification Technology, 333*, Article 125816. https://doi.org/10.1016/j.seppur.2023.125816

Peng, G., Xiang, M., Wang, W., Su, Z., Liu, H., Mao, Y., Chen, Y., & Zhang, P. (2022). Engineering 3D graphene-like carbon-assembled layered double oxide for efficient microplastic removal in a wide pH range. *Journal of Hazardous Materials, 433*, Article 128672. https://doi.org/10.1016/j.jhazmat.2022.128672

Peng, X., Urso, M., Kolackova, M., Huska, D., & Pumera, M. (2024). Biohybrid magnetically driven microrobots for sustainable removal of micro/nanoplastics from the aquatic environment. *Advanced Functional Materials, 34*(3). https://doi.org/10.1002/adfm.202307477

Piazza, V., Uheida, A., Gambardella, C., Garaventa, F., Faimali, M., & Dutta, J. (2022). Ecosafety screening of photo-fenton process for the degradation of microplastics in water. *Frontiers in Marine Science*. Publicly Available Content Database. https://doi.org/10.3389/fmars.2021.791431

Picó, Y., & Barceló, D. (2020). Pyrolysis gas chromatography-mass spectrometry in environmental analysis: Focus on organic matter and microplastics. *TrAC Trends in Analytical Chemistry, 130*, Article 115964. https://doi.org/10.1016/j.trac.2020.115964

Qamar, M., Basharat, A., Qamar, S. A., Bilal, M., Franco, M., & Iqbal, H. M. N. (2022). Enzyme-loaded nanostructured materials for the degradation of environmental pollutants. *Current Opinion in Environmental Science & Health, 30*, Article 100400. https://doi.org/10.1016/j.coesh.2022.100400

Qaretapeh, M. Z., Kouchakipour, S., Hosseinzadeh, M., & Dashtian, K. (2024). Cuttlefish Bone-Supported $CoFe_2O_4$ nanoparticles enhance persulfate Fenton-like process for the degradation of polystyrene nanoplastics. *Chemical Engineering Journal, 490*. https://doi.org/10.1016/j.cej.2024.151833

Ren, X., Tang, J., Wang, L., & Liu, Q. (2021). Microplastics in soil-plant system: Effects of nano/microplastics on plant photosynthesis, rhizosphere microbes and soil properties in soil with different residues. *Plant and Soil, 462*(1–2), 561–576. https://doi.org/10.1007/s11104-021-04869-1

Ritchie, H., Samborska, V., & Roser, M. (2023). *Plastic pollution.* https://ourworldindata.org/plastic-pollution

Rose, P. K., Yadav, S., Kataria, N., & Khoo, K. S. (2023). Microplastics and nanoplastics in the terrestrial food chain: Uptake, translocation, trophic transfer, ecotoxicology, and human health risk. *TrAC Trends in Analytical Chemistry, 167*, Article 117249. https://doi.org/10.1016/j.trac.2023.117249

Saeedi, M. (2024). How microplastics interact with food chain: A short overview of fate and impacts. *Journal of Food Science and Technology, 61*(3), 403–413. https://doi.org/10.1007/s13197-023-05720-4

Schwaminger, S. P., Fehn, S., Steegmüller, T., Rauwolf, S., Löwe, H., Pflüger-Grau, K., & Berensmeier, S. (2021). Immobilization of PETase enzymes on magnetic iron oxide nanoparticles for the decomposition of microplastic PET. *Nanoscale Advances, 3*(15), 4395–4399. https://doi.org/10.1039/D1NA00243K

Seggio, M., Arcadio, F., Cennamo, N., Zeni, L., & Bossi, A. M. (2024). A plasmonic gold nanosurface functionalized with the estrogen receptor for fast and highly sensitive detection of nanoplastics. *Talanta, 267*, Article 125211. https://doi.org/10.1016/j.talanta.2023.125211

Seong, T., Yamamoto, S., Nakatani, H., Yagi, M., Kyozuka, Y., Satuito, G., & Kim, H.-J. (2024). Effects of microplastics on reproductive characteristics and mechanisms of the marine rotifer Brachionus plicatilis. *Scientific Reports, 14*(1), 15213. https://doi.org/10.1038/s41598-024-65047-8

Sharma, N., Kumar, V., S., V., Umesh, M., Chakraborty, P., Basheer, T., Sarojini, S., Sharma, P., Pasrija, R., & Barcelo, D. (2024). Microplastic residues in clinical samples: A retrospection on sources, entry routes, detection methods and human toxicity. *TrAC Trends in Analytical Chemistry, 173*, 117618. https://doi.org/10.1016/j.trac.2024.117618

Shen, M., Huang, W., Chen, M., Song, B., Zeng, G., & Zhang, Y. (2020). (Micro)plastic crisis: Un-ignorable contribution to global greenhouse gas emissions and climate change. *Journal of Cleaner Production, 254*, Article 120138. https://doi.org/10.1016/j.jclepro.2020.120138

Shen, M., Zhao, Y., Liu, S., Hu, T., Zheng, K., Wang, Y., Lian, J., & Meng, G. (2023). Recent advances on micro/nanoplastic pollution and membrane fouling during water treatment: A review. *Science of the Total Environment, 881*, Article 163467. https://doi.org/10.1016/j.scitotenv.2023.163467

Shi, X., Zhang, X., Gao, W., Zhang, Y., & He, D. (2022). Removal of microplastics from water by magnetic nano-Fe_3O_4. *Science of the Total Environment, 802*, Article 149838. https://doi.org/10.1016/j.scitotenv.2021.149838

Shi, Q., Guo, S., Tang, J., Lyu, H., Ri, C., & Sun, H. (2023a). Enhanced removal of aged and differently functionalized polystyrene nanoplastics using ball-milled magnetic pinewood biochars. *Environmental Pollution (Barking, Essex : 1987), 316*(Pt 1), 120696. https://doi.org/10.1016/j.envpol.2022.120696

Shi, Y., Du, J., Zhao, T., Feng, B., Bian, H., Shan, S., Meng, J., Christie, P., Wong, M. H., & Zhang, J. (2023b). Removal of nanoplastics from aqueous solution by aggregation using reusable magnetic biochar modified with cetyltrimethylammonium bromide. *Environmental Pollution (Barking, Essex : 1987), 318*, 120897. https://doi.org/10.1016/j.envpol.2022.120897

Singh, A. K., Basireddy, T., & Moran, J. L. (2023). Eliminating waste with waste: Transforming spent coffee grounds into microrobots for water treatment. *Nanoscale, 15*(43), 17494–17507. https://doi.org/10.1039/d3nr03592a

Song, Y. K., Hong, S. H., Jang, M., Han, G. M., Rani, M., Lee, J., & Shim, W. J. (2015). A comparison of microscopic and spectroscopic identification methods for analysis of microplastics in environmental samples. *Marine Pollution Bulletin, 93*(1–2), 202–209. https://doi.org/10.1016/j.marpolbul.2015.01.015

Statista. (2025a). *Global plastic production.* Statista. https://www.statista.com/statistics/282732/global-production-of-plastics-since-1950/

Statista. (2025b). *Global plastic waste generation by application 2019–2060*. Statista. https://www.statista.com/statistics/1339200/global-plastic-waste-generation-outlook-by-application/

Sun, C., Wang, Z., Chen, L., & Li, F. (2020a). Fabrication of robust and compressive chitin and graphene oxide sponges for removal of microplastics with different functional groups. *Chemical Engineering Journal, 393*, Article 124796. https://doi.org/10.1016/j.cej.2020.124796

Sun, M., Chen, W., Fan, X., Tian, C., Sun, L., & Xie, H. (2020b). Cooperative recyclable magnetic microsubmarines for oil and microplastics removal from water. *Applied Materials Today, 20*, Article 100682. https://doi.org/10.1016/j.apmt.2020.100682

Sun, T., Wang, S., Ji, C., Li, F., & Wu, H. (2022). Microplastics aggravate the bioaccumulation and toxicity of coexisting contaminants in aquatic organisms: A synergistic health hazard. *Journal of Hazardous Materials, 424*, Article 127533. https://doi.org/10.1016/j.jhazmat.2021.127533

Sun, A., & Wang, W.-X. (2023). Photodegradation of microplastics by ZnO nanoparticles with resulting cellular and subcellular responses. *Environmental Science & Technology, 57*(21), 8118–8129. https://doi.org/10.1021/acs.est.3c01307

Surette, M. C., Mitrano, D. M., & Rogers, K. R. (2023). Extraction and concentration of nanoplastic particles from aqueous suspensions using functionalized magnetic nanoparticles and a magnetic flow cell. *Microplastics and Nanoplastics, 3*(1), 2. Publicly Available Content Database. https://doi.org/10.1186/s43591-022-00051-1

Surucu, O. (2025). An electrochemical detection method for polyvinyl chloride microplastics in water bodies. *Journal of Molecular Structure, 1336*, Article 142086. https://doi.org/10.1016/j.molstruc.2025.142086

Tajkia Syeed Tofa, Ye, F., Kunjali, K. L., & Dutta, J. (2019). Enhanced visible light photodegradation of microplastic fragments with plasmonic platinum/zinc oxide nanorod photocatalysts. *Catalysts, 9*(10), 819. Publicly Available Content Database. https://doi.org/10.3390/catal9100819

Talvitie, J., Mikola, A., Koistinen, A., & Setälä, O. (2017). Solutions to microplastic pollution—Removal of microplastics from wastewater effluent with advanced wastewater treatment technologies. *Water Research, 123*, 401–407. https://doi.org/10.1016/j.watres.2017.07.005

Tang, Y., Zhang, S., Su, Y., Wu, D., Zhao, Y., & Xie, B. (2021). Removal of microplastics from aqueous solutions by magnetic carbon nanotubes. *Chemical Engineering Journal, 406*, Article 126804. https://doi.org/10.1016/j.cej.2020.126804

Ullattil, S. G., & Pumera, M. (2023). Light-powered self-adaptive mesostructured microrobots for simultaneous microplastics trapping and fragmentation via in situ surface morphing. *Small (Weinheim an der Bergstrasse, Germany), 19*(38), Article e2301467. https://doi.org/10.1002/smll.202301467

Uoginte, I., Pleskyte, S., Skapas, M., Stanionyte, S., & Lujaniene, G. (2023). Degradation and optimization of microplastic in aqueous solutions with graphene oxide-based nanomaterials. *International Journal of Environmental Science and Technology, 20*(9), 9693–9706. https://doi.org/10.1007/s13762-022-04657-z

Viswanathan, S., Pallikkara, A. A., Muhammed, F., & Kallingal, A. (2024). Graphitic carbon nitride based fluorine-free hydrophobic sponges for the mitigation of microplastics, oil spillage, and harmful microbial growth in water. *Materials Today Sustainability, 26*, Article 100751. https://doi.org/10.1016/j.mtsust.2024.100751

Walker, T. R., & Fequet, L. (2023). Current trends of unsustainable plastic production and micro(nano)plastic pollution. *TrAC Trends in Analytical Chemistry, 160*, Article 116984. https://doi.org/10.1016/j.trac.2023.116984

Wang, C., Shi, X., Bao, Z., Zhang, M., Shen, Y., & Wu, Y. (2023). Semiconductor heterojunction-AgNPs mediated surface-enhanced Raman spectroscopy (SERS) sensor for portable miniaturized detection platform. *Chemosensors, 11*(9), 490. Publicly Available Content Database. https://doi.org/10.3390/chemosensors11090490

Wegner, A., Besseling, E., Foekema, E. M., Kamermans, P., & Koelmans, A. A. (2012). Effects of nanopolystyrene on the feeding behavior of the blue mussel (*Mytilus edulis* L.). *Environmental Toxicology and Chemistry, 31*(11), 2490–2497. https://doi.org/10.1002/etc.1984

Wu, M., Lin, J., Zheng, D., Yang, Y., Li, Z., Zhu, Z., Shen, Y., Ni, G., & Zhang, M. (2023). Engineering branched Au@Ag nanostar plasmonic array for coupling electromagnetic enhancement and SERS trace detection of polystyrene in aquatic environments. *Chemosensors, 11*(10), 531. Publicly Available Content Database. https://doi.org/10.3390/chemosensors11100531

Xu, D., Su, W., Lu, H., Luo, Y., Yi, T., Wu, J., Wu, H., Yin, C., & Chen, B. (2022). A gold nanoparticle doped flexible substrate for microplastics SERS detection. *Physical Chemistry Chemical Physics, 24*(19), 12036–12042. https://doi.org/10.1039/D1CP05870C

Xue, Z., Yu, X., Ke, X., & Zhao, J. (2024). A novel route for microplastic mineralization: Visible-light-driven heterogeneous photocatalysis and photothermal Fenton-like reaction. *Environmental Science: Nano, 11*(1), 113–122. https://doi.org/10.1039/D3EN00642E

Yan, R., Lin, S., Jiang, W., Yu, X., Zhang, L., Zhao, W., & Sui, Q. (2023). Effect of aggregation behavior on microplastic removal by magnetic Fe(3)O(4) nanoparticles. *The Science of the Total Environment, 898*, Article 165431. https://doi.org/10.1016/j.scitotenv.2023.165431

Yang, L., Cao, X., Cui, J., Wang, Y., Zhu, Z., Sun, H., Liang, W., Li, J., & Li, A. (2022a). Holey Ti(3)C(2) nanosheets based membranes for efficient separation and removal of microplastics from water. *Journal of Colloid and Interface Science, 617*, 673–682. https://doi.org/10.1016/j.jcis.2022.03.055

Yang, L., Ye, X., Cao, X., Zhu, Z., Sun, H., Liang, W., & Li, A. (2022b). Deploying holey rGO-based membranes for MPs removal. *Journal of Water Process Engineering, 48*. https://doi.org/10.1016/j.jwpe.2022.102875

Yang, Z., Li, Y., & Zhang, G. (2024). Degradation of microplastic in water by advanced oxidation processes. *Chemosphere, 357*, Article 141939. https://doi.org/10.1016/j.chemosphere.2024.141939

Yu, T., Huang, X., Zhang, X. F., Li, K., Liu, S. P., Dai, N., Zhang, K., Zhang, Y. X., & Li, H. (2024). A review of nanomaterials with excellent purification potential for the removal of micro- and nanoplastics from liquid. *DeCarbon, 5*, Article 100064. https://doi.org/10.1016/j.decarb.2024.100064

Yuwendi, Y., Ibadurrohman, M., Setiadi, S., & Slamet, S. (2022). Photocatalytic degradation of polyethylene microplastics and disinfection of E. coli in water over Fe- and Ag-modified TiO_2 nanotubes. *Bulletin of Chemical Reaction Engineering & Catalysis, 17*(2), 263–277. Publicly Available Content Database. https://doi.org/10.9767/bcrec.17.2.13400.263-277

Zandieh, M., & Liu, J. (2022). Removal and degradation of microplastics using the magnetic and nanozyme activities of bare iron oxide nanoaggregates. *Angewandte Chemie, 134*(47), Article e202212013. https://doi.org/10.1002/ange.202212013

Zhang, X., Ding, J., Li, Z., Dai, J., Li, N., & Li, H. (2024). Rapid remediation of nanoplastics from water using a novel superparamagnetic absorbent. *Separation and Purification Technology, 351*, Article 128069. https://doi.org/10.1016/j.seppur.2024.128069

Zhang, Y., Duan, J., Liu, R., Petropoulos, E., Feng, Y., Xue, L., Yang, L., & He, S. (2025). Efficient magnetic capture of PE microplastic from water by PEG modified Fe_3O_4 nanoparticles: Performance, kinetics, isotherms and influence factors. *Journal of Environmental Sciences, 147*, 677–687. https://doi.org/10.1016/j.jes.2023.07.025

Zhao, H., Jiang, Z., Wang, C., & Li, Y. (2023). Preparation of N, Cl Co-doped lignin carbon quantum dots and detection of microplastics in water. *Crystals, 13*(6), 983. Publicly Available Content Database. https://doi.org/10.3390/cryst13060983

Zhou, H., Mayorga-Martinez, C. C., & Pumera, M. (2021a). Microplastic removal and degradation by mussel-inspired adhesive magnetic/enzymatic microrobots. *Small Methods, 5*(9), 2100230. https://doi.org/10.1002/smtd.202100230

Zhou, J., Wen, Y., Marshall, M. R., Zhao, J., Gui, H., Yang, Y., Zeng, Z., Jones, D. L., & Zang, H. (2021b). Microplastics as an emerging threat to plant and soil health in agroecosystems. *Science of the Total Environment, 787*, Article 147444. https://doi.org/10.1016/j.scitotenv.2021.147444

Zhou, D., Wang, L., Zhang, F., Wu, J., Wang, H., & Yang, J. (2022). Feasible degradation of polyethylene terephthalate fiber-based microplastics in alkaline media with Bi_2O_3@N-TiO_2 Z-scheme photocatalytic system. *Advanced Sustainable Systems, 6*(5), 2100516. https://doi.org/10.1002/adsu.202100516

Photocatalytic Degradation of Macro—and Microplastics Using Titanium Dioxide (TiO_2)

Odangowei Inetiminebi Ogidi, Okechukwu Igwe Oguoma, and Nwanne Ijere Dike

Abstract The growing environmental threat posed by macro- and microplastics has intensified the search for sustainable degradation methods. Photocatalysis, using titanium dioxide (TiO_2) as a catalyst, has emerged as an efficient and eco-friendly approach to breaking down plastic waste. Under ultraviolet (UV) or visible light irradiation, TiO_2 generates reactive oxygen species (ROS) that oxidize polymer chains, converting plastics into smaller, less harmful by-products such as carbon dioxide, water, and organic intermediates. This process is effective for degrading various plastic types, including polyethylene, polypropylene, and polystyrene, at both macro and micro scales. Recent advancements have focused on enhancing TiO_2 photocatalytic efficiency through doping with metals or non-metals and combining it with other semiconductors to expand light absorption into the visible spectrum. Furthermore, the integration of TiO_2 into composite systems has improved its stability and recyclability, making it more suitable for large-scale applications. Despite its potential, challenges such as slow degradation rates, the formation of secondary pollutants, and limited efficiency under natural sunlight remain. This chapter will explores the principles, applications, and advancements in TiO_2-mediated photocatalytic degradation of plastics, highlighting its role in addressing plastic pollution and promoting sustainable environmental practices. Further research and technological innovation are necessary to optimize this method for widespread implementation.

Keywords Micro-plastics · Titanium dioxide (TiO_2) · Degradation · Environment · Macro-plastics · Photocatalysts

O. I. Ogidi (✉)
Department of Biochemistry, Faculty of Basic Medical Sciences, Bayelsa Medical University, Yenagoa, Bayelsa State, Nigeria
e-mail: ogidiodangowei@gmail.com

O. I. Oguoma
Department of Forensic Science, Federal University of Technology, Owerri, Imo State, Nigeria

N. I. Dike
Department of Public Health, Faculty of Health Sciences, College of Arts, Science and Professional Studies, Aba, Abia State, Nigeria

E. Priyadarshini et al. (eds.), *Advanced Nanotechnology Applications for Environmental Sustainability*, Sustainable Environmental Waste Management Strategies,
https://doi.org/10.1007/978-3-032-12933-8_13

1 Introduction

The world is currently confronting significant issues, including environmental pollution, degradation, and the urgent need for clean water and air (Ogidi & Akpan, 2024a; Ogidi & Richard-Nwachukwu, 2024b). At the same time, the rapid increase in global population, the widespread lack of regulation in energy consumption, and the manufacturing of plastics which accounts for more than 90% of global output which have all contributed to the acceleration of industrialisation (Andrady & Neal, 2009). The unchecked release of harmful contaminants and garbage into the environment has caused serious problems like air and water pollution, as well as climate change. Over 300 distinct varieties of plastics are manufactured, with more than 60 of them being highly prevalent. The production and consumption of plastics are anticipated to rise in the near future (Chen et al., 2020).

Annually, around two hundred million tonnes of plastic and plastic-based materials are manufactured, contributing to a cumulative total of million tonnes since the mid-point of the preceding era. The carpenter initially recorded the presence of plastic pellets in the North Atlantic Ocean, focusing on communities investigating microplastic pollution (Amelia et al., 2021). The extent of trash and plastic that will accumulate in waterways, rivers, and landfills, where it will decompose into smaller fragments of rock and other debris, is uncertain. Upon ingestion by organisms, microplastics (MPs) decompose into nanoplastic particles, affecting excretion, metabolism, digestion, and reproduction processes (A. Prasad. Das & John, 2025; B. Priyadarshini & Das, 2024; S. Priyadarshini et al., 2024). The origin of MPs must be evaluated, along with prioritising transportation and viable environmental solutions, to effectively manage MPs pollution (Amelia et al., 2021). MPs have been identified in several environments, including aquatic and terrestrial settings (Zhang et al. 2020a). Atmospheric fluxes are crucial in the interplanetary cycle of elements. Elements are conveyed from distant emission sources to recipient regions, resulting in these fluxes (Camarero et al., 2017).

The environmental impacts of microplastics, defined as particles smaller than 5 mm, including their capacity to proliferate, migrate, and accumulate. Accumulations of MPs on sand, sediments, seabeds, beaches, and water surfaces (Barnes et al., 2009; Ryan et al., 2009; Thompson et al., 2009). Developing countries frequently utilise single-use plastics for applications such as mulch film and greenhouse materials (Chen et al., 2020). According to the United Nations Environment Programme (UNEP) (Giacovelli, 2018), a large amount of contemporary plastic production is now focused on single-use plastics. Plastics rely heavily on oils; each year, the industry uses over 1.6 billion litres of oil to make plastics. Hydrogen, carbon, nitrogen, oxygen, sulphur, and chlorine are only a few of the many components used in plastics manufacture. Polycarbonate, acrylonitrile butadiene styrene copolymer, polysulfone, polyvinyl chloride, and phenolic resin are among the most common plastics. The predominant plastics utilised in everyday life are polypropylene (PP) and polyethylene (PE), although polystyrene (PS) and phenolic resin are also present. The primary classifications of plastics are homo-chain plastics (e.g., PS, PP, PE) and hetero-chain plastics (e.g., PET, PA) (Gewert et al., 2015).

Homo-chain plastics possess a carbon–carbon bond in the primary chain, whereas hetero-chain plastics contain diverse bonds in the primary chain. Under ambient conditions, the degradation of homopolymer chains transpires in three phases: initiation, propagation, and termination. Initiation involves bond cleavage induced by light, leading to the formation of reactive species. Propagation involves the eventual cleavage and cross-linking of the chains. Termination transpires upon the completion of the reaction. Nonetheless, degradation in aquatic environments transpires due to photodegradation, biodegradation, and hydrolysis of hetero-chain polymers (Das et al., 2025; Reddy Koteswara et al., 2025). During the brief photocatalytic breakdown of plastics, many intermediates are generated, including long alkane/olefin chains, small acid molecules such as CH_3COOH, and hydrogen, culminating in their mineralisation (Hu et al., 2021).

However, frequently, single-use plastics end up in inappropriate locations, resulting in the wastage of important resources (Boucher et al., 2019; Wang et al., 2018). Wildlife and ecosystems are adversely affected by the accumulation of plastic waste over time (Cartraud et al., 2019; Rizzi et al., 2019). Plastics has significant insulating properties, both thermally and electrically, and may be shaped into virtually any form. The variable composition of microplastics (MPs) and nano-plastics (NPs), characterised by varying additive concentrations, shapes, sizes, and physicochemical properties, precludes generalisations on the associated risks of plastics. The dimensions of a plastic particle dictate its classification. The dimensions of megaplastics, macroplastics, mesoplastics, microplastics, and nanoplastics are defined as follows: greater than 1 m, greater than 2.5 cm, greater than 5 mm, less than 5 mm, and less than 1 μm, respectively, according to reports (Chatterjee & Sharma, 2019; Gillibert et al., 2019). The basic inquiry concerning the number of MPs in aquatic environments remains unresolved (Cai et al., 2018).

The small size of microplastics and nanoparticles poses a number of problems, such as (1) their ability to move quickly through the environment, (2) their large surface area for contaminant adsorption, (3) their ease of movement and transmission through the food chain, and (4) the possibility of transmission to animals (Peng et al., 2020). The origin of microplastics permits their classification into two categories: primary and secondary plastics (Lehtiniemi et al., 2018; Peng et al., 2020). Diverse types of primary microplastics are emitted, including beads, pellets, fibres, and others. Secondary microplastics are generated from bigger plastics through biological, physical, and chemical processes. The majority of microplastics (MPs) originate from larger polymers that have undergone degradation due to bacteria, light, oxidation, and heat (Rillig et al., 2017). The family of transition metal oxides encompasses titanium dioxide (TiO2), sometimes referred to as titanium (IV) oxide. Titanium dioxide (TiO_2) serves numerous applications, such as in sunscreens, cosmetics, photovoltaic cells, environmental cleanup, nanopaint, antibacterial agents, and sun protection (Borrás et al., 2020; Upadhyay et al., 2020).

Despite the utilisation of bismuth (Tishkevich et al., 2018; Bacha et al., 2019a, 2019b, 2020), nickel nanopillars (Vorobjova et al., 2020), and nickel nanowires

(Tishkevich et al., 2019; Vorobjova et al., 2019, 2021) in environmental applications, none of these materials have been assessed for their capacity to degrade plastics. Titanium dioxide (TiO_2), a crucial photocatalyst, has optimal efficacy within the UV light spectrum (Moma & Baloyi, 2019). This photocatalytic compound has a longstanding record of safe and benign utilisation in various applications (Skocaj et al., 2011). In photocatalytic degradation reactions, it serves as an effective oxidant for contaminants (Tsang et al., 2019). Because of its low production costs and ease of use at laboratory scale, it is a widely used semiconducting material (Hu et al., 2010; Malato et al., 2002). The extensive use of this white pigment over many years attests to its non-toxicity to people and the environment (Moma & Baloyi, 2019; Shah et al., 2017). In a typical illumination scenario, two types of photochemical reactions are triggered by TiO_2. The hydrophilic transformation of TiO_2 and the redox reactions of adsorbed molecules are two of these processes that are driven by light. In the realm of construction materials, these combinations provide innovative applications of TiO_2 (Hashimoto et al., 2005). Different categories of pollutants, such as azo dyes (Wang et al., 2020), persistent pollutants (Bacha et al., 2019b), and emerging contaminants have been degraded utilising TiO_2 (Penas-Garz'on et al., 2020).

Simultaneously, researchers employed TiO_2 to decompose polymers of both macroscopic and microscopic dimensions. In 2019, N-modified TiO_2 was first employed to breakdown commercial microplastics HDPE (Ariza-Tarazona et al., 2019). Researchers have diligently endeavoured to optimise a TiO_2 photocatalyst for the degradation of diverse micropollutants. This chapter discusses the most recent advancements in the use of TiO_2 to combat plastic pollution, its essential properties, and the various methods by which it can be synthesised. The primary process of TiO_2 and its structural variations over time are emphasised. It offers a thorough examination of the literature on the subject in order to enhance our comprehension of TiO_2's capacity to degrade macro and microplastics.

2 Methods of Plastic Waste Disposal

Traditional waste management practices, such as dumping and burning non-biodegradable plastics, have shown to be ineffective. This explains the uptick in efforts to find technologically superior answers. Technologies that recycle include physical materials, energy, and resources; biodegradation and oxo-biodegradation exemplify degradation technologies (Ogidi et al., 2024a, 2024b, 2024c). Examples of oxy-biodegradation processes include thermal degradation, mechanical-chemical degradation, and photodegradation (Zhang et al., 2021a). More effective methods to recycle this organic carbon source and produce self-decomposing polymers to reduce our environmental impact should be identified (Jing et al., 2021). Under mild circumstances, Su et al. (2022) highlighted the significance of photocatalytic conversion of plastic waste into products with a low carbon number. Photocatalytic degradable plastics (PDP) are a combination of polymers and photocatalysts that enhance the biodegradability of conventional non-biodegradable plastics. Different methods of plastic waste disposal, including their mechanisms, applications, advantages, limitations, and environmental impact are shown in Table 1.

3 Photocatalytic Degradation

The utilisation of photocatalysts alongside sunlight significantly accelerates the degradation rate of plastic trash. Ultraviolet light's tremendous intensity renders it the most potent agent for polymer degradation. Plastic waste polymers experience physical and chemical transformations during photodegradation, such as chain scission, reduction in molecular weight, and alteration of molecular structure, upon exposure to UV light and oxygen (Lee & Li, 2021). Remediating water polluted with micro- and nanoplastics via heterogeneous photocatalysis is no easy feat, it has reached the industrial pilot stage and shown efficacy in eliminating MPs from water. The dimensions, morphology, and chemical composition of microplastics, along with the properties and concentration of the photocatalyst, the light source, and ambient variables, all influence the efficacy of microplastic degradation via photocatalysis.

4 The Mechanisms of Photodegradation of Polymers

A series of chemical reactions are initiated by sunlight, which ultimately results in the decomposition of microplastics. Some of the activities that can be initiated by photon absorption include oxidation, chain disruption, branching cross-linking, and excited states of macromolecules (Zhao et al., 2007). Photodegradation is expedited by the presence of a photocatalyst. Polymer oxidation by free radicals is one form of photodegradation, while polymer oxidation by singlet oxygen is another. The first phase involves the formation of singlet oxygen from the excited triplet state of the relevant sensitisers. Singlet oxygen has the capacity to perform two additional functions after reacting with the polymer: oxidising a vinyl group ($-CH=CH_2$) and forming a hydroperoxide functional group (ROOH). The byproduct of a photochemical reaction that involves an aldehyde and a ketone may react with singlet oxygen (Zhao et al., 2007).

Additionally, ultraviolet light can generate free radicals that can disrupt C–C and C–H bonds in polymers. Oxygen reacts with the free radicals that are generated during the initial, intermediate, and final stages to generate hydroxyl (O–H) and carbonyl (C=O) groups. Producing free radicals is one of the initial stages in initiating any endeavour. The photo-initiator, a polymer with an unsaturated chromophore group, absorbs light, resulting in the production of hydrogen and polymer radicals. The polymer chain is cleaved by this process (Yousif & Haddad, 2013). These polymers are distinguished by the absence of an unsaturated chromophore group, as well as contaminants, structural defects, and carbonyl groups in the polymer backbone that trigger photolysis (Zhang et al., 2021b). Physical or chemical factors may initiate

Table 1 Methods of plastic waste disposal

Disposal method	Mechanism of action	Application	Advantages	Limitations	Environmental impact
Landfilling	Plastic waste is buried in designated landfill sites	Municipal waste management, waste containment	Low cost, simple implementation	Requires large land area, plastics degrade very slowly	Groundwater contamination, soil pollution
Open dumping	Uncontrolled disposal of plastic waste in open spaces	Common in developing countries	No infrastructure needed	High pollution, creates breeding grounds for diseases	Severe soil, water, and air pollution
Incineration	Burning plastic waste at high temperatures to generate heat and electricity	Waste-to-energy plants, industrial waste disposal	Reduces landfill burden, energy recovery	Releases toxic emissions (dioxins, furans), expensive emission control required	Air pollution, greenhouse gas emissions
Mechanical recycling	Plastics are sorted, shredded, cleaned, melted, and remolded	PET bottle recycling, HDPE plastic reuse	Reduces virgin plastic use, conserves resources	Limited recyclability, requires high-quality, uncontaminated plastic	Low environmental impact if properly managed
Chemical recycling (Depolymerization)	Plastics are chemically broken down into monomers and reused	Advanced recycling plants, circular economy initiatives	Converts plastic into high-quality raw materials, suitable for low-quality plastics	High energy cost, complex technology required	Can reduce reliance on fossil fuels but may produce chemical waste
Biodegradation	Microorganisms break down biodegradable plastics into natural compounds	Composting facilities, bioplastic waste management	Environmentally friendly, reduces plastic waste	Only works for specific bioplastics (PLA, PHA), slow degradation rate	Limited environmental impact but requires special conditions
Pyrolysis	High-temperature decomposition of plastic into fuel and chemicals	Waste-to-fuel conversion, industrial applications	Produces valuable fuel, reduces plastic waste	Requires high energy input, not suitable for all plastic types	Can reduce fossil fuel dependence but releases CO_2

(continued)

Table 1 (continued)

Disposal method	Mechanism of action	Application	Advantages	Limitations	Environmental impact
Gasification	Converts plastic waste into syngas (hydrogen and carbon monoxide) at high temperatures	Syngas production, energy generation	High energy recovery, useful in industrial settings	Expensive, requires advanced technology	CO_2 emissions, potential toxic gas formation
Plasma arc gasification	Uses high-energy plasma to break plastic waste into synthetic gas and slag	Hazardous waste treatment, energy recovery	Extremely high efficiency, minimal solid waste	Very high operational costs, requires specialized equipment	Lower emissions than incineration, but energy-intensive
Ocean cleanup	Mechanical removal of floating plastics from oceans	Ocean cleanup projects, river plastic barriers	Helps reduce marine plastic pollution	Does not address plastic production, costly	Prevents marine ecosystem damage but has limited scalability
Thermal depolymerization	Heat and pressure break down plastics into oil, gas, and char	Waste-to-energy plants, fuel production	Converts plastic into reusable fuels	Requires high energy input, expensive process	Can reduce fossil fuel use but generates CO_2 emissions
Supercritical water treatment	Uses high-pressure, high-temperature water to break down plastics	Industrial-scale plastic waste disposal	Rapid breakdown, no toxic emissions	High operational costs, extreme processing conditions required	More eco-friendly than incineration but energy-intensive
Plastic-to-road technology	Mixing shredded plastic with bitumen to pave roads	Infrastructure projects, road construction	Reduces plastic waste, improves road durability	Requires specific plastic types, long-term impact unknown	Low impact on environment, improves road quality
Composting (for bioplastics)	Organic bioplastics decompose into natural elements under controlled conditions	Industrial composting facilities	Sustainable, reduces plastic footprint	Limited to compostable plastics, requires specific conditions	Minimal impact if properly managed

(continued)

Table 1 (continued)

Disposal method	Mechanism of action	Application	Advantages	Limitations	Environmental impact
Alternative fuel production	Plastic waste is converted into fuel substitutes for cement kilns and power plants	Energy-intensive industries	Reduces fossil fuel use, efficient waste management	Requires preprocessing, potential air pollution	CO_2 emissions but reduces landfill waste
Electrochemical degradation	Electric current generates reactive species to degrade plastics	Experimental lab-based applications, future waste treatment technology	No toxic byproducts, effective at breaking polymer bonds	High energy requirement, not yet widely implemented	Promising for sustainable waste treatment, but still under research
Microbial enzyme-based degradation	Special enzymes break down plastic polymers into natural byproducts	Research labs, bioremediation projects	Non-toxic, natural, sustainable	Slow process, currently not scalable	Eco-friendly, but still under development
Artificial photosynthesis degradation	Sunlight-driven catalytic process breaks plastic down into harmless compounds	Lab-scale applications, experimental technology	Uses renewable energy, no toxic byproducts	Requires advanced materials, not commercially viable yet	Potential for carbon–neutral plastic degradation
Bacterial bioremediation	Genetically modified bacteria break down plastics into simpler molecules	Research applications, landfill bioremediation	Can be tailored to specific plastics	Ethical and safety concerns, requires controlled conditions	Low-impact if successful, but still in development
Anaerobic digestion of bioplastics	Microorganisms break down biodegradable plastics in oxygen-free environments	Composting, wastewater treatment plants	Produces biogas as an energy source	Only effective for specific bioplastics (PLA, PHA), slow process	Low impact but requires proper waste sorting
Oxidative wet combustion	Controlled water-oxygen combustion breaks down plastics	Hazardous waste treatment	Fast process, reduces landfill waste	High energy requirement, emits CO_2	CO_2 emissions but no toxic dioxins

the initial phase. Chemical initiation is characterised by unsaturated sites, catalysts, peroxides, or photolytic cleavage of carbon–carbon or carbon-hydrogen bonds, while physical initiation is characterised by ultraviolet light and photolytic cleavage of these bonds (Singh & Sharma, 2008).

In the second stage of propagation, polymer macroradicals interact with oxygen, resulting in the production of polymer peroxyl radicals (PP). Thirdly, they react with another polymer to produce hydroperoxides and polymer macroradicals. The generation of (PP) was the result of the auto-oxidation of macro-radicals within the polymer. The subsequent step involves the reaction of the two hydroperoxides (HP) to generate polymer alkoxy radicals, water, and polymer peroxide radicals (Yousif & Haddad, 2013). Additionally, polymer alkoxy radicals and hydroxyl radicals (OH) are generated during the photodecomposition of (HP). Olefins and ketones are the terminal products of chain scission during propagation. The hydroxyl radical (·OH) that is generated as a consequence of the removal of hydrogen from another polymer will produce polymer macroradicals and water (Yousif & Haddad, 2013).

A consequence of persistent oxidation is the formation of microcracks on the polymer's surface, which results in the release of oxygenated molecules with low molecular weights, including hydroxyl, peroxide, and carbonyl groups. Chain scission is the result of these molecules (Zhang et al., 2021b). The final step in the process of free radicals combining to produce inert molecules is a crosslinking reaction. Polymer peroxyl radicals react with oxygen in the presence of high oxygen pressure to generate inert compounds and oxygen. Conversely, when oxygen pressure is inadequate, polymeric macroradicals and polymeric peroxyl radicals interact in Path 11. Olefins and ketones are anticipated as the termination reaction's end products. Polymers become more brittle as a result of the ongoing process of photodegradation, which reduces their molecular weight (Ali et al., 2021).

5 Titanium Dioxide (TiO_2) as Photocatalyst

Anatase, rutile, and brookite are the three most prevalent crystalline forms of titanium (IV) oxide (TiO_2) found in nature. Alongside naturally occurring forms, commercially produced variants exist, such as P25, which comprises a blend of anatase and rutile. The proportion of the two minerals ranges from 80:20 to 70:30, contingent upon the source. Due of brookite's diminished photo-reactivity compared to other polymorphic forms of TiO_2, it is often overlooked. The predominant method of preparation for utilisation is as a composite with brookite and anatase (Lu et al., 2011). Due to its photocatalytic properties, affordability, non-toxicity, and excellent photostability, TiO_2 offers numerous potential applications (Fujishma et al., 2000).

Titanium dioxide possesses numerous potential advantages when utilised as a photocatalyst. These characteristics encompass its hydrophilicity, non-toxicity, affordability, and robust photo-reactivity; its facile immobilisation through coating; its enduring chemical stability; and its potential for photodegradation efficiency. Several disadvantages of titanium dioxide warrant consideration. The optical response of the material, particularly prominent in the UV region, and the fast recombination of generated electron–hole pairs are included. Properly altering the material structure might avert both of these issues (Ahmed & Haider, 2018; Lin & Mehrvar, 2018). A scientific illustration depicting titanium dioxide (TiO_2) as a photocatalyst for the degradation of macro- and microplastics is shown in Fig. 1.

The most widely used photocatalyst is TiO_2 because of its many desirable characteristics, including its chemical stability, high-temperature resistance, cost-effectiveness, and eco-friendliness. To what extent a photocatalyst can undergo oxidation and reduction is dependent on where its energy band lies in relation to redox potentials. The ability to oxidise water is shown by a valence band maximum (VBM) that is greater than 1.23 eV. Decreases in the location of semiconductors' valence band maxima (VBMs) are directly correlated with increases in their oxidation capacities. It is the converse: for water reduction to occur, the hydrogen reduction potential must be greater than the conduction band minimum (CBM). The position of the conduction band minimum is directly connected to a semiconductor's reduction capacity. Compared to other types of semiconductors, TiO_2 has a very high oxidation capacity. Improving light absorption requires lowering the band gap (Ahmed & Haider, 2018).

Although zinc sulphide has excellent oxidation and reduction capabilities, it is unable to absorb light efficiently due to its huge band gap. Because of its ideal oxidation–reduction characteristics and suitable band gap size, TiO_2 is a remarkable photocatalyst. There are three crystalline polymorphs of titanium dioxide: anatase,

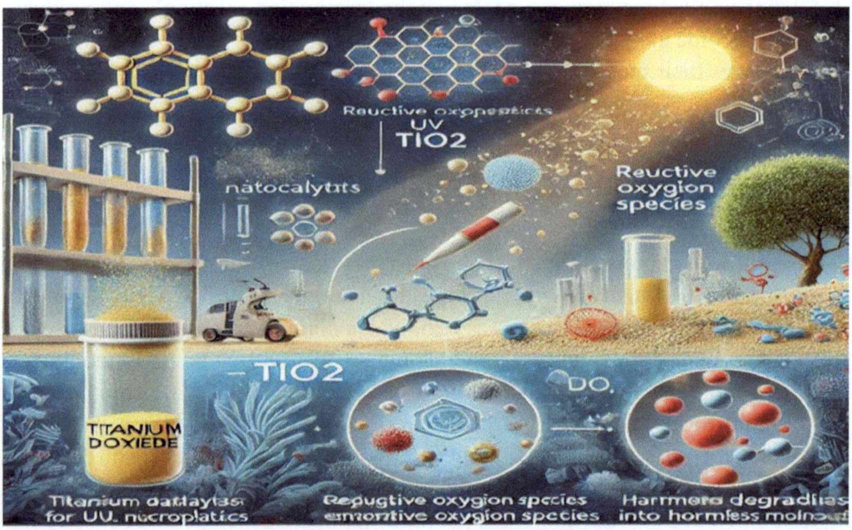

Fig. 1 Titanium Dioxide (TiO2) as Photocatalyst for the degradation of macro and microplastics. *(This is a scientific illustration depicting titanium dioxide (TiO_2) as a photocatalyst for the degradation of macro- and microplastics. It shows TiO_2nanoparticles activated by UV light, breaking down plastics into smaller, environmentally friendly compounds)*

rutile, and brookite (Bickley et al., 1991; Zhang et al., 2014, 2018). Different phases of TiO_6 show different symmetries in its core structural units, which are octahedral in shape (Zhang et al., 2014). While anatase and rutile are common photocatalysts for degradation, brookite is an unusual photocatalyst due to its structural instability. Turning anatase into rutile requires temperatures higher than 700 °C. A small subset of the many geometric forms that TiO_2 can take on are nanostructures, such as nanotubes (Mack et al., 2007), nanowires (Lee et al., 2011), and nanoparticles (Shi et al., 2013). At energy level ECB, the critical band gap (CBM) for anatase TiO_2 is 0.51 V, but for rutile TiO_2 it is 0.31 V (Yu et al., 2014). While both forms have the same valence band maximum (VBM) at an energy level of EVB = +2.69 V, the anatase form has a slightly greater bandgap (Eg = 3.2 V) than the rutile form (Eg = 3.0 V), thus showing a higher energy of photogenerated charges (Lin & Mehrvar, 2018).

Thus, anatase has a shorter absorption bandwidth and shows comparatively strong photocatalytic activity. Anatase has an indirect bandgap, in contrast to rutile and brookite which have direct ones. Photogenerated electron–hole pairs in anatase should theoretically last longer because they are not allowed to recombine directly. There is no difference in the DOS or specific energy band topology between anatase and rutile TiO_2 (Zhang et al., 2014). As a result of substantial O 2p-Ti 3d hybridisations, the valence bands consist of O 2p and multiple Ti 3d states. A stronger photo carrier transfer and wider valence bands are the outcomes of such robust p-d hybridisations. On the flip side, while O 2p and Ti 3p states do contribute to the conduction bands, Ti 3d is largely responsible for them. The band gaps predicted for anatase and rutile TiO_2 by basic density functional theory calculations are less than the values found experimentally, which should be taken into consideration (Labat et al., 2007; Park et al., 2013).

6 Mechanism of TiO_2 for Plastic Degradation

There is a small energy difference between TiO_2's conduction band (CB) and valence band (VB). When exposed to light, especially in the ultraviolet A range (315–400 nm), an electron in the valence band gains energy and moves into the conduction band (Li, 2020). The migration of electrons (e-) to and from the adsorbent can produce charged species of both positive and negative charge (Martinez & Hammer, 2011). At 388 nm, the excitation wavelength, anatase TiO2 may absorb light in the ultraviolet (UV) spectrum because to its large bandgap of 3.2 eV. The visible light spectrum includes 410 nm, which rutile TiO_2 is able to absorb due to its 3.0 eV bandgap (Pawar et al., 2018). The procedure for transferring activated TiO_2 to CB. There are two main concepts or mechanisms that govern the degrading process of plastics (Fig. 2).

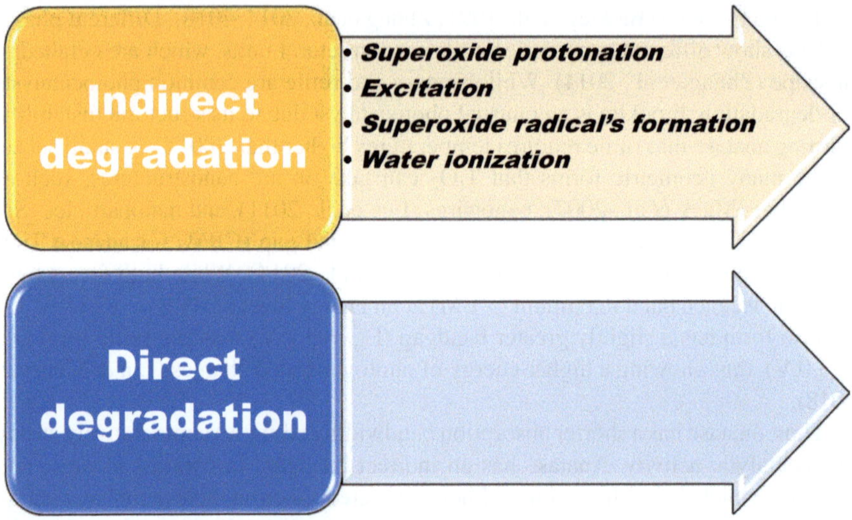

Fig. 2 Main concepts or mechanisms that govern the degrading process of plastics

7 Indirect Degradation

The subsequent techniques are the primary steps in the indirect photocatalytic degradation of plastics:

7.1 *Excitation*

When light irradiates the valence band, it excites an electron to transition to the vacant conduction band, initiating the photocatalytic event (He et al., 2020). When a photon hits TiO_2, its energy is more than or equal to its bandgap energy. The VB's hole potential (h+) and the subsequent light response are both affected by this input.

7.2 *Water Ionization*

A hydroxyl radical (OH) is generated when a water molecule interacts with a vacancy in the valence band (VB). A crucial element of photooxidation is the hydroxyl radical, a potent oxidising agent (He et al., 2020). The mineralisation of adsorbed pollutants associated with or next to the photocatalyst interface results from its nonselective attack (Santhi et al., 2020). TiO2 can photodegrade organic pollutants when subjected to both ultraviolet and visible light (Basavarajappa et al., 2020).

7.3 *Superoxide Radical's Formation*

When an oxygen molecule comes into contact with a photoexcited electron in the conduction band, the result is the superoxide radical (O2-). By oxidising and blocking the recombination of photogenerated electron–hole pairs, the O_2·-radical maintains electron neutrality with the semiconductor (Ajmal et al., 2014).

7.4 *Superoxide Protonation*

The created O_2-radical underwent protonation to form the hydroperoxyl radical (HO_2), which subsequently generates hydrogen peroxide (H_2O_2). Hydroxyl radicals are generated as hydrogen peroxide further dissociates.

8 Direct Degradation

Under visible light, TiO_2 can directly degrade plastics to a significant degree. The process involves transferring polymers from their ground state to an excited state using photons of visible light. When polymers are stimulated to this point, they can enter the catalyst's conduction band (CB) and produce semi-oxidized cation radicals. Trapped electrons degrade the target pollutant by interacting with dissolved oxygen to produce O_2, which in turn produces OH (Ajmal et al., 2014; Galindo et al., 2000). Indirect plastic degradation is more common than direct plastic degradation, because visible light reactions are less noticeable than plastic disintegration. In addition, compared to processes started by UV light, those started by visible light move at a much slower pace. Plastic breakdown procedures, including both direct and indirect approaches.

9 Application of TiO_2 for Microplastics (MPs) Degradation

TiO2 has shown significant effectiveness in breaking down MPs. In contrast to TiO_2 using ethanol (ET), which had a diameter of 177.78 nm, TiO_2 using Triton X-100 (TXT) showed a notable diameter fluctuation of about 100.07 nm. The water-based TiO2 showed a wavelength of 268.16 nm after being exposed to 365 nm UV radiation for 12 h. A combination of its band energy, surface area, particle size, charge separation, and light absorption makes it work efficiently. The degrading reaction product was determined using gas chromatography. The significantly higher generation of CO_2 relative to CO was indicative of mineralisation (Nabi et al., 2020). Degradation of PVC microplastics was achieved by electrolysis using a TiO_2/C cathode at −0.7 V relative to Ag/AgCl (Miao et al., 2020).

The reaction has significantly altered the previously smooth surface of the PVC, making it show depressions and holes. The internal components of PVC are more vulnerable to environmental conditions as the surface material deteriorates. At 100 °C, we achieved 75% dichlorination efficiency and 56% weight reduction, respectively, when we tested the system's dichlorination efficacy for PVC breakdown at various temperatures. At higher temperatures, dichlorination was most effective. A dichlorination efficiency of 29% was observed after a 6-h reaction at 100° with the graphite cathode. Using the classic Fenton procedure at 25, the effectiveness of dichlorination was found to be noticeably poor. The chromophore did not undergo degradation since there was no discernible mass loss when a photocatalyst was not present during the irradiation reaction. After 50 h of UV irradiation, HDPE-A typically loses only around 0.22% of its mass, demonstrating its remarkable resilience. That little mass loss is probably explained by the inherent characteristics of MPs and their large size in comparison to catalysts. Compared to HDPE-B, an extra component, the catalyst had smaller particles. After the 50-h procedure was finished, the catalyst caused a 4.65% decrease in total mass, as shown by the orange line. Furthermore, various LDPE diameters were tested for mass loss. The mass of 5 × 5 LDPE was around 0.97% after 50 h of reaction with the catalyst under visible light irradiation, while the mass of 3 × 3 LDPE was about 1.38%. As a result of reactive radical species reacting, the mass loss changed (Llorente- García et al., 2020).

Particle size and shape have a major impact on degradation rate and catalytic efficiency. The photocatalytic breakdown of HDPE using N-modified TiO_2 was studied under a variety of experimental settings. In HDPE, researchers have found that the photodecomposition rate is directly related to the material's porosity. The prepared sample of catalyst-modified HDPE shows the composite's typical rough surface in terms of shape. Catalysts interact better with plastic due to their large surface areas. A substantial change occurs in HDPE as a result of the catalyst's ability to produce hydroxyl radicals in a water solution. Both solid and liquid MPs can be decomposed by catalysts, according to the research.

The photolysis experiment exhibited minimal damage in visible light. Following 20 h of exposure to visible light at ambient temperature, HDPE combined with N-TiO_2 saw a mass reduction of roughly 6.40%. The solid system exhibited a kinetic rate constant that was thrice slower than that of the water system (Ariza-Tarazona et al., 2019). The decomposition experiment results indicated that, despite its smooth surface, HDPE possesses fractures and voids. The improved interaction between HDPE microplastics and C, N-TiO_2 connects the decomposition process to these morphological changes. A change in the surface composition signifies the onset of decomposition and liberates inside material for further degradation (Gewert et al., 2015).

The cumulative mass loss at different temperatures and pH levels. When exposed to visible light for 50 h in a pH 3 environment without temperature, the subjects lost the most weight (71.77%). According to the results, the rate of microplastic breakdown is affected by both temperature and pH. In 50 h of reaction at 40 °C and pH 3, the average mass loss was 12.42%, which is equivalent to 1/6 of the value achieved at pH 3 (20 °C). At a pH of 7 (20), approximately 0.47% of the bulk was diminished. At a pH of 11, the mass loss was 1.55% at 0 and 0.78% at 40. Degradation and mean weight reduction are affected by the pH of the solution and temperature (Ariza-Tarazona et al., 2020). All produced TiO_2 coatings effectively degraded MPs, irrespective of the experimental conditions. Table 2 shows the comprehensive application and additional advanced biological, chemical, and physical methods for degrading microplastics.

10 Application of TiO_2 for Macro Plastic Degradation

The degradation of macroplastic debris, a prominent form of coastal pollution, has utilised both pure and modified TiO_2. Upon exposure to light, the PVC film exhibited a significant alteration, indicating that the TiO_2 was undergoing decomposition in both its pure and vitamin C modified states. An indication that the photodegradation reaction has begun is the formation of a hollow at the PVC/catalyst contact. The PVC film that was loaded with TiO_2 and then treated with vitamin C showed more deeper penetrations and larger cavities than the pure TiO2-laden PVC film (Yang et al., 2010). Degradation of polyethylene (PE) was achieved through the use of TiO2 modified with transition metals. The creation of cavities was caused by the release of volatile chemicals and the assault of reactive oxygen species (Asghar et al., 2011).

The effectiveness of pure TiO_2 was enhanced by using bismuth oxy-iodide-modified TiO_2. Since perchlorinated iron phthalocyanine is a more stable catalyst for the oxidation of cyclohexane than FePc, even after four weeks, phthalocyanine shows immense promise as a substance to degrade pollutants (Sehlotho & Nyokong, 2004). However, the $FePcCl1_6$-TiO_2 composite with PVC showed many fragments and compact voids when subjected to the same reaction conditions. The weight decrease in PVC was more pronounced with the combination of $FePcCl1_6$ and TiO_2 than with pure TiO_2. No weight loss was seen when PVC was photolyzed (Fa et al., 2011). Titanium dioxide (TiO_2) was doped with varying amounts of copper phthalocyanine (CuPc) to improve the photodegradation of polyethylene (PE). After 160 h of irradiation, pure polyethylene showed almost little weight loss. Adding PE to TiO_2 does not significantly reduce weight, but when combined with copper phthalocyanine, it produces remarkable performance. With 0.3 monolayers of CuPc on TiO_2, the apparent breakdown rate of CuPc/TiO_2 was 0.7%. According to the findings, CuPc and TiO_2 together improve the production of active species for PE degradation, bond cleavage, and volatile product synthesis (Zhao et al., 2008).

It was also tested how well CuPc/TiO_2 degraded PS at the same time. In compared to pure TiO_2, the percentage of weight loss for PS was substantially higher when using CuPc/TiO_2. Using CuPc/TiO_2 resulted in a 6.9% decrease in PS weight, compared to a 4.1% decrease after 250 h. While the pace of weight loss slowed in the second phase, it was faster in the first. The random chain scission in polymers is hindered by an initial rapid weight drop (Shang et al., 2003). The degradation rate increased

Table 2 Applications of TiO_2 for microplastics (MPs) degradation

Degradation method	Mechanism of action	Application	Advantages	Limitations
Biodegradation (Microbial Degradation)	Enzymatic breakdown by bacteria and fungi	Bacteria such as *Ideonella sakaiensis*, *Pseudomonas sp.*, and fungi like *Aspergillus sp.* degrade MPs in soil and water environments	Environmentally friendly, sustainable	Slow degradation rate, requires specific conditions
Photo-degradation	UV radiation breaks polymer chains, leading to fragmentation	Sunlight exposure accelerates MP breakdown in marine and terrestrial environments	Natural process, cost-effective	Produces secondary pollutants, slow in deep water
Thermal degradation (Pyrolysis)	High-temperature breakdown of MPs into smaller molecules or gases	Converts MPs into energy and fuels through high-heat processing	Efficient, energy recovery potential	High energy input, expensive setup
Chemical degradation (Oxidation, Hydrolysis, Solvent Treatment)	Use of oxidants, acids, or catalysts to break down polymer structures	Fenton reaction, hydrolysis, and solvent-based degradation of MPs in industrial and wastewater treatment plants	High efficiency, controlled process	May generate toxic byproducts, requires careful handling
Nanotechnology-based degradation	Nanoparticles or nanocatalysts enhance breakdown of MPs	Nano-enzymes, photocatalysts like TiO_2, and iron-based nanoparticles used in wastewater treatment	High surface area, effective at low concentrations	Expensive, potential unknown environmental effects
Mechanical degradation	Physical forces such as grinding, milling, or ultrasound	Used in industrial recycling plants to break down MPs before further processing	Simple, scalable	Generates microplastic fragments, does not completely degrade MPs
Bioaugmentation	Introduction of genetically engineered microorganisms for enhanced degradation	Engineered bacteria or consortia used in bioremediation sites	Targeted degradation, higher efficiency than natural microbes	Ethical concerns, regulatory challenges

(continued)

Table 2 (continued)

Degradation method	Mechanism of action	Application	Advantages	Limitations
Enzymatic degradation	Specific enzymes hydrolyze plastic polymers into smaller, biodegradable molecules	PETases and MHETases used to degrade polyethylene terephthalate (PET) plastics in water treatment	Highly specific, minimal byproducts	Expensive enzyme production, requires optimized conditions
Fungal degradation	Fungi secrete enzymes that break down MPs into organic compounds	*Aspergillus niger* and *Penicillium sp.* degrade MPs in soil and composting systems	Natural, can work under varying conditions	Slow process, requires controlled conditions
Ozone treatment	Ozone oxidizes and fragments MPs, making them biodegradable	Applied in water treatment plants and industrial wastewater processing	Fast degradation, effective in water	High operational cost, possible toxic byproducts
Electrochemical degradation	Electric current generates reactive oxygen species that oxidize MP particles	Electro-Fenton and electrolysis systems used in wastewater treatment	Highly efficient, no secondary waste	High energy consumption, specialized equipment needed
Supercritical water oxidation (SCWO)	High-pressure, high-temperature water oxidizes MPs into harmless byproducts	Used in advanced industrial waste treatment systems	Rapid, complete degradation	High cost, extreme operating conditions
Bioreactor-assisted degradation	Controlled microbial and enzymatic processes break down MPs in engineered bioreactors	Used in bioremediation facilities to enhance MP degradation	Scalable, optimized for high efficiency	High initial cost, requires monitoring
Plasma treatment	High-energy plasma breaks molecular bonds in plastic particles	Used in advanced waste treatment and material recycling plants	Rapid degradation, efficient	Expensive, requires specialized infrastructure

with an increase in response time, and the maximum weight loss of PS was seen with FePc-TiO_2 (1 wt% TiO2) when applied to different films. At 1 mW/cm^2 of light intensity, PS decreased by 65% after 480 h, and at 2 mW/cm^2, it decreased by 82%. However, under the same reaction conditions, pure TiO_2 decomposed by 21% and 25% after 480 h. As the film became more fragile and pores formed throughout the reaction, it appears that there was effective deterioration, as the weight loss of FePc-TiO_2 was more than that of pure TiO_2 at low UV light intensity (Fa et al., 2008).

Notably, LDPE with 0.1 wt% TiO_2 showed effective degradation using an eco-friendly technique (Thomas & Sandhyarani, 2013). By incorporating TiO_2 particles into PVC film, the transparent structure boosts UV light absorption, leading to an increased degradation rate. This exemplifies the properties of both thin PVC film and PVC mixed with TiO_2 film. The TiO_2-integrated PS film exhibited more weight loss compared to the PS film alone. Employing 1 weight percent grafted TiO_2 resulted in a 29% reduction in PS weight, whereas PS-loaded TiO_2 demonstrated a 13% decrease upon exposure to UV radiation (Zan et al., 2004). An 18% weight loss was noted with TiO_2 (P25), whereas multiwalled carbon nanotubes impregnated with TiO_2 at concentrations of 5, 10, 20, and 40% demonstrated weight losses of 28, 29, 35, and 24%, respectively, after 180 h of irradiation (An et al., 2014). The PVC film exhibiting the most weight loss was the one subjected to treatment with micro graphite TiO_2. After 30 h, PVC with nano graphite experienced a weight reduction of 7.28%, pure TiO_2 exhibited a loss of 8.94%, and nano graphite modified TiO_2 demonstrated a decrease of 17.24% in weight. Because of its multi-layered, sheet-like monoclinic structure, nano graphite improves the catalytic performance by facilitating the efficient mixing of TiO_2 and the appropriate dispersion of PVC (Zhang et al. 2020b).

Using Fe(St)3 modified TiO_2, we were able to degrade PS film when exposed to UV radiation. The weight reduction with Fe(St)3 modified TiO_2 was 22% greater after 288 h. Pure TiO_2 diminished PS by 15%, but the photodegradation of the PS film resulted in a reduction of just 10% in PS. The photocatalytic efficacy for uniform dispersion plastic degradation can be improved by the incorporation of Fe(St)3 (Fa et al., 2013). The dispersion of polymers and catalysts is improved through functionalisation. The reduction in PVC weight was 11% with unfunctionalized TiO_2, 7% with functionalised TiO_2, and 9% with functionalised TiO_2 (Alvarado et al., 2016). Table 3 shows the application with additional macro plastic degradation methods, including advanced biological, chemical, and physical approaches.

11 Future Perspectives

One of the most important steps in photocatalytic degradation of polymeric materials is the production of oxidative species. These species help break down organic molecules when combined with oxygen, water, and high-energy photons. While a lot of research has focused on the aqueous phase and achieved photocatalytic conversion of polymeric materials into useful products, there has been a surge in study aiming

Table 3 Applications of TiO_2 for macro plastic degradation

Degradation method	Mechanism of action	Application	Advantages	Limitations
Biodegradation (Microbial Degradation)	Bacteria and fungi break down plastic polymers using enzymes	*Ideonella sakaiensis*, *Pseudomonas putida*, and *Aspergillus* species degrade plastics in soil and water	Environmentally friendly, sustainable	Slow degradation, requires optimized conditions
Photo-degradation	UV radiation breaks polymer chains, leading to fragmentation	Sunlight exposure degrades plastics in marine and terrestrial environments	Natural, cost-effective	Slow process, releases microplastics
Thermal degradation (Pyrolysis)	High-temperature breakdown of plastics into smaller hydrocarbons	Converts plastics into fuels and energy in industrial plants	Efficient, energy recovery potential	High energy input, costly setup
Chemical degradation (Oxidation, Hydrolysis, Solvent Treatment)	Chemical reactions break down plastic polymers into smaller molecules	Fenton reaction, hydrolysis, and solvent-based degradation used in recycling facilities	High efficiency, controlled process	May generate toxic byproducts, requires specialized handling
Nanotechnology-based degradation	Nanoparticles catalyze plastic breakdown at a molecular level	Nano-enzymes, photocatalysts like TiO_2 used in wastewater treatment	Highly effective at low concentrations	Expensive, potential unknown environmental effects
Mechanical degradation	Physical forces such as shredding, milling, or grinding	Used in plastic recycling plants before further processing	Simple, scalable	Generates microplastic fragments, does not fully degrade plastic
Bioaugmentation	Engineered microorganisms target specific plastic polymers	Bioremediation of plastic-contaminated sites	Targeted, higher efficiency than natural microbes	Ethical concerns, regulatory challenges
Enzymatic Degradation	Enzymes hydrolyze plastic polymers into smaller biodegradable molecules	PETase and MHETase enzymes degrade PET plastics in recycling plants	Highly specific, minimal byproducts	Expensive enzyme production, requires optimization

(continued)

Table 3 (continued)

Degradation method	Mechanism of action	Application	Advantages	Limitations
Fungal degradation	Fungi secrete plastic-degrading enzymes	*Aspergillus niger* and *Penicillium* species break down plastics in composting and landfill environments	Can operate in diverse conditions	Slow degradation, requires controlled conditions
Ozone treatment	Ozone oxidizes and weakens plastic polymer bonds	Applied in industrial waste treatment plants	Fast degradation, effective in water	High operational cost, possible toxic byproducts
Electrochemical degradation	Electric current generates reactive oxygen species that oxidize plastics	Electro-Fenton and electrolysis used in waste treatment	Highly efficient, no secondary waste	High energy consumption, requires specialized infrastructure
Supercritical water oxidation (SCWO)	High-pressure, high-temperature water oxidizes plastics	Industrial waste treatment for rapid degradation	Complete degradation, minimal residue	High cost, extreme operating conditions
Plasma treatment	High-energy plasma breaks down plastic at the molecular level	Used in advanced waste treatment and material recycling plants	Rapid degradation, highly efficient	Expensive, requires specialized equipment
Hydrothermal liquefaction	Converts plastic into liquid fuel through high-pressure water treatment	Industrial-scale recycling and waste-to-energy conversion	Produces valuable fuel products	High initial investment, requires high energy input
Photocatalytic degradation	Light-activated catalysts degrade plastic polymers into smaller molecules	Titanium dioxide (TiO_2) nanoparticles used in environmental cleanup	Can work under sunlight, highly efficient	Expensive catalysts, long processing time
Acid/base hydrolysis	Strong acids or bases break down plastic chemical bonds	Used in recycling plants to depolymerize plastics	High degradation efficiency	Corrosive chemicals require special handling
Bioreactor-assisted degradation	Controlled microbial and enzymatic processes degrade plastics in engineered reactors	Used in waste treatment and bioremediation facilities	Scalable, high efficiency	High cost, requires monitoring

(continued)

Table 3 (continued)

Degradation method	Mechanism of action	Application	Advantages	Limitations
Sonochemical degradation	Ultrasound waves create cavitation bubbles that break down plastics	Applied in wastewater treatment plants for plastic degradation	Energy-efficient, non-toxic	Not effective for all plastic types, requires specific frequencies
Thermochemical conversion	Combination of heat and chemical catalysts to convert plastics into reusable materials	Used in industrial recycling plants to convert plastic waste into synthetic fuels and chemicals	Produces valuable byproducts	High operational costs, complex process
Anaerobic digestion of bioplastics	Microorganisms break down biodegradable plastics in oxygen-free environments	Used in composting and wastewater treatment plants	Produces biogas as an energy source	Only effective for specific bioplastics (PLA, PHA), slow process
Oxidative wet combustion	Combustion in a controlled water-oxygen environment to break down plastics	Applied in hazardous waste treatment plants	Fast process, reduces landfill waste	High energy requirement, emits CO_2

at improving efficiency in the solid phase. Incorporating photosensitive components into plastic polymers, a process known as solid-state photocatalysis, improves mechanical, physical, and chemical qualities while offering several advantages for the disposal of plastic waste. Nonetheless, challenges remain to be addressed in this domain. Particle size and film thickness can also be regulated to prevent agglomeration. Nonetheless, research into semiconductor toxicity and value-added products should be prioritised in future investigations (Alvarado et al., 2016).

As complete mineralisation of plastics under visible light is currently unfeasible using TiO_2 photocatalytic methods, advancements in TiO_2 photocatalysts to enhance degradation efficiency may become a focus of future study. The breakdown of TiO_2 used in polymers can now be accomplished in a certain way. When it comes to breaking down macro and microplastics, light-induced active species including holes, ·OH, and O_2 ·-are crucial. The degradation mechanism of macro and microplastics frequently involves the breaking of polymer chain bonds by reactive species. In contrast to the well studied impacts on micro-plastics, there has been inadequate research on how temperature and solution pH affect the breakdown of macroplastics. Most modified TiO_2 does not completely degrade macro and microplastics, hence there is a need to improve its degradation efficiency. For specific polymers, TiO_2 shown some enhancement when combined with carbon, nitrogen, vitamin C, and $FePcCl1_2$. Despite extensive efforts to degrade macro and microplastics, some concerns remain unresolved, including how to enhance the efficacy of TiO_2 in degradation, improve its cost-effectiveness, and reduce its environmental toxicity (Lin & Mehrvar, 2018).

The potential material has thus far been TiO_2-based photocatalysts, predominantly limited to the UV spectrum. It is essential to discover novel methods to enhance TiO_2-based photocatalysis using visible light. A TiO_2-based photocatalytic system must efficiently degrade a diverse range of polymers, beyond those primarily examined in prior studies. The toxicity of reaction intermediates and the photocatalytic breakdown of nanoplastics using TiO_2 require additional research. Given the complexity of the photodecomposition processes affecting macro- and microplastics, it is essential to understand the role of reactive radical species in plastic degradation. To find out how effective TiO_2 is in a water-based MPs solution, we need to test it. The photoreaction chamber is an important part of this solution and has attracted a lot of attention as a potential new way to remove macro and microplastics from aquatic environments. One such solution is a floating photoreaction chamber made of TiO_2. We can effectively remove floating plastic particles from marine environments under natural sunlight by towing this photo chamber behind a boat navigating the water. This could be the basis for future research.

12 Conclusion

Due to their ubiquity and the unresolved threat they provide to living organisms, plastics have emerged as a global issue and a persistent anthropogenic contamination. With the global increase in plastic production and utilisation, various ecosystems are anticipated to suffer significant harm due to contamination from both macro and microplastics. Significant advancements have been made in the sustainable degradation of macro and microplastics due to the implementation of photocatalytic technologies in recent years. Polymers may undergo partial self-decomposition, resulting in the fragmentation into microplastic particles. The pursuit of effective degradation technologies for the removal of macro and microplastics is gaining significance. Among the several methods for addressing plastic treatment, TiO_2-based photocatalysis has frequently garnered the most focus. This chapter encompasses the latest knowledge on the breakdown of macro and microplastics with TiO_2 photocatalyst. The text also encompasses the diverse methods for preparing $TiO_2$for the degradation of macro and microplastics. Comprehensive studies have been undertaken on the physiochemical properties of TiO_2, encompassing its stability, catalytic efficacy, and applicability for plastic degradation.

Recent advancements in TiO_2 photocatalysis, including surface modifications, doping with metals and non-metals, and the development of composite materials, have enhanced its efficiency and broadened its applicability. These innovations have improved light absorption, reduced recombination rates of electron–hole pairs, and expanded the usability of TiO_2 under visible light, addressing some of its traditional limitations. Moreover, integrating TiO_2 photocatalysis with other remediation technologies offers synergistic solutions for comprehensive plastic waste management.

Despite its potential, challenges remain in scaling up TiO_2-based systems for real-world applications. Issues such as energy requirements, the formation of secondary pollutants, and the recycling of photocatalysts must be addressed to make this technology more practical and sustainable. To further guarantee that the utilisation of TiO_2 nanoparticles does not unintentionally bring new hazards, continuing research is necessary to comprehend their environmental and ecological implications. TiO_2-mediated photocatalysis stands at the forefront of innovative strategies for combating plastic pollution. By advancing material science, optimizing system designs, and fostering interdisciplinary collaboration, this technology can contribute significantly to global efforts to reduce plastic waste and protect ecosystems. Its integration into large-scale waste management frameworks will be pivotal in achieving long-term environmental sustainability.

References

Ahmed, S. N., & Haider, W. (2018). Heterogeneous photocatalysis and its potential applications in water and wastewater treatment: A review. *Nanotechnology, 29*, Article 342001.

Ajmal, A., Majeed, I., Malik, R. N., Idriss, H., & Nadeem, M. A. (2014). Principles and mechanisms of photocatalytic dye degradation on TiO2 based photocatalysts: A comparative overview. *RSC Advances, 4*, 37003–37026. https://doi.org/10.1039/C4RA06658H

Ali, S.S., Elsamahy, T., Koutra, E., Kornaros, M., El-Sheekh, M., & Abdelkarim, E.A. (2021) Degradation of conventional plastic wastes in the environment: A review on current status of knowledge and future perspectives of disposal. *Science of the Total Environment*, *771*, Article ID: 144719. https://doi.org/10.1016/j.scitotenv.2020.144719

Alvarado, J., Acosta, G., & Perez, F. (2016). Study of the effect of the dispersion of functionalized nanoparticles TiO2 with photocatalytic activity in LDPE. *Polymer Degradation and Stability, 134*, 376–382. https://doi.org/10.1016/j.polymdegradstab.2016.11.009

Amelia, T. S. M., Khalik, W. M. A. W. M., Ong, M. C., Shao, Y. T., Pan, H. J., & Bhubalan, K. (2021). Marine microplastics as vectors of major ocean pollutants and its hazards to the marine ecosystem and humans. *Progress in Earth and Planetary Science, 8*, 12. https://doi.org/10.1186/s40645-020-00405-4

An, Y., Hou, J., Liu, Z., & Peng, B. (2014). Enhanced solid-phase photocatalytic degradation of polyethylene by TiO2–MWCNTs nanocomposites. *Materials Chemistry and Physics, 148*, 387–394. https://doi.org/10.1016/j.matchemphys.2014.08.001

Andrady, A. L., & Neal, M. A. (2009). Applications and societal benefits of plastics. *Philosophical Transactions of the Royal Society B: Biological Sciences, 364*, 1977–1984. https://doi.org/10.1098/rstb.2008.0304

Ariza-Tarazona, M. C., Villarreal-Chiu, J. F., Barbieri, V., Siligardi, C., & CedilloGonzalez, E. I. (2019). New strategy for microplastic degradation: Green photocatalysis using a protein-based porous N-TiO2 semiconductor. *Ceramics International, 45*, 9618–9624. https://doi.org/10.1016/j.ceramint.2018.10.208

Ariza-Tarazona, M. C., Villarreal-Chiu, J. F., Hernandez-Lopez, J., Rivera De la Rosa, J. M., Barbieri, V., Siligardi, C., & Cedillo-Gonzalez, E. I. (2020). Microplastic pollution reduction by a carbon and nitrogen-doped TiO2: Effect of pH and temperature in the photocatalytic degradation process. *Journal of Hazardous Materials, 395*, Article 122632. https://doi.org/10.1016/j.jhazmat.2020.122632

Asghar, W., Qazi, I.A., Ilyas, H., Khan, A.A., Awan, M.A., & Rizwan Aslam, M. (2011). Comparative solid phase photocatalytic degradation of polythene films with doped and undoped TiO2 nanoparticles. *Journal of Nanomaterials*. https://doi.org/10.1155/2011/461930

Bacha, A. U. R., Cheng, H., Han, J., Nabi, I., Li, K., Wang, T., Yang, Y., Ajmal, S., Liu, Y., & Zhang, L. (2019a). Significantly accelerated PEC degradation of organic pollutant with addition of sulfite and mechanism study. *Applied Catalysis B: Environmental, 248*, 441–449. https://doi.org/10.1016/j.apcatb.2019.02.049

Bacha, A. U. R., Nabi, I., Fu, Z., Li, K., Cheng, H., & Zhang, L. (2019b). A comparative study of bismuth-based photocatalysts with titanium dioxide for perfluorooctanoic acid degradation. *Chinese Chemical Letters, 30*, 2225–2230. https://doi.org/10.1016/j.cclet.2019.07.058

Bacha, A. U. R., Nabi, I., Cheng, H., Li, K., Ajmal, S., Wang, T., & Zhang, L. (2020). Photoelectrocatalytic degradation of endocrine-disruptor bisphenol-A with significantly activated peroxymonosulfate by Co-BiVO4 photoanode. *Chemical Engineering Journal, 389*, 124–139. https://doi.org/10.1016/j.cej.2020.124482

Barnes, D. K., Galgani, F., Thompson, R. C., & Barlaz, M. (2009). Accumulation and fragmentation of plastic debris in global environments. *Philosophical Transactions of the Royal Society B: Biological Sciences, 364*, 1985–1998. https://doi.org/10.1098/rstb.2008.0205

Basavarajappa, P. S., Patil, S. B., Ganganagappa, N., Reddy, K. R., Raghu, A. V., & Reddy, C. V. (2020). Recent progress in metal-doped TiO2, non-metal doped/codoped TiO2 and TiO2 nanostructured hybrids for enhanced photocatalysis. *International Journal of Hydrogen Energy, 45*, 7764–7778. https://doi.org/10.1016/j.ijhydene.2019.07.241

Bickley, R. I., Gonzalez-Carreno, T., Lees, J. S., Palmisano, L., & Tilley, R. J. D. (1991). A structural investigation of titanium dioxide photocatalysts. *Journal of Solid State Chemistry, 92*, 178–190.

Borrás, M. C., Sluyter, R., Barker, P. J., Konstantinov, K., Bakand, S. (2020). Y2O3 decorated TiO2 nanoparticles: enhanced UV attenuation and suppressed photocatalytic activity with promise for cosmetic and sunscreen applications. *Journal of Photochemistry and Photobiology B: Biology*, 111883, https://doi.org/10.1016/j.jphotobiol.2020.111883

Boucher, J., Faure, F., Pompini, O., Plummer, Z., Wieser, O., & de Alencastro, L. F. (2019). (Micro) plastic fluxes and stocks in Lake Geneva basin. *TrAC Trends in Analytical Chemistry, 112*, 66–74. https://doi.org/10.1016/j.trac.2018.11.037

Cai, M., He, H., Liu, M., Li, S., Tang, G., Wang, W., Huang, P., Wei, G., Lin, Y., & Chen, B. (2018). Lost but can't be neglected: Huge quantities of small microplastics hide in the South China Sea. *Science of the Total Environment, 633*, 1206–1216. https://doi.org/10.1016/j.scitotenv.2018.03.197

Camarero, L., Bacardit, M., de Diego, A., & Arana, G. J. A. E. (2017). Decadal trends in atmospheric deposition in a high elevation station: Effects of climate and pollution on the long-range flux of metals and trace elements over SW Europe. *Atmospheric Environment, 167*, 542–552. https://doi.org/10.1016/j.atmosenv.2017.08.049

Cartraud, A.E., Le Corre, M., Turquet, J., & Tourmetz, J. (2019). Plastic ingestion in seabirds of the western Indian Ocean. *Marine Pollution Bulletin, 140*, 308–314. https://doi.org/10.1016/j.marpolbul.2019.01.065

Chatterjee, S., & Sharma, S. (2019). Microplastics in our oceans and marine health. *Field Actions Science Report*, 54–61. http://journals.openedition.org/factsreports/5257

Chen, Y., Awasthi, A.K., Wei, F., Tan, Q., & Li, J. (2020). Single-use plastics: production, usage, disposal, and adverse impacts. *Science of the Total Environment*, *141772*. https://doi.org/10.1016/j.scitotenv.2020.141772.

Das, A. Prasad., & John, A. (2025). *Sustainable management of urban plastic waste through circular economic approaches* (Vol. 299).

Das, A., Dey, S., Gacem, A., Yadav, K. K., Nath, H., Ram, D. K., & Das, A. P. (2025). Revolutionizing solid hydrocarbon sourcing: plastic's journey from waste to upcycled treasure. In *Environmental science and engineering, part F218*, (pp. 331–351). https://doi.org/10.1007/978-3-031-81615-4_14

Fa, W., Zan, L., Gong, C., Zhong, J., & Deng, K. (2008). Solid-phase photocatalytic degradation of polystyrene with TiO2 modified by iron (II) phthalocyanine. *Applied Catalysis B: Environmental, 79*, 216–223. https://doi.org/10.1016/j.apcatb.2007.10.018

Fa, W., Gong, C., Tian, L., Peng, T., & Zan, L. (2011). Enhancement of photocatalytic degradation of poly (vinyl chloride) with perchlorinated iron (II) phthalocyanine modified nano-TiO2. *Journal of Applied Polymer Science, 122*, 1823–1828. https://doi.org/10.1002/app.34230

Fa, W., Guo, L., Wang, J., Guo, R., Zheng, Z., & Yang, F. (2013). Solid-phase photocatalytic degradation of polystyrene with TiO2/Fe(St)3 as catalyst. *Journal of Applied Polymer Science, 128*, 2618–2622. https://doi.org/10.1002/app.37751

Fujishima, A., Rao, T. N., & Tryk, D. A. (2000). Titanium dioxide photocatalysis. *Journal of Photochemistry and Photobiology, C: Photochemistry Reviews, 1*, 1–21.

Galindo, C., Jacques, P., & Kalt, A. (2000). Photodegradation of the aminoazobenzene acid orange 52 by three advanced oxidation processes: UV/H2O2, UV/TiO2 and VIS/ TiO2: Comparative mechanistic and kinetic investigations. *Journal of Photochemistry and Photobiology, A: Chemistry, 130*, 35–47. https://doi.org/10.1016/S1010-6030(99)00199-9

Gewert, B., Plassmann, M. M., & MacLeod, M. J. E. S. P. (2015). Impacts, pathways for degradation of plastic polymers floating in the marine environment. *Environmental Science Processes & Impacts, 17*, 1513–1521. https://doi.org/10.1039/C5EM00207A

Giacovelli, C. (2018). Single-use plastics: a roadmap for sustainability. https://stg-wedocs.unep.org/handle/20.500.11822/25496

Gillibert, R., Balakrishnan, G., Deshoules, Q., Tardivel, M., Magazzu, A., Donato, M. G., Marago, O. M., Lamy de La Chapelle, M., Colas, F., Lagarde, F., & Gucciardi, P. G. (2019). Raman tweezers for small microplastics and nanoplastics identification in seawater. *Environmental Science and Technology, 53*, 9003–9013. https://doi.org/10.1021/acs.est.9b03105

Hashimoto, K., Irie, H., & Fujishima, A. (2005). TiO2 photocatalysis: A historical overview and future prospects. *Japanese Journal of Applied Physics, 44*, 8269. https://doi.org/10.1143/JJAP.44.8269

He, X., Wang, A., Wu, P., Tang, S., Zhang, Y., Li, L., & Ding, P. (2020). Photocatalytic degradation of microcystin-LR by modified TiO2 photocatalysis: A review. *Science of the Total Environment, 743*, Article 140694. https://doi.org/10.1016/j.scitotenv.2020.140694

Hu, X., Li, G., & Yu, J. C. (2010). Design, fabrication, and modification of nanostructured semiconductor materials for environmental and energy applications. *Langmuir, 26*, 3031–3039. https://doi.org/10.1021/la902142b

Hu, K., Tian, W., Yang, Y., Nie, G., Zhou, P., Wang, Y., Duan, X., & Wang, S.J.W.R. (2021). Microplastics remediation in aqueous systems: Strategies and technologies. *Water Resources*, 117144. https://doi.org/10.1016/j.watres.2021.117144

Jing, Y., Wang, Y., Furukawa, S., Xia, J., Sun, C., & Hülsey, M. J. (2021). Towards the circular economy: converting aromatic plastic waste back to Arenes over a Ru/nb2o5 catalyst. *Angewandte Chemie International Edition, 60*, 5527–5535. https://doi.org/10.1002/anie.202011063

Labat, F., Baranek, P., Domain, C., Minot, C., & Adamo, C. (2007). Density functional theory analysis of the structural and electronic properties of TiO2 rutile and anatase polytypes: Performances of different exchange-correlation functionals. *The Journal of Chemical Physics, 2007*(126), Article 154703.

Lee, J., Lee, Y., Song, H., Jang, D., & Choa, Y. (2011). Synthesis and characterization of TiO2 nanowires with controlled porosity and microstructure using electrospinning method. *Current Applied Physics, 11*, S210–S214.

Lee, Q. Y., & Li, H. (2021) Photocatalytic degradation of plastic waste: A mini review. *Micromachines, 12*, Article 907. https://doi.org/10.3390/mi12080907

Lehtiniemi, M., Hartikainen, S., Näkki, P., Engström-Öst, J., Koistinen, A., & Setälä, O. (2018). Size matters more than shape: Ingestion of primary and secondary microplastics by small predators. *Food Webs, 17*, e00097. https://doi.org/10.1016/j.fooweb.2018.e00097

Li, L. (2020). The reaction mechanism of photoelectrocatalysis on the surface of TiO2 nanotube array electrode. *Asia-Pacific Journal of Chemical Engineering*, e2511. https://doi.org/10.1002/apj.2511

Lin, Y., & Mehrvar, M. (2018). Photocatalytic treatment of an actual confectionery wastewater using Ag/TiO2/Fe2O3: Optimization of photocatalytic reactions using surface response methodology. *Catalysts, 8*, 409.

Llorente-García, B. E., Hernández-López, J. M., Zaldívar-Cadena, A. A., Siligardi, C., Cedillo-González, E. I. (2020). First insights into photocatalytic degradation of HDPE and LDPE microplastics by a mesoporous N–TiO2 coating: Effect of size and shape of microplastics. *Coatings, 10*, 69–81. https://doi.org/10.3390/coatings10070658

Lu, S. Y., Wu, D., Wang, Q. L., Yan, J., Buekens, A. G., & Cen, K. F. (2011). Photocatalytic decomposition on nano-TiO2: Destruction of chloroaromatic compounds. *Chemosphere, 82*, 1215–1224.

Mack, J. M., Tsuchiya, H., Ghicov, A., Yasuda, K., Hahn, R., Bauer, S., & Schmuki, P. (2007). TiO2 nanotubes: Self-organized electrochemical formation, properties and applications. *Current Opinion in Solid State and Materials Science, 11*, 3–18.

Malato, S., Blanco, J., Vidal, A., & Richter, C. (2002). Photocatalysis with solar energy at a pilot-plant scale: An overview. *Applied Catalysis B: Environmental, 37*, 1–15. https://doi.org/10.1016/S0926-3373(01)00315-0

Martinez, U., & Hammer, B. (2011). Adsorption properties versus oxidation states of rutile TiO2 (110). *The Journal of Chemical Physics, 134*, Article 194703. https://doi.org/10.1063/1.3589861

Miao, F., Liu, Y., Gao, M., Yu, X., Xiao, P., Wang, M., Wang, S., & Wang, X. (2020). Degradation of polyvinyl chloride microplastics via an electro-Fenton-like system with a TiO2/ graphite cathode. *Journal of Hazardous Materials, 399*, Article 123023. https://doi.org/10.1016/j.jhazmat.2020.123023

Moma, J., Baloyi, J.J.P.A. (2019). Attributes, Modified titanium dioxide for photocatalytic applications. *Photocatalytics Applications and Attributes, 18*, 15–32. https://doi.org/10.5772/intechopen.79374

Nabi, I., Bacha, A. U., Li, K., Cheng, H., Wang, T., Liu, Y., Ajmal, S., Yang, Y., Feng, Y., & Zhang, L. (2020). Complete photocatalytic mineralization of microplastic on TiO2 nanoparticle film. *iScience, 23*, 101326, https://doi.org/10.1016/j.isci.2020.101326

Ogidi, O. I., & Akpan, U. M. (2024a). Microbial and phyto-remediation of crude oil contaminated soil. In *Eco-restoration of the environment*. CRC Press, Routledge Taylor and Francis Group.

Ogidi, O. I., & Richard-Nwachukwu, N. (2024b). Microbe assisted remediation of pesticide residues from soil and water. In *Eco-restoration of the environment*. CRC Press, Routledge Taylor and Francis Group.

Ogidi, O. I., Angaye, T. C. N. & Carbom, H. C. (2024a). Microorganisms in the monitoring of terrestrial and aquatic ecosystems: Applications, challenges and future prospects. In *Biomonitoring of Pollutants in the Global South.* Springer Nature Singapore Pte Ltd.

Ogidi, O. I., Onwuagba, C. G., & Richard-Nwachukwu, N. (2024b). Biomonitoring tools, techniques and approaches for environmental assessments. In *Biomonitoring of Pollutants in the Global South.* Springer Nature Singapore Pte Ltd.

Ogidi, O. I., Ajoko, I. T., & Tawariowei, A. M. (2024c). Sustainable application of genetic ecotoxicological techniques in biomonitoring for environmental sustainability. In *Biomonitoring of pollutants in the global south.* Springer Nature Singapore Pte Ltd.

Park, H., Park, Y., Kim, W., & Choi, W. (2013). Surface modification of TiO2 photocatalyst for environmental applications. *Journal of Photochemistry and Photobiology, C: Photochemistry Reviews, 15*, 1–20.

Pawar, M., Topcu Sendogdular, S., & Gouma, P. (2018). A brief overview of TiO2 photocatalyst for organic dye remediation: Case study of reaction mechanisms involved in Ce-TiO2 photocatalysts system. *Journal of Nanomaterials, 2018*, 5953609. https://doi.org/10.1155/2018/5953609

Peñas-Garzón, M., Gómez-Avilés, A., Belver, C., & Rodriguez, J. (2020) Bedia, degradation pathways of emerging contaminants using TiO2-activated carbon heterostructures in aqueous solution under simulated solar light. *Chemical Engineering Journal*, 124867, https://doi.org/10.1016/j.cej.2020.124867

Peng, L., Fu, D., Qi, H., Lan, C. Q., Yu, H., & Ge, C. (2020). Micro- and nano-plastics in marine environment: Source, distribution and threats - a review. *Science of the Total Environment, 698*, Article 134254. https://doi.org/10.1016/j.scitotenv.2019.134254

Priyadarshini, B., & Das, A. P. (2024). The Adverse impact of synthetic microplastics released from disposable face masks pollutants on aquatic microbiota. In *Aquatic species and humans: A review. environmental science and engineering, part F2827*, (pp. 67–81). https://doi.org/10.1007/978-3-031-50840-0_5

Priyadarshini, S., Jagatee, S., & Das, A. P. (2024). Synthetic fabrics and microfiber pollution–an assessment of their global impact. In *Environmental science and engineering, part F2333* (pp. 137–157). https://doi.org/10.1007/978-3-031-51792-1_8

Reddy Koteswara, G., Bhargavi, S., Yamuna, D., Anjani Devi, M., Jeevana Lakshmi Kumar, G., Manas, T., Adarsh, S., Bhavya Sai, V., Nath, H., Alreshidi, M. A., Togun, H., Ram, D. K., & Das, A. P. (2025). Recent advancements in the remediation of marine hydrocarbon and plastic pollutants. In *Environmental science and engineering, part F218* (pp. 127–142). https://doi.org/10.1007/978-3-031-81615-4_7

Rillig, M. C., Ingraffia, R., & de Souza Machado, A. A. (2017). Microplastic incorporation into soil in agroecosystems. *Frontiers in Plant Science, 8*, 1805. https://doi.org/10.3389/fpls.2017.01805

Rizzi, M., Rodrigues, F. L., Medeiros, L., Ortega, I., Rodrigues, L., Monteiro, D. S., Kessler, F., & Proietti, M. C. (2019). Ingestion of plastic marine litter by sea turtles in southern Brazil:

Abundance, characteristics and potential selectivity. *Marine Pollution Bulletin, 140*, 536–548. https://doi.org/10.1016/j.marpolbul.2019.01.054

Ryan, P. G., Moore, C. J., van Franeker, J. A., & Moloney, C. L. (2009). Monitoring the abundance of plastic debris in the marine environment. *Philosophical Transactions of the Royal Society B: Biological Sciences, 364*, 1999–2012. https://doi.org/10.1098/rstb.2008.0207

Santhi, K., Navaneethan, M., Harish, S., Ponnusamy, S., & Muthamizhchelvan, C. (2020). Synthesis and characterization of TiO2 nanorods by hydrothermal method with different pH conditions and their photocatalytic activity. *Applied Surface Science, 500*, Article 144058. https://doi.org/10.1016/j.apsusc.2019.144058

Sehlotho, N., & Nyokong, T. (2004). Catalytic activity of iron and cobalt phthalocyanine complexes towards the oxidation of cyclohexene using tert-butylhydroperoxide and chloroperoxybenzoic acid. *Journal of Molecular Catalysis A: Chemical, 209*, 51–57. https://doi.org/10.1016/j.molcata.2003.08.014

Shah, S. N. A., Shah, Z., Hussain, M., & Khan, M. (2017). Hazardous effects of titanium dioxide nanoparticles in ecosystem. *Bioinorganic Chemistry and Applications, 2017*, 4101735. https://doi.org/10.1155/2017/4101735

Shang, J., Chai, M., & Zhu, Y. (2003). Photocatalytic degradation of polystyrene plastic under fluorescent light. *Environmental Science and Technology, 37*, 4494–4499. https://doi.org/10.1021/es0209464

Shi, H., Magaye, R., Castranova, V., & Zhao, J. (2013). Titanium dioxide nanoparticles: A review of current toxicological data. *Particle and Fibre Toxicology, 2013*(10), 15.

Singh, B., & Sharma, N. (2008). Mechanistic implications of plastic degradation. *Polymer Degradation and Stability, 93*, 561–584. https://doi.org/10.1016/j.polymdegradstab.2007.11.008

Skocaj, M., Filipic, M., Petkovic, J., & Novak, S. (2011). Titanium dioxide in our everyday life; is it safe? *Radiology and Oncology, 45*, 227–247. https://doi.org/10.2478/v10019-011-0037-0

Su, K., Liu, H., Zhang, C., & Wang, F. (2022). Photocatalytic conversion of waste plastics to low carbon number organic products. *Chinese Journal of Catalysis, 43*, 589–594. https://doi.org/10.1016/s1872-2067(21)63885-8

Thomas, R. T., & Sandhyarani, N. (2013). Enhancement in the photocatalytic degradation of low density polyethylene–TiO2 nanocomposite films under solar irradiation. *RSC Advances, 3*, 14080–14087. https://doi.org/10.1039/C3RA42226G

Thompson, R. C., Swan, S. H., Moore, C. J., Vom Saal, F. S. (2009). Our plastic age, The Royal Society Publishing. *Philosophical Transactions of the Royal Society B: Biological Sciences, 3*, 123–137. https://doi.org/10.1098/rstb.2009.0054

Tishkevich, D. I., Grabchikov, S. S., Lastovskii, S. B., Trukhanov, S. V., Zubar, T. I., Vasin, D. S., Trukhanov, A. V., Kozlovskiy, A. L., & Zdorovets, M. M. (2018). Effect of the synthesis conditions and microstructure for highly effective electron shields production based on Bi coatings. *ACS Applied Energy Materials, 1*, 1695–1702. https://doi.org/10.1021/acsaem.8b00179

Tishkevich, D. I., Vorobjova, A. I., & Vinnik, D. A. (2019). Template assisted Ni nanowires fabrication. *Materials Science Forum, 946*, 235–241. https://doi.org/10.4028/www.scientific.net/MSF.946.235

Tsang, C. H. A., Li, K., Zeng, Y., Zhao, W., Zhang, T., Zhan, Y., Xie, R., Leung, D. Y. C., & Huang, H. (2019). Titanium oxide based photocatalytic materials development and their role of in the air pollutants degradation: Overview and forecast. *Environment International, 125*, 200–228. https://doi.org/10.1016/j.envint.2019.01.015

Upadhyay, G. K., Kumar, V., & Purohit, L. (2020). Optimized CdO: TiO2 nanocomposites for heterojunction solar cell applications. *Journal of Alloys and Compounds, 157453*. https://doi.org/10.1016/j.jallcom.2020.157453

Vorobjova, A. I., Shimanovich, D. L., Sycheva, O. A., Ezovitova, T. I., Tishkevich, D. I., & Trykhanov, A. V. (2019). Studying the thermodynamic properties of composite magnetic material based on anodic alumina. *Russian MicroElectronics, 48*, 107–118. https://doi.org/10.1134/S1063739719020100

Vorobjova, A., Tishkevich, D., Shimanovich, D., Zdorovets, M., Kozlovskiy, A., Zubar, T., Vinnik, D., Dong, M., Trukhanov, S., Trukhanov, A., & Fedosyuk, V. (2020). Electrochemical behaviour of Ti/Al2O3/Ni nanocomposite material in artificial physiological solution: Prospects for biomedical application. *Nanomaterials, 10*, 45–61. https://doi.org/10.3390/nano10010173

Vorobjova, A., Tishkevich, D., Shimanovich, D., Zubar, T., Astapovich, K., Kozlovskiy, A., Zdorovets, M., Zhaludkevich, A., Lyakhov, D., & Michels, D. J. R. A. (2021). The influence of the synthesis conditions on the magnetic behaviour of the densely packed arrays of Ni nanowires in porous anodic alumina membranes. *RSC Advances, 11*, 3952–3962. https://doi.org/10.1039/D0RA07529A

Wang, L., Lin, T., Yan, C., He, W., Wang, J., & Tang, Q. (2018). Effects of different film residue and irrigation quota on nutrient and water use efficiency of cotton under drip irrigation. *Journal of Plant Nutrition, 24*, 122–133.

Wang, B., Cui, B. B., Zhang, G. Y., Li, V., Ma, H. X., Wang, W., Zhang, F. Y., & Ma, X. X. (2020). Novel axial substituted subphthalocyanine and its TiO2 photocatalyst for degradation of organic water pollutant under visible light. *Optical Materials, 109*, Article 110202. https://doi.org/10.1016/j.optmat.2020.110202

Yang, C., Gong, C., Peng, T., Deng, K., & Zan, L. (2010). High photocatalytic degradation activity of the polyvinyl chloride (PVC)-vitamin C (VC)-TiO2 nano-composite film. *Journal of Hazardous Materials, 178*, 152–156. https://doi.org/10.1016/j.jhazmat.2010.01.056

Yousif, E. & Haddad, R. (2013). Photodegradation and photostabilization of polymers, especially polystyrene: review. *SpringerPlus, 2*, Article No. 398. https://doi.org/10.1186/2193-1801-2-398

Yu, Y., Zhang, P., Guo, L., Chen, Z., Wu, Q., Ding, Y., Zheng, W., & Cao, Y. (2014). The design of TiO2 nanostructures (nanoparticle, nanotube, and nanosheet) and their photocatalytic activity. *Journal of Physical Chemistry C, 118*, 12727–12733.

Zan, L., Tian, L., Liu, Z., & Peng, Z. (2004). A new polystyrene–TiO2 nanocomposite film and its photocatalytic degradation. *Applied Catalysis, A: General, 264*, 237–242. https://doi.org/10.1016/j.apcata.2003.12.046

Zhang, J., Zhou, P., Liu, J., & Yu, J. (2014). New understanding of the difference of photocatalytic activity among anatase, rutile and brookite TiO2. *Physical Chemistry Chemical Physics: PCCP, 16*, 20382–20386.

Zhang, J., Tian, B., Wang, L., Xing, M., & Lei, J. (2018). *Photocatalysis: fundamentals, materials and applications*. Springer Nature.

Zhang, Y. Sun, T., Zhang, D., Shi, Z., Zhang, X., Li, C., Wang, L., Song, J., & Lin, Q. (2020a) Enhanced photodegradability of PVC plastics film by codoping nano-graphite and TiO2. *Polymer Degradation and Stability,181,* https://doi.org/10.1016/j.polymdegradstab.2020.109332

Zhang, Y., Kang, S., Allen, S., Allen, D., Gao, T., Sillanpaä, M. (2020b). Atmospheric microplastics: A review on current status and perspectives. *Earth Science Reviews, 203*, 103118, https://doi.org/10.1016/j.earscirev.2020.103118

Zhang, F., Zhao, Y., Wang, D., Yan, M., Zhang, J., Zhang, P. (2021a). Current technologies for plastic waste treatment: A review. *Journal of Cleaner Production, 282*, Article ID: 124523. https://doi.org/10.1016/j.jclepro.2020.124523

Zhang, K., Hamidian, A. H., Tubić, A., Zhang, Y., Fang, J. K. H., & Wu, C. (2021b) Understanding plastic degradation and microplastic formation in the environment: A review. *Environmental Pollution, 274*, Article ID: 116554. https://doi.org/10.1016/j.envpol.2021.116554

Zhao, X., Li, Z., Chen, Y., Shi, L., & Zhu, Y. (2008). Enhancement of photocatalytic degradation of polyethylene plastic with CuPc modified TiO2 photocatalyst under solar light irradiation. *Applied Surface Science, 254*, 1825–1829. https://doi.org/10.1016/j.apsusc.2007.07.154

Zhao, X.u., Li, Z., Chen, Y., Shi, L., & Zhu, Y. (2007). Solid-phase photocatalytic degradation of polyethylene plastic under UV and solar light irradiation. *Journal of Molecular Catalysis A: Chemical, 268*, 101–106. https://doi.org/10.1016/j.molcata.2006.12.012

Wastewater Management and the Role of Nanomaterials

Prasamita Samantaray, Sailaja Priyadarsini, Eepsita Priyadarshini, and Alok Prasad Das

Abstract Since industrial, agricultural, and household waste may seriously harm ecosystems and human health by contaminating water supplies, wastewater management is a crucial environmental concern. Using traditional wastewater treatment methods like biological, chemical, and physical procedures to handle new pollutants including medications, toxic metals, and microplastics is usually challenging. Recent developments in nanotechnology have created new opportunities to improve wastewater treatment's efficacy and sustainability. This article explores the use of nanoparticles in wastewater treatment, emphasizing how they can eliminate pollutants, enhance treatment procedures, and lessen environmental damage. Large area of surface, reactivity, and selectivity are just a few of the unique properties that make nanomaterials—including micron-sized particles nanocomposites, and nanostructured materials—great choices for getting rid of pollutants. Heavy metals, microbes, organic colours, and pharmaceuticals are just a few of the pollutants that these materials can effectively adsorb, degrade, or catalyse the conversion of. Nanoparticles' versatility allows for the development of specialized treatment processes for wastewater that target specific contaminants in different water matrices. The article talks about the various kinds of nanomaterials used in wastewater treatment, such as carbon-based nanomaterials, metallic and aluminium oxide nanoparticles, and polymeric nanocomposites. Additionally, it looks at their capabilities, drawbacks, and possible environmental hazards such toxicity and secondary pollution. The study also emphasizes how wastewater management systems can be optimized by integrating nanomaterials with current treatment methods like filtration, adsorption, and advanced oxidation processes. To sum up, nanoparticles offer a novel and promising way to enhance wastewater treatment; however, more study and regulation are required to guarantee their safe and efficient widespread use.

P. Samantaray · S. Priyadarsini · A. P. Das (✉)
Department of Life Sciences, Rama Devi Women's University, Bhubaneswar, Odisha, India
e-mail: alok1503@gmail.com

E. Priyadarshini
Department of Biomedical Sciences, School of Biosciences and Technology, Galgotias University, Greater Noida, Uttar Pradesh, India

E. Priyadarshini et al. (eds.), *Advanced Nanotechnology Applications for Environmental Sustainability*, Sustainable Environmental Waste Management Strategies,
https://doi.org/10.1007/978-3-032-12933-8_14

1 Introduction

Drinking water is today a valuable resource in many regions of the world due to the world's fast-growing population, the depletion of water supplies, and climate change that causes protracted droughts and floods (Mishra & Das, 2023; Mishra et al., 2024a). Reusing or recycling water in any way will therefore help to lessen this difficulty. One of the primary concerns and a popular area of contemporary research is the prudent management of water and wastewater (Mishra & Das, 2021; Mishra et al. 2022b). A significant volume of wastewater has been released into lakes, rivers, and coastal areas during the last century. As a result, the aquatic ecosystem has significant pollution issues. The environment and public health are seriously endangered by the massive amounts of liquid pollutants and effluents produced by natural, municipal, and industrial activities. Therefore, to appropriately process and handle water and wastewater, the right technique must be devised (Kumar et al., 2023; Mishra & Das, 2022). Pharmacological procedures, the adsorption flocculation, and filtering, and so on are examples of indicative/conventional processes. With particular focus to the mechanical partitioning of solid particles and the discharge of large amounts of hazardous substances such as nitrogen, phosphorus, and other electrically charged substances into the environment, these conventional technologies only address the initial stage of wastewater treatment (Mishra et al. 2022a). As a result, the latest nanotechnology greatly improves wastewater treatment by employing nanomaterials such as nanoparticles, nano-adsorbents, nanocomposites, and nanofiltration (Das et al., 2015a). To improve the elimination of specific wastewater components and boost productivity, these microscopic particles have been used in the creation of adsorbent materials, catalysts, and separation membranes. Nanotechnology can accomplish all of the preceding stated goals. Furthermore, it has been demonstrated that traditional methods of eliminating metals such as mercury, phosphate, nitrogen, and other pollutants from contaminated wastewater are not totally effective (Kumar & Das, 2017). Each has its benefits and drawbacks, but despite their cost and time commitment, they all help to some degree to lower the levels of different contaminants (Rajasulochana & Preethy, 2016). The core of nanotechnology is the manipulation and control of materials at the minuscule and atomic levels in order to create new systems, devices, and structures with improved conductive, optical, magnetic, mechanical, and electric characteristics (Sadasivuni et al., 2019). Nanotechnology has shown impressive results in a number of sectors, including wastewater treatment, and is being investigated as a possible technique. A vast range of nearly totally organic materials can be found in agricultural and culinary effluent, just like in other wastewaters. Compounds primarily composed of elements like oxygen, hydrogen, and carbon are known as organic materials. Known as carbonaceous molecules, the carbon atoms that make up organic matter may undergo oxidation chemically or biologically to produce CO_2 and energy (Ghosh & Das, 2015). A limited number of potential pollutants may be present in the various effluent streams from specific food processing operations in a processing plant (Das & Mishra, 2008; Mallick et al., 2023). However, it is almost impossible to identify all the components in the plant's discharges because various wastewaters might occasionally mix to form distinct sewage flows from the same work site (Kanungo et al., 2023a, 2023b). The properties of food and agricultural wastewater can therefore be seen as a collection of distinct physicochemical and biological criteria that are essential for the planning and operation of these types of treatment facilities (Dey et al., 2023, 2024). Nanomaterials can also be used in ex-situ waste management technologies. These methods usually use membrane filtration and separation, adsorption, and photocatalysis. Nanoparticles have a lot of surface area due to their size, which makes it easier for pollutants to adsorb onto them. This technique has garnered a lot of interest recently, much like membrane filtration and separation. The membrane's ability to act as a barrier depends on the size of the contaminating particle and the size of its holes Nanomaterials can be embedded in the membrane matrix or deposited on the surface to generate nanocomposite membranes. Nanomaterials like carbon nanotubes are added to polymeric membranes to enhance their mechanical characteristics and water permeability. Another typical metal-based nanomaterial at the nanoscale is titanium dioxide. Because of their high photosensitivity, these minuscule particles are mostly employed as photocatalysts, while they can also be used as heavy metal adsorbents (Das et al., 2015b; Ghosh & Das, 2018; Ghosh et al., 2016b; Mohanty et al., 2016). This nanomaterial can oxidize pollutants by producing extremely reactive hydroxyl radicals whenever exposed to ultraviolet (UV) light, or sunshine. Advanced oxidation techniques, as they are commonly known, use hydroxyl radicals to purify water (Das et al., 2025; Reddy Koteswara et al., 2025). Because of its large band gap energy, titanium dioxide needs UV light—not simply visible light—to start photocatalysis. These nanoparticles act as catalysts and stay unaltered during the degrading process. They are insoluble in water and chemically stable, which allows them to go through a variety of reduction processes. Carbon nanotubes (CNTs) are the most prevalent type of carbon-based nanomaterial. These are cylindrical nanostructured carbon allotropes that are notable for their high assessable adsorption sites and huge specific surface area. Their outside modification ability and variable surface chemical composition, which both increase their sorption capacity, are mainly due to their tubular structure. As a result, CNTs are primarily employed as adsorbents for polar and non-polar organic molecules, oils, and heavy metals (Priyadarsini et al., 2025; Priyadarsini & Das, 2026). CNTs' main advantage over other kinds of nanomaterials and, undoubtedly, over other traditional carbonaceous products like activated carbon is their capacity for regrowth and reuse (Application of Nanotechnology for Waste Management: Challenges and Limitations, n.d.).

2 Overview of Wastewater and Its Characteristics

To meet the growing demand for water, utilities are always searching for the most cost-effective solutions to expand their water supplies. This might have required moving surface water farther or drilling greater depths into freshwater aquifers in the recent past, but the onset of climate change has greatly raised the potential for error in these cost–benefit evaluations (Das & Mishra, 2009; Nawaz et al., 2023). Therefore, alternatives—especially to the recycling of municipal wastewater—have been becoming more and more popular since they are resistant to drought. Significantly, water-rich regions are also putting planned drinking water treatment into place to comply with stricter wastewater disposal regulations, circumvent geographic limitations on water storing capacity, protect natural resources, and boost water resource independence (Young et al., n.d). The remainder of 13–18% of flow is wastewater from the municipality reuse concentrate, which has been condensed five–six times and contains every one of the parts that are supplied to and discarded by the RO (Mishra et al., 2024b; Priyadarsini & Das, 2025). It contains increased concentrations of metallic substances, nutrients, and micropollutants, including microplastics, endocrine disrupting chemicals, medicines, personal care items, and polyfluoroalkyl substances, among other contaminants of growing concern. In addition, it contains more process chemicals such biocides and total dissolved solids (4,000–9,000 mg/L). It is a result of the growing prevalence of water reuse.

The food itself defines wastewater from food manufacturing procedures. Debris from mechanical food processing, including peeling, trimming, and hydrodynamic effects during washing and transportation, as well as dissolved organic materials from numerous activities are present in food and agricultural wastewater (Mishra et al., 2020). Large volumes of water are inevitably used to clean and, in some situations, cool food products in used for farming and food processing processes. Canning wastewater is essentially the same as household kitchen waste because it is a consequence of several canning processes, such as slicing, measuring, juice making, pureeing, blanching, and heating. Blanching and chilling crops also require large amounts of water. Cleaning plant floors, equipment, and processing areas is a common task in almost all food or agricultural processing operations. Detergents are frequently used in these operations, and they can also occasionally serve as lubricants for the equipment used in food processing. Water used in specific processing procedures is frequently reused, either with or without treatment, when doing so is both cost-effective and permitted (Kanungo et al. 2023a, 2024). Reusing water is frequently viewed as essential for pragmatic reasons since fresh water supplies are scarce in many regions of the world. Water utilization may be significantly decreased by recycling and reusing water. However, one should bear in mind that any food safety concerns resulting from the reused water should be carefully and properly investigated if it is going to be utilized for edible food items. The dairy business, which has grown steadily in recent years, is one of the largest and most important farming and processing industries in the United States. Two main dairy processing facilities that generate compressed milk, milk in powder form, compressed whey, which and other products are the source of wastewater. These wastewaters are added to the fluid milk at bottling facilities and receiving stations. The importance of these amenities is growing. Milk has a BOD of 1,00,000 mg/L, however effluent from facilities that wash cheese as well as butter may have a BOD of 1,500–3,000 mg/L. Ice creams, yogurts, sweetened the milk, and frozen desserts are made with non-dairy ingredients such aromas, sugar, which makes it fruits, nuts, and sauces. Various than milk and products related to milk, dairy manufacturing employs various raw resources.

Spills, leaks, and by-product waste are some of the ways that contaminants can get into wastewater systems. Most dairy effluent streams are neutral or slightly alkaline, with the exception of whey, which is acidic. However, the lactic acid created when lactose ferments cause these streams to quickly turn acidic (Baumann, 1974). In order to extract edible oil, oil-bearing seeds or animal fats must be solvent extracted (olive and sesame oils can be removed mechanically) and refined to remove impurities. Apart from cleaning and Wastewater streams in edible oil production facilities are produced by a number of activities, including washing operations that consume water. To eliminate these contaminants, numerous standard wastewater treatment procedures may be used in addition to the sludges produced by primary or secondary treatment procedures. Adsorption or steam/air stripping can be used to get rid of trace amounts of solvent, such hexane. Pervaporation is an additional eco-friendly technique for eliminating hexane from wastewater (Peng et al., 2003) (Table 1).

3 Applications of Nanomaterials in Wastewater Management

An exceptionally substantial number of effluents has resulted from the development of innovative and creative technologies in recent decades. Industrial effluent, which originates from many industries, is one of the primary causes of water contamination (Das & Ghosh, 2022). Wastewater contains a wide range of impurities, including organic and inorganic pollutants, heavy metals, and non-disintegrating compounds (Bal et al., 2018; Ghosh et al., 2017; Mohanty et al., 2018; Priyadarsini & Das, 2024; Sanket et al., 2016). This contamination poses a major hazard to the environment. Therefore, in order to eliminate them, innovative and novel technology and approaches must be used (Ghosh & Das, 2017; Ghosh et al., 2020; Mohanty et al., 2017). In recent years, nanomaterials have become a viable alternative to conventional techniques for the removal of pollutants. Nowadays, a wide range of reasonably priced nanomaterials with unique properties are available. In this sense, nano-absorbents are excellent materials. Heavy metal contamination of surface and underground streams is severe. An exceptionally high number of effluents has been produced as a result of the development of innovative and creative technology in recent decades (Das & Bissoyi, 2011; Das & Ghosh, 2023; Ghosh et al., 2016a; Paul et al., 2023). Industrial effluent from a range of industries is one of the primary causes of water contamination. Wastewater contains a wide range of impurities, including organic and inorganic pollutants, heavy metals, and non-disintegrating compounds. This contamination poses a major hazard to the environment. Therefore, in order to get rid of them, new and innovative techniques and equipment must be created. In recent years, nanomaterials have become a viable alternative to conventional techniques for the removal of pollutants (Das & Mishra, 2010; Mishra & Das, 2008; Mishra et al., 2009). Nowadays, a wide range of reasonably priced nanomaterials with unique properties are available and a lot of new research has been done on the chemical elimination of heavy metals.

Nanomaterials are characterized by their small size, favourable aspect to volume ratio, filtering capability, and well-organized structure. Effects on the surface area, implications for vast basic tunnels, effects on microscopic dimensions, and quantum effects are some of the unique characteristics of nanomaterials at the nanoscale (Tripathy et al., 2026). These characteristics enhance their reactivity and adsorption, making them unparalleled and excellent in removing heavy metal ions. The degradation of the natural world, the fluids we consume the air we breathe, and the land we live on is mostly caused by industrial pollutants. In addition to consuming vast amounts of agricultural land, the expanding force of industrialization degrades the environment and soil. Liquids from industrial wastewater are utilized to extract components for human activity-related production and raw material management. It contributes to environmental water contamination. According to a recent survey, one of the main causes of lake, beach, and stream contamination is industrial effluent. Due to entanglements and contamination issues in the nearby water, the ecosystem finally resurfaces as a detrimental output that adversely affects human living. Human populations are impacted by industrial effluent, which also commonly contributes to climate change. These wastewaters have a high suppressive potential, are inherently very solid, biodegradable, or highly inorganic.

The constant release of pesticides and water dyes from industry calls for the creation of complex innovations, such as the filtering or separation of chemical mixtures in combination with multi-technique detection methods, particularly at low concentrations. These estimations are time-consuming because sample preparation involves multiple middle-level handling steps (*Sadik: Nanotechnology Applications for Clean Water:...–Google Scholar*, n.d.). For example, Raman spectroscopy can be used to minimize operations in water sampling. The integration of the Raman scattering technique into a "lab-on-a-chip" device is made possible by the design of the specific structured pores silicon graphene alloys and the thick covering that is produced. On surface of these silicon graphene coatings, water samples can be readily stored, and contaminants can be found using Raman spectroscopy. It is possible for one to render such devices affectable by using molecular imprinting techniques. This material approach allows for the sub-atomic detection of extremely tiny voids in porous silica structures. Molecular templating techniques may be used to produce thick layers for dye detection utilizing silicon graphene. The maximum analytical enhancement factor, 13.98, is shown by rhodamine 6G dyes at a concentration of 9−3 M. The identical group also created permeable SG templates films to detect paraoxon, an organophosphate pesticide, at 11−4 M. Silicon graphene composites not only facilitate identification but also function as absorbents for the analyte's corresponding molecules. Studies have been performed on the attachment of organ phosphorus pesticides (OPPs) on composites of silicon grapheme (Lu & Astruc, 2020). Adsorption was encouraged by the strong π bond formed between the hybrids and the phenyl ring, as well as by the interaction among the combination and the functional groups. The removal capability and efficacy of silicon graphene molecule absorption were examined for various herbicides. The expansion of nanoparticles with magnetic properties in silicon graphene composites is essential in this instance to facilitate the composites' recovery, as was previously mentioned. With every change to the hydrophobic surface, the number of pesticide adsorption sites also increases. The physicochemical properties of the sorbent determine the range of physicochemical adsorption routes that favor the adsorption of dye onto silicon graphene composites. Strong π bonds created between the compounds and the phenyl ring, in addition to interactions between the combination of molecules and the functional groups, promoted adsorption. For a variety of herbicides, the removal potential and effectiveness of silicon graphene monomer absorption were investigated. As previously noted, the expansion of magnetic nanoparticles in silicon graphene alloys is crucial in this case to aid in the recovery of the composites. The quantity of insecticide adsorption sites rises in tandem with each modification to the hydrophobic surface. The variety of physicochemical adsorption pathways that promote dye adsorption upon silicon graphene composites is determined by the sorbent's physicochemical characteristics. Kubo et al. produced silicon graphene nanocomposites as of the multifunctional compound by incorporating a super paramagnetic nanoparticle substance into the mesoporous silica encasing graphene oxide. The nanoparticles' functionalization allowed the composites to be readily retrieved using an external magnet. But when Fe3O4 was added, the surface area was significantly reduced by 71%, and the pore diameters decreased by around 16%, from 308 mg/g to 124 mg/g, in terms of the removal ability of methylene blue. In light of this, the functionalization of the amino groups on the surface of multipurpose silicon graphene nanocomposites as may help to improve the interactions for the removal of contaminants.

Additionally, wastewater from the paint and enameling industry contains nickel. Using photocatalysts, like titanium dioxide nanoparticles, to lower the dangerous metal ions in water has been studied in detail. Dioxide of titanium Apart from publicly available particles of titanium dioxide that demonstrated nearly exceptional arsenic removal efficacy at a comparatively neutral pH value, a survey found that the nanocrystalline was a viable photocatalyst and adequate in eliminating different types of arsenic (Pearson, 1990). Heavy metal-containing wastewater is commonly cleaned using a variety of inexpensive, recently developed adsorbents. Natural resources, agricultural waste, and wastes from manufacturing processes are the sources of these adsorbents. Adsorption is the process by which a material moves from a fluid form to the outermost layer of a solid, where it is retained by chemical and physical forces. A three-step treatment procedure is one of the traditional techniques for eliminating metal: Adsorption on the particle's surface and internal transit inside of the sorbent particle move the pollutant from the majority of the solvent to the surface of the absorbing material. These methods are incredibly economical: Bio-sorption, mineralization on industrial waste products, and sorption on modified natural materials (Peralta-Videa et al., 2011).

One important area of study in the environmental sciences is the creation of low-cost adsorbent materials for pesticide maintenance. Because of their low cost and accessibility locally, industrial wastes such as fly ash-like substance sewage and carbon sludge are classified as simple materials. They can also be used as adsorbents for pesticides. Basic adsorbents that have shown a high adsorption limitation with regard to organic pollutants include fly ash-like stuff lignite, which is a solid waste from coal-fired, thermal ones, and nuclear plant operations. Some studies described fly ash's ability to absorb pesticides and proposed using it to remove insecticides from water and wastewater samples. Mercury, atrazine, and metribuzin have high maintenance limits when combined with coal fly ash. The likelihood that exposure to hazardous and toxic chemicals in drinking water would negatively impact one's health is a widely accepted definition of the potential health hazard associated with these substances. The most intricate aquatic systems are found in stream courses, where the minute elements interact and transport heavy metals in particular. Fertilizing agents pesticides, wastewater from livestock farms and poultry processing factories, and other sources are the primary causes of pollution of water in agricultural areas. Pesticide pollution caused by fertilizer heavy metals and nitrate contamination is especially hazardous. Water mineralization shows that soil quality has improved in locations where contamination of groundwater has occurred. Degradation of water quality is often caused by massive extraction of minerals storage in the surrounding area that includes extracts and strip mines.

4 Challenges and Limitations of Nanomaterial Use

In the past, many wastewater treatment methods have resulted in unexpected outcomes. Trihalomethanes and N-nitrosodimethylamine are carcinogenic byproducts of conventional wastewater treatment methods like chlorination, which were once thought to extend life expectancy worldwide. This may have something to do with the application of nanotechnology in wastewater treatment. Nanoparticles are responsible for any potential harm because of the same properties that make them useful. The components' size and toxicity end point are determined by their molecular composition, which also regulates cellular absorption. The brain, heart, spleen, marrow in the bones, and neurological system are just a few of the organs and tissues that NPs can reach due to their small size, which allows them to enter through the circulatory and the lymph system and to cross endothelium and epithelial barriers. It has been demonstrated that the toxicity of Ag NPs, CNTs, and many other metal NPs changes with their size and shape. Since NPs' diameters (which range from 2 to 102 nm) are similar to those associated with protein granules (3–11 nm), DNA spirals (2.5 nm), and cellular membrane thicknesses (9–10 nm), they can readily penetrate cells and cell organelles. Research indicates that nanoparticles of gold smaller than 7 nm can effectively penetrate the nucleus, while bigger NPs between 11 and 17 nm can only enter the cytoplasm after passing through the cell membrane.

In the end, the cost and efficacy of nanotechnology for wastewater treatment will determine its acceptability. Treatment facilities for wastewater in developed countries use advanced methods to eliminate a variety of contaminants, whereas in developing countries, it usually only takes care of the most fundamental needs (disinfection, for example). Since increasingly complicated pollutant combinations need to be treated in order to enhance water quality at a lower cost, these scenarios put a strain on the capabilities of present wastewater treatment models. This financial barrier is therefore significant but solvable. A large amount of the expense of creating nanomaterials is related to separation and purification. The cost of high-purity, consistent-quality research-grade nanomaterials has been essentially constant over the last 20 years, and it is unlikely to drop significantly without increased demand and production scale. Additionally, producing nanoparticles with a lesser purity can make using them to treat wastewater more logical. For example, fullerene soot was used in place of ultrapure C60 to produce amino-fullerene photocatalysts, which reduced costs by 90% while maintaining minimal efficiency. Furthermore, nanomaterials' cost-effectiveness may rise due to their long-term reusability. Examples include photocatalysts that sustain their activity by regenerating nano adsorbents and magnetically detachable multifunctional nanomaterials that permit many reuse cycles.

The fate of nanoparticles depends on abiotic factors, dispersibility, and the way biotic and nanomaterials interact. NPs can absorb more dirt and silt particles due to their large surface area, even though they usually settle less quickly than larger particles. Furthermore, water columns can be used to extract CNTs and fullerenes because they are very insoluble in water. Engineered nanoparticles are occasionally used for certain purposes. Oxidation as well as reduction may sometimes be favored, and light can cause photochemical changes. According to Westerhoff et al., these modifications alter the interactions between nanomaterials and their environment, which in turn affects how contaminants adhere and dissolve in water. The likelihood that physical, biological, or chemical stressors will disturb the ecosystem is known as ecotoxicity. When nanoparticles are produced, used, and disposed of, there is a chance that they will get into the treated water. Additionally, novel mineral nanomaterials like Ni and TiO2 are being produced at an increasing rate each year. It is believed that some naturally formed surface waters contain nano/micro g/L of NP. As more NP is created, these concentrations increase. Because of their antibacterial qualities, metallic silver, titanium as a metal and nanoparticles of zinc dioxide are utilized as chemical disinfectants to filter water in the majority of developing nations. Ag NP-impregnated ceramic filters are also employed. The measurable quantities of Ag NPs in the water that had been cleaned with tiny silver covered filter sheets were well below the WHO's recommended threshold. Numerous investigations have shown that rats who consumed TiO2 in NP-contaminated water suffered harm to their DNA and essential organs (De Matteis: Cellular and Molecular Toxicology of…–Google Scholar, n.d.). It is believed that some natural surface waters have NP values of nano/micro g/L. As NP production increases, so do these concentrations. Ag NP-impregnated ceramic filters are also employed. The detected quantities of Ag NPs were well below the WHO-recommended threshold during water filtration utilizing nano Ag-coated filter sheets. Numerous investigations have shown that rats who consumed TiO2 in NP-contaminated water suffered harm to their DNA and essential organs (De Matteis: Cellular and Molecular Toxicology of…–Google Scholar, n.d.).

5 Future Potential Applications of Nanomaterials in Wastewater Treatment

Even if there is currently no denying the efficacy of employing nanoparticles in the treatment of industrial wastewater, there are a number of serious disadvantages since they can tolerate high hazards for a long time and may leak into the environment while being treated. This indicates that in order to lessen environmental toxicity, more studies and surveys are required. For instance, it is difficult to create a sustainable technique for eliminating environmental contaminants since titanium dioxide, a widely used metal oxide nanoparticle, has significant drawbacks and detrimental effects on both the environment and human health. Numerous scientists and analysts are always striving to overcome these obstacles, and initiatives for new discoveries and research should be examined in this manner to overcome these obstacles. To use solar energy more efficiently, several researchers have recently found a more efficient method of narrowing the band gap. To find practical domains of use for assessing material proficiency and to develop practical techniques for producing nanomaterials, more research is needed. The market demands that the procedure's costs be maintained to a low because the aforementioned nanoparticles are environmentally friendly. For this purpose, considerable attention should be paid to green technologies derived from agricultural waste. Much effort needs to be done in this area to get better. Their application in the study has not yet been completed, which is the main emphasis of the current research. Few research are accessible for rapid economic and market analysis. Improving the industrial treatment process is the main market objective for the future, and this requires a significant financial and technological investment.

Nanomaterial treatment for water pollutants is becoming more and more common due to the awful state of water and the demand for freshwater everywhere in the universe. Progressive innovation in water treatment is desperately needed to guarantee high-quality drinking water and to eliminate toxins and micro and macro pollutants. We must use creatively replaceable water treatment solutions to improve industrial production advancements. Nanotechnology has enabled incredible breakthroughs in water purification, problem solutions, and future developments. Despite the fact that nanomaterial techniques such as nanostructured catalytic membranes, nanosorbents, and others are very efficient in terms of time, energy, and the environment, they are all quite costly and are not yet commonly employed in the industrial wastewater purification process. Because of their high rate of reactivity, nanomaterials are very efficient. In any case, there are certain disadvantages that should be avoided. Digital, computerized monitoring methods are unable to accurately quantify the prevalence of nanoparticles, which are found in water in trace proportions. Additionally, international research institutions and research institutes should create legal terms and conditions to address these situations in order to reduce health risks. Furthermore, there aren't many examples that aren't noticeable with mild treatment methods in the current situation, and the mechanical barrier that nanoengineered water approaches is inconsistently adaptive to mass growths. Furthermore, there are amazing criteria to incorporate various changes of nanomaterials, such as high productivity, cost effectiveness, ease of handling, and environmental friendliness. It's also critical to comprehend the economic obstacles and commercialization of these developments in wastewater treatment. Many uses for nanomaterials can offer a huge idea for flexible drinking water to the entire planet.

Table 1 An overview of the fundamental characteristics of nanomaterials, along with any possible hazards or difficulties

Properties of nanomaterials	An explanation of risk	References
Dissociation of contaminants	One of the main danger factors is thought to be the contamination of the NP by remaining contaminants. For instance, depending on the precursor utilized to produce iron oxide nanoparticles (FeCl3 or Fe2(SO4)3), sulfur impurities may be present. Similarly, carbon nanotubes (CNTs) deposited on the CNT surface may contain metal impurities such as nickel, yttrium, or rubidium	Bortoleto et al. (2007), Chen et al. (2004)
Impurity by nature	Because of their high reactivity, NPs interact with contaminants. Encapsulation is therefore a crucial requirement for solution-based NP synthesis (chemical method). To stabilize the NPs, nonreactive species encapsulate the reactive nano-entities during the encapsulation process	Jeevanandam et al. (2018)
The process of aggregation or mass	Due to their high solubility, phase shift, and poor corrosion resistance, weakly bound (agglomeration) and fused particles are important risk factors for NMs. This causes more degradation and makes maintaining the structure difficult	Kennedy et al. (2008)
Recycling and getting rid of things	No strict safe disposal regulations apply to NMs. The possible toxicity of NP exposure is yet unknown, and experimental findings are not yet available. Therefore, strategies for permanent disposal and recycling have not yet been devised to address the uncertainty of a nanomaterial's effect	Jeevanandam et al. (2018)
Dimensions	The particle size of NPs largely determines their reactivity and aggregation. The fact that agglomeration occurs more slowly in smaller particles is well recognized. It is impossible for the NPs to return to their initial size after synthesis. Encapsulation in NP synthesis is hence absolutely necessary. By classifying NPs based on their particle size, their unique size-dependent chemistry is differentiated from classical colloid chemistry	Ae et al. (2009)
Reactivity or release	NPs can be charged through spontaneous degradative reactions or functionalization. The specific functionality and bioavailability of NMs will be significantly influenced by chemical species and the key functional groups associated with their charges	Ae et al. (2009)

NP-Nanoparticles, NM-Nanomaterials

References

Ae, T. T., Linkov, I., José, A. E., Figueira, R., Jeffery, A. E., Ae, S., Ae, M. C., Merad, M., Tervonen, T., Linkov, I., Figueira, J. R., Steevens, J., Chappell, Á. M., Chappell, M., & Merad, M. (2009). Risk-based classification system of nanomaterials. Springer. https://doi.org/10.1007/s11051-008-9546-1. (Tervonen T, Linkov I, Figueira JR, Steevens J, Chappell M, Merad M (2009) Journal of Nanoparticle Research, 11(4), 757–766, Springer)

Bal, B., Ghosh, S., & Das, A. P. (2018). Microbial recovery and recycling of manganese waste and their future application: A review. *Geomicrobiological Journal, 36*(1), 85–96. https://doi.org/10.1080/01490451.2018.1497731

Baumann, F. J. (1974). Dichromate reflux chemical oxygen demand. Proposed method for chloride correction in highly saline wastes. *Analytical Chemistry*, *46*(9), 1336–1338. https://doi.org/10.1021/AC60345A039

Bortoleto, G. G., Borges, S. S. de O., & Bueno, M. I. M. S. (2007). X-ray scattering and multivariate analysis for classification of organic samples: A comparative study using Rh tube and synchrotron radiation. *Analytica Chimica Acta, 595*(1–2 SPEC. ISS), 38–42. https://doi.org/10.1016/J.ACA.2006.11.067

Chen, Q., Saltiel, C., Manickavasagam, S., Schadler, L. S., Siegel, R., & Yang, H. (2004). Aggregation behavior of single-walled carbon nanotubes in dilute aqueous suspension. *Journal of Colloid and Interface Science, 280*(1), 91–97. https://doi.org/10.1016/j.jcis.2004.07.028. (Q Chen, C Saltiel, S Manickavasagam, LS Schadler, RW Siegel, H Yang, Elsevier).

Das, A., Dey, S., Gacem, A., Yadav, K. K., Nath, H., Ram, D. K., & Das, A. P. (2025). Revolutionizing solid hydrocarbon sourcing: Plastic's journey from waste to upcycled treasure. In *Environmental science and engineering, part F218*, (pp. 331–351). https://doi.org/10.1007/978-3-031-81615-4_14

Das, A., & Ghosh, S. (2022). Role of microorganisms in extenuation of mining and industrial wastes. *Geomicrobiology Journal, 39*, 173–175. https://doi.org/10.1080/01490451.2022.2038953

Das, A., & Mishra, S. (2010). Biodegradation of the metallic carcinogen hexavalent chromium Cr(VI) by an indigenously isolated bacterial strain. *Journal of Carcinogenesis*, *9*. https://doi.org/10.4103/1477-3163.63584

Das, A. P., Bal, B., & Mahapatra, P. S. (2015). Chromogenic biosensors for pathogen detection. In P. Prokopovich (Ed.), *Biological and pharmaceutical applications of nanomaterials* (1st ed.). CRC Press, Taylor & Francis.

Das, A. P., Ghosh, S., Mohanty, S., & Sukla, L. B. (2015b). Consequences of manganese compounds: A review. *Toxicological & Environmental Chemistry, 96*(7), 981–997. https://doi.org/10.1080/02772248.2015.1005428

Das, A. P., & Bissoyi, A. (2011). Computational approach for comparative phylogenetic analysis of isolated chromium resistant strain Brevibacterium casei. *Journal of Engineering and Technology Research, 3*(3), 82–87. https://doi.org/10.5897/JETR.9000020

Das, A. P., & Ghosh, S. (2023). Mixed culture bioleaching: An insight into manganese biomining process and efficacy. *Manganese Mining Microorganisms*, 129–141. https://doi.org/10.1016/B978-0-12-822147-1.00001-7

Das, A. P., & Mishra, S. (2008). Hexavalent Chromium (VI): Health hazards and environmental pollutant. *J Environ Res Dev, 2*(3), 386–392.

Das, A. P., & Mishra, S. (2009). Hexavalent chromium [Cr (VI)]: Yellow water pollution and its remediation. *Sarovar Saurabh ENVIS Newsl Wetl Ecosyst, 5*(2).

Dey, S., Bissoyi, A., Koteswara Reddy, G., & Das, A. P. (2024). Synthetic microfibers: An urban emerging pollutant demanding for sustainable environmental solutions. *Sustainable Microbial Technology for Synthetic and Cellulosic Microfiber Bioremediation*, 107–121. https://doi.org/10.1007/978-3-031-62600-5_7

Dey, S., Tripathy, B., Kumar, M. S., & Das, A. P. (2023). Ecotoxicological consequences of manganese mining pollutants and their biological remediation. *Environmental Chemistry and Ecotoxicology*. https://doi.org/10.1016/J.ENCECO.2023.01.001

Ghosh, S., Bal, B., & Das, A. P. (2017). Enhancing manganese recovery from low-grade ores by using mixed culture of indigenously isolated bacterial strains. *Geomicrobiology, 35*(3), 242–246. https://doi.org/10.1080/01490451.2017.1362080

Ghosh, S., & Das, A. P. (2015). Modified titanium oxide (TiO2) nanocomposites and its array of applications: A review. *Toxicological & Environmental Chemistry, 97*(5), 491–514. https://doi.org/10.1080/02772248.2015.1052204

Ghosh, S., & Das, A. P. (2017). Bioleaching of manganese from mining waste residues using Acinetobacter sp. *Geology, Ecology, and Landscapes, 1*(2), 77–83. https://doi.org/10.1080/24749508.2017.1332847

Ghosh, S., & Das, A. P. (2018). Metagenomic insights into the microbial diversity in manganese-contaminated mine tailings and their role in biogeochemical cycling of manganese. *Scientific Reports 2018 8:1, 8*(1), 1–12. https://doi.org/10.1038/s41598-018-26311-w

Ghosh, S., Gandhi, M., van Hullebusch, E. D., & Das, A. P. (2020). Proteomic insights into Lysinibacillus sp.-mediated biosolubilization of manganese. *Environmental Science and Pollution Research 2020 28:30, 28*(30), 40249–40263. https://doi.org/10.1007/S11356-020-10863-4

Ghosh, S., Mohanty, S., Akcil, A., Sukla, L. B., & Das, A. P. (2016a). A greener approach for resource recycling: Manganese bioleaching. *Chemosphere, 154*, 628–639. https://doi.org/10.1016/J.CHEMOSPHERE.2016.04.028

Ghosh, S., Mohanty, S., Nayak, S., Sukla, L. B., & Das, A. P. (2016b). Molecular identification of indigenous manganese solubilising bacterial biodiversity from manganese mining deposits. *Journal of Basic Microbiology, 56*(3), 254–262. https://doi.org/10.1002/JOBM.201500477

Jeevanandam, J., Barhoum, A., Chan, Y. S., Dufresne, A., & Danquah, M. K. (2018). Review on nanoparticles and nanostructured materials: History, sources, toxicity and regulations. *Beilstein Journal of Nanotechnology, 9*(1), 1050. https://doi.org/10.3762/BJNANO.9.98

Kanungo, J., Sahoo, T., Bal, M., & Behera, I. D. (2023a). Performance of bioremediation strategy in waste lubricating oil pollutants: A review. *Geomicrobiology Journal*. https://doi.org/10.1080/01490451.2023.2245395

Kanungo, J., Sahoo, T., Swain, L. P., & Behera, I. D. (2023b). *Toxicity of persistent hydrocarbon pollutants, sources and sustainable remediation process*. (pp. 39–65). https://doi.org/10.1007/978-3-031-48220-5_2

Kanungo, J., Sahoo, T., Bal, M., & Behera, I. D. (2024). Performance of Bioremediation Strategy in Waste Lubricating Oil Pollutants: A Review. *Geomicrobiology Journal, 41*(4), 360–373. https://doi.org/10.1080/01490451.2023.2245395

Kennedy, A. J., Hull, M. S., Steevens, J. A., Dontsova, K. M., Chappell, M. A., Gunter, J. C., & Weiss, C. A. (2008). Factors influencing the partitioning and toxicity of nanotubes in the aquatic environment. *Environmental Toxicology and Chemistry, 27*(9), 1932–1941. https://doi.org/10.1897/07-624.1

Kumar, M., Pandit, S., Patel, V., Khanna, N., Nag, M., Lahiri, D., Ray, R. R., Das, A., & Das, D. (2023). Maximization of energy recovery from starch processing wastewater by thermophilic dark fermentation coupled with microbial fuel cell technology. *Geomicrobiology Journal*, 1–12. https://doi.org/10.1080/01490451.2023.2209555

Kumar, M. S., & Das, A. P. (2017). Emerging nanotechnology based strategies for diagnosis and therapeutics of urinary tract infections: A review. *Advances in Colloid and Interface Science, 249*, 53–65. https://doi.org/10.1016/J.CIS.2017.06.010

Lu, F., & Astruc, D. (2020). Nanocatalysts and other nanomaterials for water remediation from organic pollutants. *Coordination Chemistry Reviews, 408*. https://doi.org/10.1016/J.CCR.2020.213180

Mallick, K., Sahu, A., Dubey, N. K., & Das, A. P. (2023). Harvesting marine plastic pollutants-derived renewable energy: A comprehensive review on applied energy and sustainable approach. *Journal of Environmental Management, 348*, Article 119371. https://doi.org/10.1016/J.JENVMAN.2023.119371

Mishra, S., & Das, A. P. (2008). Hexavalent chromium reduction and 16S rDNA identification of bacteria isolated from a Cr (VI) contaminated site. *The Internet Journal of Microbiology, 7*(1).

Mishra, S., & Das, A. P. (2021). Current treatment technologies for removal of microplastic and microfiber pollutants from wastewater. *Wastewater Treatment: Cutting-Edge Molecular Tools, Techniques and Applied Aspects*, 237–251. https://doi.org/10.1016/B978-0-12-821881-5.00011-8

Mishra, S., & Das, A. P. (2022). Treatment of the Wastewater Polluted with Synthetic Microfiber Released from Washing Machine. In B. B. Das, H. Hettiarachchi, P. K. Sahu, & S. Nanda (Eds.), *Lecture Notes in Civil Engineering* (Vol. 207, pp. 109–117). Springer Science and Business Media Deutschland GmbH. https://doi.org/10.1007/978-981-16-7509-6_9/COVER

Mishra, S., & Das, A. P. (2023). Microbial remediation of synthetic microfiber contaminated wastewater. In M. P. Shah (Ed.), *Microbial technologies in industrial wastewater treatment* (pp. 337–350). Springer, Singapore. https://doi.org/10.1007/978-981-99-2435-6_17

Mishra, S., Das, A. P., & Seragadam, P. (2009). Microbial remediation of hexavalent chromium from chromite contaminated mines of Sukinda Valley, Orissa (India). *The J of Environment Research & Development*, 1122–1127.

Mishra, S., Dash, D., & Das, A. P. (2022a). Detection, characterization and possible biofragmentation of synthetic microfibers released from domestic laundering wastewater as an emerging source of marine pollution. *Marine Pollution Bulletin, 185*, Article 114254. https://doi.org/10.1016/J.MARPOLBUL.2022.114254

Mishra, S., Singh, R. P., Rout, P. K., & Das, A. P. (2022b). Membrane bioreactor (MBR) as an advanced wastewater treatment technology for removal of synthetic microplastics. *Development in Wastewater Treatment Research and Processes*, 45–60. https://doi.org/10.1016/B978-0-323-85583-9.00022-3

Mishra, S., Dash, D., & Das, A. P. (2024a). Isolation and molecular characterization of polyester-degrading bacterial strains from municipal wastewater and their application in microplastic degradation in contaminated wastewater. *Geomicrobiology Journal*. https://doi.org/10.1080/01490451.2024.2376101

Mishra, S., Dash, D., & Das, A. P. (2024b). Microbial degradation of polyester microfibers using indigenously isolated bacterial strain exiguobacterium Sp. *CLEAN – Soil, Air, Water, 52*(12), e202300343. https://doi.org/10.1002/CLEN.202300343

Mishra, S., Singh, R. P., Rath, C. C., & Das, A. P. (2020). Synthetic microfibers: Source, transport and their remediation. *Journal of Water Process Engineering, 38*, Article 101612. https://doi.org/10.1016/J.JWPE.2020.101612

Mohanty, S., Ghosh, S., Bal, B., & Das, A. P. (2018). A review of biotechnology processes applied for manganese recovery from wastes. *Reviews in Environmental Science and Bio/Technology 2018 17:4, 17*(4), 791–811. https://doi.org/10.1007/S11157-018-9482-1

Mohanty, S., Ghosh, S., Nayak, S., & Das, A. P. (2016). Isolation, Identification and screening of manganese solubilizing fungi from low-grade manganese ore deposits. *Geomicrobiological Journal, 34*(4), 309–316. https://doi.org/10.1080/01490451.2016.1189016

Mohanty, S., Ghosh, S., Nayak, S., & Das, A. P. (2017). Bioleaching of manganese by Aspergillus sp. isolated from mining deposits. *Chemosphere, 172*, 302–309. https://doi.org/10.1016/J.CHEMOSPHERE.2016.12.136

Nawaz, H., Asghar, N., Ali, A., Ameer, A., Shehzad, M. A., Nawaz, F., Mehmood, A., Iqbal, M. S., Iqbal, U., Kaleem, M., Khan, A., & Ahmad, K. S. (2023). Optimization of citric acid and EDTA levels under Ni stress using rapeseed brassica napus L. for phytoremediation. *Water, Air, and Soil Pollution, 234*(4), 1–18. https://doi.org/10.1007/S11270-023-06229-4/METRICS

Paul, A., Dey, S., Ram, D. K., & Das, A. P. (2023). Hexavalent chromium pollution and its sustainable management through bioremediation. *Geomicrobiology Journal*, 1–11. https://doi.org/10.1080/01490451.2023.2218377

Peng, M., Vane, L. M., & Liu, S. X. (2003). Recent advances in VOCs removal from water by pervaporation. *Journal of Hazardous Materials, 98*(1–3), 69–90. https://doi.org/10.1016/S0304-3894(02)00360-6

Peralta-Videa, J. R., Zhao, L., Lopez-Moreno, M. L., de la Rosa, G., Hong, J., & Gardea-Torresdey, J. L. (2011). Nanomaterials and the environment: A review for the biennium 2008–2010. *Journal of Hazardous Materials, 186*(1), 1–15. https://doi.org/10.1016/J.JHAZMAT.2010.11.020

Priyadarsini, S., & Das, A. P. (2024). Analytical and structural characterization of waste lithium-ion batteries for their effective recycling strategy. *Environmental Chemistry and Ecotoxicology.* https://doi.org/10.1016/j.enceco.2024.12.004

Priyadarsini, S., & Das, A. P. (2025). Lithium bioleaching: A review on microbial-assisted sustainable technology for lithium bio-circularity. *Journal of Water Process Engineering, 69*, Article 106744. https://doi.org/10.1016/j.jwpe.2024.106744

Priyadarsini, S., Alok, &, Das, P., Journal, G., & Das, A. P. (2025). Isolation and Identification of Native Lithium-Resistant Bacteria from Electronic Waste Contaminated sites: A focused Investigation on Tolerance Assessment with Potential Lithium Bio-recovery Application. *Geomicrobiology Journal.* https://doi.org/10.1080/01490451.2025.2539509

Priyadarsini, S., & Das, A. P. (2026). Lithium in waste water: Impacts on pollution and human health. *Decarbonization of Wastewater Pollutants as a Sustainable Solution,* 353–364. https://doi.org/10.1016/B978-0-443-33062-9.00010-2

Reddy Koteswara, G., Bhargavi, S., Yamuna, D., Anjani Devi, M., Jeevana Lakshmi Kumar, G., Manas, T., Adarsh, S., Bhavya Sai, V., Nath, H., Alreshidi, M. A., Togun, H., Ram, D. K., & Das, A. P. (2025). Recent advancements in the remediation of marine hydrocarbon and plastic pollutants. In *Environmental science and engineering, part F218,* (pp. 127–142). https://doi.org/10.1007/978-3-031-81615-4_7

Sadasivuni, K. K., Rattan, S., Waseem, S., Brahme, S. K., Kondawar, S. B., Ghosh, S., Das, A. P., Chakraborty, P. K., Adhikari, J., Saha, P., & Mazumdar, P. (2019). Silver nanoparticles and its polymer nanocomposites—synthesis, optimization, biomedical usage, and its various applications. In *Lecture notes in bioengineering,* (pp. 331–373). https://doi.org/10.1007/978-3-030-04741-2_11

Sadik: Nanotechnology applications for clean water–Google Scholar. (n.d.). Retrieved January 29, 2025, from https://scholar.google.com/scholar_lookup?title=Nanotechnology%20Applications%20for%20Clean%20Water&author=M.A.%20Omole&author=I.%20K%E2%80%99Owino&author=O.A.%20Sadik&publication_year=2009

Sanket, A. S., Ghosh, S., Sahoo, R., Nayak, S., & Das, A. P. (2016). Molecular Identification of Acidophilic Manganese (Mn)-Solubilizing Bacteria from Mining Effluents and Their Application in Mineral Beneficiation. *Geomicrobiological Journal, 34*(1), 71–80. https://doi.org/10.1080/01490451.2016.1141340

Tripathy, B., Singh, S., Behera, I. D., Mishra, S., & Das, A. P. (2026). Metagenomic profiling of diversified marine microbiome across microplastic-contaminated niches of Bay of Bengal, India. *Marine Pollution Bulletin, 222,* 118872. https://doi.org/10.1016/J.MARPOLBUL.2025.118872

Innovative Nanotechnology Applications for Sustainable Waste Management and Recycling

Vivek Raj, Santosh Kumar Karn, Moni Kumari, Rakesh Kumar, Vandana, Aman Prakash, Arun Kumar, and Arun Lal Srivastav

Abstract Nanotechnology provides transformative solutions in waste management and recycling, presenting new approaches for pollutant remediation, resource recovery, and sustainable waste processing. The sole properties of nanomaterials, including nano-adsorbents, nanocatalysts, and magnetic nanoparticles, facilitate the elimination of heavy metals, organic contaminants, and other hazardous substances from water and soil, owing to their high surface area, reactivity, and selectivity. Additionally, nanotechnology supports efficient recycling processes by enabling the recuperation of valuable materials from electronic and industrial waste, promoting a circular economy, and reducing dependency on landfills. Advanced applications, such as nanostructured membranes, filters, and sensors, offer precise contaminant separation, detection, and removal while minimizing energy use and operational costs, enhancing waste treatment efficiency. Innovations in bio-nanotechnology and

V. Raj (✉)
Department of Biotechnology, Patna Womens College (Autonomus) Patna University, Patna, Bihar, India
e-mail: dr.vivekraj07bt@gmail.com

S. K. Karn · M. Kumari
Department of Biochemistry and Biotechnology, Sardar Bhagwan Singh University, Balawala, Dehradun, Uttarakhand, India

R. Kumar
School of Pharmaceutical Sciences, Lovely Professional University, Phagwara, Punjab, India

Vandana
Indus University, Rancharda, Thaltej, Ahmedabad, Gujarat, India

A. Prakash
Department of Marine Biology, Leon H. Charney School of Marine Sciences, University of Haifa, Haifa, Israel

A. Kumar
Mahavir Cancer Sansthan & Research Centre, Patna, Bihar, India

A. L. Srivastav
Chitkara University School of Engineering and Technology, Chitkara University, Baddi, Himachal Pradesh, India

E. Priyadarshini et al. (eds.), *Advanced Nanotechnology Applications for Environmental Sustainability*, Sustainable Environmental Waste Management Strategies,
https://doi.org/10.1007/978-3-032-12933-8_15

nanocomposites further broaden sustainable recycling methods, offering effective alternatives to conventional approaches. However, issues regarding potential toxicity, impacts on environment, and regulatory challenges need to be addressed carefully to guarantee the safe integration of nanotechnology in waste management. This chapter explores nanotechnology's latest applications, emerging challenges, and future prospects in waste management, highlighting its role in achieving efficient, sustainable practices that contribute to global environmental conservation.

Keywords Nanotechnology · Waste management · Recycling · Nanomaterials · Environmental sustainability

1 Introduction

Water is a fundamental resource that underpins the survival and development of all living organisms, surpassing the importance of other necessities. The Earth, often called the "blue planet," derives its name from the fact that approximately 70% of its surface is made up of water. However, the majority of this water, 97.5%, is saline, leaving only 2.5% classified as freshwater. Of this limited freshwater supply, approximately 68.9% exists in the form of glaciers, ice caps, and permanent snow, while 30.8% is found as groundwater. Alarmingly, a mere 0.3% of this freshwater is readily obtainable for human use (Jain et al., 2021). Access to clean and safe water is essential for fostering both societal well-being and economic growth. Unfortunately, the ever-increasing global population, rapid industrialization, urbanization, and intensive agricultural practices have significantly contributed to the degradation of water quality, posing a serious threat to global water security (Singha & Mishrab, 2020). Currently, around 1.2 billion individuals have inaccessible to safe drinking water, while 2.6 billion people face challenges in meeting basic sanitation needs. The consequences of this water crisis are dire, with millions of people, particularly children, succumbing to diseases caused by unsafe and contaminated water each year. Diarrheal diseases, primarily linked to the consumption of polluted water, claim the lives of approximately 1.8 million children annually (Kumar et al., 2014). To address the growing challenge of wastewater contamination, various physicochemical and biological treatment methods have been employed. Conventional techniques include coagulation, flocculation, precipitation, adsorption, ion exchange, electrodialysis, membrane separation, and aerobic, anoxic, or anaerobic oxidation processes. These methods often rely on chemical agents such as chlorine compounds, ammonia, permanganate, alum, sodium hydroxide, hydrochloric acid, ozone, and ferric salts, along with filtration aids and ion exchange resins. However, these approaches are energy-intensive, operationally complex, and require significant engineering expertise and infrastructure. Moreover, conventional methods are often inadequate in fully removing pollutants such as toxins, nitrogen, phosphorus, and heavy metals from adulterated wastewater. While these methods reduce pollutant levels to some extent, their limitations high costs, time consumption, and partial effectiveness, highlight the urgent need for more efficient and sustainable solutions (Kumar et al., 2014).

In recent years, nanotechnology has emerged as a protagonists alternative in wastewater treatment, offering innovative solutions to address the shortcomings of traditional methods. Nanotechnology involves manipulating matter at the molecular and atomic scales to create structures, devices, and systems with superior properties, including enhanced electronic, optical, magnetic, conductive, and mechanical characteristics (Jain, 2019). This cutting-edge field has demonstrated significant achievements across various disciplines, including environmental remediation. In the area of wastewater treatment, nanostructures provide unique advantages, such as their small size, having larger surface area, and simplicity of functionalization. These properties enable the development of highly effective catalysts and redox-active media for water purification. Nanomaterials have been proven to efficiently remove a wide range of pollutants from wastewater, like heavy metals, organic and inorganic solvents, dyes, biological toxins, and pathogenic microorganisms responsible for diseases such as cholera and typhoid. Their ability to target and eliminate specific contaminants makes them a powerful tool for addressing the limitations of conventional treatment methods. By leveraging the potential of nanotechnology, it is possible to develop more efficient, cost-effective, and sustainable strategies for guaranteeing access to clean and safe water. This technological innovation not only holds promise for addressing current water challenges but also offers a pathway toward achieving long-term global water security (Kumar et al., 2014).

2 Wastewater: Sources and Composition

Wastewater comprises a variety of hazardous and dangerous substances originating from a variety sources, including domestic sewage, industrial and commercial effluents, and agricultural runoff. These sources can be characterized based on their physical appearance, chemical composition, and the abundance of microorganisms present (Ritter et al., 2002). Broadly, wastewater is defined as any water contaminated through human activities, resulting from everyday living processes or industrial operations. The primary contributors to wastewater include domestic households, agricultural practices, industrial facilities, and commercial establishments. While industries, in particular, demand high-quality water for their operations, they also generate significant volumes of adulterated and polluted wastewater, which is frequently discharged into major water bodies, exacerbating pollution levels. Wastewater is a complex matrix, consisting predominantly of water (99.9%), with the remaining 0.1% comprising a wide range of contaminants. This 0.1% includes suspended solids (ranging from 350–1200 mg/L), organic compounds such as fecal matter, food residues, and toilet paper, as well as dissolved biodegradable organics like proteins, carbohydrates, and lipids. It also contains inorganic solids, including sediments, salts, and metals, as well as particulate matter with a COD (chemical oxygen demand) of 250–1000 mg/L. Furthermore, wastewater harbours numerous

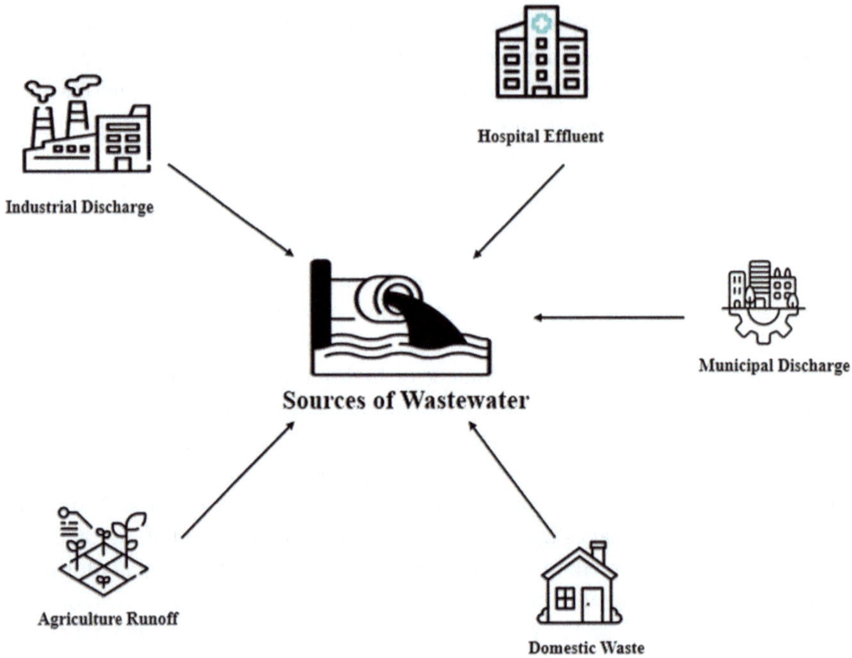

Fig. 1 Different sources of water pollution

microorganisms (up to 10^9 per mL), heavy metals, micropollutants, and nutrients (Jain et al., 2021; Warwick et al., 2013). Of particular concern is the presence of phosphate compounds, with approximately 63% of these compounds existing in soluble forms within wastewater (Warwick et al., 2013).

The composition and complexity of wastewater underscore the environmental challenges posed by its management and treatment. Each source of wastewater contributes a unique mix of contaminants, requiring tailored treatment approaches to minimize its impact on water bodies and ecosystems. Figures 1 and 2 visually illustrate the primary sources of wastewater and its typical composition, emphasizing the need for innovative and effective strategies to address this pressing environmental issue.

3 Common Steps in Wastewater Treatment

In recent years, the rapid growth of industrialization and population has led to a corresponding increase in wastewater generation, raising significant environmental and ecological concerns. Effective wastewater treatment has thus become imperative to enable water reuse or its safe discharge back into the environment (Ahmadi et al., 2020). Wastewater treatment refers to the process of removing pollutants

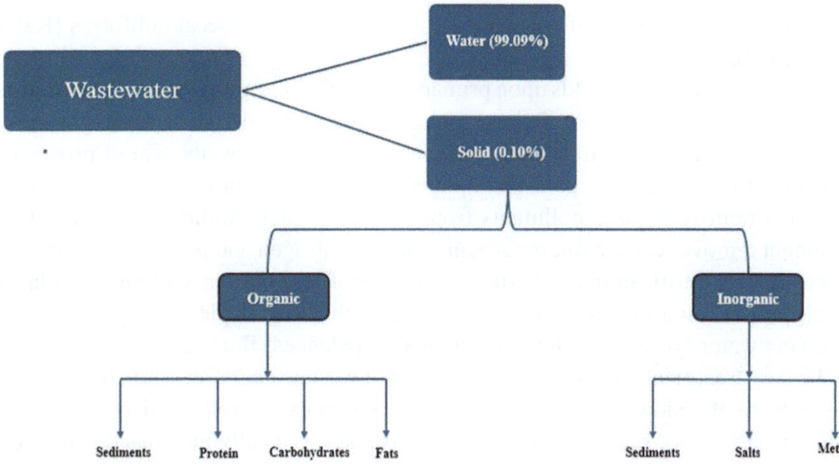

Fig. 2 Typical composition of wastewater

or contaminants from wastewater through physical or chemical methods before releasing it into the ecosystem. The primary objectives of this process are not only to eliminate contaminants to permissible levels but also to recover water and valuable micronutrients, thereby reducing threats to the environment and public health (Aboelfetoh et al., 2021). Biological wastewater treatment of domestic effluents is primarily categorized into two aerobic processes: suspended growth and fixed-film processes. Among suspended growth systems, activated sludge is the most extensively employed, whether in its conventional configuration or in modified forms such as oxidation ditches or sequential batch reactors. Anaerobic treatment, on the other hand, relies on anaerobic bacteria to convert organic matter in wastewater into biogas, primarily composed of methane and carbon dioxide. This method is particularly suited for wastewater with high concentrations of dissolved organic matter (Deshpande et al., 2020). The number of treatment stages required depends largely on the extent of pollutant removal and the specific mechanisms involved (Runguphan & Kitpichai., 2020).

The treatment process begins with preliminary treatment, which takes out the big or heavy debris. This phase is typically divided into two steps: screening and grit removal. Screening eliminates floating debris such as rags (~60% of debris), paper (~25%), and plastics (~5%) using mechanical screens. Grit removal follows, targeting inorganic substances like gravel, sand, and heavy particulates (e.g., bone fragments, coffee grounds, and corn kernels) through sedimentation in grit channels (Jain et al., 2021). After preliminary treatment, the wastewater undergoes primary treatment, where suspended solids are separated through sedimentation. In this stage, the wastewater is retained in sedimentation tanks to allow suspended solids to either settle at the bottom as sludge or rise to the surface as scum, which is subsequently skimmed off. Primary treatment removes near about 40% of biological oxygen demand (BOD), 80–90% of suspended solids, and 55% of faecal coliforms (Kalfa et al., 2020).

Secondary treatment builds upon primary treatment by employing activated sludge systems, such as oxidation ditches, trickling filters, biofilters, or turning around biological contactors, to treat organic-rich municipal wastewater. These processes rely on millions of microorganisms, including bacteria and protozoa, which metabolize and remove organic pollutants from the wastewater. Additionally, secondary treatment removes certain micronutrients, such as nitrogen and phosphorus, through processes like nitrification and luxury cell uptake. These nutrients, often assimilated by algae and fungi during eutrophication, can otherwise deplete oxygen levels in receiving water bodies, leading to ecological imbalances (Bora & Dutta, 2014).

Tertiary treatment represents the final stage of wastewater processing, focusing on the removal of residual organic and inorganic materials, as well as microorganisms, from the effluent of secondary treatment. This phase typically includes disinfection methods such as chlorination (using chlorine, chlorine dioxide, sodium hypochlorite, or chloramines), ultraviolet (UV) radiation, or ozone treatment. These measures ensure that treated wastewater is safe for environmental discharge (Bora & Dutta, 2014; Kalfa et al., 2020). It is essential to note that the aforementioned stages, preliminary, primary, secondary, and tertiary treatment, represent the traditional wastewater treatment framework. While these processes effectively address many contaminants, advanced techniques, particularly those leveraging nanotechnology, offer significant potential for improved efficiency and broader pollutant removal capabilities. These advanced methodologies will be discussed in detail in subsequent sections.

4 Nanotechnology in Wastewater Management

Numerous kinds of nanomaterials have been identified for their potential application in wastewater treatment. These include polymeric nanoparticles (NPs), metal-based nanoparticles, carbon-derived nanomaterials, zeolites, self-assembled monolayers on mesoporous supports (SAMMS), biopolymers, and various others (Baruah et al., 2019). Nanotechnology offers innovative pathways for wastewater remediation through several mechanisms, like as adsorption and biosorption, nanofiltration, photocatalysis, disinfection and pathogen control, as well as sensing and monitoring, among other advanced techniques. These approaches have demonstrated significant promise in addressing water contamination challenges effectively and efficiently.

4.1 Adsorption and Biosorption

Adsorption is an heat releasing surface phenomenon involving the transfer of molecules or ions, known as the adsorbate, from a liquid or gaseous phase onto the surface of a solid, known as the adsorbent. This process forms a monomolecular layer on the adsorbent's surface through physicochemical or chemical interactions under specific conditions (Gusain et al., 2020). A specialized form of adsorption, biosorption, utilizes biological materials such as bacteria, algae, or fungi as adsorbents due to their inherent capacity to bind and accumulate heavy metals, even from highly diluted aqueous solutions. Biosorption can occur via metabolically mediated processes requiring ATP or spontaneous physicochemical mechanisms without ATP expenditure. Key mechanisms of biosorption include microprecipitation, ion exchange, and cell-surface complexation (Derco & Vrana, 2018). Yang et al. studied the biosorption of hexavalent chromium (Cr (VI)) from synthetic wastewater using algal–bacterial aerobic granular sludge. Their results highlighted the pH dependence of Cr (VI) biosorption, with a maximum biosorption capacity of 51.0 mg/g observed at pH 2. The removal process was primarily driven by biosorption, complemented by bioreduction. Furthermore, desorption using $NaHCO_3$ recovered 64–73% of the adsorbed chromium, primarily in the form of Cr (III). Compared to conventional bacterial aerobic granular sludge, thc algal–bacterial variant exhibited superior biosorption capacity and enhanced granular stability (Yang et al., 2020).

4.1.1 Carbon-Based Nano-Adsorbents

Carbon nanostructures, encompassing various low-dimensional allotropes such as activated carbon, carbon nanotubes (CNTs), buckyballs (C60), graphite, and graphene, have been extensively studied due to their promising applications in wastewater treatment. Their widespread usage stems from their abundant availability, cost-effectiveness, chemical and thermal stability, high surface area, and environmentally friendly nature (Sayed et al., 2020). Among these, activated carbon, with its high porosity and large surface area, has long been a common adsorbent. However, its high cost has spurred the exploration of alternative carbon allotropes and functionalized carbon nanomaterials.

CNTs, cylindrical molecules formed by rolling single or multiple graphene sheets, exhibit exceptional mechanical flexibility, large pore volumes, and tunable surface chemistry. These properties make them highly effective in wastewater remediation (Verma & Balomajumder, 2020). For instance, Yadav and Srivastava demonstrated that CNTs efficiently adsorb Mn^{7+} ions, reducing concentrations from 150 to 3 ppm in synthetic wastewater (Yadav & Srivastava, 2017).

Graphene, with its two-dimensional structure of sp^2 hybridized carbon atoms, also excels as a nano adsorbent due to its high surface area, porosity, and chemical stability. Functionalized graphene derivatives, such as graphene oxide (GO), have gained attention for their ability to remove pollutants like dyes and heavy metals from wastewater. Chen et al. demonstrated the synergy of GO and aminated lignin in enhancing dye adsorption efficiency (Chen et al., 2020). Other advancements include GO functionalized with thiosemicarbazone for methylene blue removal (Bu et al., 2020) and GO hydrogels integrated with silver nanoparticles and porphyrin complexes for efficient dye adsorption (Zheng et al., 2020). Recent innovations, like

GO-chitosan hydro composites, highlight the continued development of graphene-based materials for sustainable water treatment (Sirajudheen et al., 2020).

4.1.2 Metal-Based Nano Adsorbents

Metal-based nanomaterials are increasingly attracting attention as highly effective adsorbents for wastewater treatment. Nanometals and their oxides, including Fe_3O_4, TiO_2, MnO_2, MgO, ZnO, and CdO, have been largely utilised to remove heavy metals, ions, and dyes from wastewater due to their small size, high surface area, and exceptional adsorption capabilities (Sadegh et al., 2017). Their performance surpasses traditional adsorbents like activated carbon, particularly in removing heavy and radioactive metals, as the sorption process involves strong complexation between dissolved metals and the oxygen in metal oxides (El-Sayed, 2020).

4.1.3 Polymer-Based Nanoadsorbents

Polymer nanocomposites have garnered prominent attention in current years for their capacity to enhance environmental sustainability and improve wastewater treatment processes. These materials offer advantages such as high surface area, rapid decontamination, remarkable stability, cost-effectiveness, and selectivity for removing diverse pollutants. Common polymeric adsorbents include polysaccharides like chitosan (CS) and cyclodextrin, nano-magnetic polymers, covalent organic polymers, and extracellular polymeric substances (Alaba et al., 2018). Nanocellulosics, derived from cellulose, are nontoxic, highly adsorptive, and easily modified, making them a prominent candidate for wastewater remediation (Abdi & Abedini, 2020). Lignin-derived nanomaterials have recently emerged, demonstrating efficacy in the catalytic degradation of dyes, nitroarenes, and removal of heavy metal (Chen et al., 2019).

4.1.4 Zeolites

Zeolites are three-dimensional, crystalline, microporous materials characterized by a well-defined framework of voids and conduits composed of aluminum, silicon, and oxygen. Their molecular-sized pores facilitate accessibility and enable high selectivity for adsorbing various pollutants (Ali et al., 2020). Zeolites occur naturally as silicate minerals but can also be synthesized into magnetically modified zeolites and bio-zeolites, among other forms (Nassar et al., 2017). Due to their stability in aqueous environments, cost-effectiveness, high surface area, and compatibility with the natural environment, zeolitic nanomaterials have gained considerable attention in wastewater treatment and other environmental applications (Margeta et al., 2013). Zeolites' ability to regenerate and reuse further enhances their appeal for industrial and environmental use (Mohseni-Bandpi et al., 2016).

4.2 Nanofilters

Water filtration involves the removal or reduction of particulate matter, microorganisms, and harmful biological or chemical contaminants from polluted water, making it safe for applications such as drinking, pharmaceuticals, and medical use (Koyuncu et al., 2015). Among recent advancements, nanofiltration (NF) membranes have become a prominent candidate due to their molecular weight cut-off (MWCO) in the nanometer range, enabling efficient filtration (Charcosset, 2012). NF membranes are a pressure-driven filtration technology operating between ultrafiltration (UF) and reverse osmosis (RO), with pressures ranging from 5 to 20 bars and pore sizes between 0.5 and 2.0 nm (Mulyanti & Susanto, 2018). These membranes demonstrate high rejection of divalent and multivalent ions, low rejection of monovalent ions, high flux, and lower energy consumption compared to RO, making them ideal for drinking water and wastewater treatment (Nagy, 2019). NF membranes are highly versatile and cost-effective, capable of removing organic and inorganic contaminants, including harmful microbes, from wastewater (Mänttäri et al., 2013; Shon et al., 2013). Two primary types of NF membranes are widely used: polymeric and ceramic. While polymeric membranes are cost-effective, they suffer from short lifespans due to high fouling rates and poor chemical resistance. Conversely, ceramic membranes exhibit superior mechanical, chemical, and thermal stability, enhancing their durability (Mostafavi et al., 2009; Parham et al., 2013).

4.3 Photocatalysis

Photocatalysis, derived from the Greek terms "photo" (light) and "catalysis" (a substance that alters the rate of a chemical reaction without being used), is a process where a solid material, the photocatalyst, absorbs light and accelerates a chemical reaction (Bethi et al., 2016). As a subset of Advanced Oxidation Processes (AOPs), photocatalysis produces powerful oxidants, such as hydroxyl radicals (·OH), which are capable of breaking down complex organic compounds (Rueda-Marquez et al., 2020). These processes have been widely studied for their effectiveness in removing Contaminants of Emerging Concern (CECs) from wastewater, including organic materials, pharmaceuticals, pesticides, heavy metals (e.g., mercury), and harmful microbes, including chlorine-resistant pathogens (McLain, 2018). Nanomaterials, due to their superior physical properties and quantum effects, have emerged as highly effective photocatalysts, with significant research interest focusing on their application in wastewater treatment (Hot et al., 2017). Photocatalysis operates through five key stages: (1) diffusion of pollutants to the photocatalyst surface, (2) adsorption of pollutants onto the surface, (3) photochemical reaction of adsorbed pollutants, (4) desorption of reaction products, and (5) removal or diffusion of the products away from the interface (Colmenares et al., 2009). Common nanostructured photocatalysts include titanium dioxide (TiO_2), iron oxide (Fe_2O_3), zinc oxide (ZnO), zirconium dioxide (ZrO_2), zinc sulfide (ZnS), tungsten trioxide (WO_3), and cadmium sulfide (CdS) (Mahmoudian-Boroujerd et al., 2019).

4.4 *Disinfection and Pathological Control*

The World Health Organization (WHO) reports that 80% of diseases in developing countries are associated with the consumption of the adulterated water constituted by various pathogens such as bacteria, viruses, prions, fungi, protozoa, and rickettsia. This contamination poses a significant health risk and causes millions of deaths annually worldwide (Yao et al., 2020). Disinfection, the process of reducing microbial counts to acceptable levels through physical or chemical means, is essential for protecting public health (Zhao et al., 2020). Conventional disinfection methods in wastewater treatment, including the use of free chlorine, hypochlorite, chlorine dioxide, chloramines, ozone, reverse osmosis, and peracetic acid, have been widely adopted (Zhang et al., 2020). However, these methods are often limited as they are not energy efficient, the need for costly equipment, and the formation of disinfection by-products (DBPs), which raise additional concerns (Zhang et al., 2016). Therefore, there is an urgent need for more efficient, sustainable, low-cost, and less time-consuming disinfection technologies (Xiu et al., 2012). Nanomaterials, with their exceptional surface area and specific reactivity, offer promising solutions for pathogen inactivation in water, surpassing the capabilities of conventional methods (Ivanova et al., 2013). The inactivation mechanisms of these nanomaterials include electrostatic interactions on their surface, photochemical reactions that generate reactive oxygen species (ROS), and the disruption of microbial cell walls, all of which contribute to effective disinfection (Moustafa, 2017). The unique properties of nanomaterials enable the easy inhibition of pathogens in water (Ahamed et al., 2014). Nanomaterials such as silver (Ag), copper oxide (CuO), zinc oxide (ZnO), titanium dioxide (TiO_2), polymeric nanoparticles (NPs), and carbon nanotubes (CNTs) have been explored for wastewater disinfection (Li et al., 2008).

Silver (Ag) nanoparticles, known for their antimicrobial properties for over a thousand years, remain one of the most widely used nanomaterials for microbial disinfection (Das et al., 2017; Schmidt-Mende & MacManus-Driscoll, 2007). For instance, Ag@ZnO core–shell nanoparticles have been developed for disinfecting pathogenic bacteria, including *Escherichia coli* and *Staphylococcus aureus*. These nanoparticles showed complete elimination of both bacterial strains within 60–90 min of solar photocatalysis at 35 °C, with enhanced efficiency observed at higher temperatures (Wong et al., 2019). Zinc oxide (ZnO) nanoparticles are also noted for their biocompatibility and environmental friendliness. In one study, ZnO nanoparticles synthesized under various pH conditions displayed significant antibacterial activity against *E. coli* (Gul et al., 2019). Additionally, ZnO nanocrystals doped on three-dimensional macro-mesoporous silicon (Si) wafers exhibited even stronger antibacterial effects, highlighting the potential of ZnO in water disinfection (He et al., 2019). Copper

(Cu) nanoparticles, particularly copper oxide (CuO), are among the most effective antimicrobial agents.

4.5 Sensing and Monitoring

Environmental contamination arises from diverse sources, including and non-living substances such as pathogenic microorganisms, municipal and industrial waste, sewage discharge, animal excrement, and heavy metals (Bhattacharyya et al., 2017). Monitoring water quality on both large and small scales poses significant challenges due to the low concentrations of contaminants, as well as the complexity and variability of wastewater matrices (Farahi et al., 2012). To address these challenges, the development of fast and efficient detection techniques is essential.

Recent advancements have focused on nanomaterial-based sensors for water quality monitoring. These nanosensors offer exceptional properties, such as high sensitivity for trace contaminant detection and rapid analysis (Aboelfetoh et al., 2021). Nanosensors can be defined as devices or materials capable of detecting changes in environmental stimuli, including heat, chemical, and mechanical stress, as well as changes in concentration, gravitational and magnetic forces, and electrical properties. They convey physical, chemical, or biological information from the nanoscale level to the macroscopic level (Kurbanoglu & Ozkan, 2018). Structurally, nanosensors consist of three main components: a recognition element (e.g., nanometals, nanotubes, nanowires, or nanoparticles), a transducer (e.g., voltammetric, amperometric, conductometric, or spectrophotometric), and a display for real-time monitoring (Talari et al., 2021). One pressing issue in wastewater is the rise of antibiotic-resistant pathogens and their associated antibiotic resistance genes (ARGs), which pose a severe threat to public health. For instance, methicillin-resistant *Staphylococcus aureus* (MRSA) harbors the *mecA* gene, which confers antibiotic resistance. Riquelme et al. developed gold nanosensors functionalized with oligonucleotides to monitor the *mecA* ARG in environmental samples. These nanosensors demonstrated high selectivity for the *mecA* gene in wastewater effluent, with a detection limit of 70 ppm (Atar et al., 2015).

Organophosphorus (OP) compounds, widely used as pesticides, have become prevalent in agricultural runoff and industrial wastewater. Due to their neurotoxicity, even trace amounts of these compounds in water are harmful. To detect diazinon (an OP pesticide) in tap water, agricultural runoff, and rivers, Talari et al. developed aptamer-based optical nanosensors using reduced graphene oxide quantum dots (rGQDs) and multi-walled carbon nanotubes (MWCNTs). The nanosensors exhibited high selectivity and accuracy for diazinon detection (Hu et al., 2018). Another contaminant of concern is triclosan (TCS), an antibacterial and antifungal agent commonly found in household cleaning products. Its widespread release into wastewater and natural water sources leads to chronic toxicity in aquatic organisms and potential health risks to humans.

Pharmaceutical and personal care products (PPCPs), such as caffeine, are another category of contaminants. Caffeine is released into the environment through pharmaceutical wastewater, beverages like tea and coffee, and energy drinks. Chromium (Cr) contamination is another significant environmental issue due to its extensive use in steel manufacturing, painting, leather tanning, and as a catalyst. While trivalent chromium [Cr (III)] is a trace nutrient, hexavalent chromium [Cr (VI)] is highly toxic, with mutagenic and carcinogenic effects.

5 Barriers and Risks Associated with Nanotechnology

Despite the promising potential of nanotechnology in wastewater treatment, several critical challenges hinder its widespread application and practical implementation. Key obstacles include concerns over the toxicity of nanomaterials, issues related to cost-effectiveness, and the degree of social acceptance. Furthermore, the associated risks encompass the transformation of nanomaterials in the environment, the ecotoxicological impacts of engineered nanomaterials, and their potential contribution to water pollution (Pardhi et al., 2018).

5.1 Nanomaterial Toxicity

Historically, various wastewater treatment methods have led to unintended consequences. For example, chlorination, a widely used conventional treatment technique initially believed to significantly enhance global life expectancy, was later found to produce carcinogenic byproducts such as N-nitrosodimethylamine and trihalomethanes (Raza et al., 2016). Similarly, nanotechnology, despite its potential benefits in wastewater treatment, presents challenges due to the inherent properties of nanomaterials that contribute to their toxicity. The molecular structure of nanomaterials, which dictates toxicity endpoints, and their size, which influences cellular uptake, are key determinants of their toxic effects. Due to their nanoscale dimensions, nanoparticles (NPs) can penetrate epithelial and endothelial barriers, allowing them to enter the lymphatic and circulatory systems and reach various organs, including the brain, liver, kidneys, spleen, and bone marrow (Sukhanova et al., 2018).

The toxicity of nanomaterials has been observed to vary based on size and shape. For instance, NPs ranging from 1 to 100 nm, similar in size to protein globules (2–10 nm) and DNA helices (2 nm), can easily enter cells and organelles. A demonstrated that gold NPs smaller than 6 nm can effectively enter the cell nucleus, while larger NPs (10–16 nm) are restricted to the cytoplasm (Huo et al., 2014). Similarly, titanium dioxide (TiO_2) NPs are known to cause conformational changes in tubulin, disrupting intracellular transport, cell division, and migration. The shape of NPs also influences their toxicity; for instance, spherical NPs are more readily internalized through endocytosis than nanotubes or nanofibers. Single-walled carbon nanotubes

(SWCNTs) have shown a greater capacity to block calcium channels compared to spherical NPs. Moreover, hydroxyapatite NPs with plate-like and needle-like shapes have been found to induce higher cytotoxicity in non-tumorigenic lung epithelial cells (BEAS-2B) than their spherical or rod-like counterparts (Bonelli et al., 2020). The mechanisms by which NPs induce cellular damage vary and require further investigation. Although extensive toxicological data exist for bulk materials and their shared constituents with nanomaterials, certain nanomaterials, such as fullerenes and carbon nanotubes (CNTs), lack bulk counterparts, complicating risk assessments. This underscores the necessity for more comprehensive toxicity studies and risk evaluations. A robust risk assessment framework should encompass all stages of the nanomaterial life cycle to ensure their safe and sustainable application in wastewater treatment.

5.2 Cost Effectiveness

The effectiveness and cost-efficiency of nanotechnology for wastewater treatment significantly influence its adoption. In developed countries, advanced technologies are utilized to address a broad range of pollutants, while in developing nations, wastewater treatment often focuses on basic necessities, such as disinfection. In both scenarios, there is an increasing need to address complex contaminant mixtures to achieve higher water quality at reduced costs, thereby challenging existing treatment models. Although cost remains a critical barrier, it is not insurmountable. A significant portion of the expense associated with nanomaterials stems from separation and purification processes. Research-grade nanomaterials with high purity and consistent properties have maintained relatively stable prices over the past two decades. These prices are unlikely to decrease substantially unless there is increased demand and large-scale production (Aboelfetoh et al., 2021; Pan et al., 2020). However, the economic feasibility of employing nanomaterials in wastewater treatment can be enhanced by using materials with lower purity. For instance, replacing ultrapure C60 with fullerene soot to create amino-fullerene photocatalysts has been shown to achieve a similar level of effectiveness while reducing costs by approximately 90% (Aboelfetoh et al., 2021; Pan et al., 2020). Furthermore, cost-efficiency can be improved through the long-term reusability of nanomaterials. Examples include photocatalysts that maintain their activity through regeneration of nanoadsorbents and magnetically separable multifunctional nanomaterials, which allow for multiple reuse cycles, thus enhancing their practicality and reducing overall costs (Kokkinos et al., 2020).

5.3 Nanomaterial Transformation Risk in Water

The interaction dynamics between biotic systems, nanomaterials, and abiotic factors, as well as the dispersibility and solubility of nanomaterials, play a crucial role in determining their environmental fate. Nanoparticles (NPs) typically exhibit slower settling rates compared to larger particles. However, due to their substantial surface area, they can adsorb more sediment particles and soil. Additionally, their high water insolubility, such as in carbon nanotubes (CNTs) and fullerenes, facilitates their removal through water column filtration processes (Westerhoff et al., 2013). Engineered nanomaterials are often designed for specific applications. Pristine-engineered nanomaterials, for example, are susceptible to transformation into various forms, including product-modified, product-weathered, and environmentally transferred engineered nanomaterials. External factors such as light can induce photochemical transformations, and oxidation–reduction reactions may also occur under certain conditions. These transformations significantly influence the interactions between nanomaterials and their environment, ultimately affecting the adsorption and desorption behavior of contaminants in water systems (Westerhoff et al., 2013).

5.4 Ecotoxicity Associated with Nanomaterials

Ecotoxicity refers to the potential of chemical, physical, or biological stressors to disrupt ecosystems. Nanomaterials pose a significant risk of leaching into treated water during their production, use, and disposal. This risk is exacerbated by the increasing annual production rates of emerging metal nanoparticles (NPs), such as silver (Ag) and titanium dioxide (TiO_2) (Robichaud et al., 2009). These emerging nanomaterials are of particular concern for aquatic environments, as demonstrated in a study investigating the inhibitory effects of Ag and TiO_2 nanoparticles on the growth of duckweed, an aquatic plant (Kim et al., 2011). Nanoparticle concentrations in natural surface waters have been estimated to range from nano- to micrograms per liter, with these concentrations expected to rise alongside increased NP production. Metal nanoparticles, including Ag, TiO_2, and zinc oxide (ZnO), are widely used for their antibacterial properties to purify water, particularly in developing countries where they often replace chemical disinfectants (Abramenko et al., 2018). Ceramic filters impregnated with Ag nanoparticles, for instance, have been employed for water purification. Although water filtered through nano-Ag-coated filter papers contained detectable levels of Ag nanoparticles, these levels were well below the limits established by the World Health Organization (WHO) guidelines (Albukhari et al., 2019). However, there is limited research on the toxicity of leached nanoparticles in treated water, particularly their effects on experimental animals. Some studies have reported damage to vital organs and DNA in rats exposed to TiO_2 nanoparticles in contaminated water. Additionally, engineered nanomaterials have been associated with various health risks, including pulmonary inflammation, genotoxicity, carcinogenicity, and circulatory effects (De Matteis & Rinaldi, 2018).

5.5 Water Pollution

Water pollution is a critical environmental issue, with numerous laboratories currently investigating the toxicological impacts of nanomaterial-contaminated water. The toxicological effects are further compounded by the dispersants used to stabilize nanomaterials in aqueous environments, as these materials often demonstrate poor solubility. For instance, tetrahydrofuran (THF), a commonly employed dispersant for carbon nanotubes (CNTs) and C60 fullerenes, has raised substantial concerns due to its inherent toxicity. The utilization of nanotechnology in wastewater treatment introduces both opportunities and challenges, necessitating systematic research into its potential biological and ecological impacts. Addressing these concerns is crucial to ensuring the safe and effective implementation of nanotechnology for environmental remediation. Initial studies have highlighted the remarkable efficacy of nanotechnology in treating wastewater, showcasing its potential to improve water quality significantly. However, the safe application of nanotechnology requires careful optimization of nanomaterials to mitigate their toxicological risks while enhancing their functionality (Das et al., 2018). Furthermore, the development of cost-effective strategies is imperative to facilitate large-scale applications, making such advanced technologies more accessible and feasible. Achieving this balance between efficacy, safety, and cost will determine the long-term success of nanotechnology in addressing water pollution. As research progresses, understanding the interaction between nanomaterials, dispersants, and aquatic ecosystems will provide invaluable insights into minimizing environmental and human health risks. The integration of interdisciplinary efforts, encompassing toxicology, materials science, and environmental engineering, is essential to advance the field and address the multifaceted challenges posed by nanomaterial-based wastewater treatment technologies.

6 Conclusions

The global demand for clean and safe water is surging due to rapid population growth, industrialization, urbanization, and intensive agricultural practices. Existing water purification techniques, while effective, are often chemical-intensive, energy-demanding, and reliant on sophisticated infrastructure and expertise, presenting significant challenges in implementation. In response, the development of innovative and efficient wastewater treatment technologies has become imperative. Among the emerging solutions, nanotechnology holds immense promise due to its unique potential for nanomaterial-based water treatment. Nanostructured materials exhibit remarkable properties, including high surface-to-volume ratios, enhanced sensitivity and reactivity, superior adsorption capacities, and ease of functionalization. These attributes enable them to address many limitations of traditional methods and facilitate advanced contaminant removal. However, the application of nanotechnology is not without risks. Due to their nanoscale dimensions, engineered nanomaterials can penetrate biological systems, potentially causing toxicity in humans and aquatic organisms. Factors such as pH, nanoparticle concentration, and exposure duration critically influence their toxicological effects. Concerns regarding ecotoxicity and human health risks highlight the need for comprehensive research into the behavior, interactions, and long-term impacts of nanomaterials in various environmental contexts. Despite these challenges, nanotechnology represents a transformative opportunity for wastewater management, offering potential breakthroughs in achieving high-efficiency, cost-effective, and scalable water treatment solutions. A concerted effort by scientists, policymakers, and industry stakeholders is essential to advance research, develop safe applications, and establish regulatory frameworks. With sustained commitment, nanotechnology could revolutionize wastewater treatment, ensuring access to clean water and contributing significantly to global sustainability goals. The future of this field hinges on innovative strategies that balance efficacy with safety and environmental stewardship.

References

Abdi, J., & Abedini, H. (2020). MOF-based polymeric nanocomposite beads as an efficient adsorbent for wastewater treatment in batch and continuous systems: Modelling and experiment. *Chemical Engineering Journal, 400*, 125862.

Aboelfetoh, E. F., Elabedien, M. E. Z., & Ebeid, E. Z. M. (2021). Effective treatment of industrial wastewater applying SBA-15 mesoporous silica modified with graphene oxide and hematite nanoparticles. *Journal of Environmental Chemical Engineering, 9*(1), 104817.

Abramenko, N. B., Demidova, T. B., Abkhalimov, EV., Ershov, B. G., Krysanov, E. Y., & Kustov, L. M. (2018). Ecotoxicity of different-shaped silver nanoparticles: Case of zebrafish embryos. *Journal of Hazardous Materials, 347*, 89–94.

Ahamed, M., Alhadlaq, H. A., Khan, M. M., Karuppiah, P., & Al-Dhabi, N. A. (2014). Synthesis, characterization, and antimicrobial activity of copper oxide nanoparticles. *Journal of Nanomaterials, 2014*(1), 637858.

Ahmadi, F., Zinatizadeh, A. A., Asadi, A., McKay, T., & Azizi, S. (2020). Simultaneous carbon and nutrients removal and PHA production in a novel single air lift bioreactor treating an industrial wastewater. *Environmental Technology & Innovation, 18*, 100776.

Alaba, P. A., Oladoja, N. A., Sani, Y. M., Ayodele, O. B., Mohammed, I. Y., Olupinla, S. F., & Daud, W. M. W. (2018). Insight into wastewater decontamination using polymeric adsorbents. *Journal of Environmental Chemical Engineering, 6*(2), 1651–1672.

Albukhari, S. M., Ismail, M., Akhtar, K., & Danish, E. Y. (2019). Catalytic reduction of nitrophenols and dyes using silver nanoparticles@ cellulose polymer paper for the resolution of waste water treatment challenges. *Colloids and Surfaces A: Physicochemical and Engineering Aspects, 577*, 548–561.

Ali, M. E., Hoque, M. E., Safdar Hossain, S. K., & Biswas, M. C. (2020). Nanoadsorbents for wastewater treatment: Next generation biotechnological solution. *International Journal of Environmental Science and Technology, 17*(9), 4095–4132.

Atar, N., Eren, T., Yola, M. L., & Wang, S. (2015). A sensitive molecular imprinted surface plasmon resonance nanosensor for selective determination of trace triclosan in wastewater. *Sensors and Actuators B: Chemical, 216*, 638–644.

Baruah, A., Chaudhary, V., Malik, R., & Tomer, V. K. (2019). Nanotechnology based solutions for wastewater treatment. In *Nanotechnology in water and wastewater treatment* (pp. 337–368). Elsevier.

Bethi, B., Sonawane, S. H., Bhanvase, B. A., & Gumfekar, S. P. (2016). Nanomaterials-based advanced oxidation processes for wastewater treatment: A review. *Chemical Engineering and Processing-Process Intensification, 109*, 178–189.

Bhattacharyya, S., Bennett, J., Short, L. C., Theisen, T. S., Wichman, M. D., White, J. C., & Wright, S. (2017). Nanotechnology in the Water Industry, Part 1. *Journal (American Water Works Association), 109*(11), 30–37.

Bonelli, B., Freyria, F. S., Rossetti, I., & Sethi, R. (Eds.). (2020). *Nanomaterials for the detection and removal of wastewater pollutants*. Elsevier.

Bora, T., & Dutta, J. (2014). Applications of nanotechnology in wastewater treatment—A review. *Journal of Nanoscience and Nanotechnology, 14*(1), 613–626.

Bu, J., Yuan, L., Zhang, N., Liu, D., Meng, Y., & Peng, X. (2020). High-efficiency adsorption of methylene blue dye from wastewater by a thiosemicarbazide functionalized graphene oxide composite. *Diamond and Related Materials, 101*, 107604.

Charcosset, C. (2012). *Membrane processes in biotechnology and pharmaceutics*. Elsevier.

Chen, B., Chen, S., Zhao, H., Liu, Y., Long, F., & Pan, X. (2019). A versatile β-cyclodextrin and polyethyleneimine bi-functionalized magnetic nanoadsorbent for simultaneous capture of methyl orange and Pb (II) from complex wastewater. *Chemosphere, 216*, 605–616.

Chen, H., Liu, T., Meng, Y., Cheng, Y., Lu, J., & Wang, H. (2020). Novel graphene oxide/aminated lignin aerogels for enhanced adsorption of malachite green in wastewater. *Colloids and Surfaces A: Physicochemical and Engineering Aspects, 603*, 125281.

Colmenares, J. C., Luque, R., Campelo, J. M., Colmenares, F., Karpiński, Z., & Romero, A. A. (2009). Nanostructured photocatalysts and their applications in the photocatalytic transformation of lignocellulosic biomass: An overview. *Materials, 2*(4), 2228–2258.

Das, R., Leo, B. F., & Murphy, F. (2018). The toxic truth about carbon nanotubes in water purification: A perspective view. *Nanoscale Research Letters, 13*, 1–10.

Das, S., Ranjana, N., Misra, A. J., Suar, M., Mishra, A., Tamhankar, A. J., Lundborg, C. S., & Tripathy, S. K. (2017). Disinfection of the water borne pathogens Escherichia coli and Staphylococcus aureus by solar photocatalysis using sonochemically synthesized reusable Ag@ ZnO core-shell nanoparticles. *International Journal of Environmental Research and Public Health, 14*(7), 747.

De Matteis, V., & Rinaldi, R. (2018). Toxicity assessment in the nanoparticle era. In *Cellular and molecular toxicology of nanoparticles* (pp.1–19).

Derco, J., & Vrana, B. (Eds.). (2018). *Biosorption*. BoD–Books on Demand

Deshpande, B. D., Agrawal, P. S., Yenkie, M. K. N., & Dhoble, S. J. (2020). Prospective of nanotechnology in degradation of waste water: A new challenges. *Nano-Structures & Nano-Objects, 22*, 100442.

El-Sayed, M. E. (2020). Nanoadsorbents for water and wastewater remediation. *Science of the Total Environment, 739*, 139903.

Farahi, R. H., Passian, A., Tetard, L., & Thundat, T. (2012). Critical issues in sensor science to aid food and water safety. *ACS Nano, 6*(6), 4548–4556.

Gul, S., Khan, S. A., Rehan, Z. A., Akhtar, K., Khan, M. A., Khan, M. I., Rashid, M. I., Asiri, A. M., & Khan, S. B. (2019). Antibacterial CuO-PES-CA nancomposite membranes supported Cu 0 nanoparticles for water permeability and reduction of organic pollutants. *Journal of Materials Science: Materials in Electronics, 30*, 10835–10847.

Gusain, R., Kumar, N., & Ray, S. S. (2020). Recent advances in carbon nanomaterial-based adsorbents for water purification. *Coordination Chemistry Reviews, 405*, 213111.

He, X., Yang, D. P., Zhang, X., Liu, M., Kang, Z., Lin, C., Jia, N., & Luque, R. (2019). Waste eggshell membrane-templated CuO-ZnO nanocomposites with enhanced adsorption, catalysis and antibacterial properties for water purification. *Chemical Engineering Journal, 369*, 621–633.

Hot, J., Topalov, J., Ringot, E., & Bertron, A. (2017). Investigation on parameters affecting the effectiveness of photocatalytic functional coatings to degrade NO: TiO2 amount on surface, illumination, and substrate roughness. *International Journal of Photoenergy, 2017*(1), 6241615.

Hu, R., Tang, R., Xu, J., & Lu, F. (2018). Chemical nanosensors based on molecularly-imprinted polymers doped with silver nanoparticles for the rapid detection of caffeine in wastewater. *Analytica Chimica Acta, 1034*, 176–183.

Huo, S., Jin, S., Ma, X., Xue, X., Yang, K., Kumar, A., Wang, P. C., Zhang, J., Hu, Z., & Liang, X. J. (2014). Ultrasmall gold nanoparticles as carriers for nucleus-based gene therapy due to size-dependent nuclear entry. *ACS Nano, 8*(6), 5852–5862.

Ivanova, E. P., Hasan, J., Webb, H. K., Gervinskas, G., Juodkazis, S., Truong, V. K., Wu, A. H., Lamb, R. N., Baulin, V. A., Watson, G. S., & Watson, J. A. (2013). Bactericidal activity of black silicon. *Nature Communications, 4*(1), 2838.

Jain, K. (2019). Nanohybrids of dendrimers and carbon nanotubes: A benefaction or forfeit in drug delivery? *Nanoscience & Nanotechnology-Asia, 9*(1), 21–29.

Jain, K., Patel, A. S., Pardhi, V. P., & Flora, S. J. S. (2021). Nanotechnology in wastewater management: A new paradigm towards wastewater treatment. *Molecules, 26*(6), 1797.

Kalfa, A., Shapira, B., Shopin, A., Cohen, I., Avraham, E., & Aurbach, D. (2020). Capacitive deionization for wastewater treatment: Opportunities and challenges. *Chemosphere, 241*, 125003.

Kim, E., Kim, S. H., Kim, H. C., Lee, S. G., Lee, S. J., & Jeong, S. W. (2011). Growth inhibition of aquatic plant caused by silver and titanium oxide nanoparticles. *Toxicology and Environmental Health Sciences, 3*, 1–6.

Kokkinos, P., Mantzavinos, D., & Venieri, D. (2020). Current trends in the application of nanomaterials for the removal of emerging micropollutants and pathogens from water. *Molecules, 25*(9), 2016.

Koyuncu, I., Sengur, R., Turken, T., Guclu, S., & Pasaoglu, M. E. (2015). Advances in water treatment by microfiltration, ultrafiltration, and nanofiltration. In *Advances in membrane technologies for water treatment* (pp. 83–128). Woodhead Publishing.

Kumar, S., Ahlawat, W., Bhanjana, G., Heydarifard, S., Nazhad, M. M., & Dilbaghi, N. (2014). Nanotechnology-based water treatment strategies. *Journal of Nanoscience and Nanotechnology, 14*(2), 1838–1858.

Kurbanoglu, S., & Ozkan, S. A. (2018). Electrochemical carbon based nanosensors: A promising tool in pharmaceutical and biomedical analysis. *Journal of Pharmaceutical and Biomedical Analysis, 147*, 439–457.

Li, Q., Mahendra, S., Lyon, D. Y., Brunet, L., Liga, M. V., Li, D., & Alvarez, P. J. (2008). Antimicrobial nanomaterials for water disinfection and microbial control: Potential applications and implications. *Water Research, 42*(18), 4591–4602.

Mahmoudian-Boroujerd, L., Karimi-Jashni, A., Hosseini, S. N., & Paryan, M. (2019). Optimization of rDNA degradation in recombinant Hepatitis B vaccine production plant wastewater using visible light excited Ag-doped TiO2 nanophotocatalyst. *Process Safety and Environmental Protection, 122*, 328–338.

Mänttäri, M., Van der Bruggen, B., & Nyström, M. (2013). Nanofiltration. In *Separation and purification technologies in biorefineries* (pp.233–258).

Margeta, K., Logar, N. Z., Šiljeg, M., & Farkaš, A. (2013). Natural zeolites in water treatment–how effective is their use. *Water Treatment, 5*, 81–112.

McLain, A. A. (2018). Photocatalytic properties of Zinc Oxide and graphene nanocomposites. In *Proceedings of the Wisconsin Space Conference*.

Mohseni-Bandpi, A., Al-Musawi, T. J., Ghahramani, E., Zarrabi, M., Mohebi, S., & Vahed, S. A. (2016). Improvement of zeolite adsorption capacity for cephalexin by coating with magnetic Fe3O4 nanoparticles. *Journal of Molecular Liquids, 218*, 615–624.

Mostafavi, S. T., Mehrnia, M. R., & Rashidi, A. M. (2009). Preparation of nanofilter from carbon nanotubes for application in virus removal from water. *Desalination, 238*(1–3), 271–280.

Moustafa, M. T. (2017). Removal of pathogenic bacteria from wastewater using silver nanoparticles synthesized by two fungal species. *Water Science, 31*(2), 164–176.

Mulyanti, R., & Susanto, H. (2018). Wastewater treatment by nanofiltration membranes. In *IOP conference series: Earth and environmental science* (Vol. 142, p. 012017). IOP Publishing. (Mar 2018).

Nagy, E. (2019). Nanofiltration. In *Basic equations of mass transport through a membrane layer* (pp. 417–428).

Nassar, M. Y., Abdelrahman, E. A., Aly, A. A., & Mohamed, T. Y. (2017). A facile synthesis of mordenite zeolite nanostructures for efficient bleaching of crude soybean oil and removal of methylene blue dye from aqueous media. *Journal of Molecular Liquids, 248*, 302–313.

Pan, Y., Liu, X., Zhang, W., Liu, Z., Zeng, G., Shao, B., Liang, Q., He, Q., Yuan, X., Huang, D., & Chen, M. (2020). Advances in photocatalysis based on fullerene C60 and its derivatives: Properties, mechanism, synthesis, and applications. *Applied Catalysis b: Environmental, 265*, 118579.

Pardhi, V. P., Verma, T., Flora, S. J. S., Chandasana, H., & Shukla, R. (2018). Nanocrystals: An overview of fabrication, characterization and therapeutic applications in drug delivery. *Current Pharmaceutical Design, 24*(43), 5129–5146.

Parham, H., Bates, S., Xia, Y., & Zhu, Y. (2013). A highly efficient and versatile carbon nanotube/ceramic composite filter. *Carbon, 54*, 215–223.

Raza, M. A., Kanwal, Z., Rauf, A., Sabri, A. N., Riaz, S., & Naseem, S. (2016). Size-and shape-dependent antibacterial studies of silver nanoparticles synthesized by wet chemical routes. *Nanomaterials, 6*(4), 74.

Ritter, K. S., Sibley, P., Hall, K., Keen, P., Mattu, G., & Beth Linton, L. (2002). Sources, pathways, and relative risks of contaminants in surface water and groundwater: A perspective prepared for the Walkerton inquiry. *Journal of Toxicology and Environmental Health Part A, 65*(1), 1–142.

Robichaud, C. O., Uyar, A. E., Darby, M. R., Zucker, L. G., & Wiesner, M. R. (2009). Estimates of upper bounds and trends in nano-TiO2 production as a basis for exposure assessment.

Rueda-Marquez, J. J., Levchuk, I., Ibañez, P. F., & Sillanpää, M. (2020). A critical review on application of photocatalysis for toxicity reduction of real wastewaters. *Journal of Cleaner Production, 258*, 120694.

Runguphan, T., & Kitpichai, J. (2020). Coaction of bio-sorption and bio-filtration for the remediation of domestic and agricultural wastewater contaminated with heavy metal. In *IOP Conference Series: Materials Science and Engineering* (Vol. 965, No. 1, p. 012010). IOP Publishing. (Nov 2020)

Sadegh, H., Ali, G. A., Gupta, V. K., Makhlouf, A. S. H., Shahryari-Ghoshekandi, R., Nadagouda, M. N., Sillanpää, M., & Megiel, E. (2017). The role of nanomaterials as effective adsorbents and their applications in wastewater treatment. *Journal of Nanostructure in Chemistry, 7*, 1–14.

Sayed, E. T., Alawadhi, H., Elsaid, K., Olabi, A. G., Adel Almakrani, M., Bin Tamim, S. T., Alafranji, G. H., & Abdelkareem, M. A. (2020). A carbon-cloth anode electroplated with iron nanostructure for microbial fuel cell operated with real wastewater. *Sustainability, 12*(16), 6538.

Schmidt-Mende, L., & MacManus-Driscoll, J. L. (2007). ZnO–nanostructures, defects, and devices. *Materials Today, 10*(5), 40–48.

Shon, H. K., Phuntsho, S., Chaudhary, D. S., Vigneswaran, S., & Cho, J. (2013). Nanofiltration for water and wastewater treatment–a mini review. *Drinking Water Engineering and Science, 6*(1), 47–53.

Singha, I., & Mishrab, P. K. (2020). Nano-membrane filtration a novel application of nanotechnology for waste water treatment. *Materials Today: Proceedings, 29*, 327–332.

Sirajudheen, P., Karthikeyan, P., Ramkumar, K., & Meenakshi, S. (2020). Effective removal of organic pollutants by adsorption onto chitosan supported graphene oxide-hydroxyapatite composite: A novel reusable adsorbent. *Journal of Molecular Liquids, 318*, 114200.

Sukhanova, A., Bozrova, S., Sokolov, P., Berestovoy, M., Karaulov, A., & Nabiev, I. (2018). Dependence of nanoparticle toxicity on their physical and chemical properties. *Nanoscale Research Letters, 13*, 1–21.

Talari, F. F., Bozorg, A., Faridbod, F., & Vossoughi, M. (2021). A novel sensitive aptamer-based nanosensor using rGQDs and MWCNTs for rapid detection of diazinon pesticide. *Journal of Environmental Chemical Engineering, 9*(1), 104878.

Verma, B., & Balomajumder, C. (2020). Surface modification of one-dimensional Carbon Nanotubes: A review for the management of heavy metals in wastewater. *Environmental Technology & Innovation, 17*, 100596.

Warwick, C., Guerreiro, A., & Soares, A. (2013). Sensing and analysis of soluble phosphates in environmental samples: A review. *Biosensors and Bioelectronics, 41*, 1–11.

Westerhoff, P. K., Kiser, M. A., & Hristovski, K. (2013). Nanomaterial removal and transformation during biological wastewater treatment. *Environmental Engineering Science, 30*(3), 109–117.

Wong, K. A., Lam, S. M., & Sin, J. C. (2019). Wet chemically synthesized ZnO structures for photodegradation of pre-treated palm oil mill effluent and antibacterial activity. *Ceramics International, 45*(2), 1868–1880.

Xiu, Z. M., Zhang, Q. B., Puppala, H. L., Colvin, V. L., & Alvarez, P. J. (2012). Negligible particle-specific antibacterial activity of silver nanoparticles. *Nano Letters, 12*(8), 4271–4275.

Yadav, D. K., & Srivastava, S. (2017). Carbon nanotubes as adsorbent to remove heavy metal ion (Mn+ 7) in wastewater treatment. *Materials Today: Proceedings, 4*(2), 4089–4094.

Yang, X., Zhao, Z., Yu, Y., Shimizu, K., Zhang, Z., Lei, Z., & Lee, D. J. (2020). Enhanced biosorption of Cr (VI) from synthetic wastewater using algal-bacterial aerobic granular sludge: Batch experiments, kinetics and mechanisms. *Separation and Purification Technology, 251*, 117323.

Yao, B., Luo, Z., Xiong, W., Song, B., Zeng, Z., & Zhou, Y. (2020). Disinfection techniques of human norovirus in municipal wastewater: Challenges and future perspectives. *Current Opinion in Environmental Science & Health, 17*, 29–34.

Zhang, C., Lai, C., Zeng, G., Huang, D., Yang, C., Wang, Y., Zhou, Y., & Cheng, M. (2016). Efficacy of carbonaceous nanocomposites for sorbing ionizable antibiotic sulfamethazine from aqueous solution. *Water Research, 95*, 103–112.

Zhang, T., Yu, H., Li, J., Song, H., Wang, S., Zhang, Z., & Chen, S. (2020). Green light–triggered antimicrobial cotton fabric for wastewater disinfection. *Materials Today Physics, 15*, 100254.

Zhao, J., Yan, P., Snow, B., Santos, R. M., & Chiang, Y. W. (2020). Micro-structured copper and nickel metal foams for wastewater disinfection: Proof-of-concept and scale-up. *Process Safety and Environmental Protection, 142*, 191–202.

Zheng, A. L. T., Phromsatit, T., Boonyuen, S., & Andou, Y. (2020). Synthesis of silver nanoparticles/porphyrin/reduced graphene oxide hydrogel as dye adsorbent for wastewater treatment. *FlatChem, 23*, 100174.

Nanotechnology in Waste Management and Recycling

Srijita De, Saurabh Dubey, Arijit Mohanta, Sachin Kumar Sharma, Mohd. Najmus Saquib, and Dipankar Bandyopadhyay

Abstract The advancement in industrial activities has proliferated the amount of waste produced regularly in the modern world. The three major elements of the environment, e.g., water, soil, and air, have been filled with excess dust particles, micro/nano plastics, microorganisms, organic suspensions, and heavy metal ions that have started taking a hefty toll on living entities. In this regard, among the many other pathways, the nanoscale sciences and technologies have emerged as powerful tools in mitigating the pressing challenges of waste management by converting them into 'wealth'. For example, several nanocatalysts can convert waste materials into organic solvents, acids, salts, buffers, esters, ethers, or surfactants, which are commercially important. We explore some of the nanoparticles, nanomembranes, metal oxides, and nanopolymers that can cause pollutant adsorption, catalytic degradation, separation, and material recovery processes. The nanomaterials reveal significant improvements in adsorbent specificity for heavy metals, catalytic efficacy in the breakdown of organic waste into specialty chemicals, and usage of some of them to develop nanosensors for rapid and regular monitoring of contamination. The chapter provides insights into the promises, advantages, and challenges of integrating such nanoscale technologies into existing waste management ecosystems by considering both the technical and environmental impacts. Furthermore, the scalability, economic feasibility, and the development of safe regulatory frameworks for nano-enabled waste solutions have also been briefly explored. The chapter concludes by highlighting future prospects, such as 'green' synthesis processes using sustainable nanomaterials alongside cutting-edge recovery processes. In summary, the chapter aims to understand the potential of nanotechnology to transform waste into wealth while contributing to sustainable development goals.

S. De · S. Dubey · D. Bandyopadhyay (✉)
Centre for Nanotechnology, Indian Institute of Technology Guwahati, Assam, India
e-mail: dipban@iitg.ac.in

A. Mohanta · S. K. Sharma · Mohd. N. Saquib · D. Bandyopadhyay
Department of Chemical Engineering, Indian Institute of Technology Guwahati, Assam, India

D. Bandyopadhyay
Jyoti and Bhupat Mehta School of Health Sciences and Technology, Indian Institute of Technology Guwahati, Assam, India

E. Priyadarshini et al. (eds.), *Advanced Nanotechnology Applications for Environmental Sustainability*, Sustainable Environmental Waste Management Strategies,
https://doi.org/10.1007/978-3-032-12933-8_16

Keywords Nanotechnology · Waste management · Recycling · Pollution control · Mitigation

1 Introduction

Over the past few centuries, the amount of industrial and municipal waste generation has increased exponentially with the multiplication of the world populace. The entire ecosystem of air, water, and soil has been seriously affected by the adverse effects of these wastes (Yan et al., 2022). The emission of greenhouse gases, like CO_2, NO_x, SO_x, or toxic volatile organic compounds—VOCs from industries or automobiles have contributed further to the deterioration of the surrounding air quality leading to the rapid warming of the globe. The discharge of chemicals, organic and inorganic pollutants into the water bodies, or the oil spills into the oceans have contaminated the aquatic life in a major way. Similarly, dioxins, herbicides, pesticides, agricultural waste, and chemical fertilizers have been the major causes of soil pollution.

In addition, the industrial effluents containing heavy metals like cadmium, zinc, lead, arsenic, and mercury, organic dyes, aromatic hydrocarbons, phenols, and drug residues have also been extracted and leached into the underground reservoirs during their recharge, rendering a toxic water resource (Biswas et al., 2021; Islam et al., 2023). Further, the emergence of the micro (Dubey et al., 2022; Edo et al., 2020) and nano (Urso et al., 2022) plastics have recently stirred the world with their capability to contaminate not only the soil and water but also human health. In a nutshell, with the advancement of technologies, the increase in the volume of waste has been impacting the ecosystem adversely. It has also become clear that they can neither be destroyed nor incinerated because such activities will only bring the materials back to the environment as contaminants of some other form. Therefore, it has become almost imperative to mitigate such a large-scale spread of air, water, and soil contamination by repurposing them into usable and eco-friendly products.

In this direction, among the other methodologies, nanotechnology has been emerging over the past few decades as a forte to effectively tackle environmental challenges (Manikandan et al., 2022). In particular, the novel physicochemical properties at the nanoscale, the possibility of fabricating diverse geometries, the surface functionalization capabilities, and the possibility to play around with the chemical compositions have been exploited to the fullest to mitigate the wastes at their source and produce environment-friendly and commercially viable products. For example, the use of nanoparticles (Alhalili, 2023), nanocomposites (Langer et al., 2020), nanomembranes (Maity et al., 2021a), nanosensors (Weerasinghe et al., 2025), nanobots (Jancik-Prochazkova et al., 2023a, 2023b), nanocatalysis (Wang et al., 2019), and nanostructures (Baratta et al., 2022) have been found to be effective in depolymerization and converting the solid and liquid wastes into specialty chemicals such as organic solvents, acids, salts, buffers, esters, ethers, or surfactants that are commercially important (Vanarse et al., 2025). Further, they are of extensive use to mitigate microplastic problems (Sun et al., 2020), remove toxins from water

(Lapointe et al., 2023), and eliminate heavy metals (Sugunan et al., 2005), organic pollutants, industrial dyes (You et al., 2021), pathogens, and toxic gases from the air (Mohamed et al., 2022). The prior art reveals that such advanced nanoengineering processes are rapidly becoming popular because they are highly energy-efficient, reusable, eco-friendly, and scalable (Gharpure et al., 2024; Igiebor et al., 2023).

In view of this backdrop, the present chapter explores the usage of nanotechnology for water treatment. Figure 1 pictorially summarizes the scope and objectives of this chapter. Concisely, the chapter explores the unique properties of various nanomaterials in the first section. Thereafter, the chapter explains the applications of such nanomaterials in the detection and mitigation of environmental pollutants. It then focuses on the recycling and recovery methods of pollutants using nanotechnology. Following this, the chapter will also provide economic and environmental perspectives on the use of such technology in the recovery of waste materials. Finally, the advantages of green synthesis methods for synthesizing nanomaterials have been discussed before, highlighting the methodologies to control some adverse effects of chemically synthesized nanomaterials. In summary, the chapter aims to emphasize the transformative potential of nanotechnology in waste management and environmental science, promoting innovative and sustainable approaches for mitigating contamination.

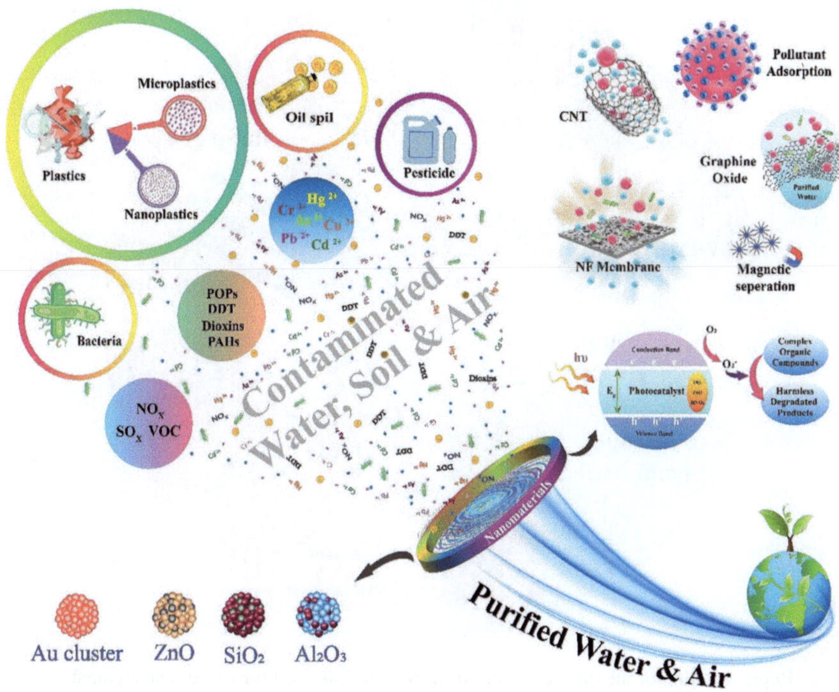

Fig. 1 Schematic of the application of nanotechnology in waste management

2 Nanomaterials for Waste Management

2.1 Classification of Nanomaterials

The nanomaterials can be broadly classified on the basis of structure, composition, and size (Elwaheidi, 2023), as represented in Fig. 2.

(a) **Zero-Dimensional (0D) Nanomaterials**: Among these are nanoparticles, which are all nanoscale (less than 100 nm) in size. Metal nanoparticles such as gold and silver are examples.
(b) **One-Dimensional (1D) Nanomaterials**: These include materials with one dimension outside the nanoscale, such as nanowires and nanotubes. One well-known example is carbon nanotubes.
(c) **Two-Dimensional (2D) Nanomaterials**: These include two-dimensional nanoscale materials like graphene and nanosheets.
(d) **Three-Dimensional (3D) Nanomaterials**: These comprise materials that are composed of nanoscale components yet have dimensions bigger than the nanoscale, such as nanocomposites and nanostructured materials.

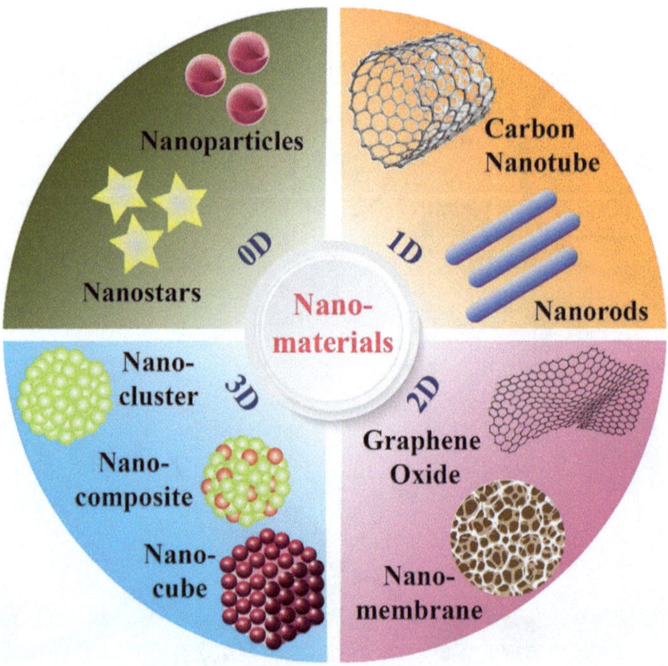

Fig. 2 Representation of the different types of nanomaterials used for waste management

2.2 Properties and Key Advantages of Nanomaterials

Nanoparticles: These are very tiny balls of less than 100 nm in size, known as nanoparticles. They are composed of a variety of substances, such as metals, non-metals, metal–metal composites, metal-non-metal composites, conjugated nanoparticles, and bioconjugated nanoparticles with various biological entities. Due to their exceptional ability to adhere to surfaces, we can use them to remove contaminants from the air or water. As potent magnets, they have the ability to draw in and retain undesirable materials. Because of their tiny size, the high surface-to-volume ratios of nanoparticles improve their adsorption and reactivity, and this is used for their specificity and selectivity for various applications. Metals, metal oxides, or organic compounds can be used to synthesize them. As they can absorb organic contaminants and heavy metals, they are effective at removing pollutants from water. They are also employed in catalytic procedures that capture the breakdown of various organic wastes (Langer et al., 2020; Talapin et al., 2016).

Nanomembranes: Nanomembranes involve very thin layers of material with small openings, making it easier to select and filter material. It is like a thin sheet with tiny holes when there is a squad of feathers. The nanomembrane, or sheet, functions as an excellent filter. This will be especially suitable for water and air purification since the system can filter small particles within the latter. They work just like a magic mesh that retains the good things in while keeping the bad things out. It is used in the treatment of water to do away with impurities, heavy metals, and microscopic plastics. They are highly selective for filtration procedures and come with high efficiency (Maity et al., 2021a, 2021b).

Nanofilms: Thin films of nanomaterials are coatings of thickness less than 100 nm, which develop special colloidal properties like SERS—surface enhanced Raman scattering, higher dielectric permittivity, GMR—giant magnetoresistance, among others. In this regard, thin films of metal oxides are considered to be the special agents that possess the ability to reduce environmental pollution. They can convert pollutants into harmless compounds using sun radiation. For instance, metal oxide such as titanium dioxide can help improve the process of cleaning water by breaking down organic substances in the water when exposed to light. Further, the thin films of the photocatalysts that embrace metal oxide properties like ZnO and TiO_2 can help reduce the concentration of organic matter in water under light. They are effective in wastewater management as they help in degrading organic pollutants. They are also used in air purification systems (Jancik-Prochazkova et al., 2023a, 2023b; Mayorga-Burrezo et al., 2023; Ullattil et al., 2023).

Nanostructures: Surfaces with nanopatterns often develop special features such as the availability of high surface-to-volume ratio suitable for catalytic actions. Such substances consisting of various materials such as carbon, metals or metal oxides have shown their capability to mitigate various air–water-coil pollutants. In particular, active carbon allotropes in the form of fullerene, graphene, and carbon nanotubes are found to be durable while at the same time high chemical adsorption capabilities.

These properties can be employed in adsorption processes to eliminate organic pollutants and heavy metals from water. Also, they enhance the other physical properties of the composite materials (Baby et al., 2019; Mohapatra et al., 2023; Nasrollahzadeh et al., 2021; Ratnam et al., 2023).

Nanopolymers: These are plastics that are safe to be in contact with living things, that is, organisms such as animals and human beings. They are made in a way that degrades themselves into the environment within a certain period of time without polluting the environment. They can be thought of as green bags that can be used instead of regular black plastic bags in the sense that they do not pollute the environment. These polymers are non-toxic and biocompatible, which fits them well with various applications in regions that involve the use of tender-age citizens or living organisms. These are used in the manufacture of biodegradable packaging in some medical facilities. In waste management, biomaterials can be applied to produce materials that break down plastics used in society (Satchanska et al., 2024; Smith et al., 2008).

Nanobiomaterials: This is a blend of bacteria with tiny particles, which may be referred to as nanobiomaterials. It is like a team of workers where the nanomaterials enhance the ability of the bacteria to break down the toxins. The cleaning staff could be bacteria, and the tools could be nanoparticles to design a better and cleaner working environment for all. This brings together the metabolic activities of bacteria to the advantages of nanomaterials. It can increase the efficiency of several specialized functions, such as enhancing the removal of pollutants as well as the biodegradation processes (Mishra et al., 2024; Singh Jadaun et al., 2022; Tripathy et al., 2024). It incorporates the catalytic natures of nanomaterials with the metabolic activities of bacteria, likely enhancing bioremediation process efficiency (Hidangmayum et al., 2023; Mandeep et al., 2020). Iron Oxide Nanoparticles are applied in bioremediation of lead and cadmium utilizing *Halomonas sp.* (Singh et al., 2022). Carbon-based nanomaterials with bacteria improve the biodegradation capability of wastewater treatment (de Carvalho Lima et al., 2024). The bioavailability of cadmium is improved with Selenium Nanoparticles together with *L. casei*. Utilization of the aforementioned nanomaterials plays a crucial role in enhancing waste management processes by improving pollutant removal, catalytic degradation, and material recovery, contributing to more sustainable waste management practices (Singh et al., 2022).

3 Applications of Nanotechnology in Waste Management

Nanotechnology addresses the complex challenges associated with waste management. In this section, the applications of wastewater treatment and pollution management have been discussed. Here, nanomaterials and nano-enabled sensors provide novel alternatives to age-old issues of environmental threat. Figure 3 represents the various nanotechnology-based mechanisms for waste management.

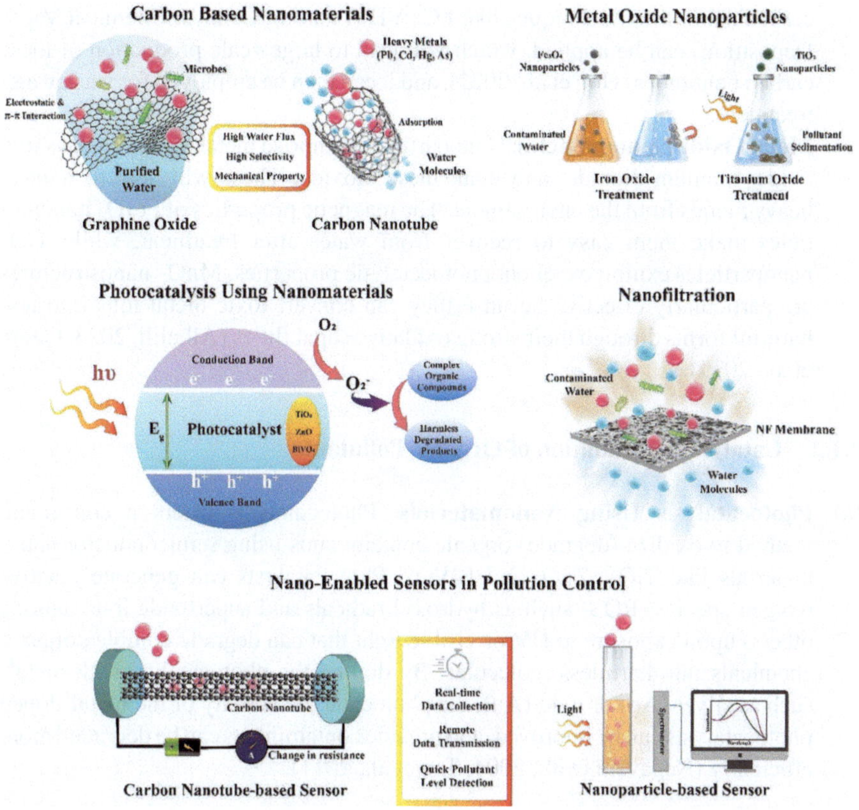

Fig. 3 Schematic of the different methods applicable for waste management via various nanomaterials

3.1 Nano-Enabled Wastewater Treatment

3.1.1 Removal of Heavy Metals

Heavy metals, including lead (Pb), cadmium (Cd), arsenic (As), and mercury (Hg), are among the most hazardous pollutants found in wastewater. These toxic metals pose risks to human health and aquatic life, even at minimal concentrations.

(a) **Carbon-Based Nanomaterials**: Two of the abundant nanomaterials for wastewater treatment are carbon nanotubes (CNTs) and graphene oxide (GO). While graphene oxide has a high surface reactivity and can form stable complexes with metal ions through electrostatic and π-π interactions, CNTs have a large surface area and a hollow structure that provide plenty of adsorption sites. Heavy metals like lead and chromium can be effectively removed with CNTs, while lead and arsenic can be effectively removed with functionalized graphene oxide (Baratta et al., 2022; Wang et al., 2021). For the continuous production of carbon nanotubes, techniques like FCCVD (Floating Catalyst Chemical Vapor Deposition) can be applied, which will lead to large-scale production of these carbon nanotubes (Hou et al., 2022), and it can then be employed for wastewater treatment.

(b) **Metal Oxide Nanoparticles**: Nanoparticles comprise metal oxides such as iron oxide, titanium dioxide, and manganese dioxide. These will help to remove heavy metals from the environment. The magnetic properties of Fe_3O_4 nanoparticles make them easy to recover from water after treatment, while TiO_2 nanoparticles exhibit excellent photocatalytic properties. MnO_2 nanostructures are particularly effective because they can convert toxic metal ions into less harmful forms through their strong oxidative capabilities (Alhalili, 2023; Gebre et al., 2019).

3.1.2 Catalytic Degradation of Organic Pollutants

(a) **Photocatalysis Using Nanomaterials**: Photocatalysis offers a convenient method to oxidize (degrade) organic contaminants using semiconductor nanomaterials like TiO_2, ZnO, and $BiVO_4$. Photocatalysts can generate reactive oxygen species (ROS) such as hydroxyl radicals and superoxide ions (among others) upon exposure to UV or visible light that can degrade complex organic chemicals into harmless molecules. By doping the photocatalyst with metals such as silver (Ag) or gold (Au), the photocatalytic activity of the metal-doped photocatalysts can be improved, and organic contaminants can be degraded more efficiently (Nagaveni et al., 2004; Tan et al., 2011).

(b) Biogenic Nanocatalysts: Using plant extracts, microorganisms, or biopolymers can potentially provide a new method for the green synthesis of nanoparticles and an environmentally friendly option to create nanomaterials. Biogenic nanoparticles exhibit not only excellent catalytic activity but are also shown to be biodegradable and non-toxic. They have exhibited great potential to degrade organic contaminants, including dyes, pesticides, and pharmaceutical residues, without harmful byproducts or secondary waste (Susan Punnoose et al., 2021).

(c) Micro/nano Plastic Degradation: Researchers are inspired by nature to utilize fractional spray technology to develop a graded purification material that deals with heavy metals, oils, pathogens, and micro/nano plastics through percolation (Zhang et al., 2022), as shown in Fig. 4a. Another group has developed hydrophobic Fe nanoparticle-based magnetic recovery of microplastics by reaching 93% seawater recovery as well as 84–78% recovery from freshwater and sediments (Grbic et al., 2019). Zinc oxide nanoparticles derived from Piper longum leaf extract have demonstrated their capacity to break down microplastics (PS and PE) through UV-induced photocatalysis at concentrations between 75–100 μL as per the research (Vas et al., 2024).

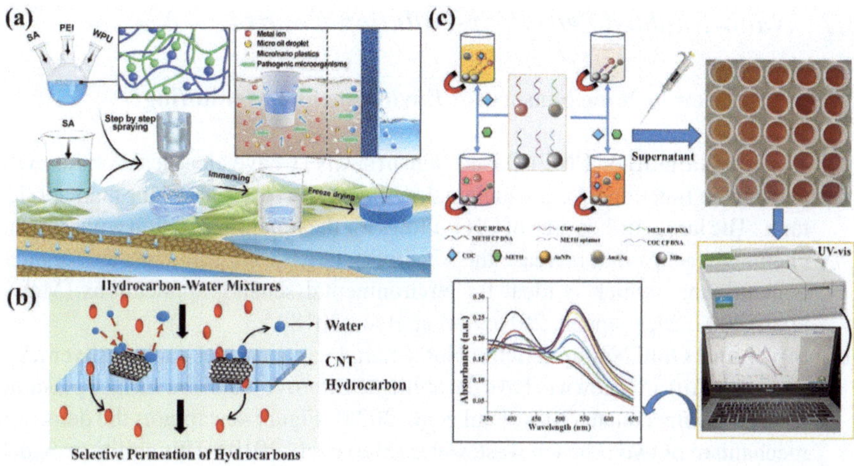

Fig. 4 Image **a** shows the use of sodium alginate (SA), branched polyethylenimine (PEI), and waterborne polyurethane (WPU)-based spraying technology for natural groundwater purification (Zhang et al., 2022), **b** portrays the superhydrophobic CNT membrane for removing trace amounts of water (Paul et al., 2022), **c** represents the mechanism of colorimetric detection of illicit drugs, methamphetamine and cocaine, in wastewater (Mao et al., 2019)

3.1.3 Membrane Filtration Techniques

(a) **Nanofiltration (NF) Membranes**: The pore size of NF membranes is intermediate. As a result, they are very good at rejecting divalent and trivalent ions, dissolved organic molecules, and microbial contaminants. They are more energy-efficient for large-scale water treatment as they operate at lower pressures than reverse osmosis membranes (Dai et al., 2023; Han et al., 2013).

(b) **Carbon Nanotube-Enhanced Membranes**: CNTs increase permeability, mechanical strength, and foul resistance of membrane systems. They have higher filtration rates and improved selectivity, which are suitable for difficult wastewater treatment applications. Furthermore, their unique structure reduces the buildup of contaminants on the membrane surface, improving the overall efficiency and longevity of the filtration system (Gethard et al., 2011; S. Paul et al., 2022). Figure 4b portrays a type of CNT membrane, which is superhydrophobic in nature and removes trace amounts of water.

3.2 *Nano-Enabled Sensors in Pollution Control*

3.2.1 Overview of Nano Sensors for Environmental Monitoring

(a) **Carbon Nanotube (CNT)-Based Sensors**: CNTs have electrical conductivity and can be functionalized, which makes them effective sensors for many pollutants. The large surface area of CNT allows for the functionalization with specific chemical groups that increase the sensitivity to gaseous pollutants and organic contaminants, which is ideal for environmental sensing applications (Mahar et al., 2007; Meyyappan, 2016; Sarkar et al., 2018).

(b) **Silver and Gold Nanoparticles**: Noble metal nanoparticles such as silver (Ag) and gold (Au) are known to have the ability to improve localized surface plasmon (LSPR) (Ahmed et al., 2024; Paul et al., 2024). Figure 4c captures the detection mechanism of two drugs in wastewater (Mao et al., 2019). This ability provides organic toxins, heavy metals, and other contaminants in the environment with the ability to be detected in very low concentrations (Mao et al., 2019; Weerasinghe et al., 2025).

3.2.2 Detecting Pollutants in Water, Air, and Soil

Nanosensors are increasingly being applied across various environmental matrices to detect and quantify pollutants with unprecedented sensitivity. These nanosensors can monitor water, air, and soil contamination items at the same time, offering comprehensive solutions for environmental health.

(a) **Water Pollution Monitoring**: Nanosensors, particularly sensitive to heavy metals (e.g., lead ($Pb2^{+}$), mercury ($Hg2^{+}$), and arsenic ($As3^{+}$)), as well as organic contaminants (e.g., pesticides, pharmaceuticals, and industrial compounds), are useful to environmentally monitor contamination due to their high capability to quantify contaminants in aquatic environments. Many of these sensor methodologies use functionalized graphene oxide (GO) and CNT sensors for real-time electrochemical sensors for pollutants in natural settings (Gupte et al., 2022; Mustafa et al., 2021; Vikesland, 2018).

(b) **Air Pollution Monitoring**: The monitoring of atmospheric volatile organic compounds (VOCs), carbon dioxide (CO_2), and nitrogen oxides (NOx), along with particulate matter (PM) depends heavily on nanosensors because of their gas-sensitive features. Metal oxide semiconductors (MOS) and nanosensors (e.g. SnO_2 and WO_3) display remarkable sensitivity for atmospheric pollution measurements at low concentrations (ppm levels) (Mohamed et al., 2022; Mushtaq et al., 2024; Santos et al., 2020). The nanocomposite uses different materials like graphene and aluminum doping into a Zinc oxide matrix for the detection of ammonia gas, which exhibits different sensing properties (Sankar ganesh et al., 2017).

Table 1 Summarization of some other nanomaterials and their applications in waste management

Nanomaterial type	Applications in waste management	References
$ZnO - TiO_2$ nano-photocatalyst	Degradation of reactive dyes from wastewater	(Haghighizadeh et al., 2023)
Nano-zero-valent iron and nano-copper	Adsorption of Levofloxacin and antibiotics	(Hamad et al., 2023)
$ZnS - NPs$ using extract *Artemisia Herba Alba*	Degradation of Methylene blue dye	(Ouni et al., 2023)
Magnetic carbon nanocomposite	Adsorption of Dyes and Pesticides	(Nille et al., 2023)
Modified nano-Zero-Valent Iron (nZVI)	Removal of Phenols from industrial wastewater	(Ding et al., 2023)
Silica nanoparticles	Removal of acid-orange	(Mohd Nazri et al., 2020)
Nano-Zero valent Ferrous	Decomposition of Nitrobenzene	(Mu et al., 2004)
Porous silica nanoparticles	CO_2 capture in bottom and fly ash stream	(Song et al., 2019)
Rhenium (Re^{+3})-doped nano-TiO_2	Decomposition of Carbofuran Pesticide	(Zeng et al., 2010)
MXene-derived γ-Fe_2O_3/Pt/TiO_2	Capture of nanoplastics	(Urso et al., 2022)

(c) **Soil Pollution Monitoring**: The application of nanosensors extends to detect different types of soil pollutants by monitoring heavy metals in combination with nitrates together with persistent organic pollutants (POPs) (Linley et al., 2021). Nanoscale zero-valent iron (nZVI) is the most recent addition to soil remediation. It benefits the removal of organic and metalloid soil pollutants (Xue et al., 2018). Other examples of nanomaterials and their applications in waste management are summarized in Table 1.

3.2.3 Advances in Sensitivity and Real-Time Monitoring

With the advancement in nanosensor technology, excellent features like real-time monitoring and remote detection have been brought forward. It makes a large contribution to environmental pollution control strategy innovations and speeding up environmental response.

(a) **Real-Time Data Collection**: IoT-based nanosensors allow the monitoring of the environment continuously and in real time. The transmission of data remotely and rapidly, as well as remote, in situ, pollutant-level disclosure, can be performed by large, intelligent nanosensor networks, which can also trigger automated responses. This is very helpful in remote locations where there are prohibitive costs associated with traditional types of monitoring. This allows for scalable, distributed environmental monitoring using nanosensors (Moretti et al., 2024; Salem et al., 2022).

(b) **Molecularly Imprinted Polymers (MIPs)**: MIPs are synthetic polymeric materials that mimic natural biological receptors with high selectivity for specific pollutants. MIPs are incorporated into nanosensors to enhance detection sensing and specificity, and tailored approaches for uncovering environmental contamination (Huang et al., 2015; Murray et al., 2012; Wu et al., 2022).

4 Recycling and Recovery Using Nanotechnology

4.1 Nano-Enabled Recycling Methods

Nanotechnology-Enabled Plastic Recycling: Plastic degradation at a slow pace combined with increasing solid waste amounts has emerged as a major environmental threat. Improving plastic recycling efficiency serves as one of the main purposes for which nano-enabled technologies are employed. This approach produces valuable products from waste plastic with the help of nanomaterials, which include carbon nanotubes (CNTs) along with graphene and nanocomposites. Scientists manage to transform polyethylene terephthalate (PET) waste using three distinct nanostructures, including photoluminescent carbon nanoparticles and carbon nanostructures. The method constitutes a dual advantage through plastic waste reduction, and at the same time, it generates valuable products that serve industries such as electronics, energy storage, and biomedicine. Nano-enabled de-polymerization technology helps break down complex polymers to obtain reusable monomers. The mechanical and thermal properties of recycled plastics containing nano-additives are implemented, which results in enhanced durability for high-performance applications (Igiebor et al., 2023).

Nano-Catalysts for Recycling Metal Waste: Metal conservation is necessary for the sustainable use of resources. Therefore, metal recycling is essential. Nanotechnology is helpful in the efficient extraction and purification of valuable metals from e-waste and industrial scrap (Manikandan et al., 2022). Gold (Au), silver (Ag), and TiO_2 nanoparticles help in efficient recovery by binding to metal ions. Graphene oxide and metal oxide nano-adsorbents can adsorb heavy metals such as lead, cadmium, and mercury (Roy et al., 2024). Electrochemical nanosensors reduce waste and increase recycling efficiency by real-time monitoring of metal recovery. The implementation of these devices leads to enhanced industrial metal recovery operations which operate with fewer harmful environmental effects. Plasma-assisted nano-processing makes the valuable recovery of rare earth elements from industrial waste possible. The environmentally friendly and energy-efficient approach serves as a key recycling operation for industries (Manikandan et al., 2022).

Nanomembrane for Waste Separation: The implementation of nano-enabled membranes produces more effective recycling methods. Due to their high selectivity in separating impurities, polymeric, along with ceramic nanomembranes, prove beneficial in pharmaceuticals, food processing, and water treatment operations (Shariful Islam et al., 2024).

4.2 *Material Recovery with Enhanced Efficiency*

Nano Technology in Metal Recovery: Conservation of rare earth materials, like platinum and palladium, is crucial for economic purposes. Different nanoscale adsorbents and advanced extraction techniques are utilized to make the resource conservation process eco-friendly. The absorption of metal ions becomes efficient using graphene and carbon nanotubes (CNTs). This technology helps in the effective recovery of metals from sludge and e-waste. Magnetic nanoparticles improve metal recovery (Mishra et al., 2022). Their surfaces are functionalized to bind to specific metal ions, which can then be easily separated with the help of magnetic fields. Nanoparticle-mediated solvent extraction enhances the efficiency of metal recovery by separating unwanted elements. Hybrid nano-filtration systems expedite and boost metal recovery efficiency by integrating nanomaterials into a single system (Manikandan et al., 2022).

Enhancing biological waste recovery with nanotechnology: The recycling of agricultural waste is of utmost importance for reducing waste. Nanocatalysts and nanosensors come to the rescue by converting organic waste into biofuels, bioplastics, and harmless fertilizers. While nanocatalysts, like TiO_2 and ZnO, decompose complex organic compounds rapidly, nano-sensors quantify different compounds and help optimize the energy conversion processes of organic waste. Bio-nano hybrid systems have proven to be effective in accelerating the rate of biodegradation, making easy recovery of rare earth elements from industrial waste (Hemalatha et al., 2024). The crop production and soil fertility have been improved by increasing the stability of bioactive compounds in fertilizers, facilitated by nano-encapsulation techniques.

5 Economic and Environmental Perspectives

5.1 *Feasibility and Cost-Effectiveness of Nano-Resources*

Cost efficiency of nano-based recycling: Nanotechnology opens a new direction to recycling, maximizing resource recovery, energy conservation, and waste reduction. Its major impacts are as follows (Igiebor et al., 2023):

(a) **Reduction in energy and temperature requirements**: Nano-catalysts enable lower temperatures and energy requirements, decreasing production costs because of their improved functionality.

(b) **Higher efficiency of material reuse**: Nanostructured catalysts combined with nano-based technologies enhance recyclability, so both essential materials can be recovered, and waste reduction remains at a minimum.
(c) **Improved resource utilization**: Nano-enabled systems can minimize water and chemical consumption. It leads to a reduced ecological footprint and efficient management of resources.
(d) **Eco-friendly solutions**: The implementation of automated nano-based recycling systems results in increased efficiency with reduced power usage, thus advancing environmental sustainability and green technological adoption.

Scalability Challenges and Solutions: Advancement in nano-enabled recycling technologies comes with the limitation of large-scale scalability. There are some key aspects that need to be addressed for effective waste management. While nanomaterials improve production efficiency, they present multiple manufacturing barriers that raise production and manufacturing complexity. Research and innovation are essential for cost-effective solutions. For nano-enabled waste management to succeed, the regulatory framework must include strict safety standards, and these frameworks should receive support from the government (Igiebor et al., 2023). Effective business strategies coupled with awareness campaigns are required to ensure the rapid demand for nano-enabled technologies in the electronics, manufacturing, and biotechnology sectors (Khan et al., 2014).

Industrial Applications and Market Potential: Nano-enabled recycling technologies are accepted in various industries to enhance efficiency and productivity. Nano-enabled electrolytic processes are employed for the recovery of rare and valuable metals (Khan et al., 2014). For the use of clear water in the agricultural sector, nano-coated and nano-porous membrane technologies are utilized in the recycling of wastewater (Shariful Islam et al., 2024). The plastic industry also deploys these technologies to convert plastics into reusable, biocompatible, and high-quality products (Roy et al., 2024). Nano-structured reclaimed materials provide more durable and robust building components for the construction industry, thereby increasing the durability and sustainability of structures (Das & Ghosh, 2022; Mishra & Das, 2023). Lightweight and robust nano-enhanced materials reduce the weight of vehicles, resulting in better fuel efficiency and decreasing environmental impact (Shariful Islam et al., 2024).

5.2 *Environmental Impact Assessment of Nanomaterials*

Every good thing comes with a cost. Similarly, nanotechnology has enabled waste management, but concerns remain about its environmental and biological impacts. Therefore, continuous monitoring of their effects is necessary (Manikandan et al., 2022).

Ecological Advantages of Nano-Enabled Recycling: Nano-enabled recycling reduces solid waste by increasing the recovery rate of materials, thereby controlling land and water pollution. These processes are more efficient than conventional processes. Recycling technologies require less energy, thereby reducing greenhouse gas emissions and mitigating the adverse effects of climate change (Mohanty et al., 2024; Sahu et al., 2024). Nanomaterials reduce environmental pollution by degrading non-biodegradable wastes, including micro-nano plastic, heavy metals, and other organic pollutants. Advanced nano-filtration systems can eliminate toxins from industrial wastewater, paving the way for preserving clean water sources.

Regulatory Challenges and Mitigation Strategies: Governments and international bodies need to establish exact environmental regulations for nanomaterials across their life cycle to safeguard ecosystems and block environmental pollution (Igiebor et al., 2023; Manikandan et al., 2022). The industry needs to concentrate on making sustainable nanomaterials that degrade naturally to decrease environmental consequences (Manikandan et al., 2022; Shariful Islam et al., 2024). Life Cycle Assessment evaluation enables the evaluation of long-term consequences for nano-enabled recycling technologies by following their complete destruction cycle. The evaluation will assist businesses in uncovering environmental hazards to innovate sustainable operational practices (Roy et al., 2024). Special methods need to be developed to implement safe nanomaterial waste treatment protocols. The proper management and disposal techniques will stop substances from accumulating uncontrollably while reducing environmental damage in the ecosystem (Khan et al., 2014; Shariful Islam et al., 2024). It is essential to educate industries, manufacturers, and consumers about the potential risks and safety standards associated with nanomaterials. This will promote responsible use of nanomaterials and facilitate the widespread adoption of safe nanotechnology (Igiebor et al., 2023; Khan et al., 2014).

6 Future Trends in Nano-Enabled Waste Management

The unique properties of nanomaterials and their uses in fields like pharmaceuticals, energy, electronics, and environmental rehabilitation compelled researchers to synthesize them. The current traditional synthesis techniques have many drawbacks, such as the high requirement of energy, harmful byproducts, and the use of toxic and potentially carcinogenic solvents. Therefore, these limitations should be addressed while the waste generation should be minimized or nullified.

In order to address this emerging problem, green synthesis techniques have been put into practice as environmentally friendly substitutes that stick with the 12 principles of green chemistry to solve the above problems. The use of safe solvents and natural and renewable resources reduces their adverse consequences on the environment while preserving the unique properties of nanomaterials. The extracts of plants contain several natural reducing and capping agents like flavonoids, polyphenols, alkaloids, and tannins (Stozhko et al., 2019). These bioactive agents reduce the ions of the metal to synthesize nanoparticles and stabilize their growth. Microorganisms like bacteria (e.g., *Bacillus cereus, Lactobacillus Kimchicus*) and fungi (e.g., *Aspergillus fumigatus, Fusarium oxysporum*) employ enzymatic systems to reduce the metal ions into nanoparticles. These nanoparticles are developed for the sensitive detection of hormones, pollutants, and toxic detergents (Li et al., 2011). Some of the algal florae can also be used for the green synthesis of nanoparticles due to their dense composition of polysaccharides and enzymes.

On the other hand, water serves as a universal solvent in green nanomaterial synthesis because it is non-toxic, readily available, and inexpensive. It favors the reduction of metal ions, enables particle nucleation, and favors uniformity, leading to the synthesis of stable nanostructures. Supercritical fluids like water and carbon dioxide are used for the synthesis of metal oxide nanoparticles like tungsten oxide nanoparticles (Kim et al., 2016). Green synthesis of nanomaterials (Arfa et al., 2024; Parmar et al., 2023) for waste management is the future of this field as that leads to sustainable solutions. The technology will be better, more frugal, biocompatible, and more eco-friendly for the environment by producing less toxic and valuable products. Furthermore, a circular economy can be developed, and this will aim to design out waste and pollution, make durable products, and regenerate natural systems. The green synthesis technique eliminates the probability of waste generation. The raw materials used are purely natural and regenerative, decreasing the carbon footprint. With negligible waste generation in synthesis, the nanomaterials are deployed for waste remediation, leading to a better environment.

7 Summary

This chapter summarizes the potential of nanotechnology in pollution control and waste management, providing advanced technologies to monitor and mitigate polluting agents. The chapter explores various nanomaterials, namely, nanoparticles, nanoclusters, nanocomposites, and CNT membranes, to name a few. The primary degradation mechanisms include photocatalytic degradation, adsorption, nanofiltration membrane, graphene oxide membrane, and surface plasmon properties of nanoparticles. Nano-enabled materials and sensors have proven to be highly efficient in the treatment of wastewater, catalytic reduction of pollutants, and environmental monitoring. These methods remove not only water contaminants, like organic pollutants, dyes, illicit drugs, inorganic substances, and heavy metals but also toxic gases and soil pollutants, such as agricultural wastes. Further, the recycling of waste materials utilizing nanomaterials and their recovery are discussed. This chapter also lights upon the economic and environmental perspectives of recycling waste. Green synthesis methods, like bioactive agents, enzymes, and algae, are employed to synthesize nanomaterials to reduce their adverse effects on the environment. As advancement continues in nanotechnology, these novel technologies are poised to revolutionize pollution control and reduce the impacts of pollution in our environment. The growing use of nanotechnology in environmental stewardship will help sustainability today and, in the future, globally as we try to maintain a cleaner and healthier planet.

References

Ahmed, M. M., Basak, M., Paul, A., Singha, D., Sinha, S., & Bandyopadhyay, D. (2024). Plasmonic nanobiosensor for detection of in aqueous samples. *IEEE Sensors Letters*, *8*(1). https://doi.org/10.1109/LSENS.2023.3337938

Alhalili, Z. (2023). Metal oxides nanoparticles: general structural description, chemical, physical, and biological synthesis methods, role in pesticides and heavy metal removal through wastewater treatment. *Molecules*, *28*(7). https://doi.org/10.3390/molecules28073086

Arfa, N., Parmar, P. R., & Bandyopadhyay, D. (2024). Self-organized synthesis of nanoserpent clusters on templated starch for sensitive detection of pancreatic α-amylase. *ACS Sustainable Chemistry and Engineering*, *12*(13), 5075–5085. https://doi.org/10.1021/ACSSUSCHEMENG.3C06403/ASSET/IMAGES/LARGE/SC3C06403_0009.JPEG

Baby, R., Saifullah, B., & Hussein, M. Z. (2019). Carbon nanomaterials for the treatment of heavy metal-contaminated water and environmental remediation. *Nanoscale Research Letters*, *14*(1). https://doi.org/10.1186/s11671-019-3167-8

Baratta, M., Mastropietro, T. F., Bruno, R., Tursi, A., Negro, C., Ferrando-Soria, J., Mashin, A. I., Nezhdanov, A., Nicoletta, F. P., De Filpo, G., Pardo, E., & Armentano, D. (2022). Multivariate metal-organic framework/single-walled carbon nanotube Buckypaper for selective lead decontamination. *ACS Applied Nano Materials*, *5*(4). https://doi.org/10.1021/acsanm.2c00280

Biswas, S., Fatema, J., Debnath, T., & Rashid, T. U. (2021). Chitosan–clay composites for wastewater treatment: A state-of-the-art review. *ACS ES and T Water*, *1*(5). https://doi.org/10.1021/acsestwater.0c00207

Dai, R., Zhou, H., Wang, T., Qiu, Z., Long, L., Lin, S., Tang, C. Y., & Wang, Z. (2023). Nanovehicle-assisted monomer shuttling enables highly permeable and selective nanofiltration membranes for water purification. *Nature Water*, *1*(3). https://doi.org/10.1038/s44221-022-00010-3

Das, A., & Ghosh, S. (2022). Role of microorganisms in extenuation of mining and industrial wastes. *Geomicrobiology Journal, 39*, 173–175. https://doi.org/10.1080/01490451.2022.2038953

de Carvalho Lima, E. N., Souza, L. H. M., Aguiar, E. M., Octaviano, A. L. M., Justo, J. F., & Piqueira, J. R. C. (2024). Can bacteria and carbon-based nanomaterials revolutionize nanoremediation strategies for industrial effluents? *Frontiers in Nanotechnology, 6*, 1389107. https://doi.org/10.3389/FNANO.2024.1389107/PDF

Ding, T., & Chen, J. (2023). Evaluating supply-demand matching of ecosystem services considering water-energy-food nexus and synergies/trade-offs in the Hangzhou of China. *Environmental Science and Pollution Research, 30*(19), 54568–54585. https://doi.org/10.1007/S11356-023-26055-9/FIGURES/8

Dubey, S., Parmar, P. R., Vanarse, V. B., & Bandyopadhyay, D. (2022). Instability facilitated micro nano-plastic disintegration under thermal influence. *On "Circular Economy on Sustainable Basis: The Role of Chemical Engineers,"* 39–40.

Edo, C., González-Pleiter, M., Leganés, F., Fernández-Piñas, F., & Rosal, R. (2020). Fate of microplastics in wastewater treatment plants and their environmental dispersion with effluent and sludge. *Environmental Pollution*, *259*. https://doi.org/10.1016/j.envpol.2019.113837

Elwaheidi, M. M. (2023). Waste management using nanotechnology. *Handbook of Research on Safe Disposal Methods of Municipal Solid Wastes for a Sustainable Environment*. https://doi.org/10.4018/978-1-6684-8117-2.ch001

Gebre, S. H., & Sendeku, M. G. (2019). New frontiers in the biosynthesis of metal oxide nanoparticles and their environmental applications: an overview. *SN Applied Sciences*, *1*(8). https://doi.org/10.1007/s42452-019-0931-4

Gethard, K., Sae-Khow, O., & Mitra, S. (2011). Water desalination using carbon-nanotube-enhanced membrane distillation. *ACS Applied Materials and Interfaces, 3*(2). https://doi.org/10.1021/am100981s

Gharpure, A., Kowalik, M., Vander Wal, R. L., & van Duin, A. C. T. (2024). Upcycling plastic waste into graphite using graphenic additives for energy storage: yield, graphitic quality, and interaction mechanisms via experimentation and molecular dynamics. *ACS Sustainable Chemistry and Engineering, 12*(11), 4565–4575. https://doi.org/10.1021/ACSSUSCHEMENG.3C07783/ASSET/IMAGES/LARGE/SC3C07783_0008.JPEG

Grbic, J., Nguyen, B., Guo, E., You, J. B., Sinton, D., & Rochman, C. M. (2019). Magnetic extraction of microplastics from environmental samples. *Environmental Science and Technology Letters, 6*(2). https://doi.org/10.1021/acs.estlett.8b00671

Gupte, T., & Pradeep, T. (2022). Nanosensors for water quality monitoring. *Separation Science and Technology (New York), 15*, 37–53. https://doi.org/10.1016/B978-0-323-90763-7.00010-X

Haghighizadeh, A., Aghababai Beni, A., Haghmohammadi, M., Adel, M. S. S., & Farshad, S. (2023). Green synthesis of ZnO-TiO_2 nano-photocatalyst doped with Fe(III) ions using bitter olive extract to treat textile wastewater containing reactive dyes. *Water, Air, and Soil Pollution, 234*(6), 1–17. https://doi.org/10.1007/S11270-023-06374-W/FIGURES/10

Hamad, M. T. M. H., & El-Sesy, M. E. (2023). Adsorptive removal of levofloxacin and antibiotic resistance genes from hospital wastewater by nano-zero-valent iron and nano-copper using kinetic studies and response surface methodology. *Bioresources and Bioprocessing, 10*(1), 1–29. https://doi.org/10.1186/S40643-022-00616-1/FIGURES/13

Han, Y., Xu, Z., & Gao, C. (2013). Ultrathin graphene nanofiltration membrane for water purification. *Advanced Functional Materials, 23*(29). https://doi.org/10.1002/adfm.201202601

Hemalatha, P. K., Kanagaraj, K., Raman, K., Nangan, S., Ramesh, M. D., Aggarwal, S., & Kumar, R. (2024). Nano-innovations for a greener future: exploring the intersection of nanotechnology and environmental sustainability. *E3S Web of Conferences, 588*, 01018. https://doi.org/10.1051/E3SCONF/202458801018

Hidangmayum, A., Debnath, A., Guru, A., Singh, B. N., Upadhyay, S. K., & Dwivedi, P. (2023). Mechanistic and recent updates in nano-bioremediation for developing green technology to alleviate agricultural contaminants. *International Journal of Environmental Science and Technology, 20*(10). https://doi.org/10.1007/s13762-022-04560-7

Hou, P. X., Zhang, F., Zhang, L., Liu, C., & Cheng, H. M. (2022). Synthesis of carbon nanotubes by floating catalyst chemical vapor deposition and their applications. *Advanced Functional Materials, 32*(11). https://doi.org/10.1002/adfm.202108541

Huang, D. L., Wang, R. Z., Liu, Y. G., Zeng, G. M., Lai, C., Xu, P., Lu, B. A., Xu, J. J., Wang, C., & Huang, C. (2015). Application of molecularly imprinted polymers in wastewater treatment: a review. *Environmental Science and Pollution Research, 22*(2). https://doi.org/10.1007/s11356-014-3599-8

Igiebor, F. A., Asia, M., & Ikhajiagbe, B. (2023). Green Nanotechnology: A modern tool for sustainable agriculture—a review. *International Journal of Horticultural Science and Technology, 10*(3). https://doi.org/10.22059/ijhst.2022.344790.571

Islam, M. D., Uddin, F. J., Rashid, T. U., & Shahruzzaman, M. (2023). Cellulose acetate-based membrane for wastewater treatment—a state-of-the-art review. *Materials Advances, 4*(18). https://doi.org/10.1039/d3ma00255a

Jancik-Prochazkova, A., Jašek, V., Figalla, S., Pumera, M., Jancik-Prochazkova, A., Pumera, M., Jašek, V., & Figalla, S. (2023b). Photocatalytic microplastics "on-the-fly" degradation via motile quantum materials-based microrobots. *Advanced Optical Materials, 11*(22), 2300782. https://doi.org/10.1002/ADOM.202300782

Jancik-Prochazkova, A., Jašek, V., Figalla, S., & Pumera, M. (2023). Photocatalytic microplastics "on-the-fly" degradation via motile quantum materials-based microrobots. *Advanced Optical Materials, 11*(22). https://doi.org/10.1002/adom.202300782

Khan, W. S., Asmatulu, R., Davuluri, S., & Dandin, V. K. (2014). Improving the economic values of the recycled plastics using nanotechnology associated studies. *Journal of Materials Science and Technology*, *30*(9). https://doi.org/10.1016/j.jmst.2014.07.006

Kim, M., Lee, B. Y., Ham, H. C., Han, J., Nam, S. W., Lee, H. S., Park, J. H., Choi, S., & Shin, Y. (2016). Facile one-pot synthesis of tungsten oxide (WO_{3-x}) nanoparticles using sub and supercritical fluids. *The Journal of Supercritical Fluids, 111*, 8–13. https://doi.org/10.1016/J.SUPFLU.2016.01.011

Langer, J., de Aberasturi, D. J., Aizpurua, J., Alvarez-Puebla, R. A., Auguié, B., Baumberg, J. J., Bazan, G. C., Bell, S. E. J., Boisen, A., Brolo, A. G., Choo, J., Cialla-May, D., Deckert, V., Fabris, L., Faulds, K., Javier García de Abajo, F., Goodacre, R., Graham, D., Haes, A. J., & Liz-Marzán, L. M. (2020). Present and future of surface-enhanced Raman scattering. *ACS Nano, 14*(1). https://doi.org/10.1021/acsnano.9b04224

Lapointe, M., Kurusu, R. S., Hernandez, L. M., & Tufenkji, N. (2023). Removal of classical and emerging contaminants in water treatment using super-bridging fiber-based materials. *ACS ES and T Water, 3*(2), 377–386. https://doi.org/10.1021/ACSESTWATER.2C00443/ASSET/IMAGES/LARGE/EW2C00443_0007.JPEG

Li, X., Xu, H., Chen, Z. S., & Chen, G. (2011). Biosynthesis of nanoparticles by microorganisms and their applications. *Journal of Nanomaterials, 2011*. https://doi.org/10.1155/2011/270974

Linley, S., & Thomson, N. R. (2021). Environmental applications of nanotechnology: nano-enabled remediation processes in water, soil and air treatment. *Water, Air, and Soil Pollution, 232*(2). https://doi.org/10.1007/s11270-021-04985-9

Mahar, B., Laslau, C., Yip, R., & Sun, Y. (2007). Development of carbon nanotube-based sensors—a review. *IEEE Sensors Journal*, *7*(2). https://doi.org/10.1109/JSEN.2006.886863

Maity, S., Bhuyan, T., Bhattacharya, R., & Bandyopadhyay, D. (2021a). Self-organized implanting of micro/nanofiltration membranes in advanced flow μ-reactors. *ACS Applied Materials and Interfaces*, *13*(16). https://doi.org/10.1021/acsami.1c01078

Maity, S., Bhuyan, T., Pattanayak, J. P., Ghosh, S. S., & Bandyopadhyay, D. (2021b). Real-time transport kinetics of drug encapsulated nanoparticles into apoptotic cancer cells inside microchannels. *Nanotechnology*, *32*(50). https://doi.org/10.1088/1361-6528/ac2391

Mandeep, & Shukla, P. (2020). Microbial nanotechnology for bioremediation of industrial wastewater. *Frontiers in Microbiology,11*. https://doi.org/10.3389/fmicb.2020.590631

Manikandan, S., Subbaiya, R., Saravanan, M., Ponraj, M., Selvam, M., & Pugazhendhi, A. (2022). A critical review of advanced nanotechnology and hybrid membrane based water recycling, reuse, and wastewater treatment processes. *Chemosphere*, *289*. https://doi.org/10.1016/j.chemosphere.2021.132867

Mao, K., Ma, J., Li, X., & Yang, Z. (2019). Rapid duplexed detection of illicit drugs in wastewater using gold nanoparticle conjugated aptamer sensors. *Science of the Total Environment*, *688*. https://doi.org/10.1016/j.scitotenv.2019.06.325

Mayorga-Burrezo, P., Mayorga-Martinez, C. C., & Pumera, M. (2023). Photocatalysis dramatically influences motion of magnetic microrobots: Application to removal of microplastics and dyes. *Journal of Colloid and Interface Science*, *643*. https://doi.org/10.1016/j.jcis.2023.04.019

Meyyappan, M. (2016). Carbon nanotube-based chemical sensors. *Small,12*(16). https://doi.org/10.1002/smll.201502555

Mishra, S., Dash, D., Al-Tawaha, A. R. M. S., & Das, A. P. (2022). A review on heavy metal ion adsorption on synthetic microfiber surface in aquatic environments. *Applied Biochemistry and Biotechnology, 194*(10), 4639–4654. https://doi.org/10.1007/S12010-022-04029-W/METRICS

Mishra, S., Dash, D., & Das, A. P. (2024). Aquatic microbial diversity on plastisphere: colonization and potential role in microplastic biodegradation. *Geomicrobiology Journal, 41*(4), 312–323. https://doi.org/10.1080/01490451.2023.2209750

Mishra, S., & Das, A. P. (2023). Microbial remediation of synthetic microfiber contaminated wastewater. In M. P. Shah (Ed.), *Microbial Technologies in Industrial Wastewater Treatment* (pp. 337–350). Springer, Singapore. https://doi.org/10.1007/978-981-99-2435-6_17

Mohamed, E. F., & Awad, G. (2022). Development of nano-sensor and biosensor as an air pollution detection technique for the foreseeable future. *Comprehensive Analytical Chemistry, 99*, 163–188. https://doi.org/10.1016/BS.COAC.2021.11.003

Mohanty, J., Dey, S., Bala, K., & Das, A. P. (2024). Recent advances in microplastic pollution for its sustainable management. *Sustainable Management of Urban Plastic Waste Through Circular Economic Approaches*, 27–48. https://doi.org/10.1201/9781003465447-3

Mohapatra, L., Cheon, D., & Yoo, S. H. (2023). Carbon-based nanomaterials for catalytic wastewater treatment: A review. *Molecules, 28*(4). https://doi.org/10.3390/molecules28041805

Mohd Nazri, M. K. H., & Sapawe, N. (2020). Removal of methyl orange over low-cost silica nanoparticles extrated from bamboo leaves ash. *Materials Today: Proceedings, 31*, A54–A57. https://doi.org/10.1016/J.MATPR.2020.10.969

Moretti, A., Ivan, H. L., & Skvaril, J. (2024). A review of the state-of-the-art wastewater quality characterization and measurement technologies. Is the shift to real-time monitoring nowadays feasible? *Journal of Water Process Engineering, 60*. https://doi.org/10.1016/j.jwpe.2024.105061

Mu, Y., Yu, H. Q., Zheng, J. C., Zhang, S. J., & Sheng, G. P. (2004). Reductive degradation of nitrobenzene in aqueous solution by zero-valent iron. *Chemosphere, 54*(7), 789–794. https://doi.org/10.1016/J.CHEMOSPHERE.2003.10.023

Murray, A., & Örmeci, B. (2012). Application of molecularly imprinted and non-imprinted polymers for removal of emerging contaminants in water and wastewater treatment: A review. *Environmental Science and Pollution Research, 19*(9). https://doi.org/10.1007/s11356-012-1119-2

Mushtaq, Z., Bangotra, P., & Sajad, S. (2024). Nanosensors to detect and quantify air pollutants. *Nanotechnology to Monitor, Remedy, and Prevent Pollution*, 169–201. https://doi.org/10.1016/B978-0-443-15660-1.00008-3

Mustafa, G., Shahzeb Khan, M., Asif, M. I., Ullah, A., Khan, I., & Ullah, I. (2021). Water pollutants and nanosensors. *Aquananotechnology: Applications of Nanomaterials for Water Purification*, 105–133. https://doi.org/10.1016/B978-0-12-821141-0.00018-5

Nagaveni, K., Sivalingam, G., Hegde, M. S., & Madras, G. (2004). Photocatalytic degradation of organic compounds over combustion-synthesized nano-TiO_2. *Environmental Science and Technology, 38*(5). https://doi.org/10.1021/es034696i

Nasrollahzadeh, M., Sajjadi, M., Iravani, S., & Varma, R. S. (2021). Carbon-based sustainable nanomaterials for water treatment: State-of-art and future perspectives. *Chemosphere,263*. https://doi.org/10.1016/j.chemosphere.2020.128005

Nille, O. S., Patel, R. S., Borate, B. Y., Babar, S. S., Kolekar, G. B., & Gore, A. H. (2023). One-step in-situ sustainable synthesis of magnetic carbon nanocomposite from corn comb (MCCC): Agricultural biomass valorisation for pollutant abatement in wastewater. *Environmental Science and Pollution Research, 30*(13), 38425–38442. https://doi.org/10.1007/S11356-022-24847-Z/FIGURES/13

Ouni, S., Mohamed, N. B. H., Haouari, M., Elaissari, A., Errachid, A., & Jaffrezic-Renault, N. (2023). A novel green synthesis of zinc sulfide nano-adsorbents using Artemisia Herba alba plant extract for adsorption and photocatalysis of methylene blue dye. *Chemistry Africa, 6*(5), 2523–2535. https://doi.org/10.1007/S42250-023-00667-7/FIGURES/12

Parmar, P. R., Mahanta, J., Dubey, S., Mandal, T. K., & Bandyopadhyay, D. (2023). Real-time monitoring of glycemic starch using biosynthesized gold nanoparticle-amylase composites. *ACS Sustainable Chemistry and Engineering, 11*(28), 10509–10519. https://doi.org/10.1021/ACSSUSCHEMENG.3C02292/ASSET/IMAGES/LARGE/SC3C02292_0008.JPEG

Paul, A., Basak, M., Ahmed, M. M., Singha, D., & Bandyopadhyay, D. (2024). Decoration of nanoenabled optoplasmonic sensors on flexible micropatterned surfaces. *IEEE Sensors Journal, 24*(17), 27152–27159. https://doi.org/10.1109/JSEN.2024.3429528

Paul, S., Bhoumick, M. C., Roy, S., & Mitra, S. (2022). Carbon nanotube enhanced membrane filtration for trace level dewatering of hydrocarbons. *Separation and Purification Technology, 292*. https://doi.org/10.1016/j.seppur.2022.121047

Ratnam, M. V., Akilamudhan, P., Kumar, K. S., Reddy, S. N., Rao, K. N., Shaik, F., & Prasad, D. M. R. (2023). Carbon-based nanoadsorbents for the removal of emerging pollutants. *Adsorption Science and Technology, 2023*. https://doi.org/10.1155/2023/3579165

Roy, D., Mandal, S., & Mukhopadhyay, K. (2024). Mitigation strategies using nano-enabled technologies for sustainable development goals. *Proceedings of the National Academy of Sciences India Section A—Physical Sciences, 94*(1). https://doi.org/10.1007/s40010-023-00854-8

Sahu, A., Dey, S., Mallick, K., Priyadarsini, S., Behera, I. D., & Das, A. P. (2024). A review on advanced technology for sustainable management of synthetic microplastic waste. *Sustainable Management of Urban Plastic Waste Through Circular Economic Approaches*, 262–279. https://doi.org/10.1201/9781003465447-15

Salem, R. M. M., Saraya, M. S., & Ali-Eldin, A. M. T. (2022). An industrial cloud-based IoT system for real-time monitoring and controlling of wastewater. *IEEE Access, 10*. https://doi.org/10.1109/ACCESS.2022.3141977

Sankar ganesh, R., Navaneethan, M., Mani, G. K., Ponnusamy, S., Tsuchiya, K., Muthamizhchelvan, C., Kawasaki, S., & Hayakawa, Y. (2017). Influence of Al doping on the structural, morphological, optical, and gas sensing properties of ZnO nanorods. *Journal of Alloys and Compounds, 698*. https://doi.org/10.1016/j.jallcom.2016.12.187

Santos, J. P., Sayago, I., & Aleixandre, M. (2020). Air quality monitoring using nanosensors. *Nanomaterials for Air Remediation*, 9–31. https://doi.org/10.1016/B978-0-12-818821-7.00002-6

Sarkar, B., Mandal, S., Tsang, Y. F., Kumar, P., Kim, K. H., & Ok, Y. S. (2018). Designer carbon nanotubes for contaminant removal in water and wastewater: A critical review.*Science of the Total Environment, 612*. https://doi.org/10.1016/j.scitotenv.2017.08.132

Satchanska, G., Davidova, S., & Petrov, P. D. (2024). Natural and synthetic polymers for biomedical and environmental applications. *Polymers, 16*(8), 1159. https://doi.org/10.3390/POLYM16081159

Shariful Islam, M., Jakaria, M., Masum Rabbi Mim, M., Raihan, J., Khan, A., & Shafiur Rahman, G. M. (2024). Application of nanotechnology in plastic waste management and recycling: Bangladesh perspective. *Publishing.Emanresearch.OrgM Shariful, MM Rabbi, GM ShafiurBiosensors and Nanotheranostics, 2024•publishing.Emanresearch.Org*. https://doi.org/10.25163/biosensors.319952

Singh, Y., & Saxena, M. K. (2022). Insights into the recent advances in nano-bioremediation of pesticides from the contaminated soil. *Frontiers in Microbiology, 13*. https://doi.org/10.3389/fmicb.2022.982611

Singh Jadaun, J., Bansal, S., Sonthalia, A., Rai, A. K., & Singh, S. P. (2022). Biodegradation of plastics for sustainable environment. *Bioresource Technology, 347*, Article 126697. https://doi.org/10.1016/J.BIORTECH.2022.126697

Smith, A. M., Duan, H., Mohs, A. M., & Nie, S. (2008). Bioconjugated quantum dots for in vivo molecular and cellular imaging. *Advanced Drug Delivery Reviews, 60*(11). https://doi.org/10.1016/j.addr.2008.03.015

Song, T., Zhao, H., Hu, Y., Sun, N., & Zhang, H. (2019). Facile assembly of mesoporous silica nanoparticles with hierarchical pore structure for CO_2 capture. *Chinese Chemical Letters, 30*(12), 2347–2350. https://doi.org/10.1016/J.CCLET.2019.07.024

Stozhko, N. Y., Bukharinova, M. A., Khamzina, E. I., Tarasov, A. V., Vidrevich, M. B., & Brainina, K. Z. (2019). The effect of the antioxidant activity of plant extracts on the properties of gold nanoparticles. *Nanomaterials, 9*(12). https://doi.org/10.3390/nano9121655

Sugunan, A., Thanachayanont, C., Dutta, J., & Hilborn, J. G. (2005). Heavy-metal ion sensors using chitosan-capped gold nanoparticles. *Science and Technology of Advanced Materials, 6*(3–4 SPEC. ISS.). https://doi.org/10.1016/j.stam.2005.03.007

Sun, C., Wang, Z., Chen, L., & Li, F. (2020). Fabrication of robust and compressive chitin and graphene oxide sponges for removal of microplastics with different functional groups. *Chemical Engineering Journal, 393*. https://doi.org/10.1016/j.cej.2020.124796

Susan Punnoose, M., Bijimol, D., & Mathew, B. (2021). Microwave assisted green synthesis of gold nanoparticles for catalytic degradation of environmental pollutants. *Environmental Nanotechnology, Monitoring and Management, 16*. https://doi.org/10.1016/j.enmm.2021.100525

Talapin, D. V., & Shevchenko, E. V. (2016). Introduction: Nanoparticle chemistry. *Chemical Reviews, 116*(18). https://doi.org/10.1021/acs.chemrev.6b00566

Tan, Y. N., Wong, C. L., & Mohamed, A. R. (2011). An overview on the photocatalytic activity of nano-doped-TiO_2 in the degradation of organic pollutants. *International Scholarly Research Notices, 2011*(1), Article 261219. https://doi.org/10.5402/2011/261219

Tripathy, B., Sahoo, P. P., Sundaray, H., & Das, A. P. (2024). In-vitro biodegradation of discarded marine microplastics across the eastern coast of the Bay of Bengal, India using Exiguobacterium sp. *Environmental Chemistry and Ecotoxicology, 6*, 236–247. https://doi.org/10.1016/J.ENCECO.2024.07.003

Ullattil, S. G., & Pumera, M. (2023). Light-powered self-adaptive mesostructured microrobots for simultaneous microplastics trapping and fragmentation via in situ surface morphing. *Small, 19*(38). https://doi.org/10.1002/smll.202301467

Urso, M., Ussia, M., Novotný, F., & Pumera, M. (2022). Trapping and detecting nanoplastics by MXene-derived oxide microrobots. *Nature Communications, 13*(1), 1–14. https://doi.org/10.1038/s41467-022-31161-2

Vanarse, V. B., Ahmed, M. M., Singha, D., Paul, A., & Bandyopadhyay, D. (2025). Depolymerization: A catalyst for environmental sustainability. *ACS Symposium Series*, 103–129. https://doi.org/10.1021/BK-2025-1498.CH006

Vas, N. V., Sundari, S. K., & Jeyachandran, S. (2024). Degradation of microplastics by zinc oxide nanoparticles synthesized using piper longum leaf extract. *Cureus, 16*(9), Article e69876. https://doi.org/10.7759/CUREUS.69876

Vikesland, P. J. (2018). Nanosensors for water quality monitoring. *Nature Nanotechnology, 13*(8). https://doi.org/10.1038/s41565-018-0209-9

Wang, L., Kaeppler, A., Fischer, D., & Simmchen, J. (2019). Photocatalytic TiO_2 micromotors for removal of microplastics and suspended matter. *ACS Applied Materials and Interfaces, 11*(36). https://doi.org/10.1021/acsami.9b06128

Wang, Z., Xu, W., Jie, F., Zhao, Z., Zhou, K., & Liu, H. (2021). The selective adsorption performance and mechanism of multiwall magnetic carbon nanotubes for heavy metals in wastewater. *Scientific Reports, 11*(1). https://doi.org/10.1038/s41598-021-96465-7

Weerasinghe, M., Jayathilaka, K., & Vithanage, M. (2025). Sensors for detection and monitoring of contaminants in wastewater. *Current Opinion in Environmental Science & Health*, 100609. https://doi.org/10.1016/J.COESH.2025.100609

Wu, H., Lin, G., Liu, C., Chu, S., Mo, C., & Liu, X. (2022). Progress and challenges in molecularly imprinted polymers for adsorption of heavy metal ions from wastewater. *Trends in Environmental Analytical Chemistry, 36*. https://doi.org/10.1016/j.teac.2022.e00178

Xue, W., Huang, D., Zeng, G., Wan, J., Cheng, M., Zhang, C., Hu, C., & Li, J. (2018). Performance and toxicity assessment of nanoscale zero valent iron particles in the remediation of contaminated soil: A review. *Chemosphere, 210*. https://doi.org/10.1016/j.chemosphere.2018.07.118

Yan, C., Qu, Z., Wang, J., Cao, L., & Han, Q. (2022). Microalgal bioremediation of heavy metal pollution in water: Recent advances, challenges, and prospects. *Chemosphere, 286*. https://doi.org/10.1016/j.chemosphere.2021.131870

You, X., Wang, R., Zhu, Y., Sui, W., & Cheng, D. (2021). Comparison of adsorption properties of a cellulose-rich modified rice husk for the removal of methylene blue and aluminum (III) from their aqueous solution. *Industrial Crops and Products, 170*. https://doi.org/10.1016/j.indcrop.2021.113687

Zeng, R., Wang, J., Cui, J., Hu, L., & Mu, K. (2010). Photocatalytic degradation of pesticide residues with RE3+ -doped nano-TiO_2. *Journal of Rare Earths, 28*(SUPPL. 1), 353–356. https://doi.org/10.1016/S1002-0721(10)60329-8

Zhang, H., Wang, F., Akakuru, O. U., Wang, T., Wang, Z., Wu, A., & Zhang, Y. (2022). Nature-inspired polyethylenimine-modified calcium alginate blended waterborne polyurethane graded functional materials for multiple water purification. *ACS Applied Materials and Interfaces, 14*(15). https://doi.org/10.1021/acsami.2c02059

[illegible]

Next-Generation Green Nanomaterials for Eco-Friendly Photocatalysis: Enhancing Efficiency in Environmental Sustainability Through Solar Energy

R. Venkatesh

Abstract Years of ever-increasing demand for environmental sustainability have been forcing significant research progress in the development of next-generation green nanomaterials, specifically in photocatalysis. Ranging from resolving serious challenges in the environment, the nanomaterials, designed to efficiently harvest solar energy, can solve problems of water purification, reduction of carbon dioxide, and generation of solar hydrogen. With the growing concern over climate change and pollution, the time has never been so crucial for technologies that can reduce harm to the environment while finding sustainable energy solutions. It discusses new developments in the field of green photocatalytic nanomaterials, with a focus on fresh design approaches that aim to make solar energy conversion more effective using these photocatalysts. Such advances, of course, are crucial for optimizing photocatalysts for particular applications. The design of both the materials themselves and their accompanying synthesis and application systems must be oriented toward sustainability so they contribute to water decontamination and reductions in greenhouse gas emissions while minimizing environmental impacts. In addition, the chapter highlights the most crucial challenges in scaling up the production of materials under focus and integrating them into practical applications. The large-scale production needs to be ensured not to undermine environmental advantages. Finally, the next-generation photocatalysts are envisioned as advanced nanotechnology shaped by green chemistry principles, offering a clean route toward sustainable environmental solutions that bring the planet closer to global ambitions toward a cleaner, greener future.

Keywords Green nanomaterials · Photocatalysis · Environmental sustainability · Nanotechnology · CO_2 reduction · Green chemistry

R. Venkatesh (✉)
Department of Physics, PSNA College of Engineering and Technology (Autonomous), Dindigul, Tamil Nadu, India
e-mail: sukatesh@gmail.com

E. Priyadarshini et al. (eds.), *Advanced Nanotechnology Applications for Environmental Sustainability*, Sustainable Environmental Waste Management Strategies,
https://doi.org/10.1007/978-3-032-12933-8_17

1 Introduction to Next-Generation Green Nanomaterials

The quest for clean energy options has propelled nanomaterial research, especially for photocatalysis, a high-tech process for harnessing the sun's power to power chemical reactions for a cleaner environment and power generation. Research now targets designing future-generation environmentally friendly nanomaterials that are efficient as well (Hussain et al., 2021). These novel materials provide superior photocatalytic activity due to their large surface area, distinctive optical properties, and nanoscale engineering capacity for tailored purposes. Metal oxide and carbon nanomaterials, among others, have proven particularly promising in photocatalysis with major uses in water cleaning, CO_2 reduction, and solar hydrogen production (Fatima et al., 2020). But numerous materials that have been developed still have issues of poor stability, limited absorption of visible light, and low efficiency in actual applications. Green nanomaterials aim to break these limitations by enhancing performance with new designs while staying within environmental sustainability and green chemistry principles.

2 Evolution of Nanomaterials in Photocatalysis

Nanomaterials have been employed in photocatalysis for many decades, and important developments occurred towards the end of the twentieth century. Initial work centered on semiconductor materials such as titanium dioxide (TiO_2), which exhibited photocatalytic behavior under UV illumination for reactions such as water splitting and degradation of organic pollutants. Due to TiO_2's narrow UV absorption and poor quantum efficiency, however, researchers searched for other materials and innovative strategies to improve its performance (Tahir et al., 2020).

Recent developments in nanomaterials have profoundly enhanced photocatalysis by tapping into new features like high surface area, quantum effects, and tunable structure. Methods like nanostructuring, metal or nonmetal doping, and mixing disparate nanomaterials have all helped to augment photocatalytic efficiency (Nasirian et al., 2017). Doping, for instance, enhances visible light absorption, while combining materials like TiO_2-graphene improves electron transport and charge separation. Bandgap engineering additionally maximizes light absorption, further increasing overall efficiency. Though these advances have significantly enhanced the activity of photocatalysts, there are still challenges in terms of scalability, environmental concerns, and stability over extended periods (Nasirian et al., 2018). Present research is directed towards developing "green" nanomaterials that are not only effective but also sustainable and eco-friendly. This table (Table 1) gives a concise summary of the major developments in nanomaterials for photocatalysis over the years, with materials and innovations during each stage and their effect on the efficiency and sustainability of photocatalytic processes.

Table 1 Advancements in nanomaterials for photocatalytic applications

Development	Materials	Advancements	Impact on photocatalysis
Early years (Late twentieth century)	Titanium Dioxide (TiO_2)	Photocatalytic activity under UV light for water splitting and organic pollutant degradation	Limited UV absorption and low quantum efficiency
Recent developments (twenty-first century)	Nanostructured materials (e.g., nanoparticles, nanowires, nanorods)	Tailored shapes and sizes to increase surface area and improve charge carrier separation	Improved photocatalytic efficiency by increasing surface area
Nonmetals (e.g., nitrogen, sulfur)	Doped materials for visible light absorption	Introduction of dopants to extend light absorption into the visible range and enhance photocatalytic activity	Enhanced light absorption and photocatalytic performance
Composite materials	TiO_2 with graphene or carbon-based materials	Hybrid systems formed by combining different nanomaterials to exploit synergistic effects	Improved charge separation, electron mobility, and efficiency
Bandgap engineering	Advanced Nanomaterials with tunable electronic structures	Tuning of electronic structures to optimize light absorption and charge transfer for photocatalysis	Broader spectrum of solar energy utilization, enhanced efficiency
Green nanomaterials (current focus)	Environmentally friendly and sustainable materials	Focus on scalability, environmental impact, and long-term stability	Transition to sustainable, efficient photocatalysts

3 Green Chemistry Principles in Nanomaterial Design

Green chemistry principles (Fig. 1), which emphasize minimizing environmental footprint and resource consumption, are central to the design of next-generation nanomaterials for photocatalysis (Yadav & Ahmaruzzaman, 2022). The materials are intended to be effective, sustainable, and safe throughout their life cycle, from synthesis to end-of-life disposal.

Some of the most important green chemistry principles informing environmentally friendly nanomaterials are:

Atom Economy and Resource Efficiency: Environmentally friendly synthesis techniques minimize waste and utilize resources effectively, with processes such as sol-gel, hydrothermal, and bio-inspired synthesis minimizing toxic chemicals.

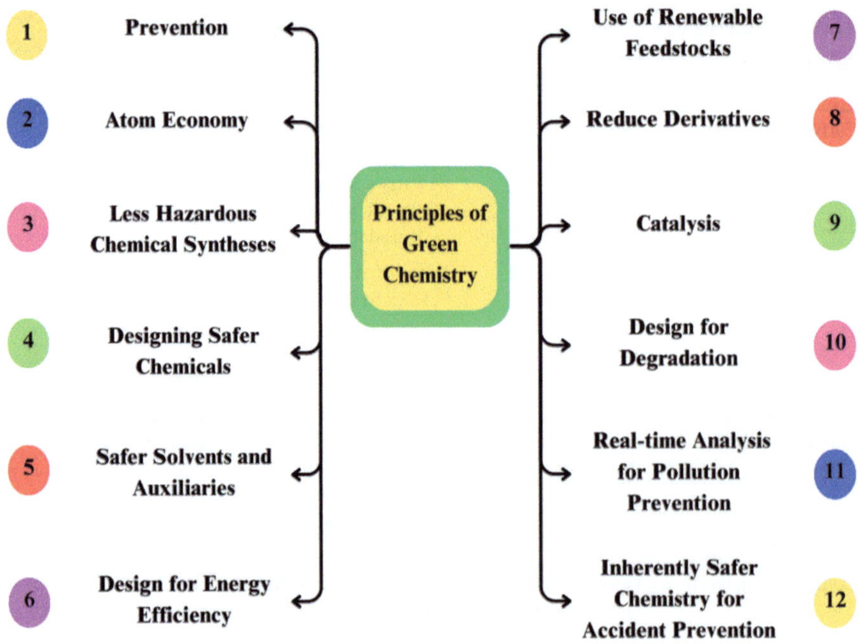

Fig. 1 Essential principles of sustainable chemistry

Non-toxic and Biocompatible Materials: Nanomaterials are synthesized from non-toxic, plentiful, and biodegradable sources such as zinc oxide, copper, iron, graphene, and carbon nanotubes, making them safe to dispose of and having low environmental footprint.

Energy Efficiency and Green Synthesis Routes: Green chemistry favors energy-efficient, low-temperature, or solvent-free synthesis routes, usually powered by renewable energy sources such as solar energy.

Minimization of Hazardous Chemicals and Waste: The application of non-toxic solvents and reagents lowers harmful byproducts, with strategies such as bio-based reduction and mechanochemistry minimizing environmental impact.

End-of-Life Considerations: Green chemistry promotes the design of nanomaterials to be recyclable, reusable, or biodegradable, which avoids environmental build-up.

Through these principles, green nanomaterials enhance performance while boosting sustainability. The incorporation of green chemistry into the formulation of photocatalytic nanomaterials is a milestone toward effective, environmentally friendly options in solar energy and environmental sustainability (Anaya-Rodríguez et al., 2023).

4 Advancements in Nanomaterial Design for Photocatalysis

The evolution of next-generation nanomaterials for photocatalysis has emerged as a vibrant and multidisciplinary field of research, focusing on optimizing the efficiency of solar energy conversion towards environmental and energy applications (Priyadarshini & Das, 2024; Singh et al., 2024). With growing demand for cleaner and greener technologies, the engineering of photocatalytic materials that optimize light absorption, enhance charge carrier dynamics, and improve stability under practical conditions has become essential (Thongam & Chaturvedi, 2021).

5 Design Strategies for Enhanced Photocatalytic Efficiency

Photocatalyst performance relies on a number of parameters, such as light absorption characteristics, charge separation efficiency, and reactivity at the surface. To develop the efficiency of nanomaterials in photocatalytic processes thus necessitates the use of strategic design principles that enhance these parameters (Kaus et al., 2021). A number of significant approaches have been developed to enhance photocatalytic efficiency, such as structural engineering, doping, heterojunction construction, and surface modification.

Structural Engineering and Morphological Control: Nanomaterials of certain shapes, sizes, and structures can significantly enhance photocatalytic activity. Tuning the nanomaterial structure makes it possible to optimize surface area, light absorption, and electron transport properties. High surface-to-volume ratio nanomaterials, including nanoparticles, nanorods, nanosheets, and hollow spheres, offer enhanced active surfaces for reactions, leading to improved efficiency. One-dimensional (1D) and two-dimensional (2D) nanostructures, such as nanowires and nanosheets, enable improved electron flow, minimizing recombination and enhancing charge separation, and also facilitating light trapping for enhanced absorption (Vu et al., 2019). Hollow and core–shell nanostructures enhance efficiency by providing room for reactants and improving charge separation and preventing performance loss due to aggregation.

Bandgap Engineering: Bandgap engineering is significant in tailoring semiconducting nanomaterials for photocatalytic purposes by adjusting their light-absorbing capability and converting this into chemical energy. Through bandgap modification, scientists can adjust the nanomaterials to have the ability to absorb more sunlight, including visible light. One of the general approaches is to narrow the bandgap via doping or material alteration, making it possible for photocatalysts to absorb visible light more effectively, which is essential for use in solar hydrogen generation (Ayyub & Rao, 2020). The second strategy is photonic design, in which plasmonic nanomaterials such as gold and silver nanoparticles are utilized to absorb light within the visible to infrared region. Plasmonic materials amplify photocatalytic reactions by delivering energy to neighboring semiconductors, which increases overall light-harvesting efficiency.

Charge Separation and Electron Transport: Effective charge separation and electron transport are critical to the optimization of photocatalytic activity, as electron-hole recombination can decrease the process efficiency. One method to enhance this is by the creation of heterojunctions and composite materials, where the juxtaposition of semiconductors with complementary band structures maximizes charge separation and reduces recombination. For instance, the combination of n-type semiconductors such as TiO_2 with p-type semiconductors such as Cu_2O, or the introduction of carbon-based compounds such as graphene, can enhance electron transfer and photocatalytic efficiency (Chen et al., 2021). Furthermore, cocatalysts and electron mediators such as noble metals (Pt, Au) or non-metallic compounds such as N-doped carbon can be loaded on nanomaterials to serve as electron sinks, suppressing recombination and enhancing the overall rate of reaction.

Surface Modification and Functionalization: Surface modification of nanomaterials is an important factor in the improvement of photocatalytic performance through the change of their chemical reactivity and light interaction. Methods such as doping, molecular group functionalization, and surface coating can tailor these properties. Metal (e.g., platinum or palladium) or non-metal (e.g., nitrogen or sulfur) doping can adjust the electronic features of semiconductors, broadening the range of light absorption and improving catalytic efficiency. For instance, nitrogen doping of TiO_2 broadens its absorption to the visible light region (Molinari et al., 2000). Surface plasmon resonance (SPR) in metal nanoparticles also increases light absorption and enables charge transfer between the nanomaterials and the photocatalytic surface, enhancing overall efficiency.

6 Synthesis Methods and Material Innovation

The synthesis techniques employed to synthesize photocatalytic nanomaterials play the central role in defining their structure, composition, and, in turn, their efficiency. Classical top-down methods like ball milling or lithography were complemented with more sophisticated bottom-up techniques to enable increased control over material design and the properties of the resultant photocatalyst. The invention of new, eco-friendly, and scalable synthesis techniques has been a central field of innovation in photocatalysis.

Sol-Gel and Hydrothermal Synthesis: Sol-gel processing and hydrothermal synthesis are two of the most popular methods of synthesizing nanomaterials, providing flexibility, affordability, and fine-tuned control over particle size, shape, and morphology.

The sol-gel process is a process of converting a solution or sol into a gel phase, from which nanomaterials are derived by drying and calcination. The method enables the formation of homogeneous nanostructures, including mesoporous materials, metal oxide nanoparticles, and composites, which are especially well-suited for photocatalytic purposes because of their high surface area and adjustable properties (Obregón & Rodríguez-González, 2021).

The hydrothermal synthesis employs high-pressure and high-temperature conditions in aqueous media to create nanomaterials. The technique is particularly useful for the synthesis of high-purity, crystalline materials and intricate nanostructures. It is widely employed to synthesize semiconducting nanomaterials such as TiO_2, ZnO, and CdS, which are critical for numerous photocatalytic applications (Gan et al., 2020).

Green Synthesis and Bio-Inspired Approaches: In accordance with the green chemistry principles, increasingly there has been a focus placed on green synthetic approaches that lower the utilization of toxic chemicals and solvents. Bio-inspired as well as biogenic strategies using natural processes in nanomaterial synthesis have risen to prominence since they are easy, sustainable, and economical.

Biomolecule-assisted synthesis is one of the promising approaches, wherein plant extracts, microbes, or biomolecules such as proteins, peptides, and enzymes are employed for synthesizing nanomaterials (Samanta et al., 2020). These biological molecules can assist in regulating the size, shape, and crystallinity of nanomaterials along with lowering the environmental burden normally linked with conventional synthesis routes.

Biomimetic and self-assembly methods also promise a lot. Biomimetic methods deal with imitating natural phenomena such as mineralization or crystallization in order to synthesize nanomaterials with desired properties (Bigham et al., 2024). Self-assembly methods, where nanoparticles of their own accord form desired structures, allow complex photocatalysts to be synthesized without the use of aggressive chemicals, providing a greener option for material synthesis (Baskar et al., 2021).

Plasma-Enhanced and Microwave-Assisted Synthesis: Sophisticated synthesis methods such as plasma-enhanced chemical vapor deposition (PECVD) and microwave-assisted processes provide fast, energy-saving, and highly regulated pathways for nanomaterial synthesis.

Plasma-based techniques employ ionized gases to alter the surface characteristics or dope materials under gentle conditions. These methods work effectively to produce high-quality films or composite materials to improve photocatalytic efficiency, which suits them well for use in energy and environmental applications (Kundu et al., 2019).

Microwave-assisted synthesis involves the use of microwave radiation to speed up chemical reactions, allowing for the quick fabrication of nanomaterials with precise size and morphology. This process minimizes reaction time and energy requirements compared to traditional heating processes, thus constituting a more efficient and environmentally friendly method for nanomaterial fabrication (Kadam et al., 2014).

Nanomaterials development has greatly enhanced the efficiency of photocatalysts in environmental and energy applications. Optimizing their structure, bandgap, and charge separation, as well as employing novel, scalable, and sustainable synthesis processes, can enhance the efficiency and sustainability of photocatalysts. Future innovation in the synthesis and fabrication of materials will be crucial in overcoming the existing challenges and realizing the full potential of solar-powered environmental remediation and clean energy production.

7 Applications of Green Nanomaterials in Environmental Sustainability

With increasing environmental issues worldwide, green nanomaterials have been seen as a promising solution to tackle numerous ecological problems. Their extraordinary properties, such as high surface areas, optoelectronic properties that can be tuned, and catalytic functionality, make them promising materials for functions that seek to enhance environmental sustainability.

7.1 Water Purification and Pollutant Degradation

Water scarcity and water pollution are among the biggest environmental issues of the twenty-first century. Green nanomaterials have been shown to possess great potential for improving water purification and the breakdown of harmful pollutants and thus are crucial in initiatives focusing on supplying clean water and preventing environmental pollution (Behera & Das, 2025; Priyadarshini et al., 2024).

Photocatalytic Water Purification: Nanomaterials, especially semiconductors such as titanium dioxide (TiO_2), zinc oxide (ZnO), and cerium oxide (CeO_2), have demonstrated excellent capability in photocatalytic water purification. These nanomaterials employ solar energy to break down toxic organic pollutants and pathogens in water. The photocatalytic reaction produces reactive oxygen species (ROS) such as hydroxyl radicals (OH•) and superoxide anions (O_2•–) upon illumination, which degrade organic pollutants into harmless byproducts, usually transforming them into carbon dioxide and water. TiO_2 is among the most researched photocatalysts because of its high stability, low toxicity, and capacity to degrade a broad spectrum of organic pollutants under UV light (Sonu et al., 2021). Scientists have enhanced its visible light absorption via methods such as doping with non-metal elements (such as nitrogen or sulfur) and co-coupling with other semiconductors. Composite photocatalysts such as TiO_2/graphene or TiO_2/CuO have also greatly increased photocatalytic activity. Graphene enhances charge separation and decreases recombination, and CuO serves as a co-catalyst to further enhance the degradation process.

Heavy Metal Ion Removal: Green nanomaterials, including nanoscale zero-valent iron (nZVI), graphene oxide (GO), and bio-based materials, are efficient at removing heavy metal ions such as lead, mercury, and arsenic from polluted water. These nanomaterials exhibit high surface reactivity, making them able to adsorb and immobilize hazardous metal ions by processes such as reduction, complexation, or electrostatic interactions. nZVI, especially, has come into focus for its capacity to lower metal ions, including chromium (VI) and lead (II), to less toxic forms because of its huge surface area and high reactivity. This renders nZVI a favorite for the scale-up of water cleaning (Dutta et al., 2022). Graphene and graphene derivatives, such as reduced graphene oxide, are also synthesized as adsorbents for water cleaning. Due to their high surface area and functionalized edges, they are very effective in removing organic and inorganic contaminants. Graphene-based adsorbents are also capable of being reused and regenerated, making them a sustainable option for long-term water treatment.

Removal of Pathogens and Microorganisms: Green nanomaterials are also efficient in targeting and eliminating dangerous pathogens, such as bacteria, viruses, and protozoa, from water. The antimicrobial and antiviral activities of silver nanoparticles (AgNPs), copper oxide (CuO) nanoparticles, and carbonaceous materials such as graphene oxide have been extensively documented. Silver nanoparticles have, in particular, been found to have great antimicrobial effects from the release of silver ions (Ag^+), which interfere with the bacterial and viral cell walls. AgNPs have been incorporated into composite materials or applied as coatings on water purification filters, offering a viable means of removing a wide spectrum of pathogens (Ahmadi et al., 2021). Graphene materials, particularly graphene oxide, have been promising in the area of water filtration by creating antimicrobial membranes that filter water and kill microorganisms upon contact. Additional functionalization of graphene with antimicrobial agents can make it more capable of degrading organic pollutants and disinfecting water efficiently (Mohanty et al., 2025; Satapathy et al., 2024).

8 Solar Hydrogen Generation and CO_2 Reduction

The international transition towards renewable energy sources has increased the role of solar energy in sustainable energy generation. Green nanomaterials are leading the way to harness solar energy into valuable fuels like hydrogen and to prevent the buildup of greenhouse gases like carbon dioxide (CO_2). Specifically, nanomaterials have demonstrated great potential in photocatalytic water splitting to generate hydrogen and photocatalytic CO_2 reduction to generate valuable chemicals and fuels.

Solar Hydrogen Generation: Hydrogen, an efficient and clean fuel, is regarded as a potential solution to the world energy crisis. Photocatalytic water splitting, through which solar energy is utilized to split water into hydrogen ($H_2O \rightarrow H_2 + O_2$), is a sustainable method for hydrogen generation. Nanomaterials, with their high surface area, tunable characteristics, and photocatalytic activity, are perfect candidates to upgrade photocatalysts for this process. Semiconductor nanomaterials such as TiO_2, CdS, and g-C_3N_4 are highly researched for this. Although TiO_2 is a popular photocatalyst, it is usually limited to being activated by UV light, hence its efficiency (Gopinath & Nalajala, 2020). New developments, including doping TiO_2 with metals (e.g., platinum) or non-metals (e.g., nitrogen), have broadened its light absorption to the visible region, making it more effective in solar hydrogen production. In addition, composite and hybrid photocatalysts, which integrate semiconductors with materials such as graphene, CNTs, or CQDs, have exhibited better charge separation and photocatalytic activity. The composite systems enhance electron transport and conductivity, reducing electron–hole recombination and increasing the total hydrogen production rate.

CO_2 Reduction to Value-added Chemicals: Reduction of CO_2 into useful chemicals, for example, methane (CH_4), ethanol, or carbon monoxide (CO), is an excellent climate change mitigation approach to recycle CO_2 back into fuels and chemicals. Nanomaterials play an imperative role in fueling the reduction process with the utilization of solar power in order to decrease CO_2 under benign conditions. Metal oxide photocatalysts, like Cu_2O, Cu-based catalysts, and transition metal composite systems such as palladium and copper have been found to be promising in the reduction of CO_2 to methane, ethylene, and other hydrocarbons. Graphene, carbon nanotubes (CNTs), and carbon quantum dots are also found to be suitable materials for CO_2 reduction because of their high electrical conductivity and high surface area (Bhatia et al., 2021). These materials can enhance effective electron transfer during the reduction process, enhancing product selectivity and yield as catalysts or support materials. Plasmonic nanomaterials, including gold and silver nanoparticles, can also facilitate CO_2 reduction via localized surface plasmon resonance. This enhances the local electromagnetic field, enhancing light absorption and enabling the conversion of CO_2 into valuable chemical fuels.

Integration into Solar Fuel Devices: Improved progress in the incorporation of nanomaterials in solar fuel devices, including photocatalytic reactors and photoelectrochemical cells (PECs), has largely enhanced the efficiency and scalability of CO_2 reduction under sun-driven conditions and hydrogen production. The devices bring together photocatalysts and semiconducting materials to develop efficient systems that transform solar energy into chemical energy. Green nanomaterials are at the forefront of achieving environmental sustainability with their applications in water cleaning, pollutant destruction, solar hydrogen production, and CO_2 reduction (Bramhaiah & Bhattacharyya, 2021). Through the utilization of solar power, green nanomaterials present a sustainable means of responding to essential global issues such as access to clean water and climate change. With continued research and development, green nanomaterials promise to be very important during a shift towards an eco-friendlier and more sustainable future.

9 Challenges and Future Directions

Although the prospect for green nanomaterials for sustainable photocatalysis is tremendous, the field still has tremendous challenges for it to gain popularity and reach commercialization. Solving these challenges, together with determining areas for future research directions, will be critical for harnessing the complete potential of the materials towards environmental sustainability.

9.1 Scalability and Commercialization of Green Nanomaterials

The scale-up from laboratory studies to large-scale application poses various challenges for green nanomaterials. Despite considerable advances in the synthesis of new photocatalysts, scaling them up requires surmounting technical, economic, and regulatory hurdles.

Synthesis and Production Challenges: Scaling up the synthesis of green nanomaterials for mass production is still challenging, although their potential photocatalytic activity has been demonstrated in the laboratory. Present approaches are generally energy-consumptive, costly, and equipment-intensive, thus not ideal for cost-effective and environmentally friendly large-scale production. To meet this, more reproducible, standardized, and energy-efficient synthesis routes are needed, e.g., employing renewable feedstocks, less harmful solvents, and low-energy processes such as bio-inspired or room-temperature reactions (Arun et al., 2022). Moreover, batch-to-batch consistency in nanomaterials is important since differences in size, morphology, and surface chemistry can influence their photocatalytic activity. Enhancing regulation of synthesis factors such as temperature, pH, and precursor concentration is crucial to achieving consistency and efficiency in its use in industrial processes.

Cost and Economic Feasibility: In spite of the readiness of raw materials for most of the green nanomaterials, their fabrication is still an expensive process, particularly when employing precious precursors or catalysts. More advanced materials, including precious metal-containing materials and sophisticated composites, add to the final expense. To render these materials economical, minimizing the use of precious metals like platinum, gold, and silver is essential. This can be done by employing substitute materials or doping with lower-cost metals such as iron or copper (Bahadur et al., 2020). Moreover, large-scale production from the laboratory to industrial levels poses logistical issues, such as maintaining large-scale production without compromising quality or photocatalytic activity. Pilot-scale experiments and life cycle assessment (LCA) will be crucial in determining the economic viability and environmental sustainability of large-scale production of green nanomaterials.

Regulatory and Environmental Considerations: The safety and environmental issues of nanomaterials are significant impediments to their large-scale application. Although green nanomaterials are made to be environmentally friendly, their long-term performance in terms of stability, toxicity, and environmental effects needs to be thoroughly researched. There is no one-size-fits-all regulatory system for the safe application of nanomaterials. To facilitate large-scale application, there should be clear guidelines on safety and standardized tests for toxicity, biodegradability, and environmental effects (Dhall et al., 2024). Public opinion and acceptance also come into play in commercialization. Open up research and risk studies to educate the public regarding the possible risks of even "green" nanoparticles, and this will be essential in winning over support and trust.

10 Future Prospects in Eco-Friendly Photocatalysis

The future of photocatalysis will be centered on innovations that make it more efficient, expand material utilization, and incorporate photocatalytic systems into practical uses.

Development of Novel Photocatalysts with Broader Light Absorption: The principal drawback of existing photocatalysts such as TiO_2 is that they are UV-light dependent, a fraction of the solar spectrum. In order to become more efficient, photocatalysts must absorb light over a wider spectrum that includes visible and infrared light. Improvements in bandgap engineering, including doping, forming heterojunctions, or employing novel materials such as metal–organic frameworks (MOFs) and perovskite-like frameworks, can enable photocatalysts to absorb a broader spectrum of solar radiation, enhancing performance under practical conditions (Yadav et al., 2020). Moreover, plasmonic nanomaterials such as gold and silver nanoparticles are capable of boosting the light absorption rate via localized surface plasmon resonance (LSPR), which has enormous potential to enhance the light-harvesting efficiency, particularly in processes such as solar hydrogen production and CO_2 reduction.

Hybrid Photocatalytic Systems and Integrated Devices: Future work will be directed towards the incorporation of photocatalysts into multi-functional platforms that consolidate photocatalysis with other technologies for energy harvesting or environmental remediation, resulting in more efficient and scalable devices for practical applications. One of the promising directions is photoelectrochemical cells (PECs), which synergistically combine photocatalysis with electrochemical methods towards applications such as solar hydrogen evolution and CO_2 reduction. By maximizing the semiconductor-electrode interface, PEC systems would be able to have greater efficiencies in solar energy conversion to chemical fuels (Devi et al., 2023). Another new development is photocatalytic membranes, which integrate nanomaterials into water purification filtration devices. These membranes provide continuous, self-cleaning, and energy-efficient water treatment, allowing for the removal of contaminants, pathogens, and heavy metals from wastewater, offering sustainable solutions to water shortages.

Artificial Photosynthesis and Solar Fuels: Artificial photosynthesis for mimicking the natural process and generating renewable fuel such as hydrogen, methane, or liquid hydrocarbons using water and CO_2 is an area of emphasis for photocatalysis. This is being dominated by green nanomaterials that allow energy from sunlight to be efficiently translated into chemical bonds. Advancements in photocatalytic materials and integrated systems for sunlight-to-solar fuels, for example, hydrogen or carbon-based fuels, are anticipated to speed up, with research aimed at maximizing reaction pathways, enhancing stability, and reducing costs for rendering these systems more economically viable for large-scale power generation (Nair et al., 2023). Moreover, conversion of CO_2 to high-value chemicals such as ethylene, methanol, and complex hydrocarbons is an important area of research. Green nanomaterials, when used in artificial photosynthesis systems, present a potential sustainable solution to carbon mitigation as well as renewable chemical production.

Integration into Circular Economy and Sustainability: One of the directions for the development of photocatalysis in the future will be the inclusion of green nanomaterials within circular economy approaches, in which waste products become raw materials. Photocatalytic systems to transform waste CO_2, plastics, and other contaminants into desirable fuels or chemicals will be the key to diminishing the environmental impact of industrial activities. Incorporating photocatalytic systems within technologies for waste recycling is a big plus environmentally. For instance, photocatalytic CO_2 reduction can be integrated with plastic recycling to convert plastic waste into valuable chemicals or fuels, enabling a more circular and sustainable waste management practice (Jha & Dubey, 2024).

Commercialization and future development of green nanomaterials for environmentally friendly photocatalysis have tremendous potential to address environmental issues. However, challenges such as scalability, cost, and regulation will need to be addressed through continued innovation in material synthesis and manufacturing techniques. Future advancements will aim at better light absorption, greater material functionality, and systems that interface solar energy with pollution remediation. As these technologies become advanced, they will help in fulfilling the goals of sustainability, lowering the utilization of fossil fuels, and mitigating pollution of the environment.

11 Conclusion

Green nanomaterials are exhibiting tremendous potential in environmentally friendly photocatalysis with sustainable solutions to energy and environmental issues. Their special characteristics, such as large surface area and effective catalytic activity, make them capable of utilizing solar power for applications ranging from water treatment to pollutant degradation, solar hydrogen generation, and CO_2 reduction. The development of materials through bandgap engineering and hybrid systems has enhanced efficiency and scalability. But scalability, cost, and environmental concerns must first be overcome before they can be implemented at a large scale. Future studies hope to improve light absorption, charge separation, and stability while maintaining long-term environmental safety. As these technologies continue to mature, they could help propel us toward a low-carbon, resource-efficient future.

References

Ahmadi, Y., Bhardwaj, N., Kim, K., & Kumar, S. (2021). Recent advances in photocatalytic removal of airborne pathogens in air. *The Science of the Total Environment, 794*, Article 148477. https://doi.org/10.1016/j.scitotenv.2021.148477

Anaya-Rodríguez, F., Durán-Álvarez, J. C., Drisya, K. T., & Zanella, R. (2023). The challenges of integrating the principles of green chemistry and green engineering to heterogeneous photocatalysis to treat water and produce green H_2. *Catalysts, 13*(1), 154. https://doi.org/10.3390/catal13010154

Arun, J., Nachiappan, S., Rangarajan, G., Alagappan, R. P., Gopinath, K. P., & Lichtfouse, E. (2022). Synthesis and application of titanium dioxide photocatalysis for energy, decontamination and viral disinfection: A review. *Environmental Chemistry Letters, 21*(1), 339–362. https://doi.org/10.1007/s10311-022-01503-z

Ayyub, M. M., & Rao, C. (2020). Design of efficient photocatalysts through band gap engineering. In *Elsevier eBooks* (pp. 1–18). https://doi.org/10.1016/b978-0-12-817836-2.00001-6

Bahadur, N., Das, P., & Bhargava, N. (2020). Improving energy efficiency and economic feasibility of photocatalytic treatment of synthetic and real textile wastewater using bagasse fly ash modified TiO_2. *Chemical Engineering Journal Advances, 2*, Article 100012. https://doi.org/10.1016/j.ceja.2020.100012

Baskar, A. V., Benzigar, M. R., Talapaneni, S. N., Singh, G., Karakoti, A. S., Yi, J., Al-Muhtaseb, A. H., Ariga, K., Ajayan, P. M., & Vinu, A. (2021). Self-assembled fullerene nanostructures: Synthesis and applications. *Advanced Functional Materials, 32*(6). https://doi.org/10.1002/adfm.202106924

Behera, I. D., & Das, A. P. (2025). *Environmental hydrocarbon pollution and zero waste approach towards a sustainable waste management*. 436.

Bhatia, P., Dharaskar, S., & Unnarkat, A. P. (2021). CO_2 reduction routes to value-added oxygenates: A review. *Environmental Science and Pollution Research, 28*(44), 61929–61950. https://doi.org/10.1007/s11356-021-16003-w

Bigham, A., Zarepour, A., Safarkhani, M., Huh, Y., Khosravi, A., Rabiee, N., Iravani, S., & Zarrabi, A. (2024). Inspired by nature: Bioinspired and biomimetic photocatalysts for biomedical applications. *Nano Materials Science*. https://doi.org/10.1016/j.nanoms.2024.02.006

Bramhaiah, K., & Bhattacharyya, S. (2021). Challenges and future prospects of graphene-based hybrids for solar fuel generation: Moving towards next generation photocatalysts. *Materials Advances, 3*(1), 142–172. https://doi.org/10.1039/d1ma00748c

Chen, F., Ma, T., Zhang, T., Zhang, Y., & Huang, H. (2021). Atomic-level charge separation strategies in semiconductor-based photocatalysts. *Advanced Materials, 33*(10). https://doi.org/10.1002/adma.202005256

Devi, L., Dalal, K., Manchala, S., Behera, K., Sharma, K. N., & Das, A. (2023). Nanostructured photoelectrochemical biosensors. In *Elsevier eBooks* (pp. 265–284). https://doi.org/10.1016/b978-0-323-88431-0.00005-3

Dhall, S., Nigam, A., Harshavardhan, M., Mukherjee, A., & Srivastava, P. (2024). A comprehensive overview of methods involved in nanomaterial production and waste disposal from research labs and industries and existing regulatory guidelines for handling engineered nanomaterials. *Environmental Chemistry and Ecotoxicology, 6*, 269–282. https://doi.org/10.1016/j.enceco.2024.06.002

Dutta, S., Dixit, R., & Sharma, R. (2022). Remediation of heavy metals with nanomaterials. In *Separation science and technology* (pp. 97–138). https://doi.org/10.1016/b978-0-323-90763-7.00009-3

Fatima, R., Warsi, M. F., Zulfiqar, S., Ragab, S. A., Shakir, I., & Sarwar, M. I. (2020). Nanocrystalline transition metal oxides and their composites with reduced graphene oxide and carbon nanotubes for photocatalytic applications. *Ceramics International, 46*(10), 16480–16492. https://doi.org/10.1016/j.ceramint.2020.03.213

Gan, Y. X., Jayatissa, A. H., Yu, Z., Chen, X., & Li, M. (2020). Hydrothermal synthesis of nanomaterials. *Journal of Nanomaterials, 2020*, 1–3. https://doi.org/10.1155/2020/8917013

Gopinath, C. S., & Nalajala, N. (2020). A scalable and thin film approach for solar hydrogen generation: A review on enhanced photocatalytic water splitting. *Journal of Materials Chemistry A, 9*(3), 1353–1371. https://doi.org/10.1039/d0ta09619a

Hussain, C. M., Kecili, R., & Hussain, C. G. (2021). *Sample Preparation with Nanomaterials.* https://doi.org/10.1002/9783527685622

Jha, H., & Dubey, B. K. (2024). Challenges and opportunities in enabling circular economy for sustainable wastewater treatment. In *Earth and environmental sciences library* (pp. 483–507). https://doi.org/10.1007/978-3-031-63046-0_20

Kadam, A., Dhabbe, R., Kokate, Gaikwad, Y., & Garadkar, K. (2014). Preparation of N doped TiO_2 via microwave-assisted method and its photocatalytic activity for degradation of Malathion. *Spectrochimica Acta Part A Molecular and Biomolecular Spectroscopy, 133*, 669–676. https://doi.org/10.1016/j.saa.2014.06.020

Kaus, N. H. M., Rithwan, A. F., Adnan, R., Ibrahim, M. L., Thongmee, S., & Yusoff, S. F. M. (2021). Effective strategies, mechanisms, and photocatalytic efficiency of semiconductor nanomaterials incorporating RGO for environmental contaminant degradation. *Catalysts, 11*(3), 302. https://doi.org/10.3390/catal11030302

Kundu, D., Banerjee, D., Ghosh, S., Das, N., Thakur, S., Das, B., & Chattopadhyay, K. (2019). Plasma enhanced chemical vapour deposited amorphous carbon coating for hydrophobicity enhancement in commercial cotton fabrics. *Physica E Low-Dimensional Systems and Nanostructures, 114*, Article 113594. https://doi.org/10.1016/j.physe.2019.113594

Mohanty, M., Mohanty, J., Priyadarsini, S., Rodríguez-Díaz, J. M., Yadav, S., Cabral-Pinto, M. M. S., Nath, H., Ram, D. K., Koteswara Reddy, G., & Das, A. P. (2025). Environmental hydrocarbon pollutants: Sources, transport, and effect on human health. *Environmental Science and Engineering, Part, F218*, 21–44. https://doi.org/10.1007/978-3-031-81615-4_2

Molinari, A., Amadelli, R., Antolini, L., Maldotti, A., Battioni, P., & Mansuy, D. (2000). Phororedox and photocatalytic processes on Fe (III)–porphyrin surface modified nanocrystalline TiO_2. *Journal of Molecular Catalysis a Chemical, 158*(2), 521–531. https://doi.org/10.1016/s1381-1169(99)00446-x

Nair, P., Das, B., & Kuriakose, T. (2023). Artificial photosynthesis using nanotechnology. In *Springer eBooks* (pp. 639–667). https://doi.org/10.1007/978-3-031-31111-6_25

Nasirian, M., Lin, Y. P., Bustillo-Lecompte, C. F., & Mehrvar, M. (2017). Enhancement of photocatalytic activity of titanium dioxide using non-metal doping methods under visible light: A review. *International Journal of Environmental Science and Technology, 15*(9), 2009–2032. https://doi.org/10.1007/s13762-017-1618-2

Nasirian, M., Bustillo-Lecompte, C. F., Lin, Y. P., & Mehrvar, M. (2018). Optimization of the photocatalytic activity of N-doped TiO_2 for the degradation of methyl orange. *Desalination and Water Treatment, 110*, 193–208. https://doi.org/10.5004/dwt.2018.22219

Obregón, S., & Rodríguez-González, V. (2021). Photocatalytic TiO_2 thin films and coatings prepared by sol–gel processing: A brief review. *Journal of Sol-Gel Science and Technology, 102*(1), 125–141. https://doi.org/10.1007/s10971-021-05628-5

Priyadarshini, B., & Das, A. P. (2024). The adverse impact of synthetic microplastics released from disposable face masks pollutants on aquatic microbiota, aquatic species and humans: A review. *Environmental Science and Engineering, Part, F2827*, 67–81. https://doi.org/10.1007/978-3-031-50840-0_5

Priyadarshini, S., Jagatee, S., & Das, A. P. (2024). Synthetic fabrics and microfiber pollution–an assessment of their global impact. *Environmental Science and Engineering, Part, F2333*, 137–157. https://doi.org/10.1007/978-3-031-51792-1_8

Samanta, D., Basnet, P., Chanu, T. I., & Chatterjee, S. (2020). Biomolecule assisted morphology-controllable synthesis of Zinc Sulphide nanomaterials for efficient photocatalytic activity under solar irradiation. *Journal of Alloys and Compounds, 844*, Article 155810. https://doi.org/10.1016/j.jallcom.2020.155810

Satapathy, S., Kar, D., Priyadarsini, S., Das, A. P., & Kuanar, A. (2024). Application of microorganisms in emerging waste management strategies and remediation of synthetic and cellulosic microfibers. *Sustainable Microbial Technology for Synthetic and Cellulosic Microfiber Bioremediation*, 183–206. https://doi.org/10.1007/978-3-031-62600-5_10

Singh, D., Chauhan, U., Maurya, S., & Gupta, G. (2024). Exploring the potential of grasses in the phytoremediation of various xenobiotic compounds. *Sustainable Management of Environmental Pollutants through Phytoremediation*, 120–134. https://doi.org/10.1201/9781003442295-6

Sonu, K., Puttaiah, S. H., Raghavan, V. S., & Gorthi, S. S. (2021). Photocatalytic degradation of MB by TiO_2: Studies on recycle and reuse of photocatalyst and treated water for seed germination. *Environmental Science and Pollution Research, 28*(35), 48742–48753. https://doi.org/10.1007/s11356-021-13863-0

Tahir, M. B., Nawaz, T., Nabi, G., Sagir, M., Shehzad, M. A., Yasmin, A., Hussain, S., Bhatti, M. P., Ahmed, A., & Gilani, S. S. A. (2020). Recent advances on photocatalytic nanomaterials for hydrogen energy evolution in sustainable environment. *International Journal of Environmental & Analytical Chemistry*, 1–19. https://doi.org/10.1080/03067319.2019.1691188

Thongam, D. D., & Chaturvedi, H. (2021). Advances in nanomaterials for heterogeneous photocatalysis. *Nano Express, 2*(1), Article 012005. https://doi.org/10.1088/2632-959x/abeb8d

Vu, N., Kaliaguine, S., & Do, T. (2019). Critical aspects and recent advances in structural engineering of photocatalysts for sunlight-driven photocatalytic reduction of CO_2 into fuels. *Advanced Functional Materials, 29*(31). https://doi.org/10.1002/adfm.201901825

Yadav, G., & Ahmaruzzaman, M. (2022). New generation advanced nanomaterials for photocatalytic abatement of phenolic compounds. *Chemosphere, 304*, Article 135297. https://doi.org/10.1016/j.chemosphere.2022.135297

Yadav, S. K., Grandhi, G. K., Dubal, D. P., De Mello, J. C., Otyepka, M., Zbořil, R., Fischer, R. A., & Jayaramulu, K. (2020). MHP@MOF hybrids: metal halide perovskite@Metal-organic framework hybrids: Synthesis, design, properties, and applications (Small 47/2020). *Small, 16*(47). https://doi.org/10.1002/smll.202070258

New Advancement of Nanobiosensors in Antibiotic Residual Detections for Management of Public Health Hazards

Sonali Sandipta Dash, Alaka Sanghamitra, Yostnamayee Panda, and Dhananjay Soren

Abstract From the onset of 1940s, antimicrobial agents are widely used to treat a variety of infectious diseases including Urinary Tract Infections, Tuberculosis, Pneumoniae, etc. As a result, overconsumption of broad-spectrum antibiotics has led to an increase in drug resistant strains beside environmental pollution, both of which pose serious threat to the surrounding and human health. Despite their excellent properties, biosensors are not even established as a stable device under some circumstances. Thus, to ensure public health and to minimise antibiotic resistance, nano-biosensors i.e., biosensors with nanomaterial conjugation are being developed for effective eradication of drug residues from the biota. As compared to conventional detection techniques, nano-biosensors are the promising future platforms for easy and quick monitoring of residual drugs in a variety of samples viz., water bodies even at trace element levels. Due to their selectivity, specificity and sensitivity, nano-biosensors are often considered as biocompatible biosensors with integration of nanomaterials having low detection limits and high speed of response. Furthermore, recent developments and modifications for practical applications of nano-biosensors in antibiotic residual detections are suggested to overcome the existing challenges and outlook for future perspectives in this field. Hence, new ideas towards technological advancements will be the key players to prevent antimicrobial resistance and manage public health and hazards for the future generation.

Keywords Nano-biosensors · Antibiotic residuals · Public health · Electrochemical · Aptamer · Enzyme

---

S. S. Dash · A. Sanghamitra · Y. Panda · D. Soren (✉)
Enzymology and Cancer Biology Laboratory, Department of Zoology, Ravenshaw University, Cuttack, Odisha, India
e-mail: dsoren@ravenshawuniversity.ac.in

Centre of Excellence in Environment and Public Health, Department of Zoology, Ravenshaw University, Cuttack, Odisha, India

E. Priyadarshini et al. (eds.), *Advanced Nanotechnology Applications for Environmental Sustainability*, Sustainable Environmental Waste Management Strategies,
https://doi.org/10.1007/978-3-032-12933-8_18

1 Introduction

Public health is seriously threatened by the rising levels of antibiotic residues in the environment, particularly in food and water. Antibiotic residues are persistent in a variety of ecosystems as a result of overconsumption and misleading usage of antibacterial drugs in agriculture and industrial practices, veterinary sciences and human medicine. These residues directly endanger human health, affect microbial populations and consequently lead to antibiotic resistance. The emergence of such a serious concern directly poses toxicological risk factors thus making the development of efficient monitoring systems essential.

The extensive use of antibiotics either natural or synthetic, is quite effective for treating several infectious diseases as well as in agro-industries and animal care facilities (Saini et al., 2012; Fair and Tor, 2014; Abedalwafa et al., 2019; Khan, 2022; Cardoso et al., 2021; Nepfumbada et al., 2024). More or less, antibiotic contamination is paving its way to enter the biota via wastewater effluents, plant debris, animal and human faecal matter, dietary supplements, agrochemicals, medicinal and industrial wastes and so on (Dehghani et al., 2013; Robati et al., 2016).

Due to unhealthy lifestyle, there is an exponential increase of infections at a cost of life-threatening condition. Inflammation, hypersensitivity or allergic reactions, cancer, diabetes, teratogenesis and many more can't even be cured due to antibiotic resistance (Jin et al., 2022; Abedi-Firoozjah et al., 2024). Being one of the saviours, antibiotics also proved to be the leading organic pollutant and can be highly toxic whenever consumed unknowingly. Additionally, the occurrence of antibiotic residues is commonly observed in sources of nutrients, being altered to contaminants in human body system (Abedalwafa et al., 2019). This instance is so far considered as a slow poison case in deteriorating the human health completely as well as the surrounded environment (Nepfumbada et al., 2024).

The antibacterial drugs are simply characterised on the basis of their molecular arrangements, structural complexities and functionalities, and mode of action (Nepfumbada et al., 2024). Some of the different classes of antibiotics including beta lactams, aminoglycosides, tetracyclines, glycopeptides, fluoroquinolones, sulfonamides, macrolides, etc. serve for inhibition of cell wall synthesis, protein and nucleic acid synthesis accordingly (Nepfumbada et al., 2024). These drugs can get accumulated and mixed in the environmental samples, drinking water and food supplements and finally into human gut leading to low metabolism and high health risks (Cuprys et al., 2018; Gaudin et al., 2017; Gou et al., 2021; Kraemer et al., 2019; Lan et al., 2017; Nepfumbada et al., 2024). The commercially available broad spectrum of antibiotic classes is often consumed illegally without any suggestions and valid prescriptions from physicians. In fact, these are even used in higher dosages so as to meet the market value of nutritional supplements, dairy business, animal husbandry, therapies and other industrial alliances, etc. (Abedi-Firoozjah et al., 2024). Even low levels of antibiotic residues in milk items can kill or slow down the bacteria responsible for fermentation of dairy products and other such associated processes involved. This phenomenon creates a major challenge in shaping the economy of the industry and to cope up with the recent technological issues hampering community health, improper sanitation, etc. (Das et al., 2024b, 2024c). It is recommended to ban the unauthorised sales and distribution of drugs in the market with reference to proper implementation of Food and Drug Administrations (FDA) and World Health Organisations (WHO) approved notions and reports (Abedalwafa et al., 2019; Abedi-Firoozjah et al., 2024).

In this outlook, the national administration and government policy makers should take a stand to validate the exact percentage of antibiotic consumption in aquaculture, dairy industry, agriculture and other industrial sectors so as to safeguard human's health at stake. The conflict is to be resolved jointly by WHO and FDA with a compensated MRL (i.e., Maximum Residue Limit) accompanied with the Acceptable Daily Intake (or ADI) of the desired antibacterial drugs for an optimum use (Abedalwafa et al., 2019; Nepfumbada et al., 2024).

Several spectrophotometry and chromatography methods of antibiotic residual detections are applied in government laboratories and other research and development authorities (Abedalwafa et al., 2019; Nepfumbada et al., 2024). Such conventional applications comprise capillary electrophoresis (CE), Liquid chromatography-mass spectroscopy (LC–MS), High performance liquid chromatography (HPLC), etc. However, they are not easy to handle, need professionals and technicians to operate, require monetary support as well as long period of time to wind up the procedure. No rapid and routine analysis is carried out thus switching towards modern solutions apart from traditional ones (Kharewal et al., 2020; Wu et al., 2016).

Thus, new techniques of sensing methods are introduced to counteract the drawbacks of the ancestral ways. For on-site monitoring, high sample throughput, rapid detection, cost effective, high sensitivity and specificity, reliable and flexible usage are the basic characteristics worked tremendously for antibiotic residual identification. Henceforward, the advanced techniques outreached the development of detection assays highlighting the importance of functionalities of state-of-the-art sensing patterns (Jahanbani & Benvidi, 2016; Nepfumbada et al., 2024). Some of them are electrochemical detection, immunological and microbiological tests, Raman spectroscopy, colorimetric sensors, fluorometric detection, surface plasmon resonance (SPR) sensors, ELISA or Enzyme Linked Immuno-Sorbent Assay, immunochemical detection, etc. (Abedalwafa et al., 2019; Abedi-Firoozjah et al., 2024).

In recent years, the occurrence of human illness and death has reached a milestone due to antimicrobial resistance (AMR). It has been estimated that about 10 million deaths are expected by the year 2050 which itself is a major concern for socio-economic and public health management (Xia et al., 2024). The widely reported data claim the pattern of resistance among numerous bacterial strains for their strong association with biofilm as a defence mechanism. In order to strategize their survival rate, they often form biofilm to revert back the drug penetration into the cellular system thus exhibiting resistance towards antibacterial medications (Furst and Francis 2019; Xia et al., 2024).

With the rapid inclination of molecular biology and biotechnological techniques, the field of nanotechnology has also been risen so far to meet today's expectations in science. The advanced approaches being applied in terms of healthcare facilities, environmental community safeguarding, biomedical applications, food safety monitoring, industrial effluent management, etc. serve outstanding illustrations (Mohanty et al., 2024b; Priyadarshini & Das, 2024; Priyadarshini et al., 2024; Xia et al., 2024).

The cutting-edge analytical tools otherwise known as biosensors, make use of biorecognition substances in close proximity to a transducer. Antibodies, DNA and enzymes are the most commonly utilised recognising elements in biosensor development (Sánchez-Paniagua López et al., 2017; Yazdanparast et al., 2019). Their performance is significantly influenced by a recognition element's binding affinity with a target analyte. There is no doubt that developing biosensors present several obstacles, chief among them relies with the effective acquisition of biorecognition signals and their conversion into optical, acoustic, gravimetric, electrical or electrochemical signals (Nepfumbada et al., 2024; Abedi-Firoozjah et al., 2024). The limitations of these regulatory devices consist inaccurate analyte detection, inappropriate range of target analyte identification, low specificity and sensitivity which can directly affect the test results by interpreting negative and complicated false-positive results. Moreover, the amplification of transducer for better reproducibility and low frequency of LOD (i.e., Lower Detection Limit) is to be avoided along with fake miniature of sensing devices with usage of unnecessary micro-elements (Abedi-Firoozjah et al., 2024). In order to overcome the limitations, nanomaterials are integrated with sensing tools for advanced biosensor fabrication, altogether termed as nano-biosensors. These are the combinational approach of nanomaterials (NMs)-based biosensors accountable for amplifying signals of transducers with high specificity, selectivity and sensitivity. Higher stability and ideal conductivity aids for better nano-biosensors in association with high surface to volume ratio, shock-enduring capabilities and tunable for both structure and colour (Abedi-Firoozjah et al., 2024; Kumar & Neelam, 2016; Lawal, 2018; Nepfumbada et al., 2024). The nanomaterials that are often used in such context involves metal nanoparticles, metal organic frameworks, carbon nanotubes, nanowires, quantum dots, polymeric nanoparticles, nano-rods, DNA, enzymes, dendrimers, antibodies, etc. (Abedi-Firoozjah et al., 2024).

Thus, the nano-biosensors for antibiotic residual detection are an improved biosensing system influencing charge transfer and lower detection limit (LOD) effortlessly (Nepfumbada et al., 2024). The conjugation of certain nano bio-materials with a carrier or receptor is simply awful in sensing applications. Nowadays, specific binding of the analytes represents reliability and accuracy with proper standardised protocols (Barros et al., 2023). Furthermore, the detection of antibiotic residues even in trace amount ensure validation of rapid screening methods with low or no limitations. Use of microfluidic chips, electrodes and a wide range of biomolecules as an interaction partner imprints successfully with hybrids of nanomaterials and biomolecules and thus termed as nano-biosensors.

2 Antibiotic Residual Detections

2.1 Beta-Lactams

Beta-lactams mainly penicillin and cephalosporins produced from *Penicillium notatum* and *Cephalosporium* respectively show different structural complexities whereas share similar mechanism of action for inhibition of cell wall synthesis. Being water soluble, inexpensive and mostly prior antibiotics, penicillin are prescribed for the treatment of gonorrhoea, urinary and respiratory disorders, skin and dental infections. They are mostly found in food and water contaminants as a result of which WHO has announced its range of ADI as 0–50 μg/kg human body weight and MRL as 100 μg/kg. On the other hand, bronchitis, tonsilitis, pneumoniae and many more are treated by cephalosporins (Abedalwafa et al., 2019; Masoudyfar & Elhami, 2019) (Table 1).

Colorimetric detection assays are carried out in presence of functionalised gold nanoparticles as a metal nanoparticle, polythiophene and a colorimetric agent such as TMB (Tetramethylbenzidine). Apart from that, cetyl trimethyl ammonium bromide (CTMB) is capped with enzymatic influenced metal NPs especially gold nanoparticles to detect beta-lactam residues at a certain UV wavelength. The required limit of detection (LOD) for the functionalised recognition is 0.007 mg/ml. Even, it is about 1 ng/ml in case of antibody incorporated platinum nanoparticles as a colorimetric sensing method for Penicillin-G (PEN-G) identification which is meant to have more sensitivity and specificity. To detect ampicillin, functionalised gold nanoparticles with aptamers are designed at a lower detection limit of 10 ng/ml (Song et al., 2012). All these biosensor-based nanomaterials are made visibly receptive and adequate enough to possibly detect antibiotics whether by TMB oxidation, receptor-mediated detection via colorimetric fluoride ion detection and visible colour change which clearly depicts antibiotic recognition in the desired samples (Abedalwafa et al., 2019).

2.2 Aminoglycosides

Aminoglycosides are majorly introduced from *Streptomyces* having a chemical structure of streptomine ring. Kanamycin, streptomycin, neomycin, amikacin, gentamicin, tobramycin are some of the bactericidal drugs for inhibiting protein synthesis. Being cost effective, aminoglycosides are used predominantly but sometimes are banned for their harsh side effects in kidney and inner ear damage. But still their usage in food industries is demandable which ultimately cause toxic effects and residual drug products even in low concentration of 35 and 40 μg/ml in animal-derived food items. As per the suggestions of WHO, the MLR range of 200 μg/kg body weight is selected and hence high-sensitive techniques for bio-sensing systems are to be allotted (Abedalwafa et al., 2019).

Table 1 Antibiotic residual detections by different biosensing components

Representative antibiotics	Biosensor elements	Samples	Detection mechanism	LOD	References
Pencillin-G	Enzyme conjugated gold nanoclusters, Tetramethylbenzidine (TMB) and silver doped gold nanoparticles	Food supplements	Amine groups interacted for aggregation	30 μM	Zhang et al. (2014)
Amoxicillin	TMB, aptamer mediated gold nanoparticles	Milk products and aquaculture samples	Non-polar structures formed due to colorimetric signals amplified and nanomaterials aggregated	0.54 μM	Abedalwafa et al. (2019)
Kanamycin	TMB, gold and silver doped nanoparticles	Cattle farms	Fabricated metallic nanoparticles aggregate and often enhance TMB for oxidation	0.1 nM	Wang et al. (2017)
Gentamycin	Lipid infused capsules of gold nanocomposites	Pharmaceutical and water samples	Phosphatidylcholine interacted via hydrogen bonding	7.4 nM	Liu et al. (2016)
Neomycin	Melamine and cyclam incorporated gold nanoparticles	Animal derived food products	Hydrogen bonding between neomycin and biosensor elements	30 pM	Xiao et al. (2015)
Tetracycline	TMB, aptamer mediated gold and nano clusters	Milk products	Binding of aptamers initiate transducers to amplify signals	266 pM	Ramezani et al. (2015)
Doxycycline	Gold nanoparticles	Water samples	Integration of positive amino acids on the surface of nanomaterials due to strong electrostatic interactions	48 nM	Abedalwafa et al. (2019)

(continued)

Table 1 (continued)

Representative antibiotics	Biosensor elements	Samples	Detection mechanism	LOD	References
Oxytetracycline	Gold capped nanomaterials with aptamers, and TMB supremacy	Dairy products and wastewater samples	Complexation of graphene aptamers induce TMB redox reaction	25 nM	Abedalwafa et al. (2019)
Azithromycin	Silver nanoparticles	Water samples	Dispersion of silver nanoparticles coated with citrate, cause electrostatic repulsion reactions	0.2 μM	Chavada et al. (2017)
Ciprofloxacin	Silver nanomaterials doped with glassy carbon electrode	Milk, rain, tap and river water samples	Cyclic Voltammetry and Differential Pulse Voltammetry	0.2 μM	Gissawong et al. (2021)
Ofloxacin	Copper oxide coating and carbon printed electrode	Milk and medicine	Differential Pulse Voltammetry	1.9 μM	Taherizadeh et al. (2023)
Norfloxacin	Metal ions particularly iron	Aquaculture samples and pharmaceutical wastes	Cyclic Voltammetry and complexation with Fe ions	0.15 μM	Vu et al. (2023)
Enrofloxacin	Enzyme conjugation	Dairy products and water samples	Immuno-reaction between enrofloxacin and conjugated enzyme	0.041 μM	Lu et al. (2021)
Sulfadimethoxine	Aptamer coated gold nanoparticles	Medicinal Wastes	Gold nanoparticles stability is decreased upon addition of Nacl and colour formation leading to sulfadimethoxine-aptamer target binding	0.1 pg/ml	Abedalwafa et al. (2019)
Chloramphenicol	TMB	Animal derived food supplements	TMB oxidation with the presence of hydrogen peroxide	0.03 nM	Chang et al. (2017)

Moreover, incorporation of liposomes and metal nanoparticles especially gold and silver, microbiological assays to recognise aminoglycosides are observed colorimetrically. The drug kanamycin is detected by the use of TMB which on Fenton reaction is oxidised to form hydrogen peroxide with 0.1 nm LOD (Liu et al., 2018). Likewise, gold nanoparticles functionalised aptamers and nucleic acids i.e., double stranded DNA and strongly initiated for identifying aminoglycosides with a developed yet simple bio-sensing technique. This is followed by the binding of aptamers to antibiotics so as to aggregate with desired metal nanoparticles for visible change of colour from red to black due to pH variation (Abedalwafa et al., 2019).

2.3 *Tetracyclines*

Another potent antibacterial agent is being extracted from *Streptomyces*is tetracycline. The name itself denotes its structural annotations having four rings alike aminoglycosides and even similar mechanism of action i.e., inhibition of protein synthesis. Although, it is categorised to be a safe alternative with an outstanding therapeutic approach, it unfortunately lacks behind this concept of treating parasitic infections and other deadly diseases. Tetracycline residues are uncontrollably found in environmental samples, food and beverages even in plant sources besides which antibiotic resistance is consequently developed (Zhou et al., 2018; Abedalwafa et al., 2019). Some of the major health issues are related to liver, kidney and inner ear damage. Toxic allergic reactions, hyper pigmentation of skin and teeth, etc. are all because of unnecessary deposits of tetracyclines in the biota. Therefore, the MRL and ADI ranges of the drug are suggested to be 100 μg/kg and 0–30 μg/kg, respectively. Accordingly, the bio-sensing techniques are introduced as per the requirements which results in direct and rapid recognition of tetracyclines, doxycycline and oxytetracycline, etc. in modified direct enzyme-linked single-stranded deoxyribonucleic acid (ssDNA) aptamer assay scheme. Tetramethylbenzidine (TMB) oxidation is mediated by Fenton reaction, where aptamer coated gold and silver nanoparticles are the aggregation agents interacting with tetracyclines for their accelerated detection. Furthermore, the electrostatic binding affinity of the reducing agents are employed to confirm the significant aggregation of nanoparticles with the required drug molecules (Abedalwafa et al., 2019).

2.4 *Macrolides*

Macrolides are the bacteriostatic drugs inhibiting bacterial ribosomal subunits for protein synthesis. Erythromycin and azithromycin are the most widely used macrolides for skin and gut infections and even respiratory disorders. Due to unknown range of MRL, azithromycin is often debarred from veterinary uses. But cattle farms are observed to have macrolide residues, whenever the drug is conjugated with silver nanoparticles and the electrostatic interaction between them results in a visibly colour change (Chavada et al., 2017; Abedalwafa et al., 2019).

2.5 Fluoroquinolones

Fluoroquinolones or quinolones are the DNA synthesis inhibitors specifically discovered for respiratory and skin inflammation and UTIs. However, norfloxacin is characterised as a toxic organic pollutant for animal-derived food supplements. These commercially available drugs are detected in fish culture farms in the presence of iron coated in PVC layer. Additionally, enrofloxacin is detected by the complexation of sandwich ELISA and immunochromatography assay with the help of an immune-strip biosensor (Abedalwafa et al., 2019).

2.6 Sulfonamides

The synthetic antibacterial drugs mainly sulfonamides are modified to behave like anticonvulsants, antiglaucoma and diuretic agents which initiate the production of folate in bacterial systems. Sulfafurazole and sulfadiazine are ingested safely due to low price yet involved in forming residues in the environment by the release of urinal and faecal matters. This potentially results in antibiotic resistance, cancer, lithiasis, anaphylaxis and thyroid (Ji et al., 2017). Thus, ultrasensitive aptamer-based biosensors are capped with gold and silver nanoparticles for modified identification of the drugs.

2.7 Phenicols

Phenicols are inhibitors of protein synthesis which are well utilised to slow down the growth of bacteria and in preventing further pathogenic infections. Among all, chloramphenicol is well known for its efficacy and increased demand over industrial applications (Huang et al., 2019). Somewhat, its overconsumption can be accumulated in milk and meat, out of which diarrhoea, kidney dysfunction, hypersensitivity, nausea, bone marrow damage and gray body syndrome can be observed. According to WHO, it has no MRL range thus restricting its usage mandatorily (Miao et al., 2016). Similar techniques are introduced for phenicol detection as well. Sandwich aptamers are functionalised on indirect ELISA in the presence of biotin, streptavidin and silver nanoparticles, leading to visible colour change. Likewise, TMB oxidation carried out with the help of hydrogen peroxide, to detect the drug residuals calorimetrically (Abedalwafa et al., 2019).

3 Nano-Biosensing Techniques

Nanomaterials are considered to be an ideal option to be fabricated as transducers for identifying antibiotic residuals. Their unique and appealing nature to amplify signals by adhering to biorecognition components is appreciated. The special attributes of nanomaterials comprise high specificity, selectivity and sensitivity, high response rate, budget and travel friendly, better electrical conductance, high surface area, stability and distinctive magnetic properties which altogether make it designable (Seth and Rathinasabapathi, 2022). Additionally, nanomaterial-based biosensors typically hold physical, chemical, electrochemical, and optical characteristics employed in designing several nano-biosensing systems as per the requirements (Abedalwafa et al., 2019).

3.1 *Optical Nano-Biosensors*

It is known that to detect low levels of antibiotic residuals, some sort of reliable and reproducible sensors are to be considered. With prior efficiency, optical properties are highlighted with diverse biorecognition elements and visualise the analytes remarkably. Both fluorometric and colorimetric aspects of sensing technologies are put forward for clear indication of antibiotic concentrations (Hong et al., 2023).

3.1.1 Colorimetric Nano-Biosensors

Nowadays, colorimetric nano-biosensors are paying attention for being a sophisticated device with all the conventional qualities and modern analytical technologies (Abedalwafa et al., 2019). A simple change in colour can determine several antibiotic deposits by using dyes, enzymes, polymers, nanoparticles, etc. In order to have comprehensible aggregations, the biosensors are composed of spherical and colloidal nanoparticles and often require biorecognition components for dispersion (Abedi-Firoozjah et al., 2024).

Magnetic Nanoparticles

The potential of magnetic nanoparticles in antibiotic residual, colorimetric analysis has been demonstrated by their constant ability to be disintegrated in the presence of magnetic field. Nevertheless, Van der Waals' forces of interaction cause them to readily combine, which restricts their catalytic capabilities. It can only be regained by surface alteration and structure fabrication that lessens the likelihood of aggregation. Fe_3O_4Cu nanoparticles and "sandwich" oligonucleotide hybridisation were used to develop a novel colorimetric aptamer-based biosensor exhibiting enzymatic activities like that of peroxidase (Wang et al., 2022). It has a lower detection limit of 9 ng/ml.

It is even considered to be a risk free, quickly responding detection analysis in case of tetramethylbenzidine (Abedi-Firoozjah et al., 2024).

Metallic Nanoparticles

Metallic nanoparticles are superior nanomaterials in terms of optical features, tunability, high coefficients of extinction, easy synthesis and outstanding biological applications. The implementation of organic dyes for visible colour change at certain wavelength is appreciated with a simple target analyte by customising shape, size and surface area of nanoparticles. Gold and silver nanoparticles are functionally introduced for calorimetrically identification of antibiotic residuals, as when colour changes from pale yellow to bright pink (El-Said et al., 2020). Apart from that, morphological analysis also employed in recent years for simple and quick detection in water samples. Excellent durability, high-throughput performance, low cost and maintenance, and higher stability made metallic nanoparticles, a good to go option for colorimetric biosensing procedure. A new bio-sensing technique using a nanoprobe of gold nanoparticles and cDNA strand for chloramphenicol detection is on demand. The aptamer is conjugated with cDNA to create a hybrid with no drug thus acting as a biosensor. It often identifies the residual concentration in dairy products LOD and LDR of 0.13 pg/ml and 1.0 to 10 ng/ml, respectively. On the basis of hydrogen bonding between lysine and antibiotics, tetracyclines are aggregated with gold and silver nanoparticles and represent a LOD score of 22.6 nm. But on the other hand, it is better to invest in silver rather than in gold because of its high cost and maintenance (Abedi-Firoozjah et al., 2024).

Fabrication of Organic Compounds

Antibiotic residues can also be detected by a variety of organic compounds being fabricated with nanomaterials. One such component which is a tetraamine macrocyclic ligand with 14 members that has a strong affinity for transition metal cations (Aghajari & Azadbakht, 2018). Silver nanoparticles incorporated with cyclam function as a colorimetric biosensor which witness a colour change from violet to blue due to hydrogen bond. In addition to being utilised as a medication, phosphatidylcholine is utilised to detect gentamycin like that of cyclam. Likewise, polymethacrylic acid, melamine and luminol, a chemiluminescence either through physicochemical reactions or oxidation and reduction mechanisms are demonstrated as new modified capping agents to identify antibacterial targets calorimetrically (Abedi-Firoozjah et al., 2024).

3.1.2 Fluorometric Nano-Biosensors

There is no doubt that fluorometric nano-biosensors are a promising detection with low analysis error (Fig. 1).

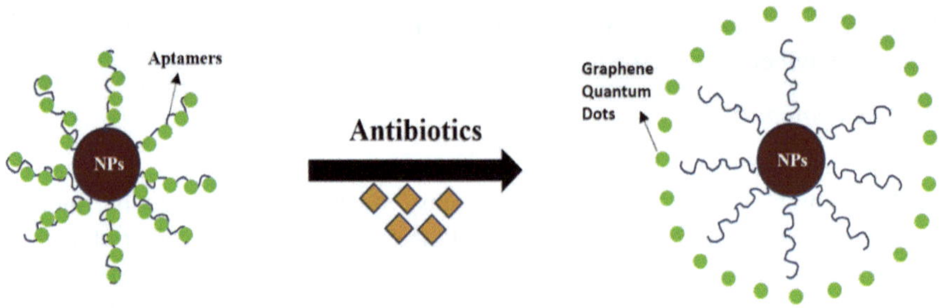

Fig. 1 Fluorometric Detection Method Using Nanocomposites

Carbon-Mediated Nanoparticles

Carbon mediated nanoparticles include mainly quantum dots, carbon nanotubes and graphene related nanomaterials. Carbon nanotubes are a common class of nanomaterial based optical biosensing systems to identify the antibiotic residues. Nowadays, quantum dots especially graphene quantum dots (GQDs) are employed as fluorescence probes. Their biocompatible and non-toxic features are quite safe to rely on. Even GQDs have the capacity to alter fluorescence, and other characteristics like water solubility, high surface area and reproducibility nature. Ratio-metric systems are initiated in association with hydrogen peroxide, Fe_3O_4 metallic nanoparticles and quantum dots, which are way more sensitive for detection purposes (Wang et al., 2018). The inner filter effect between the analytes and carbon nanotubes as well as Fe_3O_4 incorporation produce hydroxyl radicals via Fenton chemistry as a result of quenching of fluorescence probes.

Metal Organic Frameworks (MOFs)

Metal Organic Frameworks (MOFs) have been employed as fluorescent nanomaterials where amine groups are potentially utilised for better tunability and electron transport facilities. Low LOD for active sites are quite impressible having respectable recovery rates. For long and strong green fluorescence, 3D terbium-containing MOF are used ratiometrically to identify Tetrazolium chloride residue. Additionally, zeolitic imidazolate framework-8 (ZIF-8) and luminescent zirconium MOF (MIP/Zr-LMOF) are also considered to have a dual emission fluorescence probe on attachment with gold and silver nanoparticles (Abedi-Firoozjah et al., 2024).

3.2 *Electrochemical Nano-Biosensors*

Antibiotic residual monitoring is encouraged by electrochemical analysis of concerned contaminants for analytical exploitation. Both chemical and biochemical, as well as electrochemical signals are exhibited in accordance to impedance, voltage, capacitance and current. The detecting analyte, the transducing element and

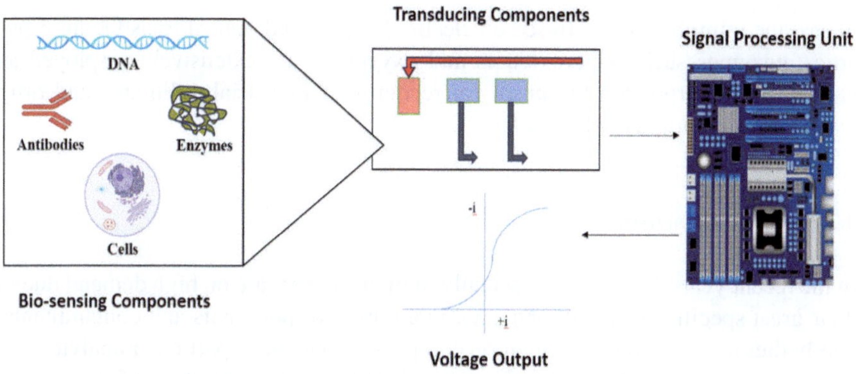

Fig. 2 Schematic representation of electrochemical bio-sensing technique

electronic components sending and receiving signals are the structural composition of an ideal biosensor. However, an efficient one requires nano cluster conjugation for overcoming limitations regarding performance and sensitivity (Fig. 2).

3.2.1 Aptamer Based Nano-Biosensors

Aptamer based nano-biosensors or simply aptasensors are synthetic and functional single stranded nucleic acids being useful for accurately identifying a wide range of targets, such as structural proteins, tiny compounds and even the entire cellular systems (Januarie et al., 2023). They tend to possess merits such as stable with low toxic and immunogenic levels, easily modified due to its tiny size. It can be done so by fabricating the activated surface non-covalently or by the bio-fabricated sensor surface directly with aptamers. A reliable and easy to perform detection tool often measure ciprofloxacin contaminants in a broad range by photoelectrochemical principle, whereas the functional nanomaterials include silver nanoparticles and carbon fibres. Others consist of ultrasensitive microfluidic aptasensors, for photosensitive silver doped copper graphene loaded copper indium disulfide as well as black anatase titanium nanomaterials adorned with reduced graphene oxide (Abedi-Firoozjah et al., 2024).

3.2.2 Enzyme Based Nano-Biosensors

Enzymes are the leading biomolecules and biocatalysts in biological systems. Depending on the target analyte, they are modified into biosensors by detecting and inhibiting substrates and enzymes at active sites respectively (Das et al., 2024a; Mohanty et al. 2024a; Priyadarshini et al., 2024). This is further accomplished by catalytic reaction and binding affinities of target analytes with respect to enzyme immobilisation. Based on electrochemical and optical transducing properties, enzymes such as peroxidases and oxygenase are extensively employed as nanocluster incorporated biosensor for recognising industrial pollutants and drug residues (Kurbanoglu et al., 2020).

3.2.3 Immunosensors

In the recent years, biosensors especially immunosensors are on high demand due to their great specificity in environmental monitoring of pollutants and contaminants. Antibodies are used as biorecognition component to identify particular analytes and toxic components. Both direct and indirect ELISA are being employed for a suitable signal transducing system. It depends on the development of antigen–antibody (Ag–Ab) complex for better immobilization (Aymard et al., 2022). Antibiotic residues such as ciprofloxacin and enrofloxacin with required LOD ranges are identified by the usage of either piezoelectric or double electrochemical immunosensors and even magnetic carbon nanocomposites which is specifically needed to prepare the biosensors (Abedi-Firoozjah et al., 2024).

4 Challenges and Limitations

Food safety depends on the identification of antibiotic residues in animal driven food products. Although the electrochemical and optical methods including fluorometric and colorimetric approaches are appreciated, there are some limitations holding back these procedures. Due to lower specificity, sensitivity and selectivity of nanofibers, it requires extended time for recognition of drug residuals. Sometimes, metallic nanoparticles are fabricated extensively over the requirement for a real time and ultrasensitive error free analysis like that of the conventional methods of detection. A variety of immunoassays including ELISA-mediated biosensors should be introduced besides antibody functionalization for effective employment of extended detection devices to enhance their capacity and stability (Abedalwafa et al., 2019).

5 Conclusions and Future Prospective

The need for nourishing ourselves with nutritious food has increased as a result of shifting dietary habits and sedentary lifestyles. Developing trustworthy health control measures are necessary to promote enough and safe supplements via diversified mechanisms with respect to financial basis.

Concern over antibiotic contamination is growing, which is why scientists are looking forward into various biosensing systems as a substitute for traditional analysis of drug residual concentrations. Some of the major benefits include affordability, usefulness, rapid response rate and high sensitivity. Towards meeting the leading expectations, nanomaterials are often incorporated with electrochemical and optical biosensing techniques. However, these effective aspects too come with drawbacks of being slow and sometimes costly. Thus, to manage all the cons, numerous detection platforms should be improved and put forward in an expected and combined manner.

Innovative identifying techniques from a futuristic perspective may be able to address this issue. It is remarkably known that the ability via different categories is absolutely possible to measure accurately and classify the antibiotic residuals accordingly, so that antibiotic resistance will not be developed. Hence, the new advancements in nano-biosensing techniques are worth saving the public health and hazards in the favour of the future generation.

Acknowledgements All the authors duly acknowledge the Department of Science and Technology, Government of India, New Delhi for infrastructure development by DST-FIST (First cycle, Year 2012, and Second Cycle, Year 2024, File No.SR/FST/LS-I/2024/1270). The Centre of Excellence in Environment and Public Health, Odisha State Higher Education Council, Government of Odisha (File No. 26913/HED/HE-PTC-WB-02-17/OHEPEE) by the World Bank Fund is duly acknowledged for the Infrastructure development for the Department of Zoology.

Author DS (principal Investigator) duly acknowledge the Science and Engineering Research Board (SERB), Government of India, New Delhi for Laboratory infrastructure facility created through sanctioned Extramural Research Grants (File No. EEQ/2016/000808).

Author AS and DS (Principal Investigator) duly acknowledge the Science and Technology Department, Government of Odisha for financial assistant to execute the Extramural Research Project on Advanced area of Biotechnology (File No. ST-BT-MISC-0009-2022-2744/ST).

Author YP duly acknowledge the Council for Scientific and Industrial Research (CSIR), New Delhi towards Junior Research Fellowship for financial assistant to pursue the Ph.D. program.

Author Contribution SSD: Draft writing and manuscript preparation; YP: manuscript editing; AS: data collection and design; DS: conceptualization, data analysis, review and editing. All authors contributed to the book chapter.

Conflict of Interest The authors declare no conflict of interest.

References

Abedalwafa, M. A., Li, Y., Ni, C., & Wang, L. (2019). Colorimetric sensor arrays for the detection and identification of antibiotics. *Analytical Methods, 11*(22), 2836–2854.https://doi.org/10.1039/C9AY00371A

Abedi-Firoozjah, R., Alizadeh-Sani, M., Zare, L., Rostami, O., Azimi Salim, S., Assadpour, E., Azizi-Lalabadi, M., Zhang, F., Lin, X., & Jafari, S. M. (2024). State-of-the-art nanosensors and kits for the detection of antibiotic residues in milk and dairy products. *Advances in Colloid and Interface Science, 328*, Article 103164. https://doi.org/10.1016/j.cis.2024.103164

Aghajari, R., & Azadbakht, A. (2018). Amplified detection of streptomycin using aptamer-conjugated palladium nanoparticles decorated on chitosan-carbon nanotube. *Analytical Biochemistry, 547*, 57–65. https://doi.org/10.1016/j.ab.2018.02.005

Aymard, C., Kanso, H., Serrano, M. J., Pagán, R., Noguer, T., & Istamboulie, G. (2022). Development of a new dual electrochemical immunosensor for a rapid and sensitive detection of enrofloxacin in meat samples. *Food Chemistry, 370*, Article 131016. https://doi.org/10.1016/j.foodchem.2021.131016

Barros, S. C., Silva, A. S., & Torres, D. (2023). Multiresidues multiclass analytical methods for determination of antibiotics in animal origin food: A critical analysis. *Antibiotics (Basel, Switzerland), 12*(2), 202. https://doi.org/10.3390/antibiotics12020202

Cardoso, A. R., Carneiro, L. P., Cabral-Miranda, G., Bachmann, M. F., & Sales, M. G. F. (2021). Employing bacteria machinery for antibiotic detection: using DNA gyrase for ciprofloxacin detection. *Chemical Engineering Journal*, *409*, 128135.https://doi.org/10.1016/j.cej.2020.128135

Chang, H., Lv, J., Zhang, H., Zhang, B., Wei, W., & Qiao, Y. (2017). Photoresponsive colorimetric immunoassay based on chitosan modified AgI/TiO_2 heterojunction for highly sensitive chloramphenicol detection. *Biosensors & Bioelectronics, 87*, 579–586. https://doi.org/10.1016/j.bios.2016.09.002

Chavada, V. D., Bhatt, N. M., Sanyal, M., & Shrivastav, P. S. (2017). Surface plasmon resonance based selective and sensitive colorimetric determination of azithromycin using unmodified silver nanoparticles in pharmaceuticals and human plasma. *Spectrochimica Acta Part A, Molecular and Biomolecular Spectroscopy, 170*, 97–103. https://doi.org/10.1016/j.saa.2016.07.011

Cuprys, A., Pulicharla, R., Brar, S. K., Drogui, P., Verma, M., & Surampalli, R. Y. (2018). Fluoroquinolones metal complexation and its environmental impacts. *Coordination Chemistry Reviews*, *376*, 46–61.https://doi.org/10.1016/j.ccr.2018.05.019

Das, A., Dey, S., Bala, K., Gnanasekaran, R., & Das, A. P. (2024a). Microbial bioremediation technology for sustainable treatment and management of synthetic microfiber waste. In *Sustainable microbial technology for synthetic and cellulosic microfiber bioremediation* (pp. 221–238). https://doi.org/10.1007/978-3-031-62600-5_12

Das, A., Dey, S., & Das, A. P. (2024b). Ecotoxicological impacts of synthetic microfiber pollutants and development of sustainable mitigation strategies. *Environmental Chemistry and Ecotoxicology*. https://doi.org/10.1016/J.ENCECO.2024.12.003

Das, A., Ray, S. K., Mohanty, M., Mohanty, J., Dey, S., & Das, A. P. (2024c). Ecotoxicity of microplastic wastes and their sustainable management: A review. *Environmental Chemistry and Ecotoxicology, 6*, 144–152. https://doi.org/10.1016/J.ENCECO.2024.05.003

Dehghani, S., Jonidi Jafari, A., Farzadkia, M., & Gholami, M. (2013). Sulfonamide antibiotic reduction in aquatic environment by application of Fenton oxidation process. *Iranian Journal of Environmental Health Science & Engineering, 10*(1), 29. https://doi.org/10.1186/1735-2746-10-29

El-Said, W. A., Abdelshakour, M., Choi, J. H., & Choi, J. W. (2020). Application of conducting polymer nanostructures to electrochemical biosensors. *Molecules (Basel, Switzerland), 25*(2), 307. https://doi.org/10.3390/molecules25020307

Fair, R. J., & Tor, Y. (2014). Antibiotics and bacterial resistance in the 21st century. *Perspectives in Medicinal Chemistry, 6*, 25–64. https://doi.org/10.4137/PMC.S14459

Furst, A. L., & Francis, M. B. (2019). Impedance-based detection of bacteria. *Chemical Reviews, 119*(1), 700–726. https://doi.org/10.1021/acs.chemrev.8b00381

Gaudin, V. (2017). Advances in biosensor development for the screening of antibiotic residues in food products of animal origin-A comprehensive review. *Biosensors & Bioelectronics, 90*, 363–377. https://doi.org/10.1016/j.bios.2016.12.005

Gissawong, N., Srijaranai, S., Boonchiangma, S., Uppachai, P., Seehamart, K., Jantrasee, S., Moore, E., & Mukdasai, S. (2021). An electrochemical sensor for voltammetric detection of ciprofloxacin using a glassy carbon electrode modified with activated carbon, gold nanoparticles and supramolecular solvent. *Mikrochimica Acta, 188*(6), 208. https://doi.org/10.1007/s00604-021-04869-z

Gou, Y., Chen, P., Yang, L., Li, S., Peng, L., Song, S., & Xu, Y. (2021). Degradation of fluoroquinolones in homogeneous and heterogeneous photo-Fenton processes: A review. *Chemosphere, 270*, Article 129481. https://doi.org/10.1016/j.chemosphere.2020.129481

Hong, J., Su, M., Zhao, K., Zhou, Y., Wang, J., Zhou, S. F., & Lin, X. (2023). A minireview for recent development of nanomaterial-based detection of antibiotics. *Biosensors, 13*(3), 327. https://doi.org/10.3390/bios13030327

Huang, W., Zhang, H., Lai, G., Liu, S., Li, B., & Yu, A. (2019). Sensitive and rapid aptasensing of chloramphenicol by colorimetric signal transduction with a DNAzyme-functionalized gold nanoprobe. *Food Chemistry, 270*, 287–292. https://doi.org/10.1016/j.foodchem.2018.07.127

Jahanbani, S., & Benvidi, A. (2016). Comparison of two fabricated aptasensors based on modified carbon paste/oleic acid and magnetic bar carbon paste/Fe3O4@oleic acid nanoparticle electrodes for tetracycline detection. *Biosensors & Bioelectronics, 85*, 553–562. https://doi.org/10.1016/j.bios.2016.05.052

Januarie, K. C., Oranzie, M., Feleni, U., & Iwuoha, E. (2023). Quantum dot amplified impedimetric aptasensor for interferon-gamma. *Electrochimica Acta, 463*, 142825.https://doi.org/10.1016/j.electacta.2023.142825

Ji, H., Wu, Y., Duan, Z., Yang, F., Yuan, H., & Xiao, D. (2017). Sensitive determination of sulfonamides in environmental water by capillary electrophoresis coupled with both silvering detection window and in-capillary optical fiber light-emitting diode-induced fluorescence detector. *Electrophoresis, 38*(3–4), 452–459. https://doi.org/10.1002/elps.201600364

Jin, Y., Dou, M., Zhuo, S., Li, Q., Wang, F., & Li, J. (2022). Advances in microfluidic analysis of residual antibiotics in food. *Food Control, 136*, 108885.https://doi.org/10.1016/j.foodcont.2022.108885

Khan, M. Z. H. (2022). Recent biosensors for detection of antibiotics in animal derived food. *Critical Reviews in Analytical Chemistry, 52*(4), 780–790. https://doi.org/10.1080/10408347.2020.1828027

Kharewal, T., Verma, N., Gahlaut, A., & Hooda, V. (2020). Biosensors for penicillin quantification: A comprehensive review. *Biotechnology Letters, 42*(10), 1829–1846. https://doi.org/10.1007/s10529-020-02970-6

Kraemer, S. A., Ramachandran, A., & Perron, G. G. (2019). Antibiotic pollution in the environment: From microbial ecology to public policy. *Microorganisms, 7*(6), 180. https://doi.org/10.3390/microorganisms7060180

Kumar, H., & Neelam, R. (2016). Enzyme-based electrochemical biosensors for food safety: A review. In *Nanobiosensors in disease diagnosis*, (pp. 29–39). https://doi.org/10.2147/NDD.S64847

Kurbanoglu, S., Erkmen, C., & Uslu, B. (2020). Frontiers in electrochemical enzymebased biosensors for food and drug analysis. *TrAC Trends in Analytical Chemistry, 124*, 115809.https://doi.org/10.1016/j.trac.2020.115809

Lan, L., Yao, Y., Ping, J., & Ying, Y. (2017). Recent advances in nanomaterial-based biosensors for antibiotics detection. *Biosensors & Bioelectronics, 91*, 504–514. https://doi.org/10.1016/j.bios.2017.01.007

Lawal, A. T. (2018). Progress in utilisation of graphene for electrochemical biosensors. *Biosensors & Bioelectronics, 106*, 149–178. https://doi.org/10.1016/j.bios.2018.01.030

Liu, Q., Mu, H., Sun, C., & Duan, J. (2016). Highly specific determination of gentamicin by induced collapse of Au–lipid capsules. *RSC Advances, 6*(18), 14483–14489.https://doi.org/10.1039/C5RA21752K

Liu, J., Zeng, J., Tian, Y., & Zhou, N. (2018). An aptamer and functionalized nanoparticle-based strip biosensor for on-site detection of kanamycin in food samples. *Analyst, 143*(1), 182–189.https://doi.org/10.1039/C7AN01476G

Lu, S., Wang, S., Wu, P., Wang, D., Yi, J., Li, L., Ding, P., & Pan, H. (2021). A composite prepared from covalent organic framework and gold nanoparticles for the electrochemical determination of enrofloxacin. *Advanced Powder Technology, 32*(6), 2106–2115.https://doi.org/10.1016/j.apt.2021.04.025

Masoudyfar, Z., & Elhami, S. (2019). Surface plasmon resonance of gold nanoparticles as a colorimetric sensor for indirect detection of Cefixime. *Spectrochimica Acta Part A, Molecular and Biomolecular Spectroscopy, 211*, 234–238. https://doi.org/10.1016/j.saa.2018.12.007

Miao, Y. B., Ren, H. X., Gan, N., Cao, Y., Li, T., & Chen, Y. (2016). Fluorescent aptasensor for chloramphenicol detection using DIL-encapsulated liposome as nanotracer. *Biosensors & Bioelectronics, 81*, 454–459. https://doi.org/10.1016/j.bios.2016.03.034

Mohanty, J., Dey, S., Bala, K., & Das, A. P. (2024a). Recent advances in microplastic pollution for its sustainable management. In *Sustainable management of urban plastic waste through circular economic approaches* (pp. 27–48). https://doi.org/10.1201/9781003465447-3/RECENT-ADVANCES-MICROPLASTIC-POLLUTION-SUSTAINABLE-MANAGEMENT-JYOTIRMAYEE-MOHANTY-SUDESHNA-DEY-KIRAN-BALA-ALOK-PRASAD-DAS

Mohanty, S., Benya, A., Hota, S., Das, A. P., & Singh, S. (2024b). Impact of COVID-19 generated emerging pollutants on aquatic environment and its effect on fish population and its edibility: A review. In *Environmental science and engineering, Part F2827*, (pp. 17–33). https://doi.org/10.1007/978-3-031-50840-0_2

Nepfumbada, C., Mthombeni, N. H., Sigwadi, R., Ajayi, R. F., Feleni, U., & Mamba, B. B. (2024). Functionalities of electrochemical fluoroquinolone sensors and biosensors. *Environmental Science and Pollution Research International, 31*(3), 3394–3412. https://doi.org/10.1007/s11356-023-30223-2

Priyadarshini, B., & Das, A. P. (2024). The adverse impact of synthetic microplastics released from disposable face masks pollutants on aquatic microbiota, aquatic species and humans: A review. In *Environmental science and engineering, Part F2827*, (pp. 67–81). https://doi.org/10.1007/978-3-031-50840-0_5

Priyadarshini, S., Jagatee, S., & Das, A. P. (2024). Synthetic fabrics and microfiber pollution–An assessment of their global impact. In *Environmental science and engineering, Part F2333*, (pp. 137–157). https://doi.org/10.1007/978-3-031-51792-1_8

Ramezani, M., Mohammad Danesh, N., Lavaee, P., Abnous, K., & Mohammad Taghdisi, S. (2015). A novel colorimetric triple-helix molecular switch aptasensor for ultrasensitive detection of tetracycline. *Biosensors & Bioelectronics, 70*, 181–187. https://doi.org/10.1016/j.bios.2015.03.040

Robati, R. Y., Arab, A., Ramezani, M., Langroodi, F. A., Abnous, K., & Taghdisi, S. M. (2016). Aptasensors for quantitative detection of kanamycin. *Biosensors & Bioelectronics, 82*, 162–172. https://doi.org/10.1016/j.bios.2016.04.011

Saini, V., McClure, J. T., Léger, D., Dufour, S., Sheldon, A. G., Scholl, D. T., & Barkema, H. W. (2012). Antimicrobial use on Canadian dairy farms. *Journal of Dairy Science, 95*(3), 1209–1221. https://doi.org/10.3168/jds.2011-4527

Sánchez-Paniagua López, M., Redondo-Gómez, E., & López-Ruiz, B. (2017). Electrochemical enzyme biosensors based on calcium phosphate materials for tyramine detection in food samples. *Talanta, 175*, 209–216. https://doi.org/10.1016/j.talanta.2017.07.033

Seth., S.S., &Rathinasabapathi, P. (2022). A short review on detection of antibiotics in milk using nanomaterial-based biosensor. *Food Analytical Methods, 15*, 2181–2192. https://doi.org/10.1007/s12161-022-02291-6

Song, K. M., Jeong, E., Jeon, W., Cho, M., & Ban, C. (2012). Aptasensor for ampicillin using gold nanoparticle based dual fluorescence-colorimetric methods. *Analytical and Bioanalytical Chemistry, 402*(6), 2153–2161. https://doi.org/10.1007/s00216-011-5662-3

Taherizadeh, M., Jahani, S., Moradalizadeh, M., & Foroughi, M. M. (2023). Synthesis of a dual-functional terbium doped copper oxide nanoflowers for high-efficiently electrochemical sensing of ofloxacin, pefloxacin and gatifloxacin. *Talanta, 255*, Article 124216. https://doi.org/10.1016/j.talanta.2022.124216

Vu, O. T., Nguyen, Q. H., Nguy Phan, T., Luong, T. T., Eersels, K., Wagner, P., & Truong, L. T. N. (2023). Highly sensitive molecularly imprinted polymer-based electrochemical sensors enhanced by gold nanoparticles for norfloxacin detection in aquaculture water. *ACS Omega, 8*(3), 2887–2896. https://doi.org/10.1021/acsomega.2c04414

Wang, L., Zhou, H., Wu, X., Song, Y., Huang, Y., Yang, X., & Chen, X. (2022). A novel colorimetric aptasensor for sensitive tetracycline detection based on the peroxidase-like activity of

Fe_3O_4@Cu nanoparticles and "sandwich" oligonucleotide hybridization. *Mikrochimica Acta, 189*(3), 86. https://doi.org/10.1007/s00604-022-05195-8

Wang, C., Chen, D., Wang, Q., & Tan, R. (2017). Kanamycin detection based on the catalytic ability enhancement of gold nanoparticles. *Biosensors and Bioelectronics, 91*, 262–267. https://doi.org/10.1016/j.bios.2016.12.042

Wang, Y., Ni, P., Jiang, S., Lu, W., Li, Z., Liu, H., Lin, J., Sun, Y., & Li, Z. (2018). Highly sensitive fluorometric determination of oxytetracycline based on carbon dots and Fe3O4 MNPs. *Sensors and Actuators B: Chemical, 254*, 1118–1124. https://doi.org/10.1016/j.snb.2017.07.182

Wu, D., Du, D., & Lin, Y. (2016). Recent progress on nanomaterial-based biosensors for veterinary drug residues in animal-derived food. *TrAC Trends in Analytical Chemistry, 83*, 95–101. https://doi.org/10.1016/j.trac.2016.08.006

Xia, Q., Jiang, H., Liu, X., Yin, L., & Wang, X. (2024). Advances in engineered nano-biosensors for bacteria diagnosis and multidrug resistance inhibition. *Biosensors, 14*(2), 59. https://doi.org/10.3390/bios14020059

Xiao, C., Liu, J., Yang, A., Zhao, H., He, Y., Li, X., & Yuan, Z. (2015). Colorimetric determination of neomycin using melamine modified gold nanoparticles. *Microchimica Acta, 182*, 1501–1507. https://doi.org/10.1007/s00604-015-1480-1

Yazdanparast, S., Benvidi, A., Abbasi, S., & Rezaeinasab, M. (2019). Enzyme-based ultrasensitive electrochemical biosensor using poly (l-aspartic acid)/MWCNT bio-nanocomposite for xanthine detection: A meat freshness marker. *Microchemical Journal, 149*, 104000. https://doi.org/10.1016/j.microc.2019.104000

Zhang, L., Jin, Y., Mao, H., Zheng, L., Zhao, J., Peng, Y., Du, S., & Zhang, Z. (2014). Structure-selective hot-spot Raman enhancement for direct identification and detection of trace penicilloic acid allergen in penicillin. *Biosensors & Bioelectronics, 58*, 165–171. https://doi.org/10.1016/j.bios.2014.02.052

Zhou, C., Zou, H., Sun, C., Ren, D., Xiong, W., & Li, Y. (2018). Fluorescent aptasensor for detection of four tetracycline veterinary drugs in milk based on catalytic hairpin assembly reaction and displacement of G-quadruplex. *Analytical and Bioanalytical Chemistry, 410*(12), 2981–2989. https://doi.org/10.1007/s00216-018-0981-2

[illegible]

Nanotechnology for Resource Conservation

Hasan Volkan Oral

Abstract Nature's resources are limitless, but human activity is destroying and polluting them daily. In addition, the effects of global warming and climate change have sped up the degradation of the ecosystem. Since the 1980s, there has been discussion about sustainability and sustainable development as a means of preventing this. These ideas are insufficient in the current technological era to offer practical solutions to the issue. As a result, creative solutions are needed to halt the world's expanding resource consumption as well as the negative effects of climate change and global warming. Resource conservation can be achieved by nanotechnology, which is the control of matter at the molecular level. The purpose of this section is to present the applications of nanotechnology in the fields of environmental impact and safety, sustainable water management, material recovery, and energy management. Each section is including a theoretical notion followed by a practical implementation. For example, for the section on sustainable water management, examples supported by nanotechnology will be tried to be described in an understandable way from nature-based solutions and circularity perspectives. Case studies were selected different geographical locations supported the information from the literature. This approach has also been applied to other chapters, and case studies have been presented from many different countries and regions, including China, Spain, Malaysia, and Brazil.

Keywords Sustainability · Nanotechnology · Nature resource management · Nature-based solutions · Circular economy

H. V. Oral (✉)
Department of Civil Engineering, Istanbul Aydin University, Istanbul, Türkiye
e-mail: volkanoral@aydin.edu.tr

E. Priyadarshini et al. (eds.), *Advanced Nanotechnology Applications for Environmental Sustainability*, Sustainable Environmental Waste Management Strategies,
https://doi.org/10.1007/978-3-032-12933-8_19

1 Introduction

Nanotechnology is characterized as a discipline within science and engineering that emphasizes the design, creation, and utilization of structures, devices, and systems through the manipulation of atoms and molecules at the nanoscale, which is generally recognized as having one or more dimensions measuring approximately 100 nm (100 millionths of a millimeter) or less (EU) (2006). Nimesh (2013) referenced Norio Taniguchi's definition from 1974, stating that "Nano-technology primarily involves the processing of materials through the separation, consolidation, and deformation at the level of individual atoms or molecules." Additionally, nanotechnology pertains to technology that functions at the nanoscale, offering numerous potential applications. Olawade et al. (2024) characterize the synergy between Artificial Intelligence (AI) and nanotechnology, often highlighted in discussions about the 2025s, as a groundbreaking intersection that promises significant advancements in both scientific and technological fields (Bala et al., 2024; Das et al., 2024a, 2024b). According to Naik and Jagtap (2024), the development of AI and nanotechnology is significant and has the potential to revolutionize many areas of human endeavor. The ability of artificial intelligence (AI) to analyze vast volumes of data, spot patterns, and learn enables the automation of a variety of procedures as well as decision-making and problem-solving tools across industries to enhance those sectors and boost their production and efficiency.

However, no matter how much technology advances and how much AI offers solutions that will make human life easier, we are observing the effects of climate change and global warming very frequently today. Nature is being rapidly consumed and polluted. The purpose of this book chapter is to present the use of nanotechnology in resource conservation. The book chapter's organizational structure is as follows: Definitions are provided so that audiences can comprehend the study's topic. Applications of nanotechnology are discussed in Sect. 2 in a number of areas, including material recovery, energy management, sustainable water management covered by nature-based solutions (NbS) and circularity perspectives, and environmental effect and safety. The conclusion is finally concluded in Sect. 3.

2 Definitions

The objective of this part is to provide definitions that help readers easily understand the overall notion of the topic.

Sustainability: The Cambridge Dictionary (2025a) defines sustainability as the ability to continue for an extended period of time while incurring little or no environmental damage.

Sustainable Development: If we want a better tomorrow, we must live in a way that promotes sustainable development, which means addressing current needs without compromising the ability of future generations to do the same (UN, 2023).

Nature Resource Management: The methods by which civilizations control the availability or accessibility of the natural resources that they depend on for their existence and advancement are known as natural resource management (Britannica, 2025).

Nature-based solutions (NbS): NbS make use of the natural world and the strength of robust ecosystems to ensure a stable and biodiverse future, optimize infrastructure, and protect people (IUCN, 2023).

Circular Economy: The circular economy is a production and consumption model that focuses on sharing, leasing, reusing, repairing, refurbishing, and recycling existing materials and products to the greatest extent possible (Das et al., 2024a, 2024b; John & Das, 2024). This prolong the product's life cycle (EU, 2015).

Environmental impact and safety: The effect that the activities of people and businesses have on the environment (Cambridge, 2025b), and environmental safety refers to the policies and practices put in place to protect people and the environment against risks and dangers in their surroundings (Roberts, 2001).

Sustainable Water Management: This term can be identified as applying the means of sustainability to the water management (Water Foundation, 2025).

Material Recovery: Materials recovered in the waste stream are restored to a beneficial application, which may be different from the original use (EEA, 2025).

Energy Management: The process involves overseeing, strategizing, enhancing, and conserving energy to achieve a system that operates with maximum energy efficiency. (Sharma et al., 2022).

The applications of nanotechnology in specified fields

The uses of nanotechnology in particular sectors are shown in Fig. 1.

3 Environmental Impact and Safety

According to Stander and Theodore (2011), nanotechnology relevant to environmental impacts is searching for the answers to the following questions:

- What possible environmental issues are connected to this new technology?
- Is it realistic for industry and society to anticipate the release of toxic or hazardous materials into the environment during the production or usage of nanoproducts?
- Could environmental deterioration result from nano applications, specifically from the bioaccumulation of nano products in living tissue?
- How are rules going to affect this new technology?

Linked with the Questions, Rezvani et al., (2019) revealed that silver and silver Nanoparticles (NPs) (AgNPs) have antimicrobial qualities against certain bacteria, fungi, and viruses, the growing use of nanosilver in consumer goods, water purification, and medical settings has sparked worries about the material's safety for the environment and public health. Hydrophobicity is one of the physical and chemical characteristics of nanomaterials that are thought to be crucial to their environmental impact. It is possible to forecast changes in hydrophobicity brought on by biotic

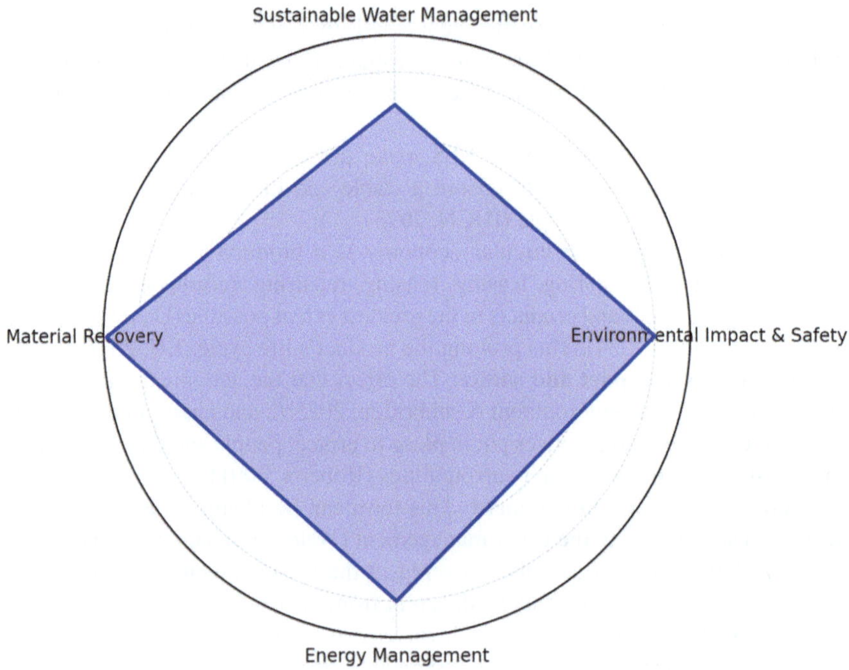

Fig. 1 Sectors relevant to Nanotechnology application

and abiotic processes. The environmental behaviors of NPs (NPs), such as bioaccumulation, toxicity, and aggregation. To assess the hydrophobicity changes brought about by sulfidation and the creation of natural organic matter (NOM) corona, the binding rate of silver (Ag) NPs on manufactured surfaces was tracked using dark-field microscopy (DFM) (Roncari et al., 2023).

Case study Examples: Roncari et al., (2023) conducted a study on the samples taken from the Suwannee River, in the southern United States that flows from south Georgia to Florida. Furthermore, using fuzzy logic on samples from the Besòs River in northeastern Spain, which empties into the Mediterranean Sea close to Barcelona, Ramirez et al. (2021) examined the environmental risk of silver NPs in aquatic ecosystems.

4 Sustainable Water Management, Nature-Based Solutions, and Circularity Perspectives

NPs, nanocomposites, and nanofiltration membranes are examples of nanomaterials with enhanced adsorption, catalytic, and filter properties. Numerous pollutants, including organic pollutants, microorganisms, and heavy metals, are effectively eliminated from water sources by these NPs. Furthermore, real-time detection of water quality indicators is made possible by nanotechnology-based sensors and monitoring systems, guaranteeing safe and clean water sources. The authors revealed that biosensors and their applications can also be noticed particularly in the fields of environmental monitoring and medicine. Biosensors with enhanced sensitivity and performance have been made possible by the unique properties of nanomaterials such as graphene, carbon nanotubes (CNTs), and nanospheres. In biosensing applications, microalgae are commonly employed as stable and adaptable receptors (Kumar et al., 2024).

Case Study Examples: Nanofiltration membranes (NF) used for water treatment is one of the NbS applications in this field. Currently, this technology has been applied in Jiaxing City, Zhejiang Province of China. The combined treatment capacity of the two water plants' phase nanofiltration systems is 300,000 cubic meters per day. They are likely to remove more than 85% of the total organic carbon (TOC), 85% of the sulfate, and less than 1.0 mg/L of COD in the permeate (DuPont, 2023). Another example of a case study comes from Canada, where an NF system was created to generate 100 gallons of water per minute for use in making drinking water (PureAqua, 2023).

5 Material Recovery

Nanostructured adsorbents are increasingly employed in wastewater treatment due to their superior efficiency and rapid response capabilities. Research is underway to develop magnetic nanoparticles for the adsorption of metals and various chemical compounds (Priyadarshini & Das, 2024; Priyadarshini et al., 2024; Sahu et al., 2024). Additionally, nanofiltration technology is applied to eliminate industrial toxins from groundwater, including bisphenol A, phthalates, and alkylphenols. The concept of green nanotechnology has recently gained prominence, with its main objective being the mitigation of health risks and environmental issues (Jangid & Inbaraj, 2021).

Case Study Example: Applications in this field are monitored in China due to its population density and large geographical area (Qingping et al., 2017).

6 Energy Management

One of the most important issues facing the twenty-first century is energy. The seriousness of this problem is increasing due to the combined risks of depleting energy supplies and rising consumption. There are many different approaches to energy consumption, from innovative technology to behavioral modifications (Asgari et al., 2024). Renewable energy sources can be enhanced by nanotechnology; for instance, rotor blades made of lighter, stronger nanomaterials can increase wind energy efficiency (Mishra et al., 2024; Mohanty et al., 2024a, 2024b). Precision farming based on nanotechnology is being employed in biomass energy to maximize the crops used to make biofuels. While nano-composites are used to increase the wear resistance of drilling machines in geothermal energy, nano-coatings can be used to stop corrosion in tidal energy equipment (Hussein, 2015).

Case Study Examples: With an emphasis on uses like smart sensors for tracking soil health and NPs for controlled-release fertilizers, nations like Brazil and Malaysia are investing in nanotechnology for agricultural sustainability (Khatoon & Velidandi, 2025). Kyoto University of Japan created a novel semiconductor that allows solar panels to convert twice as much sunlight into power (Iberdola, 2025).

7 Results and Discussion

Nanotechnology has been proposed as a sustainable solution to a variety of environmental problems, such as fuel crises, wastewater treatment, greenhouse gas emissions, and the elimination of various pollutants that may contribute to climate change. Consequently, nanotechnology is opening up new environmental applications and may hold the answer to slowing down future climate change. The possibilities of developing technology will provide us with the opportunity to use nanotechnology more in many areas. However, it is necessary to see and be aware of the negative effects of the use of nanotechnology in some fields. According to the Center for Food Safety (2025), nanotechnology is all about size reduction, and because of this drastic size reduction, materials that appear to be commonplace may act very differently than when they are in their bigger bulk or macro form. NPs (NPs) are more easily able to pass through biological membranes, cells, tissues, and organs than larger particles because of their small size. Nanomaterials can travel throughout the body after they enter the bloodstream and settle in many organs and tissues, such as the brain, liver, heart, kidneys, spleen, bone marrow, and nervous system. Once within cells, they have the potential to induce oxidative damage, disrupt regular biological processes, and even kill cells. Both biological systems and the environment may be at risk from NPs. It is also becoming obvious that the size, structure, and type of nanomaterials, such as graphene/graphene oxide with gold NPs, and carbon and carbon nitride nanotubes, have diverse effects on plants and the environment. Therefore, to evaluate the effects of NPs, long-term life cycle assessments are required (Kumar et al., 2019).

As a result, using nanotechnology responsibly and in terms of natural resource conservation, taking into account its potential negative effects on the environment, is important in reducing the effects of climate change.

References

Asgari, H., et al. (2024). The impact of nanomaterials on energy-centric form-finding of educational buildings in a semi-arid climate. *Heliyon*. https://doi.org/10.1016/j.Heliyon.2024.e39882

Bala, K., Anshul, Verma, S., Dogra, A., Das, A. P., Kumar, R., & Nandini. (2024). Sustainable management for urban plastic waste generation. *Sustainable Management of Urban Plastic Waste Through Circular Economic Approaches*, 211–236. https://doi.org/10.1201/9781003465447-13

Britannica (2025). https://www.britannica.com/money/natural-resource-management

Cambridge Dictionary (2025a). https://dictionary.cambridge.org/dictionary/english/sustainability

Cambridge Dictionary (2025b). https://dictionary.cambridge.org/dictionary/english/environmental-impact

Center for Food Safety (2025). https://www.centerforfoodsafety.org/issues/682/nanotechnology/risks-of-nanotechnology

Das, A., Dey, S., Bala, K., Gnanasekaran, R., & Das, A. P. (2024). Microbial bioremediation technology for sustainable treatment and management of synthetic microfiber waste. *Sustainable Microbial Technology for Synthetic and Cellulosic Microfiber Bioremediation*, 221–238. https://doi.org/10.1007/978-3-031-62600-5_12

Das, A. Prasad., Behera, I. Dipamitra., & Das, N. Prasad. (2024). *Renewable energy generation and value addition from environmental microfiber pollution through advanced greener solution.* 278.

Dupont (2023). https://www.dupont.com/content/dam/water/amer/us/en/water/public/documents/en/NF-FilmTec-NF270-MDW-Jiaxing-China-300MLD-CS-45-D04715-en.pdf

European Environment Agency (2025). https://www.eea.europa.eu/help/glossary/eea-glossary/material-recovery

European Union (2007). https://ec.europa.eu/health/scientific_committees/opinions_layman/en/nanotechnologies/about-nanotechnologies.htm#7

European Union (2015). https://www.europarl.europa.eu/topics/en/article/20151201STO05603/circular-economy-definition-importance-and-benefits

Hussein, A. K. (2015). Applications of nanotechnology in renewable energies—A comprehensive overview and understanding. *Renewable and Sustainable Energy Reviews, 42*, 460–476. https://doi.org/10.1016/j.rser.2014.10.027

INDERDOLA (2025). https://www.iberdrola.com/innovation/nanotechnology-applications

IUCN (2023). https://iucn.org/our-work/nature-based-solutions

Jagtap, V and Naik, G (2024). Two heads are better than one: Unraveling the potential impact of artificial intelligence in Nanotechnology. Nano TransMed. 3. https://doi.org/10.1016/j.ntm.2024.100041.

Jangid, P and Inbaraj, P. (2021). Applications of nanomaterials in wastewater treatment. *Materials Today: Proceedings, 43*. https://doi.org/10.1016/j.matpr.2021.01.126.

John, A., & Das, A. P. (2024). Government strategies for achieving circular economy for plastic waste management in southeast Asia. *Sustainable Management of Urban Plastic Waste Through Circular Economic Approaches*, 72–82. https://doi.org/10.1201/9781003465447-6

Khatoon, U. T., & Velidandi, A. (2025). An overview on the role of government initiatives in nanotechnology innovation for sustainable economic development and research progress. *Sustainability, 17*(3), 1250. https://doi.org/10.3390/su17031250

Kumar, A., Gupta, K., Dixit, S., et al. (2019). A review on positive and negative impacts of nanotechnology in agriculture. *International Journal of Environmental Science and Technology, 16*, 2175–2184. https://doi.org/10.1007/s13762-018-2119-7

Kumar, A., et al. (2024). Integrating green nanotechnology with sustainable development goals: A pathway to sustainable innovation. *Discov Sustain, 5*, 364. https://doi.org/10.1007/s43621-024-00610-x

Mishra, S., Dash, D., & Das, A. P. (2024). Microbial degradation of polyester microfibers using indigenously isolated bacterial strain exiguobacterium Sp. *CLEAN—Soil, Air, Water*, *52*(12), e202300343. https://doi.org/10.1002/CLEN.202300343

Mohanty, J., Dey, S., Bala, K., & Das, A. P. (2024). Recent advances in microplastic pollution for its sustainable management. *Sustainable Management of Urban Plastic Waste Through Circular Economic Approaches*, 27–48. https://doi.org/10.1201/9781003465447-3

Mohanty, M., Tripathy, B., Koteswara Reddy, G., & Das, A. P. (2024). Greener technologies for establishing restraint over microplastic pollution. *Sustainable Microbial Technology for Synthetic and Cellulosic Microfiber Bioremediation*, 239–258. https://doi.org/10.1007/978-3-031-62600-5_13

Nimesh, S (2013) Gene therapy, potential applications of nanotechnology, A volume in Woodhead Publishing Series in Biomedicine, Woodhead Publishing

Olawade, D. B., et al. (2024). The synergy of artificial intelligence and nanotechnology towards advancing innovation and sustainability—a mini-review

Priyadarshini, B., & Das, A. P. (2024). The adverse impact of synthetic microplastics released from disposable face masks pollutants on aquatic microbiota, aquatic species and humans: A review. *Environmental Science and Engineering, Part, F2827*, 67–81. https://doi.org/10.1007/978-3-031-50840-0_5

Priyadarshini, S., Jagatee, S., & Das, A. P. (2024). Synthetic fabrics and microfiber pollution–an assessment of their global impact. *Environmental Science and Engineering, Part, F2333*, 137–157. https://doi.org/10.1007/978-3-031-51792-1_8

PureAqua (2023). https://pureaqua.com/nanofiltration-system-for-beverage-use-100-gpm-canada/

Qingping, W., et al. (2017). Application of nanotechnology in wastewater treatment. *Nanoscience and Nanotechnology, 1*(1), 19–25.

Ramirez, R., et al. (2021). Environmental risk assessment of silver NPs in aquatic ecosystems using fuzzy logic. *Water, 14*(12), 1885. https://doi.org/10.3390/w14121885

Rezvani, E., et al. (2019). Adverse effects of nanosilver on human health and the environment. *Acta Biomaterialia, 94*. https://doi.org/10.1016/j.actbio.2019.05.042.

Roberts, M.C. (2001). Injuries and accidents: psychosocial aspects, international encyclopedia of the social and behavioral sciences, 7517–7521

Roncari, F., et al. (2023). Direct quantification of hydrophobicity: A case study of environmentally relevant silver NPs. *Front. Nanotechnol., 5*, 1271009. https://doi.org/10.3389/fnano.2023.1271009

Sahu, A., Dey, S., Mallick, K., Priyadarsini, S., Behera, I. D., & Das, A. P. (2024). A review on advanced technology for sustainable management of synthetic microplastic waste. *Sustainable Management of Urban Plastic Waste Through Circular Economic Approaches*, 262–279. https://doi.org/10.1201/9781003465447-15

Sharma, P., et al. (2022). A critical and comparative review of energy management strategies for microgrids. *Applied Energy, 327*, Article 120028. https://doi.org/10.1016/j.apenergy.2022.120028

Stander, L., & Theodore, L. (2011). Environmental implications of nanotechnology—An update. *International Journal of Environmental Research and Public Health, 8*(2), 470. https://doi.org/10.3390/ijerph8020470

United Nations (U.N.) (2023). https://www.un.org/sustainabledevelopment/blog/2023/08/what-is-sustainable-development/

Water Foundation (2025). https://waterfdn.org/sustainable-water-management-swm-profile/#:~:text=Resources,those%20needs%20in%20the%20future.

Use of Nanomaterials in Water and Air Purification: Technologies and Challenges

Triveni, Prakash Kurmi, and Siddhartha Sankar Dhar

Abstract Nanomaterials play a major role in the environmental remediation based on water and air purification. Due to their diverse and tunable properties and ease of handling, they are widely explored in the areas of wastewater treatment, photocatalytic contaminant degradation, treatment of inorganic and organic air pollutants using adsorption, separation or photoreduction. Here we will explore the use of Metal Organic Frameworks, Quantum dots and Graphene based nanomaterials and their nanocomposites in such applications involving environmental remediation of air and water. Due to high adsorption surface area, high photocatalytic activity, antimicrobial properties, particle segregation qualities and other unique optoelectronic properties make nanomaterials and their nanocomposites desirable for environmental remediation. However, limitations such as dispersion and retention of the nanomaterials and extended sustainability of their antimicrobial quality act as major setbacks as this can hinder in the process of mass scale water treatment. The stability of nanomaterials in water is explored in the context as the recent contributions by Metal Organic Frameworks to the field of nanomaterials promises improved stability. Gaseous inorganic and organic pollutants, pathogens and particulate matter are the main contributors to air pollution. Photocatalysis is one of the promising techniques to purify the air contaminants into less harmful products. Advantages such as ambient conditions, strong potential of elimination of most types of air pollutants and complete degradation to harmless final products with some minute intermediate by-products render this technology as favourable from research point.

Keywords Water and air purification · MOFs · Photocatalysis · QDs · Nano-purification technology · CNTs

Triveni (✉) · P. Kurmi · S. S. Dhar
Department of Chemistry, National Institute of Technology Silchar, Silchar, Cachar, Assam, India
e-mail: triveni21_rs@che.nits.ac.in

E. Priyadarshini et al. (eds.), *Advanced Nanotechnology Applications for Environmental Sustainability*, Sustainable Environmental Waste Management Strategies,
https://doi.org/10.1007/978-3-032-12933-8_20

1 Introduction

Nanoscience generally refers to the varying properties and size of materials in the dimension of 10 nm to 100 nm. So, nanomaterials generally are materials of which a single unit is sized in between 1 and 100 nm in any one dimension. Sedimentation, coagulation, flocculation and activated carbon are some conventional ways of water treatment which are not very effective (Theron et al., 2008; Qu et al., 2013a). Due to ease of operation the more popular water treatment methods are microfiltration, ultrafiltration and nanofiltration (Ngwenya et al., 2013; Bruggen & Vandecasteele, 2003).

As advanced water treatment and desalination technologies are of everyday interest there is continuous need for finding better advanced materials for water treatment. Nanomaterials come in four types mainly metal based, carbon based, dendrimers and composites. Due to their physicochemical properties, they are used in water purification for separation and as reactive medium.

2 Types of Nanomaterials for Water Purification

Nanomaterials can be categorised into 3D (nanoparticles, MOFs and composites), 2D (g-CN nanosheets), 1D (CNTs) and 0D (QDs) depending on their special confinement in three dimensions as depicted in Fig. 1. They can show different forms such as fused, spherical, single, hybrid, irregular and tubular assemblies. Nanowires, nanospheres, nanoribbons, nanotubes, nanofibers, nanosheets, quantum dots are some common depictions of nanomaterials. These nanomaterials in hybrid forms or composite forms possess various advantages such as minimum aggregation, better surface area, better optoelectronic properties, enhanced thermal and electrical transport and sturdy mechanical properties than their single 1D or 2D building cores.

3 Nanoparticles, MOFs and Composites (3D)

Nanoparticles offer considerable advantages in purifying potable water. Efficient uses such as removal of heavy metals, microorganisms and organic pollutants are the focus when it comes to prolonged use of water treatment. In this case noble metal nanoparticles are unique in their application of water purification but their high energy surface leads to a system attempting minimized surface area through methods like agglomeration, chemical transformation and protection. Consequently, the likely adsorption of varying species occurs over the surface of the nanoparticles. However, the same properties such as increased surface area and chemical reactivity which aid in degrading environmental contaminants can sometimes impart them as unsafe to humans. Ag nanoparticles are applied to remove pesticides, microorganisms and

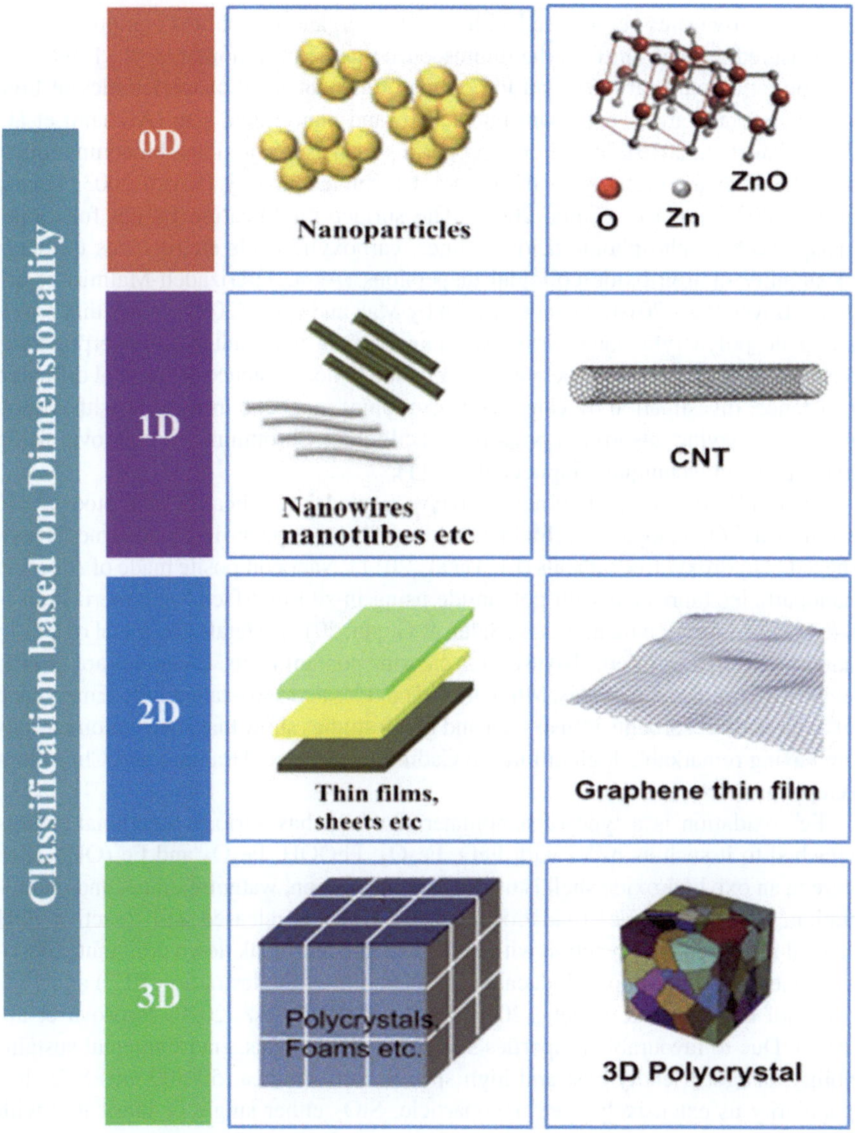

Fig. 1 Nanomaterial classification based on dimensions for the application of water and air purification

heavy metals from drinking water thus making it fit for human use while Au nanoparticles are used as effective sensors for detection of ultra-low pesticide concentrations in water through the changes in the properties of the functional group attached to the nanoparticle in the presence of the pesticide in the solution. OPH enzyme is used to modify the Au nanoparticle which act as fluorophores. The fluorophore positioned at the OPH active sites get released addition of an organophosphorus compound. This is measured through shifts in the fluorescence intensity (Simonian et al., 1999).

Iron oxide nanoparticles find importance for their novel characteristics of high specific surface area, super para-magnetism and nano range size (Afkhami et al., 2010; Pan et al., 2010). In addition, properties such as low toxicity, biocompatibility and chemical inertness make them useful in biotechnology (Roco, 2003; Huang et al., 2003; Gupta & Gupta, 2005). The surface modification by apt functional groups such as phosphonic acids, amines, carboxylic acids etc. decides the level of stability of iron oxide colloidal dispersions (Assa, Jafarizadeh-Malmiri et al., 2016; Boyer et al., 2010). Demonstration by Mak and Chen (2004) shows that novel, magnetic, polyacrylic bound, iron oxide nanoparticle can result in the adsorption and desorption of Methylene Blue within 2 min due to nonexistence of internal diffusion resistance. Investigation by Gupta et al. combines magnetic iron oxide with carbon nanotubes having adsorption properties facilitating Chromium (III) removal using pH dependent technique (Gupta et al., 2011).

Liga and his co-researchers have also investigated the application of photocatalytic Ag-doped TiO_2 nanoparticles for inactivation of bacteriophage in aqueous media with the aid of hydroxyl free radicals (Liga et al., 2011). Nanocomposite made of Alumina nanoparticles fabricated with polyamide using in-situ interfacial polymerization is used as ultrafiltration membranes (Saleh & Gupta, 2012). Metals and metal oxides in nanoscale dimensions are also useful in clearing contaminants. Examples are Fe (III) oxides, Mg oxides, Ce oxides, Mn oxides, Ti oxides and zero-valent ions. Emergence of nano scale absorbents is also seen and many studies show that such nanomaterials are having remarkable high affinity to Cadmium, Arsenic, Uranium and Chromium (Gopakumar et al., 2019).

Fe^0 oxidation is a type of nanomaterial which has various functional groups attached to it such as nZVI with FeO, Fe_2O_3, FeOOH, Fe_3O_4 and Fe $(OH)_3$. Fe^0 core in an oxo hydroxide shell is utilised for adsorption, water treatment and precipitation. Here due to Fe^0 oxidation in water H_2O_2 is released and Reactive •OH radical is formed after reaction with Fe^{2+}. The radicals break down pollutants like p-nitrophenol, hexabromocyclodecane, estrogen 17α-ethinylestradiol (EE2) and 17ß-estradiol E2 (Botes & Cloete, 2010; Pasinszki & Krebsz, 2020; Jarošová et al., 2015). Due to favourable properties such as easy synthesis, environmental sustainability, budget friendly cost and high specific surface area (50-500 nm) SiO_2 has popularity as extensively used nanoparticle. SiO_2 either single or fabricated with other materials is experimented for efficient adsorption of aqueous contaminants (Elizondo-Villarreal et al., 2024; Zeb et al., 2020). Nanomaterial TiO_2 is useful in both photocatalysis and adsorption. The specific surface area, size and crystallinity are influencing criteria of metal adsorption. This nanomaterial shows higher adsorption capabilities for Pb (II), Ni (II), Cu (II), Cd (II) and Zn (II). Lower size range of a nanoparticle exposes higher active surface sites because it presents larger saturated reactive atoms which have an affinity for the surface (Armaković et al., 2023; Naseem & Durrani, 2021; Engates & Shipley, 2011).

Now coming to MOFs there is a recent emergence for their application in environmental water cleansing research. First MOF which is really important is MOF-5. A MIL-53(Fe) MOF doped with 3.0 wt % g-CN viz. CMFe-3 reported a photocatalytic Cr (VI) reduction to Cr (III) efficiency of 2.0 and 2.1 times under visible light. A recyclable Cu-MOF/Fe_3O_4 spherical flower nanocomposite is synthesized and applied for Malachite Green and Pb (2 +) removal with an efficiency of 89.92 and 86.75%. Ragab and co-researchers has fabricated the double layer polytetrafluoroethylene microfiltration membrane with ZIF-8 for the removal of micropollutants in water purification with observed adsorption capacity of increase by 40% and doubled water permeability. The driving criteria behind this is the hydrophobicity of ZIF-8, which leads to reduced friction of the membrane surface with water, thereby increasing permeability of the MOF based membrane. Second criteria seen is the formation of hydrogen bonds by MOF nanoparticles with water which increases hydrophilicity thereby improving permeability. UiO-66 is used as an anchor on Graphene oxide layer by Ma and co-workers where the MOF UiO-66 works as a porous modifier. A range of nanocomposite membranes is synthesized with polyether sulfone (PES) where it is observed that 3.0 wt% of UiO-66 led to improved flux, rejection rate, antifouling and hydrophilicity. Furthermore, Yuan and his co-researchers designed a novel ZIF-300 membrane with the purpose of heavy metal removal. Obtained results had shown 99.21% rejection rate for $CuSO_4$, high stabilisation energy, and 39.2 L/ m^2 h bar. Organic solvent nanofiltration membranes is prepared by Sarango and team by dipping ZIF-8 and ZIF-67 on a P84 asymmetric template with reports showing repulsion of 90% dye and higher methanol selective permeability (Kiteto et al., 2024; Li et al., 2020).

g-CN fabricated nanocomposite is applied critically in water treatment and wastewater purification. Due to its photocatalytic properties, its experimentation in photodegradation of organic contaminants and toxic heavy metal removal is widespread. Mainly dyes, pharmaceuticals and pesticides are the popular organic contaminants on which photodegradation is studied (Cui et al., 2021) g-CN provides a sustainable and efficient manner of water purification thus contributing to study of safer and cleaner water treatment technologies (Ajiboye et al., 2020; Shi et al., 2019). For example, a magnetic graphene oxide fabricated with g-CN (mGCN) prepared by Dai and coresearchers has been applied in Visible LED light irradiated photoreduction of Uranium (VI). This nanocomposite is reported to show reusability, selectivity and high photocatalysis with a 2880.6 mg g^{-1} extraction output (Dai et al., 2020). While g-CN/ZnO has been synthesized hydrothermally by Gnana Prakash and co-researchers for photodegradation of Crystal Violet. This experiment reported a photocatalytic efficiency of 97% with a 1.4 better rate constant as compared to pristine ZnO and g-CN along with excellent stability. Holes were the main reactive species involved and the heterojunction between g-CN and ZnO promoted remarkable charge carrier connection (Manimozhi et al., 2021). g-CN and Porous g-CN was produced by Kumar and co-researchers through melamine pyrolysis and chemical protonation techniques. Their photocatalytic applications using organic dyes is analysed under sunlight. The protonated porous g-CN is observed to show a blue shift and showed better photocatalytic efficiency in comparison to its pristine counterpart. This is due to increased surface area which exposes more dye active sites for photodegradation (Kumaravel et al., 2021). Thus, for these g-CN-based nanocomposites versatility for diverse environmental applications is reported.

4 Graphitic Carbon Nitride Nanosheets (2D)

g-CN nanosheets are studied as Membrane Photoreactor by Zhou and team for its effect on dye degradation of Sudan Orange, Rhodamine 110 and Methylene Blue. This is possible due to remarkable surface area and space range between the g-CN nanosheets being similar with the size range of dye ultimately showing efficient degradation (Zhou et al., 2016). While R. Li and team has highlighted g-CN producing membranes with properties such as photocatalysis, self-cleaning, water-permeability ($11.7\ L\ m^2\ h^1\ bar^{1}$) and antibacterial nature of 99.83%. Screening effect of membranes leads to rejection of dyes and salts pass though the membrane surface easily which results in dye and salt separation. The self-cleaning effect is observed under visible light for 3 cycles which establishes it back to its original state (Li et al., 2019). Photocatalytic properties significantly depend on band gap energy, porosity, surface area, crystal composition and particle size. g-CN having narrow 2.7 eV band gap makes it effective for sunlight use and combined with its physicochemical properties render it ideal for photocatalytic membranes. The most relevant aspect of g-CN fabricated membranes is its ability to do photocatalysis together with hydrophilicity on the same surface simultaneously despite both participating in different mechanisms. This is also why it has self-cleaning aspect (Mulungulungu et al., 2022). 2D g-CN fabricated membranes reveals its enormous potential in removal of divalent ions and organic chemicals. It has a laminar framework with 3.11 Å regularly arranged triangular nanopores which serve as both water transport channels and molecular sieves. Chen and team thus by incorporating g-CN nanosheets in polyamide nanofiltration membranes obtained a fair water flux $37.6\ L\ m^2\ h^1$ compared to $20.9\ L\ m^2\ h^1$ of the bare membrane. An 84% salt rejection rate is also seen owing to Donnan Exclusion Effect and satisfactory antifouling properties (Chen et al., 2016).

5 Carbon Nanotubes (1D)

Carbon nanotubes (CNTs) are 1D nanomaterial having mechanical, electrical and chemical properties. Their unique sorption capacity allows them to degrade certain pollutants in wastewater. The FTIR spectra of CNT regarding temperature influence is studied to show the ability of modified CNT which aid them in adsorbing heavy metals from water. Processes such as fluorination, cycloaddition, free radical polymerization, oxidation and diazonium salt reaction is utilized for enhancing reactivity and solubility of CNT. Following up in research of researchers, multi-walled carbon nanotubes treated by microwave method is reported for removing Zn (II) with 99%

effective removal. Degradation of dangerous pollutants by CNTs is also reported in numerous instances. While some researchers report poor physical adsorption of pristine CNTs owing to poorly linked phenolic and aqueous compounds to the outer surface but it showed improved adsorption for modified CNTs. The reason being the presence of simultaneous H-bonding and stacking which led to greater phenol binding with CNT-OH (Nasrollahzadeh et al., 2021; Singh et al., 2022).

Composition of CNT consists of graphite sheets rolled in the form of cylindrical tube which appears in fence latticework. Single-walled Carbon nanotubes (SWCNTs) have a single layer of graphene in the cylindrical framework while the Multi-walled Carbon nanotubes (MWCNTs) have multiple graphene sheets layers in the cylinder lattice. Both types of CNTs find use in water desalination directly and indirectly in removal of compounds which interfere in the desalination process. This desalination is made possible due to the internal hollow cavity found in CNTs. Characteristics such as homogeneous textured hydrophobic walls, elevated aspect ratios, and internal pore dimeter make CNTs effective in highly efficient water transport. However, the fascinating effect is that CNTs facilitate low energy need in water purification. Also, these membrane types allow almost frictionless flow of water while retaining broad spectrum in water contaminants (Thostenson et al., 2001; Dai et al., 2005; Li & Zou, 2011; Nasrabadi & Foroutan, 2011; Joseph et al., 2011).

6 Quantum Dots (0D)

Quantum dots such as Carbon Quantum dots (CQDs) and Graphene Quantum Dots (GQDs) have various applications in the arena of water treatment. CQDs initially discovered by Xu and team in 2004 and named by Sun's team in 2006 indicate Carbon-based nanoparticles with at least 1 dimension lesser than 10 nm (Su, Liu, et al., 2020). They are having either planar or pseudo spherical design comprising sp^3 and sp^2 hybrid Carbon atoms or either any one hybrid form of Carbon almost like diamond, graphene or GO sheets (Somayeh et al., 2020). They show numerous functionalised surfaces and report fluorescence depending on their composition as established by Lim et al. (2015) and team in 2015 and Sun (Wang et al., 2016) and team in 2016. GQDs simply contain graphene either in single layer or few layers or a mixture of both. GQDs with their fabricated equivalents have novel opportunities in application and merits such as improved physicochemical tunability, quantum confinement bandgap opening, comparable biomolecular size, larger dispersibility and additionally numerous active areas (Nasrollahzadeh et al., 2021).

CQDs having 1.2 eV and 3.8 nm size is reported by Li and team in 2010 for its elevated up-conversion PL characteristics using a one-pot alkali assisted electrochemical technique. Methylene Blue dye is decomposed using as-synthesized TiO_2@CQDs or SiO_2@CQDs nanocomposites under stable, strong visible light. Both these nanocomposites have shown photodegradation abilities of 100% after 25-min 300W halogen lamp irradiation but in the case of pristine CQDs or mixture of both pristine CQDs and the nanocomposites found a 0% minimal photodegradation

(Li et al., 2010). Another nanocomposite CQD@ZnO with 30 nm size is prepared by Yu and team in 2012 through a one pot hydrothermal method and applied it in toxic MeOH and Benzene degradation with greater than 80% efficiency in a 24-h time scale with reaction conditions of visible light source and RT (Yu et al., 2012). GQDs are mainly applied in environmental water remediation by photocatalytic technology which focusses on organic contaminant elimination. Various procedures are experimented for studying photocatalytic efficiency which includes semiconductor GQD-incorporated photocatalysts, noble metal or metal oxide loading, structure frameworks, and synthesis of semiconductor hybrid nanocomposites (Kaur et al., 2018; Yan et al., 2019). Novel GQD designed nano-photocatalysts which include TiO_2 coupled GQDs (Bian et al., 2017; Qu et al., 2013a, 2013b), Fe_3O_4 coupled GQDs (Wu et al., 2014), Zn porphyrin coupled GQDs (Lu et al., 2015), $MnNb_2O_6$ coupled GQDs (Yan et al., 2017), ZnO coupled GQDs (Ebrahimi et al., 2016), g-CN coupled GQDs (Chen & Bai, 2020; Sahu et al., 2021) etc. are used in photodegradation experiments.

7 Technologies and Challenges

A. Photodegradation is the first technology that is observed when we survey nanomaterials for water treatment. Here the toxicity and sustainability of the technique should be evaluated in synthesis, application and by-product. Elaboration on the study in green-synthesized nanomaterials need more focus at industrial as well as commercial levels. The contribution to secondary pollution should be researched separately with critical evaluation and address (Nasrollahzadeh et al., 2021; Singh et al., 2022).
B. Membrane systems is the second technology involved in water purification by nanomaterials. Despite the nanomaterial preparation being eco-friendly and user-friendly, challenges in optimization aspects need to be analysed which include factors such as reaction parameters and material stability, nanomaterial behaviour, morphology, removal performance. Purification and extraction of the membranes is also a necessary aspect for water purification (Chen et al., 2016).
C. Filtration systems is the third technology which discusses about separation of contaminants from wastewater. More investigation is required in the design of novel multifunctional nanohybrid nanomaterials in enhancing effective use (Mulungulungu et al., 2022).
D. Osmosis methods is the fourth technology focussing on the purification of wastewater for desalination. However, cost-friendly nature of the studies needs to be addressed in incorporating suitable nanomaterials in the forward osmosis and reverse osmosis technologies (Mulungulungu et al., 2022).
E. MOFs are generally employed as a fifth technology in itself for the water treatment as they have multiple uses and are tunable. The design of the efficacy and remedial optimized performance is on lab scale with mimic exposure conditions

but realistic environmental values need to be studied and investigated (Kiteto et al., 2024; Li et al., 2020).

F. Nanocatalysis is the sixth method which turns a waste product into another useful or green byproduct without leaving toxic wastes. Some nanomaterials such as CNTs have high-cost needs relative to adsorption methods including effect on aquatic systems (Nasrabadi & Foroutan, 2011; Yuehan et al., 2020; Yan et al., 2019).

8 Types of Nanomaterials for Air Purification

For air purification nanomaterials commonly witnessed are 0D nanoparticles and nano-cubes, 1D nanorods, nanowires, nanotubes and nanobelts, 2D nano-disks, nanoplatelets, nano-sheets and nanofilms, 3D nano-dendrites, nano-flowers, nano-spheres. However, in case of air treatment mostly the wide classification found is of 3 types, mainly Metal based nanomaterials, Non-metal based nanomaterials and Mixed nanocomposites. Air pollution is introduced as changes in the original atmospheric composition on a chemical, physical or biological level by contaminants of biogenic, anthropogenic or geogenic sources. Mainly the pollutants exist in either gaseous or particulate form. Here, the former refers to gases like CO, SO_2, O_3 while latter includes minute microlevel aggregates of chemical compounds in nm-μm range and pathogens such as bacteria, viruses, fungi etc.

9 Metal Based Nanomaterials

A nanometal semiconductor co-catalyst in photocatalytic systems is considered equal to an electrochemical cathode which brings to the point that CO_2 reduction electrocatalysts apply also to photocatalytic co-catalysts in CO_2 reduction. There is a large variation in the reaction rate and selectivity among the catalysts. In CO_2 reduction Cu, Ag, Zn and Au (Group-IB) portray excellent electrocatalytic reactivity with efficient CO selectivity by Ag, Au, and Zn and hydrocarbons affinity for Cu metal. P-block mainly Hg, Bi, Cd, Sn, Tl, Pb and, In produce formate in solutions. Bi ultrathin nanosheets having single crystalline nature and large surface areas is synthesized by Han and team using in-situ topotactic electrochemical transformation of BiOI nanosheets. Application wise it reported high selectivity and high durability for formate output by CO_2 reduction in a time period of 10 h (Zhao et al., 2020). Photocatalytic nanomaterials portray performance based on experimental variables such as type, pollutant flow rate, concentration, humidity, radiation type and intensity, photocatalyst texture (powder, foam, film) and photocatalyst nature (morphology, surface area, particle size). One commercial widely used nano-catalyst Degussa P25 (TiO_2 P25) is reported decomposing 29.3% benzene, 69.3% ethylbenzene, 56.7%

toluene, 81.65% NO and 72.1% o-xylene within reaction degradation time of 3.8 min with 6W UV source, and 600 μW/cm^2 UV intensity (Ren et al., 2017).

10 Non-Metal Based Nanomaterials

CNTs can be used to control CO_2 and CH_4 exhausts from industries and vehicles. It also has ability to trap greenhouse gases emitted by coal mines and electricity plants. Furthermore, CNTs as nano-adsorbents can separate air pollutants owing to selective enhanced adsorption affinity for the particular contaminant. Such nanomaterials show criteria of enhanced pore diameter, pore volume and large surface area which influence the improvement of surface reactive sites that bind with the selective air pollutants as reported by Saleh and Gupta (2013). A study is reported where (3-aminopropyl) tri-ethoxy-silane (APTES) modified MWCNTs synthesized by Gui and team in 2013 for the study of CO_2 removal by chemisorption. This chemisorption is as a result of carbamate development through reaction between CO_2 and secondary amine groups of APTES. It has resulted in 75.4 mg CO_2 adsorbed per adsorbent maximal adsorption of CO_2 which is a superior performance as compared to adsorbents such as SBA-15 (Gui et al., 2013).

Nanomaterial membrane such as electro-spun polyacrylonitrile-fabricated carbon nanofiber which has remarkable microporosity including N-containing functionalities for efficient adsorption is reported by Lee and team for Formaldehyde removal even at less concentration (Lee et al., 2010). A trial experiment on a CNT filter using variable membrane support of 10 mm pore range has been developed for collecting various gases such as NO_x, CO_2, CO and HC. For NO_x, in 20% loadings of range 0.3–0.5 mg cm^2 the proficiencies observed is comparatively less. Here, the morphology of the CNT is multiwalled with non-linear growth because of changing diameters and longitudes. Moreover, an amorphous carbon coating is carried out on the CNTs. Properties such as fuel, brand, model conditions of the car affect the CNT adsorption. There is range of separation efficiency from 10.0% to 95.0% for the CNT loading 0.2-.6 mg cm^2. In the NO_x, CO_2, CO and HC collection efficiency percentage is from 5.0% to 60.0% while keeping same CNT loadings (Romero-Guzmán et al., 2018).

Graphene has properties such as high conductivity, lower electrical noise, 2D conjugated frameworks, large surface area including remarkable electronic qualities. The increased surface-active sites with lowered electrical noise in reduced Graphene Oxide (r-GO) is due to improved surface area to volume ratio. These above-mentioned properties render graphene and r-GO as potential candidates for gas adsorption and detection. Recently, epitaxial graphene is used in detection of NO_2 gas at ppb levels specially for single-layered graphene in contrast to bi-layered graphene. For SO_2 detection, Petryshak and team (2017) has intercalated CNTs with helical-structured and parallel-oriented liquid crystal mixture applying it for SO_2 adsorption. This adsorption technique transformed the CNT conductivity, thus generating a senser signal for SO_2 detection. Chemical and physical adsorption of target gases are also carried out by CNTs with enhanced surface-volume ratios. Zouaghi and team (2018) has also reported CNT- incorporated gas sensors on SiO_2 coated and B-doped Si materials using Chemical Vapor deposition (Saleem et al., 2022).

11 Nanocomposites and MOFs

Low and team in 2017 has prepared TiO_2-CNT (Low et al., 2017) composites where the CNTs transfer electrons effortlessly while hindering the electron–hole recombination rate simultaneously. Similarly, a g-CN/TiO_2 (Giannakopoulou et al., 2017) nano-photocatalyst with variable ratios is synthesized and experimented by Giannakopoulou in 2017 which reported that 1:4 ratio nanocomposite has most photocatalytic activity in NO oxidation relative to pristine g-CN and TiO_2. Formaldehyde is a volatile chemical whose long exposure can lead to skin irritation, and nasal tumours. A TiO_2 reduced GO prepared by hydrothermal route has observed formaldehyde removal efficiency of 10.0%, 20.0% and 60.0% by TiO_2, r-GO and TiO_2 reduced GO in 100-min irradiation as experimented by Yu in 2018. This is due to r-GO prolonging the longevity of charge carriers, which then enhances probability for photochemical reactivity and increases reactive radical number. Similarly, TiO_2-silicate nanocomposite is reported by Portela and team for formaldehyde photocatalytic oxidation.

MnO-FeO-CrO fabricated with TiO_2 nanorods has demonstrated high Hg separation efficiency of 80.0–83.0% at 100–200 C in aerobic conditions. Nitic oxide addition improved the separation efficiency. This is owing to high crystallization and larger surface area which brought about enhanced activity and uniform dispersion. This setup can be employed in coal-utilised power plants for superior Hg^0 separation (Jampaiah et al., 2019). MOFs have unique criteria such as enhanced surface-active areas, tunability, variable pore alteration etc. that render them remarkable candidates for air contaminant trapping. Due to imbalance in the metal ions' presence on the surface and defect in MOFs a positive charge is created which enhances electrostatic interactions leading to removal efficiency of Particulate matter having less than 2.5 μm size (Chen, 2017). Even Bian and team in 2018 has manufactured MOF nanofibers as filters for formaldehyde and Particulate matter (2.5 μm). When ZIF-67 is added to this electro-spun nanofiber the particulate matter filtration increased from 74.5% to 87.2%. Moreover, the filtration efficiency stayed higher than 99.0% in the time period of 30 days. For formaldehyde this nanocomposite showed a removal performance of 84.0%. Hydrothermally synthesized TiO_2/r-GO is reported for superior removal activity towards gaseous MeOH owing to its high adsorption capability, superior charge transfer and lesser band gap (Saleem et al., 2022).

12 Technologies and Challenges

Filtration (Saleem et al., 2022) and Adsorption (Romero-Guzmán et al., 2018) technology is used to trap particulate matter, pollutants such as CNTs and adsorb them from air by TiO_2, Ag nanoparticles. If the nanomaterials are Fe based it can interfere with the setup as intense magnetic combustion as well as friction disturbance affect with a major fraction. So, at industrial points and in vehicles it can wear the technology employing it (Zhao et al., 2020).

The second technology employed is photocatalysis (Ren et al., 2017) which uses sunlight or UV light for degradation of air pollutants into harmless byproducts. Sometimes some nanomaterials can remove air pathogens from air helping in air bioremediation. Particulate matter can enter human body and accumulate in the lungs causing major health effects. Likewise, CNTs (Gupta & Saleh, 2013) and other C-based minute nanomaterials have been found in alveoli of human kids. CNTs are generally confirmed as entering human lungs and thus termed as atmospheric aggressors (Saleem et al., 2022).

13 Conclusion

For water purification nanomaterials prove more useful in assisting the filter-separation techniques which are remarkably high in efficiency than conventional techniques. Metal and metal oxide nanoparticles, nano-Silica, nano-Titania, 2D g-CN (graphitic carbon nitride), MOFs (metal organic frameworks), quantum dots (QDs), carbon nanotubes (CNTs) and nanocomposites are potential helpers in water remediation. Techniques such as membrane systems, filtration, photodegradation, nano-catalysis, electrolysis, ion-exchange, and adsorption are effective for contaminant removal as highlighted in Table 1. All these nanomaterials show remarkable efficiency in treating wastewater polluted by dyes, pathogens and harmful chemicals.

On the other hand, for nanomaterial assisted air treatment focus is put on powder, film or aerosol textures of metallic, non-metallic or mixed composites. The main air contaminants are either gases like SO_2 (sulphur dioxide), NO_x (nitrogen oxides), CO (carbon monoxide), CH_4 (methane gas), H_2S (hydrogen sulphide), HC (hydrocarbons) etc. or microlevel chemical aggregates including air-borne pathogens. Filtration, adsorption and photocatalytic technologies are applied together in removing the hazard of air contaminants. In summary we can conclude that nanomaterials have tremendous potential for both air and water remediation.

Table 1 Pollutant sources for air and water purification, technology involved and challenges faced in practical use

Nanomaterials in water purification-pollutant sources, technology and challenges			
Pollutant sources	**Technology**	**Merits**	**Challenges**
Sewage and domestic waste (Soap and detergents, oils, pharmaceuticals, cosmetics, microbes and germs)	MOF system and Nanocatalysis	Degrades various contaminants, cost effective, easy operation, effective, nano amount	Nano toxicity yet to be determined, reaction conditions not set, lab level capacity, recovery of material is debatable. (Kiteto et al., 2024; Li et al., 2020)
Industrial waste (Heavy metals, dyes, VOCs, oily waste)	Photodegradation	Green method, ease in operation, generates less harmful byproducts	Rapid electron–hole recombination, limited visible light absorption, incomplete mineralization. (Naseem & Durrani, 2021; Ajiboye et al., 2020)
Agricultural waste (Pesticides, herbicides, fertilisers, toxic artificial salts)	Membrane system	Rapid and efficient, reusability of byproducts, removes variety of pollutants	High cost in operation and maintenance, non-reusability or non-regeneration, short lifespan, Large scale durability debatable. (Chen et al., 2016)
Surface waste (Mining waste, metal contamination, CO)	Filtration using adsorption and Osmosis	Rapid, cost-effective, easy and variable pollutant removable, regeneration of material	Material loss, Waste disposal a problem, utilization of waste debatable, fouling in osmosis. (Dai et al., 2005; Nasrollahzadeh et al., 2021)
Nanomaterials in air purification-pollutant sources, technology and challenges			
Contaminant sources	**Technology**	**Advantages**	**Challenges**
Locomotive sources (Vehicles, railways, metros)	Filtration and adsorption	Wider applicability, cost effective, regeneration possible, easy operation and maintenance	pH dependant and ionic strength dependence, scalability, potential for secondary pollution, longevity of adsorbents and filter materials. (Gupta & Saleh, 2013; Romero-Guzmán et al., 2018)

(continued)

Table 1 (continued)

Nanomaterials in water purification-pollutant sources, technology and challenges			
Pollutant sources	**Technology**	**Merits**	**Challenges**
Stationary sources (Household, industries, power plants, oil refineries)	Photocatalysis	Clean, affordable, renewable, ease of application, cost-effective	Product selectivity, poor durability of photocatalysts, lab scale level. (Saleem et al., 2022; Zhao et al., 2020)

References

Afkhami, A., Saber-Tehrani, M., & Bagheri, H. (2010). Modified Maghemite nanoparticles as an efficient adsorbent for removing some cationic dyes from aqueous solution. *Desalination, 263*(1–3), 240–248. https://doi.org/10.1016/j.desal.2010.06.065

Ajiboye, T. O., Kuvarega, A. T., & Onwudiwe, D. C. (2020). Graphitic carbon nitride-based catalysts and their applications: A review. *Nano-Structures & Nano-Objects, 24*, 100577, ISSN 2352-507X. https://doi.org/10.1016/j.nanoso.2020.100577

Armaković, S. J., Savanović, M. M., & Armaković, S. (2023). Titanium dioxide as the most used photocatalyst for water purification: an overview. *Catalysts, 13*(1), 26. https://doi.org/10.3390/catal13010026

Assa, F., Jafarizadeh-Malmiri, H., Ajamein, H., Anarjan, N., Vaghari, H., Sayyar, Z., Berenjian, A. A. (2016). Biotechnological perspective on the application of iron oxide nanoparticles. *Nano Research, 9*(8), 2203–2225. https://doi.org/10.1007/s12274-016-1131-9

Bian, S., Zhou, C., Li, P., Liu, J., Dong, X., & Xi, F. (2017). Graphene quantum dots deco rated titania nanosheets heterojunction: Efficient charge separation and enhanced visible-light photocatalytic performance. *ChemCatChem, 9*, 3349–3357. https://doi.org/10.1002/cctc.201601594

Botes, M., & Eugene Cloete, T. (2010). The potential of nanofibers and nanobiocides in water purification. *Critical Reviews in Microbiology, 36*(1), 68–81. https://doi.org/10.3109/10408410903397332

Boyer, C., Whittaker, M. R., Bulmus, V., Liu, J., & Davis, T. P. (2010). The design and utility of polymer-stabilized iron oxide nanoparticles for nanomedicine applications. *NPG Asia Materials, 2*(1), 23–30. https://doi.org/10.1038/asiamat.2010.6

Chen, Y., & Bai, X. (2020). A review on quantum dots modified g-C3N4-based photo catalysts with improved photocatalytic activity. *Catalysts, 10*, 142. https://doi.org/10.3390/catal10010142

Chen, J., Li, Z., Wang, C., Wu, H., & Liu, G. (2016). Synthesis and characterization of g-C3N4 nanosheet modified polyamide nanofiltration membranes with good permeation and antifouling properties. *RSC Advances, 6*(113), 112148–112157. https://doi.org/10.1039/C6RA21192E

Cui, Y., An, X., Zhang, S., Tang, Q., Lan, H., Liu, H., & Qu, J. (2021). Emerging graphitic carbon nitride-based membranes for water purification. *Water Research, 200*, 117207, ISSN 0043-1354. https://doi.org/10.1016/j.watres.2021.117207

Dai, K., Shi, L., Fang, J., Zhang, D., & Bingkun, Y. (2005). NaCl adsorption in multi-walled carbon nanotubes. *Materials Letters, 59*(16), 1989–1992. https://doi.org/10.1016/j.matlet.2005.01.042

Dai, Z., Sun, Y., Zhang, H., Ding, D., & Li, L. (2020). Photocatalytic reduction of U(VI) in wastewater by mGO/g-C3N4 nanocomposite under visible LED light irradiation. *Chemosphere, 254*, 126671, ISSN 0045-6535. https://doi.org/10.1016/j.chemosphere.2020.126671

Ebrahimi, M., Samadi, M., Yousefzadeh, S., Soltani, M., Rahimi, A., Chou, T.-C., Chen, L.-C., Chen, K.-H., & Moshfegh, A. Z. (2016). Improved solar-driven photo catalytic activity of hybrid graphene quantum dots/ZnO nanowires: A direct Z scheme mechanism. *ACS Sustainable Chemistry & Engineering, 5*, 367–375. https://doi.org/10.1021/acssuschemeng.6b01738

Elizondo-Villarreal, N., Gandara-Martínez, E., García-Méndez, M., Gracia-Pinilla, M., Guzmán-Hernández, A. M., Castaño, V. M., & Gómez-Rodríguez, C. (2024). Synthesis and characterization of SiO_2 nanoparticles for application as nano-adsorbent to clean wastewater. *Coatings, 14*(7), 919. https://doi.org/10.3390/coatings14070919

Engates, K., & Shipley, H. (2011). Adsorption of Pb, Cd, Cu, Zn, and Ni to titanium dioxide nano particles: Effect of particle size, solid concentration, and exhaustion. *Environmental Science and Pollution Research International, 18*, 386–395. https://doi.org/10.1007/s11356-0100382-3

Giannakopoulou, T., Papailias, I., Todorova, N., Boukos, N., Liu, Y., Jiaguo, Y., & Trapalis, C. (2017). Tailoring the energy band gap and edges' potentials of g-C_3N_4/TiO_2 composite photocatalysts for NOx removal. *Chemical Engineering Journal, 310*, 571–580. https://doi.org/10.1016/j.cej.2015.12.102

Gopakumar, D. A., Pai, A. R., Pasquini, D., Ben, L. S. Y., HPS, A. K., Thomas, S. (2019). Nanomaterials—state of art, new challenges, and opportunities, nanoscale materials in water purification (pp. 1–24). Elsevier. https://www.researchgate.net/publication/330439830_Nanomaterials_State_of_Art_New_Challenges_and_Opportunities

Gui, M. M., Yap, Y. X., Chai, S.-P., & Mohamed, A. R. (2013). Multi-walled carbon nanotubes modified with (3-aminopropyl) triethoxysilane for effective carbon dioxide adsorption. *International Journal of Greenhouse Gas Control, 14*, 65–73. https://doi.org/10.1016/j.ijggc.2013.01.004

Gupta, A. K., & Gupta, M. (2005). Synthesis and surface engineering of iron oxide nanoparticles for biomedical applications. *Biomaterials, 26*(18), 3995–4021. https://doi.org/10.1016/j.biomaterials.2004.10.012

Gupta, V. K., & Saleh, T. A. (2013). Sorption of pollutants by porous carbon, carbon nanotubes and fullerene- An overview. *Environmental Science and Pollution Research, 20*, 2828–2843. https://doi.org/10.1007/s11356-013-1524-1

Gupta, V. K., Agarwal, S., & Saleh, T. A. (2011). Chromium removal by combining the magnetic properties of Iron oxide with adsorption properties of carbon nanotubes. *Water Research, 45*(6), 2207–2212. https://doi.org/10.1016/j.watres.2011.01.012

Huang, S. H., Liao, M. H., & Chen, D. H. (2003). Direct binding and characterization of lipase onto magnetic nanoparticles. *Biotechnology Progress, 19*(3), 1095–1100. https://doi.org/10.1021/bp025587v

Jampaiah, D., Chalkidis, A., Sabri, Y. M., Mayes, E. L. H., Reddy, B. M., & Bhargava, S. K. (2019). Low-temperature elemental mercury removal over TiO_2 nanorods-supported MnOx-FeOx-CrOx. *Catalysis Today, 324*(2019), 174–182. https://doi.org/10.1016/j.cattod.2018.11.049

Jarošová, B., Filip, J., Hilscherová, K., Tuček, J., Šimek, Z., Giesy, J. P., Zbořil, R., & Bláha, L. (2015). Can zero-valent iron nanoparticles remove waterborne estrogens? *Journal of Environmental Management, 150*, 387–392. https://doi.org/10.1016/j.jenvman.2014.12.007

Joseph, L., Heo, J., Park, Y.-G., Flora, J. R. V., & Yoon, Y. (2011). Adsorption of bisphenol A and 17α-ethinyl estradiol on single walled carbon nanotubes from seawater and brackish water. *Desalination, 281*, 68–74. https://doi.org/10.1016/j.desal.2011.07.044

Kaur, M., Kaur, M., & Sharma, V. K. (2018). Nitrogen-doped graphene and graphene quantum dots: A review on synthesis and applications in energy, sensors and environment. *Advances in Colloid and Interface Science, 259*, 44–64. https://doi.org/10.1016/j.cis.2018.07.001

Kiteto, M., Vidija, B., Mecha, C. A., et al. (2024). Advances in metal–organic frameworks as adsorbents, photocatalysts and membranes: A new frontier in water purification. *Discover Water, 4*, 54. https://doi.org/10.1007/s43832-024-00119-4

Kumaravel, S., Manoharan, M., Haldorai, Y., & Rajendra Kumar, R. T. (2021). Enhanced visible-light degradation of organic dyes via porous g-C_3N_4. *Phosphorus, Sulfur, and Silicon and the Related Elements, 197*(3), 200–208. https://doi.org/10.1080/10426507.2021.1991343

Lee, K. J., Shiratori, N., Lee, G. H., Miyawaki, J., Mochida, I., Yoon, S.-H., & Jang, J. (2010). Activated carbon nanofiber produced from electrospun polyacrylonitrile nanofiber as a highly efficient formaldehyde adsorbent. *Carbon, 48*(15), 4248–4255. https://doi.org/10.1016/j.carbon.2010.07.034

Li, H., & Zou, L. (2011). Ion-exchange membrane capacitive deionization: A new strategy for brackish water desalination. *Desalination, 275*(1–3), 62–66. https://doi.org/10.1016/j.desal.2011.02.027

Li, H., He, X., Kang, Z., Huang, H., Liu, Y., Liu, J., Lian, S., Tsang, C. H. A., Yang, X., & Lee, S. T. (2010). Water-soluble fluorescent carbon quantum dots and photo catalyst design. *Angewandte Chemie International Edition, 49*, 4430–4434. https://doi.org/10.1002/anie.200906154

Li, R., Ren, Y., Zhao, P., Wang, J., Liu, J., & Zhang, Y. (2019). Graphitic carbon nitride (g-C3N4) nanosheets functionalized composite membrane with self-cleaning and antibacterial performance. *Journal of Hazardous Materials, 365*, 606–614. https://doi.org/10.1016/j.jhazmat.2018.11.033

Li, J., Wang, H., Yuan, X., Zhang, J., & Chew, J. W. (2020). Metal-organic framework membranes for wastewater treatment and water regeneration. *Coordination Chemistry Reviews, 404*, 213116, ISSN 0010-8545. https://doi.org/10.1016/j.ccr.2019.213116

Liga, M. V., Bryant, E. L., Colvin, V. L., & Li, Q. (2011). Virus inactivation by silver doped titanium dioxide nanoparticles for drinking water treatment. *Water Research, 45*(2), 535–544. https://doi.org/10.1016/j.watres.2010.09.012

Lim, S. Y., Shen, W., & Gao, Z. (2015). Carbon quantum dots and their applications. *Chemical Society Reviews, 44*, 362–381. https://doi.org/10.1039/C4CS00269E

Low, J., Cheng, B., & Jiaguo, Y. (2017). Surface modification and enhanced photocatalytic CO_2 reduction performance of TiO_2: A review. *Applied Surface Science, 392*(2017), 658–686. https://doi.org/10.1016/j.apsusc.2016.09.093

Lu, Q., Zhang, Y., & Liu, S. (2015). Graphene quantum dots enhanced photocatalytic activity of zinc porphyrin toward the degradation of methylene blue under visible-light irradiation. *Journal of Materials Chemistry, 3*, 8552–8558. https://doi.org/10.1039/C5TA00525F

Mak, S. Y., & Chen, D. H. (2004). Fast adsorption of methylene blue on polyacrylic acid-bound Iron oxide magnetic nanoparticles. *Dyes Pigments, 61*(1), 93–98. https://doi.org/10.1016/j.dyepig.2003.10.008

Manimozhi, R., Mathankumar, M., & Prakash, A. P. G. (2021). Synthesis of g-C3N4/ZnO heterostructure photocatalyst for enhanced visible degradation of organic dye. *Optik, 229*, Article 165548. https://doi.org/10.1016/j.ijleo.2020.165548

Mulungulungu, G. A., Mao, T., & Han, K. (2022). Two-dimensional graphitic carbon nitride-based membranes for filtration process: Progresses and challenges. *Chemical Engineering Journal, 427*, 130955, ISSN 1385-8947. https://doi.org/10.1016/j.cej.2021.130955

Naseem, T., & Durrani, T. (2021). The role of some important metal oxide nanoparticles for wastewater and antibacterial applications: A review. *Environmental Chemistry and Ecotoxicology, 3*, 59–75, ISSN 2590-1826. https://doi.org/10.1016/j.enceco.2020.12.001

Nasrabadi, A. T., & Foroutan, M. (2011). Ion-separation and water-purification using single-walled carbon nanotube electrodes. *Desalination, 277*(1–3), 236–243. https://doi.org/10.1016/j.desal.2011.04.028

Nasrollahzadeh, M., Sajjadi, M., Iravani, S., Varma, R.S. (2021). Carbon-based sustainable nanomaterials for water treatment: State-of-art and future perspectives. *Chemosphere, 263*, 128005, ISSN 0045-6535. https://doi.org/10.1016/j.chemosphere.2020.128005

Ngwenya, N., Ncube, E. J., & Parsons, J. (2013). Recent advances in drinking water disinfection: Successes and challenges. *Reviews of Environmental Contamination and Toxicology, 222*, 111–170. https://doi.org/10.1007/978-1-4614-4717-7_4

Pan, B., Qiu, H., Pan, B., Nie, G., Xiao, L., Lv, L., Zhang, W., Zhang, Q., & Zheng, S. (2010). Highly efficient removal of heavy metals by polymer-supported nanosized hydrated Fe (III) oxides: Behaviour and XPS study. *Water Research, 44*(3), 815–824. https://doi.org/10.1016/j.watres.2009.10.027

Pasinszki, T., & Krebsz, M. (2020). Synthesis and application of zero-valent iron nanoparticles in water treatment, environmental remediation, catalysis, and their biological effects. *Nanomaterials (Basel), 10*(5), 917. https://doi.org/10.3390/nano10050917

Qu, X., Alvarez, P. J. J., & Li, Q. (2013a). Applications of nanotechnology in water and wastewater treatment. *Water Research, 47*(12), 3931–3946. https://doi.org/10.1016/j.watres.2012.09.058

Qu, D., Zheng, M., Du, P., Zhou, Y., Zhang, L., Li, D., Tan, H., Zhao, Z., Xie, Z., & Sun, Z. (2013b). Highly luminescent S, N co-doped graphene quantum dots with broad visible absorption bands for visible light photocatalysts. *Nanoscale, 5*, 12272–12277. https://doi.org/10.1039/C3NR04402E

Ren, H., Koshy, P., Chen, W.-F., Qi, S., & Sorrell, C. C. (2017). Photocatalytic materials and technologies for air purification. *Journal of Hazardous Materials, 325*, 340–366. https://doi.org/10.1016/j.jhazmat.2016.08.072

Roco, M. C. (2003). Nanotechnology: Convergence with modern biology and medicine. *Current Opinion in Biotechnology, 14*(3), 337–346. https://doi.org/10.1016/S0958-1669(03)00068-5

Romero-Guzmán, L., Reyes-Gutiérrez, L. R., Romero-Guzmán, E. T., & Savedra-Labastida, E. (2018). Car- bon Nanotube Filters for Removal of Air Pollutants from Mobile Sources. *Journal of Minerals and Materials Characterization and Engineering, 6*, 105–118. https://doi.org/10.4236/jmmce.2018.61009

Sahu, R. S., Dubey, A., & Shih, Y.-H. (2021). Novel metal-free in-plane functionalized graphitic carbon nitride with graphene quantum dots for effective photodegradation of 4-bromophenol. *Carbon, 182*, 89–99. https://doi.org/10.1016/j.carbon.2021.05.033

Saleem, H., Zaidi, S. J., Ismail, A. F., & Goh, P. S. (2022). Advances of nanomaterials for air pollution remediation and their impacts on the environment. *Chemosphere, 287*(Part 2), 132083. https://doi.org/10.1016/j.chemosphere.2021.132083

Saleh, T. A., & Gupta, V. K. (2012). Synthesis and characterization of alumina nano-particles polyamide membrane with enhanced flux rejection performance. *Separation and Purification Technol., 89*, 245–325. https://doi.org/10.1016/j.seppur.2012.01.039

Shi, Y., Huang, J., Zeng, G., Cheng, W., Hu, J., Shi, L., & Yi, K. (2019). Evaluation of self-cleaning performance of the modified g-C3N4 and GO based PVDF membrane toward oil-in-water separation under visible-light. *Chemosphere, 230*, 40–50, ISSN 0045–6535. https://doi.org/10.1016/j.chemosphere.2019.05.061

Simonian, A. L., Disioudi, B. D., & Wild, J. R. (1999). An enzyme-based biosensor for the direct determination of di-isopropyl fluorophosphate. *Analytica Chimica Acta, 389*(1–3), 189–196. https://doi.org/10.1016/S0003-2670(99)00170-1

Singh, K.K., Singh, A., Rai, S. (2022). A study on nanomaterials for water purification. *Materials Today: Proceedings, 51*(Part 1), 1157–1163, ISSN 2214-7853. https://doi.org/10.1016/j.matpr.2021.07.116

Somayeh, T., Zahra, D., Kaiqiang, Z., Hadi, B., Van, L. Q., Won, J. H., & Mohammadreza, S. (2020). Carbon and graphene quantum dots: A review on syntheses characterization biological and sensing applications for neurotransmitter determination. *RSC Advances, 10*(26), 15406–15429. https://doi.org/10.1039/D0RA00799D

J. Theron, J.A. Walker, T.E. Cloete, (2008). Nanotechnology and water treatment: applications and emerging Opportunities, Crit. Rev. Microbiol. 34 (1) 43–69. https://doi.org/10.1080/10408410701710442

Thostenson, E. T., Ren, Z., & Chou, T.-W. (2001). Advances in the science and technology of carbon nanotubes and their composites: A review. *Composites Science and Technology, 61*(13), 1899–1912. https://doi.org/10.1016/S0266-3538(01)00094-X

Van Der Bruggen, B., & Vandecasteele, C. (2003). Removal of pollutants from surface water and groundwater by nano filtration: Overview of possible applications in the drinking water industry. *Environmental Pollution, 122*(3), 435–445. https://doi.org/10.1016/S0269-7491(02)00308-1

Wang, X., Sun, G., Li, N., & Chen, P. (2016). Quantum dots derived from two dimensional materials and their applications for catalysis and energy. *Chemical Society Reviews, 45*, 2239–2262. https://doi.org/10.1039/C5CS00811E

Wu, X., Zhang, Y., Han, T., Wu, H., Guo, S., Zhang, J. (2014). Composite of graphene quantum dots and Fe3O4 nanoparticles: peroxidase activity and application in phenolic compound removal. *RSC Advances, 4*, 3299-3305. https://doi.org/10.1039/C3RA44709J

Yan, M., Hua, Y., Zhu, F., Gu, W., Jiang, J., Shen, H., & Shi, W. (2017). Fabrication of nitrogen doped graphene quantum dots-BiOI/MnNb2O6 p-n junction photocatalysts with enhanced visible light efficiency in photocatalytic degradation of antibiotics. *Applied Catalysis b: Environmental, 202*, 518–527. https://doi.org/10.1016/j.apcatb.2016.09.039

Yan, Y., Gong, J., Chen, J., Zeng, Z., Huang, W., Pu, K., Liu, J., & Chen, P. (2019). Recent advances on graphene quantum dots: From chemistry and physics to applications. *Advanced Materials, 31*, 1808283. https://doi.org/10.1002/adma.201808283

Yu, H., Zhang, H., Huang, H., Liu, Y., Li, H., Ming, H., & Kang, Z. (2012). ZnO/carbon quantum dots nanocomposites: One-step fabrication and superior photo catalytic ability for toxic gas degradation under visible light at room temperature. *New Journal of Chemistry, 36*, 1031–1035. https://doi.org/10.1039/C2NJ20959D

Yuehan, S., Liu, G., Zeng, C., Yaobin, L., Luo, H., & Zhang, R. (2020). Carbon quantum dots-decorated TiO2/g-C3N4 film electrode as a photoanode with improved photoelectrocatalytic performance for 1,4-dioxane degradation. *Chemosphere, 251*, Article 126381. https://doi.org/10.1016/j.chemosphere.2020.126381

Zeb, S., Ali, N., Ali, Z., Bilal, M., Adalat, B., Hussain, S.,... & Iqbal, H. M. (2020). Silica-based nanomaterials as designer adsorbents to mitigate emerging organic contaminants from water matrices. *Journal of Water Process Engineering, 38*, 101675. https://doi.org/10.1016/j.jwpe.2020.101675

Zhao, Q., Li, X., Zhou, Q., Wang, D., & Xu, H. (2020). Nanomaterials developed for removing air pollutants. In *Micro and nano technologies, advanced nanomaterials for pollutant sensing and environmental catalysis* (pp. 203–247). Elsevier, ISBN 9780128147962. https://doi.org/10.1016/B978-0-12-814796-2.00006-X

Zhou, K. G., McManus, D., Prestat, E., Zhong, X., Shin, Y., Zhang, H. L., Haigh, S. J., & Casiraghi, C. (2016). Self-catalytic membrane photo-reactor made of carbon nitride nanosheets. *Journal of Materials Chemistry A, 4*(30), 11666–11671. https://doi.org/10.1039/C5TA09152G

Influence of Nano-Formulated Fertilizers on the Well-Being of Crops and Their Impacts on the Functional Microorganisms in the Rhizosphere

Santosh Kumar Sethi, Arati Kindo, Madhusmita Behera, and Birendra Kumar Bindhani

Abstract Contemporary farming methods heavily rely on synthetic fertilizers for supplying necessary nutrients to enhance crop health and production. While the short-term benefit, the extended and consistent use of these fertilizers proves detrimental to soil to increase its fertility and the essential nutrient rich of the rhizosphere microbiome. The adoption of nanobased technology, in the form of Nano-fertilizers, offers a novel, efficient, and environmentally friendly substitute to chemical fertilizers. Nano-fertilizers enable a gradual and sustained nutrient release, supporting plant growth and preserving the diversity of the functionally important microbes. These characteristics can aid the phytomicrobiome in effectively mitigating environmental stress variables (biotic and abiotic). However, despite their remarkable effective use, certain drawbacks, such as a complex production process, delicate transfer and liable dose effectiveness, hinder the widespread implementation of Nano-fertilizers in modern agriculture. The present comprehensive review work focusses to underscore the current trends and perspectives regarding the utilization of Nano-fertilizers to enhance soil fertility, with a particular emphasis on their impact on the beneficial functional microbiome.

Keywords Nano-formulated fertilizers · Rhizosphere · Sustainable agriculture · Microbiome · Crop health · Functional microbes · Efficacy

S. K. Sethi (✉) · A. Kindo
School of Biotechnology, Gangadhar Meher University, Amruta Vihar, Sambalpur, Odisha, India
e-mail: sksethi2k@gmail.com

B. K. Bindhani
School of Biotechnology, KIIT University, Bhubaneswar, Odisha, India

M. Behera
Science laboratory, UGHS, Khaparaganda, Dist-Ganjam, Odisha, Soroda, India

E. Priyadarshini et al. (eds.), *Advanced Nanotechnology Applications for Environmental Sustainability*, Sustainable Environmental Waste Management Strategies,
https://doi.org/10.1007/978-3-032-12933-8_21

1 Introduction

Traditional agriculture is encountering unprecedented challenges due to rapid population growth and environmental degradation (Sang et al., 2019). For years, fertilizers and insecticides have played a crucial role to supplying the demands of agricultural production (Guo et al., 2018). Nevertheless, over use of these pesticides also impede the growth of sustainable agriculture (Raliya et al., 2017; Singh et al., 2021). Expanding the application of nanotechnology can yield creative solutions to enhance sustainable agriculture and satisfy food demands (Sanzari et al., 2019). Numerous agricultural issues, including overuse of pesticides and fertilizers and plant stress from harsh weather, can be effectively addressed by nanotechnology, according to recent research (Demirer et al., 2021; Giraldo et al., 2019; Kamle et al., 2020; Kumar et al., 2019; Lahiani et al., 2013). Furthermore, the quick and simple monitoring of plant growth status is another application of nanotechnology (Kwak et al., 2017, 2019).

Nanofertilizers are basically nutrient particles reduced to a nanoscale, which increases their surface area and reactivity. This new composition and the change in size of these nutrients allow better absorption, less waste and potential benefits not previously seen in conventional fertilizers (Mehta et al., 2015). But what sets nanofertilizers apart from their traditional counterparts? This is mainly their size and mode of operation. Nanofertilizers can penetrate plant cells more efficiently, releasing nutrients in a controlled manner loss and promoting efficient use of nutrients (Zahra et al., 2022).

Nanofertilizers, their ability to deliver nutrients directly to the target area can reduce nutrient runoff, a major environmental problem with traditional fertilizers. The spread of nanofertilizers in agriculture promises many benefits, such as higher yields, better plant health and better ecological results. However, as with any new technological advancement, it is important to understand its impact not only on crops but also on the wider ecosystem. In this regard, the purpose of this review is to explain the application and importance of nanofertilizers, focusing on its effect on the well-being of the plant. The transition from traditional to nanotechnology-enhanced agricultural practices is not simply a matter of changing the type of fertilizer. This is an evolution in how we understand and approach crop nutrition. The nuances of this transition are diverse. While conventional fertilizers often release nutrients either too quickly, leading to waste, or too slowly, inhibiting plant growth, nanofertilizers, with their controlled release mechanisms, try to strike the right balance. Their size allows a closer interaction with plant cells, which promotes better uptake of nutrients and thus better plant health (Kumar et al., 2015). Plant welfare is not limited to growth and yield. This includes the plant ability to deal with stress, its nutritional profile and overall vigor. Nanofertilizers are able to address these aspects of plant protection in a more targeted way than conventional fertilizers due to their precise action. Although the prospect sounds promising, it is equally important to consider all unintended consequences. The nanoscale world behaves differently, and while certain effects are beneficial, they can have unknown consequences for plant health and the surrounding environment.

Thus, the purpose of this review work is focus not only to shed light on the obvious benefits of nanofertilizers, but also to explore areas that require research. Recent research has emphasized the potential of nanobased fertilizers in farming and covered the ongoing lack of education that must be filled in order to apply nanotechnology in modern agriculture in a sustainable manner (Adisa et al., 2019). Although nanoparticles offer numerous benefits, there are also recognized ecotoxicological consequences associated with them (Cobos et al., 2020; Rajput et al., 2018; Tourinho et al., 2012). For the most part, nanoparticles' antibacterial potential has been thoroughly investigated (Boateng et al., 2020; Cobos et al., 2020; Fan et al., 2021; Mehwish et al., 2021). Therefore, it is possible that nanofertilizers will unintentionally harm microbial populations that are good for the soil and plants (Amen et al., 2021). In light of this, the near-term success of using nanofertilizers for sustainable agriculture will depend on all of these elements combined. We have provided a thorough explanation of the impacts of numerous nanoparticles in this paper, particularly those that can be applied to soil, rhizosphere, and phyllosphere as nanofertilizers.

2 Nano Formulated Fertilizers

Materials that provide essential components or nutrients for plant growth and development are referred to as plant nutrients or fertilizers. However, the overuse of fertilizers has led to decreased soil fertility by harming beneficial microbes. This has created a demand for alternative, eco-friendly nanofertilizers (NFs), which are a key application of nanotechnology in agriculture (Abobatta et al., 2018). NFs can be created from traditional fertilizers, bulk fertilizer materials, or derived from other plants by coating them with nanomaterials (NMs). Nanoparticles can store a large amount of nutrients and release them gradually, allowing crops to absorb nutrients as needed, without the harmful effects associated with conventional fertilizers (Preetha & Balakrishnan, 2017). Because NFs deliver nutrients directly to the plants, they reduce eco-toxicity and prevent nutrient loss to soil or groundwater (Kaur et al., 2023). Adequate supply of essential minerals is crucial for plant growth. Minerals required in small amounts, such as iron (Fe), zinc (Zn), manganese (Mn), copper (Cu), boron (B), nickel (Ni), sodium (Na), and chlorine (Cl), are known as micronutrients, while those needed in larger amounts, like nitrogen (N), phosphorus (P), and potassium (K), are macronutrients. Additionally, minerals like magnesium (Mg), calcium (Ca), sulfur (S), and silica (Si) are considered secondary macronutrients (Solanki et al., 2015). Nanoparticles used in nanofertilizers fall into three categories: carbon-based, inorganic, and organic (Ealia et al., 2017). Organic nanoparticles or polymers include dendrimers, micelles, liposomes, and ferritin, which are non-toxic and biodegradable. Inorganic nanoparticles, or non-carbon-based NPs, are typically made from metals or metal oxides, with metals such as aluminum (Al), iron (Fe), lead (Pb), silver (Ag), cadmium (Cd), cobalt (Co), copper (Cu), gold (Au), and zinc

(Zn) being commonly used. These metal-based nanoparticles can be further modified into metal oxide nanoparticles, such as iron NPs that rapidly oxidize into iron oxide (Fe_2O_3), increasing their reactivity compared to iron NPs (Sajid et al., 2020). Carbon-based NPs are composed entirely of carbon (Jadoun et al., 2021).

3 NFs and Its Application on Microbial Population

Microorganisms are particularly useful for synthesizing nanoparticles (NPs) due to their fast growth, easy cultivation, and ability to thrive under ambient conditions of pressure, pH, and temperature. This "green synthesis" involves microorganisms like bacteria, fungi, yeast, and microalgae that can produce NPs (Ali et al., 2020). A wide range of microbes serve as potential biofactories to produce various metal nanoparticles (MNPs) in a cost-effective and eco-friendly manner (Mandal et al., 2005).

Plants, however, offer a readily available source and generate larger quantities of biomolecular reducing agents compared to bacteria. Plant materials like stalks, leaves, roots, flowers, and fruits are used to synthesize metallic nanoparticles (Bahrulolum et al., 2021). The process involves cleaning, drying, boiling, and filtering plant material to obtain extracts. This plant filtrate is then mixed with metallic salt solutions and incubated to produce nanoparticles (Rajeshkumar & Bharath, 2017). Phytochemicals in plant extracts, including terpenoids, flavones, quinones, ketones, and aldehydes, act as electron donors to convert metal ions into NPs in aqueous solutions (Farhadi et al., 2019).

Nanoparticles are more likely to penetrate plants if their size is smaller than the pores of leaves and roots, enhancing their absorption and effectiveness. Nanofertilizers (NFs) significantly improve pest and pathogen resistance, increase productivity, and enhance soil quality, ultimately boosting the vegetative growth and crop yields (Kumar et al., 2020) (Table 1).

4 Nanotechnology in Agriculture

Different methods of use of Nanofertilizers in agriculture, such as application of physical method, different chemical treatment method, and the biological ones, can be used to create nanoparticles (Ghidan et al., 2017). These methods have shortcomings in that they extracted and purified the nanoparticles from the micro-emulsion techniques which used as through the method of oil, surfactant, co-surfactant, and aqueous phase were difficult to scale up and require a lot of resources (Pilarska et al., 2013). Now a days it has already seen the sharp increase the interest in nanotechnology because of its extensive applications in a variety of fields, including materials science, medical medications, energy, and catalysis. Minuscule particles with

Table 1 Effect of the use of different nanoparticles by green synthesized on different crop plants through different mode of application

Sr. No	Plants used	Methods used	Mode of application	Results	References
1	*Cassia occidentalis*	FeO NPs was synthesized by using flower extract of *Cassia occidentalis*	FeO NPs applied to rice seedling at 20 & 40 mg/l	Result showed that the application of FeO NPs to *Oryza sativa* showed 50% increase in root length & dry weight	Afzal et al. (2021)
2	*Fenugreek (Trigonellafoenum-graceum) (methi)*	AgNPS biologically synthesized by seed extract of fenugreek	Foliar application of Ag NPS with different concentration of 0, 10, 20, 30, 40 μg/ml	Result showed increased maximum seed germination 76.11% at a concentration of 10 μg/ml	Kumar et al. (2012)
3	*Calotropis gigantea* (arak)	ZnO NPS was biosynthesized by using leaf of *Calotropis gigantea*	Foliar application of ZnO NPS to seedlings with the concentration of 30 mg/l	Result showed the effect of ZnO NPSon the vegetative growth and yield potential of three important tree seedlings i.e. *Azadirachta indica, Alstonia scholaris & P. pinnataby* 30%, 20% and 10% respectively	Chauduri et al. (2017)
4	*Moringa oleifera*	Ag NPs was phytologically synthesized using leaf extract of *Moringa oleifera*	Direct application to two varieties of wheat plant (Galaxy-13, Pak-13) with 4 different concentration 20, 40, 60, 80 ppm	Result showed that Ag NPs increases growth and production by 20–25%	Ehsan et al. (2022)
5	*Berberis vulgaris*	Ag NPswas phytologically synthesized using leaf and root extract of *Berberis vulgarish*	Direct applied on plant with different concentration of 1, 3 & 5 mM	Result showed the antibacterial activity against gram +ve bacteria *S. aereus* & gram −ve bacteria *E. coli*	Behravanet al. (2019)

(continued)

Table 1 (continued)

Sr. No	Plants used	Methods used	Mode of application	Results	References
6	*Coriandrum sativum* (Dhania)	ZnO NPS was green synthesized by the leaf extract of *Coriandrum sativum*	Direct application on plant with concentration of 8.6 mg/l	Result showed increased in crop productivity by 6–7%. And germination by 25 and 14% in fenugreek and tomato seed respectively	Ukidave and VIngale (2022)
7	*Aloe barbaden sismill*	ZnO NPS was green synthesized by leaf extract of *Aloebarbaden sismill*	Applied on wheat plant at different concentration of 15, 62, 125, 250, 500 mg/l	Result showed the seed grown better with 15, 62, 125, 250 & 500 mg/l & moderate amount of ZnO NPS (62 mg/l) showed significant enhancement in shoot & root, 105% and 50% respectively	Jaspreet et al. (2022)
8	*Cornelian cherry*	Fe_2O_3 NPs green synthesized by using fruit extract of *Cornelian cherry*	Direct applied on barley plant with the concentration of 10–100 mg/l	Result showed that Fe_2O_3 NPs at concentration ranging from 10 to 100 mg/l showed growth of barley seedling & root biomass by 42%	Rostamizadeh et al. (2020)
9	*Andean blueberry*	ZnO, MnONPs was synthesized by using *Andean blueberry* extract	Foliar application with concentration of 270 ppm ZnO, MnO NPs In cabbage	Result showed that at a concentration of 270 ppm the ZnO, MnO NPs in cabbage (*Brassica oleracea)* exhibited increase of 10.3% in root size, 55.1% in dry biomass, 7.1% in chlorophyll & 25.6% in leaf area	Herrera et al. (2022)
10	*Nigella sativa*	ZnO NPs was green synthesized by seed extract of black seeds of *Nigella sativa*	Foliar application was given to the seeds of Broccoli with 7 different concentration 0, 50, 100, 200, 400 and 1000 µg/l	Result showed the application of ZnO NPs Broccoli (*Brassica oleracea)* with different concentration enhanced seed germination of 35%, root length of 56%, shoot length of 16%, leaves no of 41%andleafareaby24% increased significantly	Awan et al. (2021)

(continued)

Table 1 (continued)

Sr. No	Plants used	Methods used	Mode of application	Results	References
11	*Juniperusprocera*	AgNPs was synthesized biologically using seed extract of *Juniperus procera*	Foliar application of AgNPs with different concentration 2.5, 5, 10 and 25 mg/l to tomato plant (*Solanum lycopersicum*)	Result showed that the treatment of Ag NPs with 5 mg/l has highest germination rate 96.7%, followed by 10 mg/l (89.67%), 25 mg/l (87%), and 2.5 mg/l (86%)	Sahil et al. (2022)
12	*Citrus reticulate*	ZnO NPswasgreen synthesize d using peel extract of *Citrus reticulate*	Foliar application to the seeds of *Brassica nigra* (Black mustard)	Result showed that treatment with ZnONPs Showed increased in germination percentage of seeds *of Brassica nigra* by 67%	Rafique et al., 2021
13	Clove bud	ZnNPS was green synthesized by clove bud extract	Direct application on plant at 400 and 600 ppm	Result showed significant growth at 400 and 600 ppm in *Pisum sativum* & improved productivity of crop pea plant by 15–20%	Ahmed et al. (2022)
14	*Andean blueberry*	FeO ZnO NPs was synthesizedbyusing *Andean blueberry* extract	Foliar application with concentration of 270 ppm FeOZnO NPs in *Andean lupin*	Result showed that at a concentration of 270 ppm the FeOZnO NPs in *Andean lupin* exhibited increase inheightby6%, root size by 19%, chlorophyll content by 3.5% & leaf area by 300%	Herrera et al. (2022)
15	Moringa	Z n NPS was green synthesizedby Moringa leaf extract	Foliar application with 10, 50, 100, 250, 500 mg/l	Result showed the effect of ZnNPS at10 mg/l yield maximum plant height & fresh weight with foliar application	Reshma et al. (2022)
16	*Cuminumcyminum* (jeera)	TiO_2NPs was synthesizedby using seed extract of *Cuminumcyminum*	Foliar application with 6 potential concentration 25, 50, 100, 150, 200, 250 mg/l	Result showed significant enhancement in germination of *Vigna radiata* (mung bean) by 65%	Mathew et al. (2021)

(continued)

Table 1 (continued)

Sr. No	Plants used	Methods used	Mode of application	Results	References
17	Vegetable waste (peel)	ZnO NPs was biologically synthesizedby peel waste of vegetable	Foliar application with the concentration of 20 mg/l	Result showed ZnO NPs have antimicrobial potential against pathogens in plants as well as remarkable results in improving growth of germination of seeds by 95%	Rexlin et al. (2023)
18	*Cassia fistula*	CuO NPS was biosynthesized by leaf extract of *Cassiafistula*	Foliar root application with different concentrationof 5–350 μg/ml	Result showed that exposer of CuO NPS with 300 μg/ml significantly reduces disease, shows antifungal activity against *Fusarium oxysporum* the causing agent of fusarium wilt of tomato	Ashraf et al. (2021)
19	*Citrus sinensis*	Ag NPSwas synthesizedby peel extract of *Citrussinensis*	Direct application on plant at concentration 50 μl	Result showed AgNPS exhibited good antibacterial activity against both gram negative and gram-positive bacteria but it showed higher antibacterial activity against *E. coli & P. aeruginosa* then *S.aureus*	Kaviya et al. (2011)
20	*Panicum sumatrense* (little millet)	CuO NPS was green synthesized by using grains extract of *Panicum sumatrense*	Direct applied on plant with different concentration of 60, 80 and 150 μg/ml	Result showed that CuONPS at 150 μg/ml showed antibacterial activity against *Salmonella typhi &staphylococcus* and CuONPS at 60 μg/ml showed efficient growth of *Vingna mungo* plant	Velsankar et al. (2022)
21	Allium cepa (Onion)	AgNPSwas synthesized by using onion extract	Foliar application with 5.4 mg/l	Result showed Ag NPSat 5.4 mg/l improved onion seed germination by 66%	Acharyaet al. (2019)

(continued)

Table 1 (continued)

Sr. No	Plants used	Methods used	Mode of application	Results	References
22	*Trifolium pretense (red clover)*	ZnONPs was green synthesized by using flower extract of *Trifoliumpretense*	ZnONPs applied to plant with different concentration of 1028, 516, 256 and 125 µg/ml	Result showed the antimicrobial activity of ZnO NPs against *S. aureus* (ATCC4163) *E. Coli* (ATCC25922), *P. aeruginosa* (ATCC6749)	Dobrucka and Długaszewska (2016)
23	*Moringa oleifera*	ZnO NPswas green synthesized by using leaf extract of *Moringa oleifera*	ZnO NPs was applied to plant at concentration of 200 mg/ml	Result showed the antimicrobial activity against *B. subtilis*, *P. aeruginosa, S. aureus, E. coli*	Elumalai et al. (2015)
24	*Carica papaya*	ZnO NPswas green synthesized by using milk latex	Applied with different concentration volumes 10, 15, 20, 25 & 30 ml	Result showed ZnO NPs have remarkable antibacterial activity against *Pseudomanas aeruginosa & Staphylococcus aureus*	Sharmaet al. (2016)
25	*Oryza sativa* (rice)	Ag NPS was green synthesized by using plant extract	Foliar application with different concentration of 10, 20 and 40 mg/l	Result showed increased in content of chlorophyll & carotenoid and also promote growth of rice seedlings by 15%	Guptaet al. (2018)
26	*Nyctanthes arbor-tristis*	ZnO NPSwas synthesized by using *yctanthesarbortristis flower extract*	ZnO NPS was applied to plant at concentration of 16 µg/ml	Result showed antifungal activity	Jamdagni et al. (2018)
27	*Jacaranda mimosifolia*	ZnO NPS was synthesized by using *Jacaranda mimosa folia flower* extract	ZnONPSwas applied to plant at concentration of 100 µg/ml	Result showed antimicrobial behavior against gram +ve *Enterococcus faccium* (ATCC35665) &gram –ve E. coli (ATCC25922) pathogenic strain	Deepali et al. (2015)

(continued)

Table 1 (continued)

Sr. No	Plants used	Methods used	Mode of application	Results	References
28	*Drcocephalum Moldavica (dragon head)*	Titarium oxide (TiO_2) synthesized biologically by plant extract	Foliar application with different concentration 10, and 40 ppm	Result showed thatfoliar application of 10 ppm TiO_2 NPs increased plant shoot & essential oil content	Mohammadi et al. (2016)
29	*Punicagranatum (pomegranate pomace)*	Zincnanoparticlewas synthesized biologically by *P.granatum*	Foliar applicationof Zn NPswith concentration of 300 mg/l to the leaf of *Zataria multiflora*	Result showed large essential oil production & improved quality & quantity	Bahmanzadeganet al. (2022)
30	*Eucalyptus lanceolata*	Zinc oxide (Zn NPs) was green synthesized by leaf of *Eucalyptus lanceolata*	Foliar application with 200 ppm	Result showed the antimicrobial activity against pathogen Ca using disease in maize plant & also improved stem diameter & leaf surface area	Sharma et al. (2022)

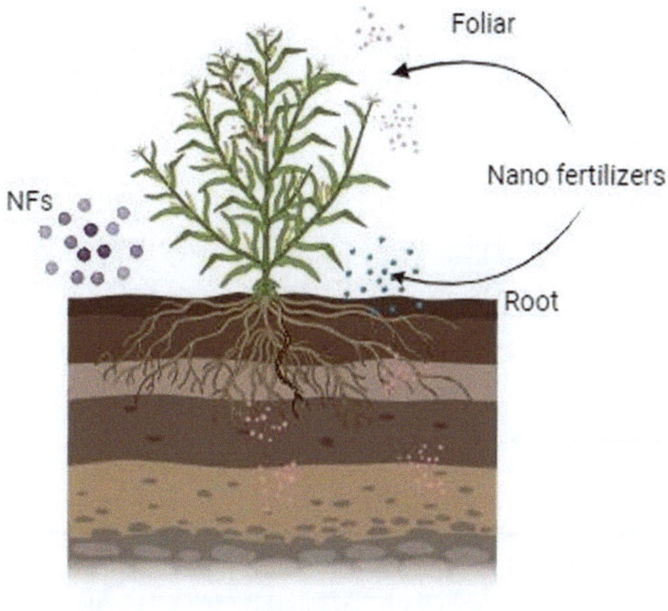

Fig. 1 NFs and its mode of application on soil and plant

medical, industrial, and agricultural fields may benefit from smaller-to-larger surface areas (1–100 nm) (Figs. 1 and 2).

There are several ways to create nanoparticles, such as chemical, biological, and physical methods such a soil emulsion, surfactant, co-surfactant, and aqueous phase, these approaches have disadvantages (Pilarska et al., 2013). Simple, practical, environmentally friendly, and requiring less reaction time are the benefits of using plant and microbial extracts in green approaches for nanoparticle manufacturing. Plant growth regulators, insecticides, and fertilization processes could all be improved by using nanomaterials made using environmentally friendly and green processes (Ghidan et al., 2017). Minimizing the amount of harmful chemical that pollutes the environment. Hence, this technology helps in reducing the environment pollutants. (Huang et al., 2015) (Figs. 2 and 3, Table 1).

5 Nano Fertilizer and Its Released Mechanism

5.1 Action Based Nanofertilizer

Nano-fertilizers can be classified into five categories: action-based, controlled release, targeted delivery, plant growth stimulants, and water and nutrient loss management. These innovative fertilizers offer several benefits, such as enhanced

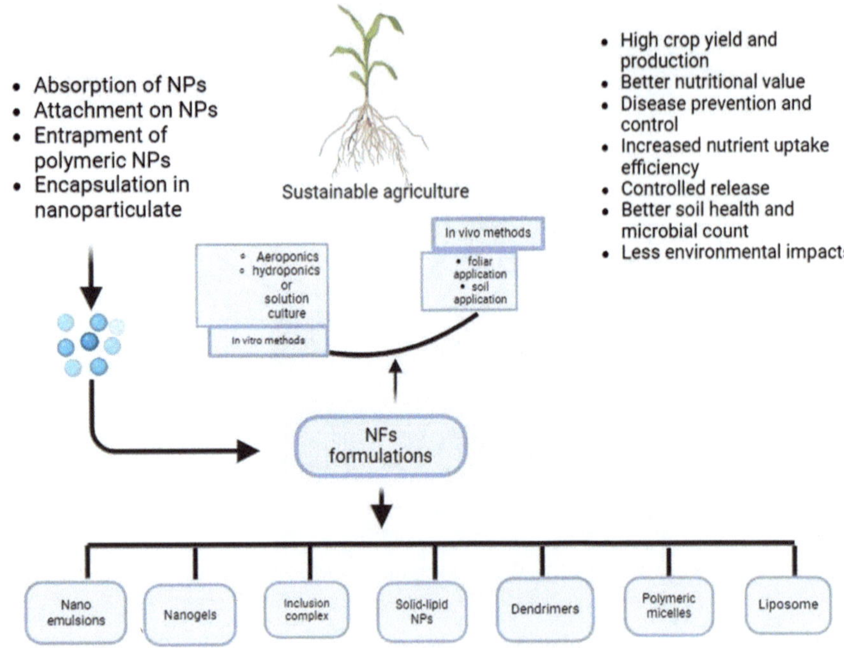

Fig. 2 Role of NFs on crop and soil health for sustainable agriculture

nutrient efficiency, controlled nutrient release, accurate nutrient targeting, improved plant growth, and minimized nutrient loss, making them crucial for sustainable agriculture.

5.2 Controlled Release Nanofertilizers

The use of Nano-fertilizers have become a promising alternative to chemical fertilizers, addressing challenges like increase the nutrient leaching processand lack of nutrient use. These advanced fertilizers, which helps to nutrients encapsulated within nanoscale carriers composed of materials mainly the different polymers and some inorganic substances (Liu & Lal, 2015). Nanoparticles in controlled-release nano-fertilizers help regulate nutrient delivery, improving absorption and minimizing environmental harm (Derosa et al., 2010; Mohanty et al., 2024; Sahu et al., 2024). The nutrients released from these carriers is affected by environmental stress factors mainly biotic and abiotic stress and by different stimuli-responsive mechanisms through different biodegradation process and by different enzymatic based degradation process is play the vital role (Ghormade et al., 2011).

Controlled-release nano-fertilizers provide targeted and stable nutrient release, helps to increase the plant nutrient absorption and decrease the use for heavy and

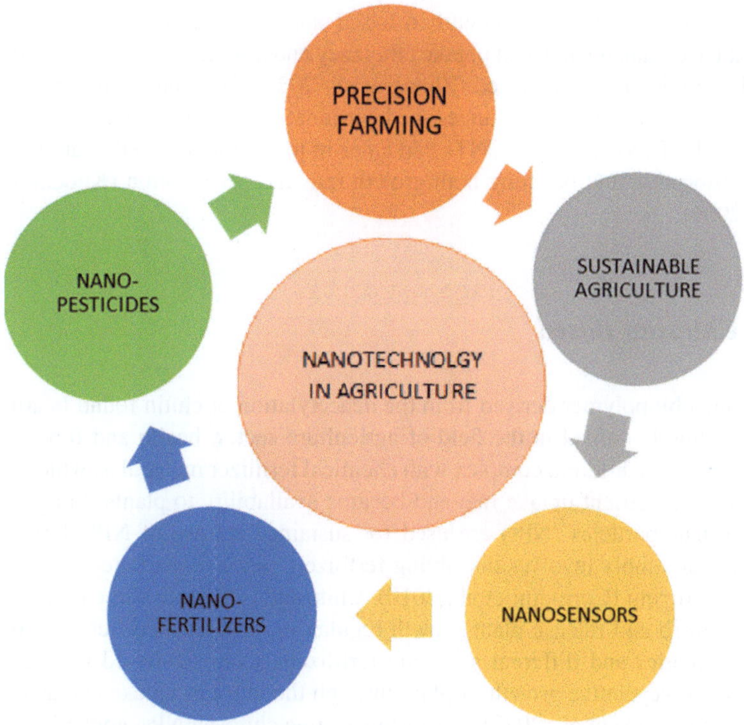

Fig. 3 Application of nanotechnology in modern agriculture

frequent dose of fertilizer applications (Liu & Lal, 2015). By decrease the nutrient leaching significantly, they help prevent water contamination and eutrophication, thereby safeguarding aquatic ecosystems (Singh et al., 2019). These fertilizers have also been shown to enhance crop productivity (Kah et al., 2018). Their potential applications span various crops, including cereals and legumes, with significant results was found in increasing crop yields and nutrient retention efficiency (Liu & Lal, 2015).

5.3 Carbon Based

Carbon is vital for sustaining life, which is the major components of all organic compounds found and is involved in numerous biochemical processes. Bio-structured carbon is especially significant for plants (Jariwala et al., 2013). Among carbon-based nano-fertilizers, carbon nanotubes (CNTs) have gained considerable attention due to their high aspect ratio, mechanical strength, and distinct chemical properties (Safdar et al., 2022). The use of CNTs in agriculture not only improves the nutrient delivery and uptake but also leading to higher crop productivity (Khodakovskaya et al., 2009). Graphene, another carbon-based nano-fertilizer, has also demonstrated potential in enhancing nutrient uptake efficiency and also helps to promoting the crop growth significantly as reported (Zhang et al., 2022). Growing plants in a nutrient medium containing CNTs enhances seed germination and also enhances the vegetative growth of crop plant, as CNTs can enter in to the tomato seed coat, improving water absorption and boosting both growth rate and germination (Khodakovskaya et al., 2009).

5.4 Chitosan Based

Chitosan, a biopolymer derived from the deacetylation of chitin found in arthropod shells, is highly valued in the field of agriculture sector, health and food industry sectors. It forms different complex with chemical fertilizer molecules, which is helps to enhancing nutrient uptake rate and became availability to plants. In agriculture, chitosan nanoparticles (NPs) are used for sustained release of NPK fertilizers, a process that simply involves dissolving fertilizers in a nanoparticle-based solution form and stirring (Corradini et al., 2010). Chitosan's structure can easily be modified to absorb and release plant growth regulators, pesticides, different herbicides and insecticides and different synthetic fertilizers. Chitosan-based nanofertilizers improve the vegetative growth of plant through the nutrient uptake and availability in plants (Mujtaba et al., 2020). For instance, zinc chitosan NPs, applied as a foliar spray, promote wheat growth (Dimkpa et al., 2017), while copper chitosan NPs have shown potential to enhance growth in corn and tomatoes (Saharana et al., 2015).

5.5 Clay Based

Nanoparticles composed of clay, possesses immense surface area and the nanostructure layer reactivity, which can be utilized to different fabricated based formulation for application (Basak et al., 2012). Nanofertilizer based on clay is an important component in formulation-based application due to its active surface of Nanoclay which provides the different mechanism of physicochemical & biochemical process (Sempeho et al., 2014).

5.6 Layer Double Hydroxide

Layered double hydroxides (LDHs) are two-dimensional materials consisting of layers with intercalated anions in the spaces between them. These anions help neutralize the effect of positive charge present in the LDH and facilitate the sustained release of various anions of LDH. LDHs has their anion exchange capacity and thermal stability properties, can differ based on the synthesis method (Rakhimol et al., 2021). For instance, Mg/Al LDHs have been reported to utilized to counteract the three differnt anionic herbicides (2,4-D, MCPA, and Picloram), and the resulting complexes were examined for controlled release in soil and water columns. The results indicate that LDHs could be effective and helps in formulate to slow-release of acidic herbicides (Cardoso et al., 2006).

5.7 Nano Capsule Based

Nano capsules are minute structures, crafted from either organic based or inorganic based materials, designed to encapsulate and deliver the formulated fertilizers (Petosa et al., 2017). These capsules can be made from different biopolymers, silica, different metal oxides and lipids or different carbon nanotubes (Kottegoda et al., 2011). By induce a protective layer, nanocapsules facilitate the sustained release of nutrients, which enhances fertilizer efficiency, allows for better application control, and ensures a steady nutrient supply to crops. This controlled release mechanism helps provide nutrients at optimal times and quantities, which reducing the risk of metal leaching (Kottegoda et al., 2011). Studies have already demonstrated that the applying of different encapsulated zinc oxide nanoparticles to cereal crop (maize) which found the significantly boosts photosynthesis, growth, and yield, even under cobalt stress (Salam et al., 2022).

5.8 Nanogel Based

Nanogels are based on different spongy materials which composed of polymers and mixed with a liquid based designed to absorb and helps to gradually release fertilizers or related other substances (Krishnani et al., 2022). Infused with essential growth elements required for plant growth mainly the nitrogen, phosphorus, potassium (NPK) (Bindra & Singh, 2021), these nanogels are commercially available with of various applications, including use in fertilizers industry, water purification and drug delivery sectors. Additionally, nanogel particles which helps to absorb and gradual release fertilizers with retain water in soil, improving the water retention capacity and helping to prevent soil erosion (Madzokere et al., 2021).

5.9 Polyurethane Based

Polyurethane-based nano-fertilizers are developed from polymers composed of organic units linked by urethane bonds. Typically, these fertilizers are synthesized using ring-opening metathesis polymerization (ROMP), a method where a monomer is created from two distinct molecules with a double bond, which is then polymerized. The resulting polymers form nanoparticles (NPs) that help deliver essential nutrients to the soil (Das et al., 2024; Dey et al., 2024). These NPs can improve the efficiency of conventional fertilizers by enhancing the delivery of key nutrients like nitrogen, phosphorus, and potassium (Thirugnanasambandan, 2021; Yuan et al., 2022).

5.10 Starch Based

Starch-based nano-fertilizers consist of starch-derived nano crystals that are easily dissolved in water and applied to plants as a liquid or aerosol (Lin et al., 2011). These nanocrystals produce no chemical waste while efficiently fertilising crops with high-quality renewable energy sources (Kumar et al., 2018). Starch-based nanofertilizers represent a significant advancement in agriculture due to their eco-friendly and efficient delivery of nutrients. Unlike traditional fertilizers, these nanofertilizers utilize starch, a biodegradable and renewable material, to enhance nutrient release and uptake. This reduces environmental impact and minimizes soil and water pollution. The nanoscale size of the particles allows for precise targeting, ensuring that plants receive nutrients more effectively (Lin et al., 2011). This method not only improves crop yields but also promotes sustainable farming practices by reducing the need for excessive chemical inputs. Overall, starch-based nanofertilizers contribute to more efficient, sustainable, and environmentally friendly agriculture.

5.11 Zeolite Based

Zeolites are a type of microporous aluminosilicate minerals that have been added to soil for many years because of their capacity to both absorb and hold onto organic molecules such as water and nutrients (Jakkula & Wani, 2018). The process of creating zeolite-based nano-fertilizers entails synthesizing zeolite nanoparticles (NPs), which are subsequently combined with different ingredients to produce fertilizers. Because these NFs are tiny, they can more effectively transport nutrients to the crop's root zone and delve deeper into the soil (Sharama et al., 2022). Zeolite based composite fertilizers are known for increasing nutrients use efficiency in plant (Salam et al., 2022). Zeolite based nitrogen fertilizers have potential of promoting growth activity in maize plant (Manikandan & Subramanian, 2016).

6 Advantages of Controlled Release Nanofertilizer

Carbon-based nanofibers improve crop growth, enhance the water and nutrient uptake, and aid during droughts (Behravan et al., 2019). Chitosan-based nanofibers are biodegradable, customizable, easily modifiable, and protect biomolecules from environmental factors (Subramanian et al., 2015). Clay-based nanofibers have a large surface area, reactive nano layers, and control anion release (Mishra et al., 2017). Nano-capsule-based nanofibers offer controlled nutrient release, efficient nutrient delivery, and reduced leaching risk (Gupta et al., 2018). Nano gel-based nanofibers are highly soluble, biodegradable, non-toxic, and improve water retention (Zhang et al., 2021). Polyurethane-based nanofibers provide controlled nutrient release, enhanced water-holding capacity, and decreased soil erosion (Gupta et al., 2018). Starch-based nanofibers are a renewable energy source, facilitate effective nutrient delivery, and generate minimal chemical waste (Mishra et al., 2019). Zeolite-based nanofibers enhance nutrient delivery, allow for tailored nutrient provision, and lower fertilization costs (AbdEl-Aziz et al., 2015).

Nano fertilizers offer several advantages in agriculture. Their small size allows for better nutrient absorption by plants, leading to enhanced growth and yield. They provide targeted delivery of nutrients, reducing waste and minimizing environmental impact (Bala et al., 2024; Das et al., 2024; Prasad. Das et al., 2024). This precision also allows for lower application rates, cutting costs and reducing the need for frequent applications. Nano fertilizers can improve soil health by promoting beneficial microbial activity and reducing nutrient leaching. Their use can enhance crop resilience to stress conditions, improving overall agricultural sustainability and productivity. As a result, nano fertilizers represent a promising advancement in modern farming practices.

7 Nanofertilizers for Targeted Delivery

Nanofertilizer technology represent an innovative approach for delivering fertilizers in agriculture. They work by targeting specific molecules in the soil and delivering nutrients or other substances directly to plants (Rameshaiah et al., 2015). These tiny molecules can bind to enzymes and plant hormones, effectively directing them to the plants. Research has demonstrated that nano-aptamers can improve plant nutrient uptake from the soil (Gupta et al., 2018). Fertilizers can be delivered to plants by attaching them to nanoparticles, like gold nanoparticles or liposomes, which protect the aptamers from degradation and ensure precise delivery to the plant cells. The aptamers bind to specific receptors on the plant cell surface, allowing the nanoparticles to enter the cell and release the fertilizer (Ghormade et al., 2011). By linking the root system with microorganisms, nano-aptamers help control the spread of fertilizer and ensure the correct amount is applied (Despande et al., 2017). The use of nanomaterials for targeted delivery of essential nutrients, pesticides, and genetic material could greatly advance the agriculture industry (Rakhimol et al., 2021). Besides clay-based fertilizers, targeted nano-fertilizers include nano-coated urea, iron oxide nanoparticles, nano hydroxyapatite, carbon-based nutrient carriers, nano emulsions, and nano-encapsulated micronutrients (Cardoso et al., 2006).

8 Plant Growth Stimulating Nano-Fertilizer

Certain nano-fertilizers, such carbon nanotubes (CNTs), interact with the root system of plants to promote hormone synthesis and thus accelerate plant development. They raise the soil's concentration of carbon and other nutrients. Carbon nanotubes (CNTs) have the ability to both absorb and release nutrients, improve soil structure, retain more water, and promote plant development. Low carbon nanotube concentrations may be beneficial for water transport, root development, and seed germination without being harmful, in contrast to the negative effects of big dose fertiliser trials. (Zhang et al., 2022). The direct addition of CNTs to the soil enhances its ability to retain water and nutrients (Fig. 3 and Table 1). Because CNTs can delve deeper into the soil, they can sustain the flora for an extended amount of time. They have the ability to move to plant systemic areas like fruits, leaves, and roots, suggesting a strong connection with plant cells (Singh et al., 2022).

Plant growth stimulating nano-fertilizers are cutting-edge agricultural products designed to enhance plant growth and crop yields. These fertilizers use nanoparticles to deliver essential nutrients directly to plant roots and foliage more efficiently than traditional fertilizers (Zhang et al., 2022). Their small size allows for better absorption and targeted delivery, minimizing nutrient loss and reducing environmental impact. Nano-fertilizers can improve soil health, increase resistance to pests and diseases, and promote faster growth. By optimizing nutrient uptake and reducing waste, they offer a sustainable alternative to conventional fertilizers, potentially revolutionizing agricultural practices and supporting global food security (Ghormade et al., 2011).

9 Water and Nutrient Loss Controlling Fertilizer

Farmers can use less fertiliser while keeping the same crop yield by using nanofertiliser, which contains NPs that can regulate the pace at which fertilisers are released into the soil. To limit water and nutrient loss in soil, urea can be coated with NPs of iron oxide, sulphur, magnesium, zinc, copper, molybdenum, boron, ammonium sulphate, and potassium (Safdar et al., 2022). Nano-beads & nano-emulsion are two prominent nanofertilizers that control soil water & nutrient loss (Salam et al., 2022).

10 Nanobeads

Nano-beads are tiny particle engineered to contain slow-release nutrients diminishing plant water loss (Krishnani et al., 2022). Iron, carbon, and other metals are among the materials used to make nanobeads. Pollution can also be cleaned up with the use of nanobeads. Commercially available nanobead fertiliser is one of the most widely used types. containing macro and micronutrients, called Nano-Fert (Yuan et al., 2022). Another popular nanofertilizer is N-Flex, containing nanosized particles that solely release nutrient over time. The slow release of nutrients helps ensuring the plant receive the nutrients they need for optimal growth without risk of over fertilization. N-Flex also contains more nitrogen than many traditional fertilizers, which helps promoting healthy growth.

11 Nano Emulsion-Based Fertilizer

A combination of nutrients that are both water soluble and in soluble is contained in small droplets called Nano emulsions. High energy mixing is used to combine a surfactant with liquid to create a stable, uniform mixture that is the basis for nanoemulsion-based fertilizers. In order for the liquid droplets to stay uniformly dispersed throughout the solution, the surfactant assists to maintain them suspended in the water-soluble combination. Droplets are easily absorbed and utilized by the plant, improving crop health and output (Corradini et al., 2010). The minuscule droplet size of nano emulsion fertilizers enhances the permeability of plant cell walls, enabling quick and thorough fertilizer absorption by the plants (Singh et al., 2022).

12 Current Status and Future Prospects

Nanoparticles in fertilizers present a promising opportunity to boost crop yields while minimizing environmental impact compared to conventional fertilizers. To fully leverage their potential, it's essential to focus on several key research and development areas. This includes creating environmentally friendly and cost-effective methods for synthesizing nanomaterials, fine-tuning their physical and chemical properties, and addressing potential risks. Future research should also aim to develop formulations that allow for controlled nutrient release and targeted delivery using "smart" nanofertilizers that respond to environmental factors like pH or temperature. Understanding the environmental and health risks associated with nanofertilizers and establishing clear regulations and guidelines for their safe use are crucial. Additionally, evaluating the environmental impact of nanofertilizers is necessary for sustainable practices. Overcoming current challenges in their delivery is vital despite their potential economic and environmental benefits. Addressing these challenges will lead to the creation of commercially viable nanofertilizers, potentially increasing agricultural productivity with less fertilizer. Continuous advancement and commercialization require thorough analysis of these smart fertilizers based on nanotechnology. Successful development of slow-release nanofertilizers could have significant benefits for environmental conservation, energy efficiency, and economic sustainability.

Nanofertilizers are an emerging technology designed to enhance nutrient delivery and reduce environmental impact. Currently, they are gaining traction due to their ability to increase nutrient use efficiency, improve crop yields, and minimize leaching and runoff. Their unique properties, such as high surface area and controlled release, make them superior to traditional fertilizers. Looking ahead, the future prospects of nanofertilizers are promising with advancements in nanotechnology likely to improve their performance and sustainability. However, challenges like cost, scalability, and environmental safety need to be addressed. Ongoing research and development are expected to drive innovations and broaden their applications in agriculture.

13 Conclusion

The growing global population demands increased food production, while nutrient deficiencies significantly impact crop yields and economic losses in agriculture. Though chemical fertilizers can provide short-term boosts in crop output, they are not a sustainable solution in the long run. Current research is heavily focused on developing eco-friendly alternatives to these chemicals (Babu et al., 2022; Bradu et al., 2023; Chaudhuri and Malodia 2017; Hasan and Jhung 2015; LeBrun et al., 1992; Sahani et al., 2021; Salih et al., 2022; Seleiman et al., 2021; Toksha et al., 2021). Nano-fertilizers, which use nanotechnology, are emerging as a promising method to enhance crop yields and soil health. This innovation presents a transformative opportunity for agriculture by promoting sustainability. To maximize their impact, ongoing research, risk management, interdisciplinary collaboration, farmer education, and ensuring affordability and accessibility are crucial. Nanomaterials, known for their efficient delivery systems, also support agricultural sustainability through applications such as Nano-sensors and Nano-pesticides. Additionally, nanotechnology enhances food stability, processing, and reduces waste. While nanoparticles like Zn, MgO, Ag, Au, ZnO, TiO_2 and SiO_2 offer unique benefits, they may pose health risks if they penetrate human and animal cells. These risks can be managed by using green synthesis methods, which involve plant extracts for producing Nano-fertilizers, ensuring both improved yields and long-term agricultural sustainability.

Acknowledgements Thanks go to the authority of Gangadhar Meher University for providing financial assistance through sanctioning a minor research project grant to Dr. Santosh Kumar Sethi through Seed Fund Project (Sanctioned order No. 4934/GMU Dt. 02-12-2023).

References

Abd El-Aziz, A. R. M., Al-Othman, M. R., Mahmoud, M. A., & Metwaly, H. A. (2015). Biosynthesis of silver nanoparticles using *Fusarium solani* and its impact on grain borne fungi. *Digest Journal of Nanomaterials and Biostructures, 10*(2), 655–662.

Acharya, P., Jayaprakasha, G. K., Crosby, K. M., Jifon, J. L., & Patil, B. S. (2019) Green-synthesized nanoparticles enhanced seedling growth, yield, and quality of onion (*Allium cepa* L.). *ACS Sustainable Chemistry & Engineering, 7*(17), 14580–14590.

Afzal, S., Sharma, D., & Singh, N. K. (2021). Eco-friendly synthesis of phytochemical-capped iron oxide nanoparticles as nano-priming agent for boosting seed germination in rice (*Oryza sativa* L.). *ESPR, 28*, 40275–40287.

Ahmed, S., Qasim, S., Ansari, M., Shah, A. A., Rehman, H. U., Shah, M. N., & Datta, R. (2022). Green synthesis of zinc nanoparticles and their effects on growth and yield of *Pisum sativum*. *Journal of King Saud University-Science, 34*(6), Article 102132.

Ashraf, H., Anjum, T., Riaz, S., Ahmad, I. S., Irudayaraj, J., Javed, S., & Naseem, S. (2021). Inhibition mechanism of green-synthesized copper oxide nanoparticles from *Cassia fistula* towards *Fusarium oxysporum* by boosting growth and defense response in tomatoes. *Environmental Science: Nano, 8*(6), 1729–1748.

Awan, S. S., Hahzadi, K., Javad, S., Tariq, A., Ahmad, A., & Ilyas, S. (2021). A preliminary studyof influence of zinc oxide nanoparticles on growth parameters of *Brassica oleracea* var italic. *Journal of the Saudi Society of Agricultural Sciences, 20*(1), 18–24.

Basak, B. B., Pal, S., Datta, S. C. (2012). Use of modified clays for retention and supply of water and nutrients. *Current Science,* 1272–1278.

Babu, S., Singh, R., Yadav, D., Rathore, S. S., Raj, R., Avasthe, R., & Singh, V. K. (2022). Nanofertilizers for agricultural and environmental sustainability. *Chemosphere, 292*, Article 133451.

Bahmanzadegan, A., Tavallali, H., Tavallali, V., & Karimi, M. A. (2022). Variations in biochemical characteristics of Zataria multiflora in response to foliar application of zinc nanocomplex formed on pomace extract of Punicagranatum. *Industrial Crops and Products, 187*, Article 115369.

Behravan, M., Panahi, A. H., Naghizadeh, A., Ziaee, M., Mahdavi, R., & Mirzapour, A. (2019). Facilegreen synthesis of silver nanoparticles using Berberis vulgaris leaf and root aqueous extract and its antibacterial activity. *International Journal of Biological Macromolecules, 124*, 148–154.

Bindra, H. S., & Singh, B. (2021). Nanofertilizers and nanopesticides: Future of plant protection. In: *Advances in nano-fertilizers and nano-pesticides in agriculture* (pp. 57–84). Woodhead Publishing.

Bradu, P., Biswas, A., Nair, C., Sreevalsakumar, S., Patil, M., Kannampuzha, S., & Gopalakrishnan, A. V. (2023). Recent advances in green technology and industrial revolution 4.0 for a sustainable future. *Environmental Science and Pollution Research, 30*(60), 124488–124519.

Cardoso, L. P., Celis, R., Cornejo, J., & Valim, J. B. (2006). Layered double hydroxides as supports for the slow release of acid herbicides. *Journal of Agricultural Food Chemistry, 54*(16), 5968–5975.

Corradini, E., DeMoura, M. R., & Mattoso, L. H. C. (2010). A preliminary study of the incorporation of NPK fertilizer into chitosan nanoparticles. *Express Polymer Letters, 4*(8).

Chaudhuri, S. K., & Malodia, L. (2017). Biosynthesis of zinc oxide nanoparticles using leaf extract of *Calotropis gigantea*: Characterization and its evaluation on tree seedling growth in nursery stage. *Applied Nanoscience, 7*(8), 501–512.

DeRosa, M. C., Monreal, C., Schnitzer, M., Walsh, R., & Sultan, Y. (2010). Nanotechnology in fertilizers. *Nature Nanotechnology, 5*, 91.

Dimkpa, C. O., Bindraban, P. S., Fugice, J., Agyin-Birikorang, S., Singh, U., & Hellums, D. (2017). Composite micronutrient nanoparticles and salts decrease drought stress in soybean. *Agronomy for Sustainable Development, 37*, 1–13.

Dobrucka, R., & Długaszewska, J. (2016). Biosynthesis and antibacterial activity of ZnO nanoparticles using Trifoliumpretense flower extract. *Saudi Journal of Biological Sciences, 23*(4), 517–523.

Ghidan, A. Y., Al-Antary, T. M., Salem, N. M., & Awwad, A. M. (2017). Facile greensynthetic route to the zinc oxide (ZnONPs) nanoparticles: Effect on green peach aphid and antibacterial activity. *The Journal of Agriculture Science, 9*(2), 131–138.

Ghormade, V., Deshpande, M. V., & Paknikar, K. M. (2011). Perspectives for nano-biotechnology enabled protection and nutrition of plants. *Biotechnology Advances, 29*(6), 792–803.

Gupta, S. D., Agarwal, A., & Pradhan, S. (2018). Phytostimulatory effect of silver nanoparticles (AgNPs) on rice seedling growth: An insight from antioxidative enzyme activities and gene expression patterns. *Ecotoxicology and Enviornmental Safety, 161*, 624–633.

Hasan, Z., & Jhung, S. H. (2015). Removal of hazardous organics from water using metal- organic frameworks (MOFs): Plausible mechanisms for selective adsorptions. *Journal of Hazardous Materials, 283*, 329–339.

Huang, S., Wang, L., Liu, L., Hou, Y., & Li, L. (2015). Nanotechnology in agriculture, livestock, and aquaculture in China. A Review. *Agronomy for Sustainable Development, 35*, 369–400.

Jadoun, S., Arif, R., Jangid, N. K., & Meena, R. K. (2021). Green synthesis of nanoparticles using plant extracts: A review. *Environmental Chemistry Letters, 19*(1), 355–374.

Jakkula, V. S., & Wani, S. P. (2018). Zeolites: Potential soil amendments for improving nutrient and water use efficiency and agriculture productivity. *SRCC, 8*(1), 1–15.

Jariwala, D., Sangwan, V. K., Lauhon, L. J., Marks, T. J., & Hersam, M. C. (2013). Carbon nanomaterials for electronics, optoelectronics, photovoltaics, and sensing. *Chemical Society Reviews, 42*(7), 2824–2860.

Jamdagni, P., Khatri, P., & Rana, J. S. (2018). Green synthesis of zinc oxide nanoparticles using flower extract of Nyctanthes arbor-tristis and their antifungal activity. *Journal of King Saud University-Science, 30*(2), 168–175.

Kaviya, S., Santhanalakshmi, J., Viswanathan, B., Muthumary, J., & Srinivasan, K. (2011). Biosynthesis of silver nanoparticles using *Citrussinensis* peel extract and its antibacterial activity. *Spectrochimica Acta Part A: Molecular and Biomolecular Spectroscopy, 79*(3), 594–598.

Kah, M., Kookana, R. S., Gogos, A., & Bucheli, T. D. (2018). A critical evaluation of nanopesticides and nanofertilizers against their conventional analogues. *Nature Nanotechnology, 13*(8), 677–684.

Kaur, H., Hussain, S. J., Mir, R. A., Verma, V. C., Naik, B., Kumar, P., & Dubey, R. C. (2023). Nanofertilizers–emerging smart fertilizers for modern and sustainable agriculture. *Biocatalysis and Agricultural Biotechnology, 54*, Article 102921.

Khodakovskaya, M., Dervishi, E., Mahmood, M., Xu, Y., Li, Z., Watanabe, F., & Biris, A. S. (2009). Carbon nanotubes are able to penetrate plant seed coat and dramatically affect seed germination and plant growth. *ACS Nano, 3*(10), 3221–3227.

Krishnani, K. K., Boddu, V. M., Chadha, N. K., Chakraborty, P., Kumar, J., Krishna, G., & Pathak, H. (2022). Metallic and non-metallic nanoparticles from plant, animal, and fisheries wastes: Potential and valorization for application in agriculture. *Environmental Science and Pollution Research, 29*(54), 81130–81165.

Kottegoda, N., Munaweera, I., Madusanka, N., & Karunaratne, V. (2011). A green slow-release fertilizer composition based on urea-modified hydroxyapatite nanoparticles encapsulated wood. *Current Science*, 73–78.

Kumar, P., Pahal, V., Gupta, A., Vadhan, R., Chandra, H., & Dubey, R. C. (2020). Effect of silver nanoparticles and Bacillus cereus LPR2 on the growth of Zea mays. *Science and Reports, 10*(1), 20409.

Kumar, R., Ashfaq, M., & Verma, N. (2018). Synthesis of novel PVA–starch formulation-supported Cu–Zn nanoparticle carrying carbon nanofibers as a nanofertilizer: Controlled release of micronutrients. *Journal of Materials Science, 53*(10), 7150–7164.

LeBrun, P., Gaffet, E., Froyen, L., & Delaey, L. (1992). Structure and properties of Cu, Ni and Fe powders milled in a planetary ball mill. *Scripta Materialia, 26*, 1743–1748.

Liu, R., & Lal, R. (2015). Potentials of engineered nanoparticles as fertilizers for increasing agronomic productions. *Science of the Total Environment, 514*, 131–139.

Mathew, S. S., Sunny, N. E., & Shanmugam, V. (2021). Green synthesis of anatase titanium dioxide nanoparticles using Cuminumcyminum seed extract; effect on Mung bean (*Vigna radiata*) seed germination. *Inorganic Chemistry Communications, 126*, Article 108485.

Manikandan, A., & Subramanian, K. S. (2016). Evaluation of zeolite-based nitrogen nano-fertilizers on maize growth, yield and quality on inceptisols and alfisols. *International Journal of Plant and Soil Science, 9*(4), 1–9.

Madzokere, T. C., Murombo, L. T., & Chiririwa, H. (2021). Nano-based slow releasing fertilizers for enhanced agricultural productivity. *Materials Today: Proceeding, 45*, 3709–3715.

Mishra, M., Dashora, K., Srivastava, A., Fasake, V. D., & Nag, R. H. (2019). Prospects, challenges and need for regulation of nanotechnology with special reference to India. *Ecotoxicology and Environmental Safety, 171*, 677–682.

Mishra, S., Singh, B. R., Naqvi, A. H., & Singh, H. B. (2017). Potential of biosynthesized silver nanoparticles using *Stenotrophomonas* sp. BHU-S7 (MTCC 5978) for management of soil-borne and foliar phytopathogens. *Scientific Reports, 7*(1), 45154.

Mohammadi, H., Esmailpour, M., & Gheranpaye, A. (2016). Effects of TiO_2 nanoparticles and water-deficit stress on morpho-physiological characteristics of dragonhead (Dracocephalum-moldavica L.) plants. *Acta Agriculturae Slovenica, 107*(2), 385–396.

Mujtaba, M., Khawar, K. M., Camara, M. C., Carvalho, L. B., Fraceto, L. F., Morsi, R. E., & Wang, D. (2020). Chitosan-based delivery systems for plants: A brief overview of recent advances and future directions. *International Journal of Biological Macromolecules, 154*, 683–697.

Preetha, P. S., & Balakrishnan, N. (2017). A review of nano fertilizers and their use and functions in soil. *International Journal of Current Microbiology and Applied Sciences, 6*(12), 3117–3133.

Rafique, M., Sohaib, M., Tahir, R., Bilal Tahir, M., & Rizwan, M. (2021). Plant-mediated green synthesis of zinc oxide nanoparticles using peel extract of citrus reticulate for boosting seed germination of brassica Nigra seeds. *Journal of Nanoscience and Nanotechnology, 21*(6), 3573–3579.

Rakhimol, K. R., Thomas, S., Kalarikkal, N., & Jayachandran, K. (2021). Nanotechnology in controlled-release fertilizers. In: *Controlled release fertilizers for sustainable agriculture* (pp. 169–181). Academic Press, Elsevier. https://doi.org/10.1016/B978-0-12-819555-0.00010-8

Rajeshkumar, S., & Bharath, L. V. (2017). Mechanism of plant-mediated synthesis of silver nanoparticles–a review on biomolecules involved, characterization and antibacterial activity. *Chemico-Biological Interactions, 273*, 219–227.

Rameshaiah, G. N., Pallavi, J., & Shabnam, S. (2015). Nano fertilizers and nano sensors–an attempt for developing smart agriculture. *International Journal of Engineering Research and General Science, 3*(1), 314–320.

Rexlin, J., Vijayakumar, S., Nilavukkarasi, M., Vidhya, E., Alharthi, N. S., Sajjad, M., & Praseetha, P. K. (2023). Bioengineered ZnO nanoparticles as a nanopriming agent in Cyamopsistetragonoloba (L). Taub. To improve yield and disease resistance. *Applied Nanoscience, 13*(9), 5993–6001.

Rostamizadeh, E., Iranbakhsh, A., Majd, A., Arbabian, S., & Mehregan, I. (2020). Green synthesis of Fe_2O_3 nanoparticles using fruit extract of *Cornus mas* L. and its growth-promoting roles in Barley. *Journal of Nanostructure in Chemistry, 10*, 125–130.

Sahani, S., & Sharma, Y. C. (2021). Advancements in applications of nanotechnology in global food industry. *Food Chemistry, 342*, Article 128318.

Safdar, M., Kim, W., Park, S., Gwon, Y., Kim, Y. O., & Kim, J. (2022). Engineering plants with carbon nanotubes: A sustainable agriculture approach. *Journal of Nanbiotechnology, 20*(1), 275.

Salam, A., Khan, A. R., Liu, L., Yang, S., Azhar, W., Ulhassan, Z., & Gan, Y. (2022). Seed priming with zinc oxide nanoparticles downplayed ultrastructural damage and improved photosynthetic apparatus in maize under cobalt stress. *Journal of Hazardous Materials, 423*, 127021.

Salih, A. M., Qahtan, A. A. A., Qurainy, F. A., & Munqedhi, B. M. (2022). Impact of biogenic ag-containing nanoparticles on germination rate, growth, physiological, biochemical parameters, and antioxidants system of tomato (*Solanum tuberosum* L.) in vitro. *Processes, 10*(5), 825.

Seleiman, M. F. A., Suhaibani, N., Ali, N., Akmal, M., Alotaibi, M., Refay, Y., & Battaglia, M. L. (2021). Drought stress impacts on plants and different approaches to alleviate its adverse effects. *Plants, 10*(2), 259.

Sempeho, S. I., Kim, H. T., Mubofu, E., & Hilonga, A. (2014). Meticulous overview on the controlled release fertilizers. *Advances in Chemistry, 63071.* https://doi.org/10.1155/2014/363071

Sharma, D., Sabela, M. I., Kanchi, S., Mdluli, P. S., Singh, G., Stenström, T. A., & Bisetty, K. (2016). Biosynthesis of ZnO nanoparticles using *Jacaranda mimosifolia* flowers extract: Synergistic antibacterial activity and molecular imulated facet specific adsorption studies. *Journal of Photochemistry Photobiology B: Biology, 162*, 199–207.

Sharma, P., Urfan, M., Anand, R., Sangral, M., Hakla, H. R., Sharma, S., & Bhagat, M. (2022). Green synthesis of zinc oxide nanoparticles using *Eucalyptus lanceolata* leaf litter: Characterization, antimicrobial and agricultural efficacy in maize. *Physiology and Molecular Biology Plants, 28*(2), 363–381.

Singh, J., Kumar, S., Alok, A., Upadhyay, S. K., Rawat, M., Tsang, D. C., & Kim, K. H. (2019). The potential of green synthesized zinc oxide nanoparticles as nutrient source for plant growth. *Journal of Cleaner Production, 214*, 1061–1070.

Solanki, P., Bhargava, A., Chhipa, H., Jain, N., & Panwar, J. (2015). Nano-fertilizers and their smart delivery system. In: *Nanotechnologies in food and agriculture* (pp. 81–101).

Saharan, V., Sharma, G., Yadav, M., Choudhary, M. K., Sharma, S. S., Pal, A., & Biswas, P. (2015). Synthesis and in vitro antifungal efficacy of Cu–chitosan nanoparticles against pathogenic fungi of tomato. *Int J Biolmacromol, 75*, 346–353.

Subramanian, K. S., Manikandan, A., Thirunavukkarasu, M., & Rahale, C. S. (2015). Nano-fertilizers for balanced crop nutrition. In *Nanotechnologies in food and agriculture* (pp. 69–80).

Thirugnanasambandan, T. (2021). Advances of engineered nanofertilizers for modern agriculture. Plant-microbes-engineered nano-particles (PM-ENPs) nexus in agro-ecosystems: Understanding the Interaction of plant, microbes and engineered nano-particles. *ENPS,* 131–152.

Toksha, B., Sonawale, V. A. M., Vanarase, A., Bornare, D., Tonde, S., Hazra, C., & Chatterjee, A. (2021). Nanofertilizers: A review on synthesis and impact of their use on crop yield and environment. *Environmental Technology and Innovation, 24*, Article 101986.

Ukidave, V., & VIngale, L. T. (2022). Green synthesis of zinc oxide nanoparticles from *Coriandrum sativum* and their use as fertilizer on Bengal gram, Turkish gram, and green gram plant growth. *International Journal of Agronomy, 2022*, 1–14.

Velsankar, K., Parvathy, G., Mohandoss, S., Kumar, R. M., & Sudhahar, S. (2022). Green synthesis and characterization of CuO nanoparticles using *Panicum sumatrense* grains extract for biological applications. *Applied Nanoscience, 12*(6), 1993–2021.

Yuan, S., Cheng, L., & Tan, Z. (2022). Characteristics and preparation of oil-coated fertilizers: A review. *JCR, 345*, 675–684.

Zhang, X., Cao, H., Wang, H., Zhao, J., Gao, K., Qiao, J., & Ge, S. (2022). The effects of graphene-family nanomaterials on plant growth: A review. *Nanomatererials, 12*(6, 936.

Zhang, P., Guo, Z., Ullah, S., Melagraki, G., Afantitis, A., & Lynch, I. (2021). Nanotechnology and artificial intelligence to enable sustainable and precision agriculture. *Nature Plants, 7*(7), 864–876.

Essential Oils and Nanotechnology: A Synergistic Approach to Environmental Sustainability

Geetanjali Raikwar, Saloni Santia, Kanchan Naagar, Fiza Fatima, Sumedha Mohan, and Praveen Dahiya

Abstract Essential oils (EOs) are volatile aromatic plant secondary metabolites and can be derived from different plant parts. EOs are renowned for their antimicrobial, antioxidant, and anti-inflammatory properties, making them valuable for diverse applications in healthcare, agriculture, and environmental sustainability. Being natural and effective, EOs exhibit significant efficacy against various pathogenic microorganisms, offering an eco-friendly alternative to synthetic chemicals/antibiotics. However, the long-term use of essential oils is limited by inherent challenges such as volatility, low water solubility, instability under environmental conditions, and rapid degradation. These limitations reduce their bioavailability and efficacy, especially in prolonged or large-scale applications. Nano-biotechnology has now emerged as a transformative approach to encounter such challenges, enhancing stability, bioactivity, and controlled release. Nano-encapsulation techniques, such as liposomes, polymeric nanoparticles, and solid lipid nanoparticles, protect essential oil components from degradation, improve their solubility, and enable sustained release for prolonged action. Additionally, green synthesis of nanoparticles using essential oils as reducing agents not only stabilizes the oils but also provides eco-friendly methods to create multifunctional nanomaterials. Hence, this chapter discusses the properties and uses of essential oils, limitations of their direct application, and the innovative role of nanobiotechnology in maintaining and enhancing their bioavailability. It also explores diverse applications of EO-based nano-formulations, including antimicrobial coatings, wound healing, cancer treatment, and sustainable agriculture. In conclusion, the integration of essential oils with nanobiotechnology enhances its long-term efficacy and environmental sustainability which can be achieved for research and implementation in various industries for long-term ecological balance and green innovation.

Keywords Essential oils · Nanotechnology · Bioavailability · Green synthesis

---

G. Raikwar · S. Santia · K. Naagar · F. Fatima · S. Mohan · P. Dahiya (✉)
Amity Institute of Biotechnology, Amity University Uttar Pradesh (AUUP), Noida, Uttar Pradesh, India
e-mail: pdahiya@amity.edu; praveen_sang@yahoo.com

E. Priyadarshini et al. (eds.), *Advanced Nanotechnology Applications for Environmental Sustainability*, Sustainable Environmental Waste Management Strategies,
https://doi.org/10.1007/978-3-032-12933-8_22

1 Introduction

Essential oils (EOs) are complex mixture of volatile and aromatic substances including terpenes, terpenoids and phenols. EOs have captivated scientific and therapeutic interest due to their complex chemistry and remarkable biological properties i.e., antioxidant, anxiolytic, antidepressant, anti-inflammatory, antimicrobial, antiviral, anticancer etc. (Chauhan & Dahiya, 2016; Raikwar et al., 2024). These natural compounds are primarily extracted from various parts of aromatic plants including leaves, flowers, seeds, bark, roots, and fruit peels. EOs serves critical ecological and physiological roles in plants, ranging from defence mechanisms to pollinator attraction.

EOs major active constituents falls predominantly into two broader classes i.e., terpenoids and phenylpropanoids. Distinct metabolic pathways such as mevalonic acid and 2C-methyl-D-erythritol-4-phosphate takes part in synthesis of precursor isoprene moiety which is crucial for biosynthesis of terpenes (Abd El-Hamid et al., 2023). Though monoterpenes (C_{10}), sesquiterpenes (C_{15}), and their oxygenated derivatives are found in predominance, other components such as ketones, esters, alcohols, and carboxylic acids each contributing uniquely to the bioactivity of EOs. These activities may influence by various factors such as plant species, growth conditions, and critically, the extraction method employed. There are several techniques available to extract essential oils. Traditional techniques (steam distillation, hydrodistillation) are most widely used due to their simplicity and efficiency. However, heat utilization in these techniques makes them limited for heat sensitive constituents. Modern techniques such as cold pressing, solvent extraction, microwave-assisted extraction, and supercritical CO_2 extraction overcoming these limitations. These techniques aim to preserve the natural profile and efficacy of EOs while offering higher extraction efficiencies, lower environmental impact, and better scalability (Keivanfar et al., 2023).

EOs such as fennel, juniper, kalonji (Purkayastha et al., 2012), *Mentha spicata* and *Mentha piperita* (Zaidi & Dahiya, 2015), tea tree, oregano, eucalyptus and clove are well known for their therapeutic antimicrobial potential against various multi drug resistant pathogens by exerting membranes disruption, protein synthesis disruption, and inhibiting viral adhesion and replication. Additionally, EOs possess notable antioxidant activity, primarily through scavenging reactive oxygen species and mitigating oxidative stress-induced cellular damage-an effect largely attributed to constituents like chamazulene and linalool. Their anti-inflammatory potential is equally significant, as they modulate pro-inflammatory cytokines and signaling pathways, thereby offering therapeutic value in managing chronic inflammatory disorders. Recent studies further highlight the cytotoxic and antiproliferative properties of various EOs against cancer cell lines, indicating promising applications in oncology either as adjuvants or standalone therapeutic agents (Brah et al., 2023). Despite their immense properties, efforts to develop new technologies, applications or functional agents based on EOs encounter challenges related to the loss of biological constituents due to high volatility, high risk of deterioration/oxidation, dependence on seasonality, limited stability and/or reduced efficacy of their biotechnological properties. However, nanotechnology offers innovative solutions to these barriers by enhancing the delivery, stability, and targeted action of EOs (Santos et al., 2023).

EO-loaded nano-carriers, including solid lipid nanoparticles (SLNs), nanoemulsions, polymeric nanoparticles, and liposomes enable controlled release, improved cellular uptake, and reduced degradation. These nano-formulations have shown enhanced efficacy in treating wound infections, inflammatory disorders, and cancer, while minimizing side effects. In wound care, for instance, combining EOs with silver nanoparticles (AgNPs) has resulted in dressings that exhibit superior antimicrobial and tissue regenerative properties. Moreover, encapsulation of EOs into biodegradable food packaging materials is transforming food preservation by offering safe, natural alternatives to synthetic additives (Kokina et al., 2019). In summary, EOs are not merely fragrant plant extracts, they are complex, bioactive agents with significant potential in medicine, agriculture, and industry. The convergence of phytochemistry, molecular biology, and nanotechnology is redefining their scope, offering novel pathways to harness their full therapeutic potential in a scientifically robust and sustainable manner. Nanotechnology, coupled with green chemistry principles and biopolymer-based delivery systems, is poised to play a pivotal role in expanding the reach and reliability of EO-based products.

Hence, current chapter discusses the properties and applications of essential oils, the limitations of their direct use, and the innovative role of nanobiotechnology in improving their stability and bioavailability. It also highlights the diverse applications of EO-based nano-formulations in areas such as antimicrobial coatings, wound healing, cancer treatment, and sustainable agriculture. Ultimately, integrating essential oils with nanobiotechnology offers a promising strategy to enhance therapeutic efficacy and promote environmental sustainability, paving the way for green innovation across various industries.

2 Essential Oils

Essential oils (EOs) are complex natural mixes of volatile, lipophilic, and odoriferous compounds that are frequently found in aromatic plants. They are sometimes referred to as essences, volatile oils, etheric oils, or aetheroleum. These oils are thought to include around 200 different constituents. These are aromatic chemicals that are abundant in oil sacs or oil glands that are located at various depths in the fruit peel, primarily in the cuticles and flavedo portion. Furthermore, EOs are aromatic oily liquids that are derived from various plant parts, including leaves, bark, seeds, flowers, and peels. In few exceptions (*cinnamon, sassafras, and vetiver*), most EOs are colorless or pale yellow, liquid at room temperature, and less thick than water (De Sousa et al., 2023). Furthermore, some of the chemical components of EOs are optically active, soluble in most organic solvents (ether, alcohol, and acetone), and insoluble in water.

EOs are used in a wide variety of consumer goods such as detergents, soaps, toilet products, cosmetics, pharmaceuticals, perfumes, confectionery food products, soft drinks, distilled alcoholic beverages (hard drinks) and insecticides. The world production and consumption of EOs and perfumes are increasing very fast. Unlike fatty oils, these "essential" oils are highly concentrated, substances extracted from flowers, leaves, stems, roots, seeds, bark, resin or fruit rinds. These EOs can be kept in glandular trichome secretory cells, cavities, and epidermal cells. The amount of EOs found in these plants can be anywhere from 0.01 to 10% of the total. That's why tons of plant material are required for just a few hundred pounds of oil.

There are often over a hundred components in each oil. Nonetheless, the two chemical families of terpenoids (monoterpenoids and sesquiterpenoids) and phenylpropanoids contain the most significant active ingredients (Fig. 1). These two categories are produced by distinct metabolic pathways. The main components of the EOs of several plant and floral species are terpenes and terpenoids (Pezantes-Orellana et al., 2024). It is possible to think of terpenes as composed of isoprene units. Terpenoids, also known as isoprenoids, class of chemicals that are still more common than terpenes. Hydrocarbons make up terpenes and can have linear, monocyclic, or polycyclic chemical structures. Terpenes can be distinguished into different class based on the number of C_5 elementary units in the skeletal structure. The monoterpenoid and sesquiterpenoid groups of terpenoids provide the most significant constituents of EOs from most plants.

Monoterpenes: Almost all EOs contain monoterpenes or monoterpenoid molecules, which are composed of ten carbon atoms and at least one double bond. These are the most basic and typical terpenes. Two isoprene units are assembled to create them (C_{10}). Acyclics, bornanes, camphanes and isocamphanes, fenchanes, thujanes, p-menthanes, caranes, and pinanes are the several subgroups of monoterpenes that can be separated based on structural variation. The biggest group of naturally occurring monoterpenes is p-methane type. Geraniol, terpineol (found in *lilacs*), limonene (citrus fruits), myrcene (in hops), linalool (*lavender*), and pinene (found in

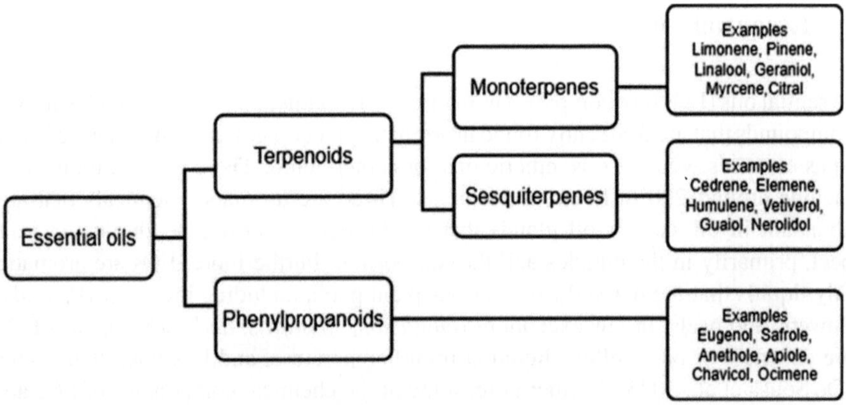

Fig. 1 Essential oils and its classes of compounds with examples

pine trees) are a few examples of monoterpenes and monoterpenoids. They respond quickly to sources of heat and air.

Sesquiterpenes: These compounds have complicated pharmacological properties, such chamazulene, which is composed of 15 carbon atoms and has the chemical formula $C_{15}H_{24}$. Three isoprene molecules (C_{15}) are coupled to form sesquiterpenes, which are found naturally in insects and higher plants. They are the most varied class of terpenoids, exhibiting a wide variety of unusual configurations with acyclic, monocyclic, bicyclic, or tricyclic frameworks.

Phenylpropanoids and Oxygenated compounds: A class of chemical substances known as phenylpropanoids is produced by plants using the shikimate pathway; their biogenetic precursor is the aromatic amino acid L-phenylalanine. A C_3 propane molecule joined to a phenyl ring forms the core of phenylpropanoids. Though less prevalent than terpenes, they are present in the kingdom of plants. Plants produce a class of chemical molecules called phenylpropanoids through the shikimate pathway. The potent alteration of a very small number of core structures is the source of the amazing diversity of phenylpropanoids. Oxygenated groups are the prevalent type of functional group found in EOs. The shikimate pathway produced aromatic chemicals such as alcohol, methoxy, phenol and aldehyde (Vogt, 2010).

Esters and Ketones: Often present in many EOs, esters are substances that are produced when an alcohol reacts with an acid (a process called esterification). They are likely to have fruity therapeutic benefits, including as sedative and antispasmodic properties, and are soothing and relaxing. When separated from other components, ketones can occasionally be neurotoxic and mucolytic. They facilitate tissue development, encourage cell regeneration, and liquefy mucus. They are mostly present in oils used for the upper respiratory system and are beneficial for ailments like dry cough, colds, flu, and dry asthma.

Alcohols and Carboxylic Acids: Organic substances with a hydroxyl group or more hydroxyl groups. It includes monohydric and polyhydric alcohols. Alcohols with a single hydroxyl group such as phenolic and tyrosol is monohydric and polyhydric include sugars with many hydroxyl groups, such as glycerol, and sugar alcohols, such as mannitol, which play important roles in metabolism. Carboxylic acid has a carboxyl group and are crucial for metabolic processes. Many plants contain benzoic acid, which is used as a preservation agent for food. Salicylic acid is known for its application in anti-inflammatory drugs and its function in plant defence.

Biosynthesis of terpenoids Isopentenyl diphosphate (IPP) and its allylic isomer dimethylallyl diphosphate (DMPP) are the universal precursors of isoprene molecules. The precursors of isoprene, isopentenyl, and dimethylallyl diphosphate are produced by two distinct biochemical pathways, which are the mevalonic acid (MVA) and 2C-methyl-D-erythritol-4-phosphate (MEP) pathways as depicted in Fig. 2. While the MEP pathway is limited to a few species, such as higher plants and algae, the MVA pathway is functional in all eukaryotes.

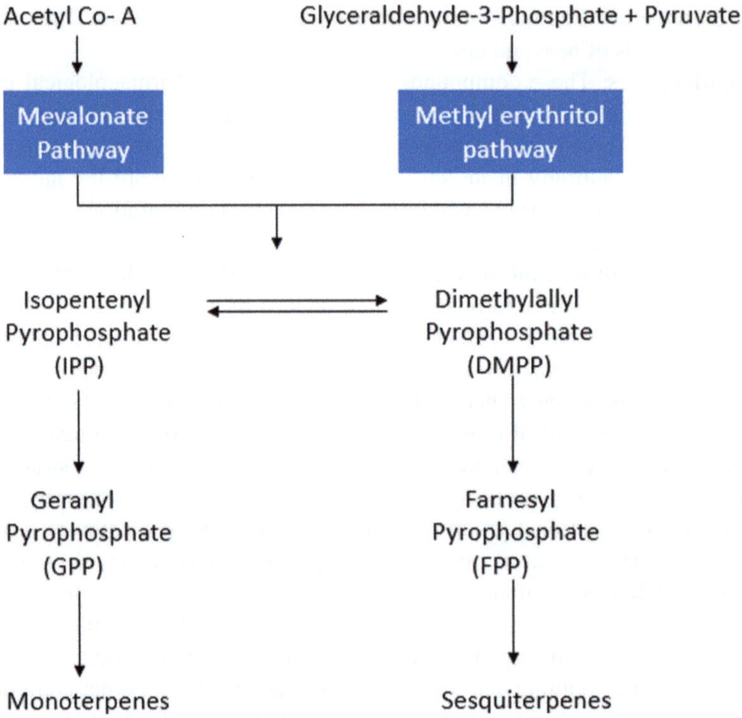

Fig. 2 Biosynthesis of terpenoids through mevalonic acid and methyl-erythritol phosphate pathways

3 Extraction Methods

There are variety of extraction procedures that are employed, with steam extraction and hydro distillation being the most traditional and conventional, straightforward, and widely applied technique. Additional procedures include microwave extraction, solvent extraction, cold pressing, supercritical CO_2 extraction etc. The plant material utilized usually determines the EOs extraction process. It is among the primary determinants of EOs quality. The yield and quality of the distillate produced, however, can be greatly impacted by the procedure selected (Souiy et al., 2023). The chemical makeup of EOs may be harmed or changed by an improper extraction process. The natural qualities and bioactivity are lost as a result. In the worst situations, this is followed by physical changes such as increased viscosity, discolouration, and an unpleasant taste or odour.

3.1 Conventional Extraction Methods

Conventional extraction methods are characterized by several drawbacks such as the degradation of unsaturated compounds and loss of certain components and are usually energy-consuming. There are various conventional extraction techniques such as steam distillation, hydro-distillation, hydro-diffusion and solvent extraction. There are on-going efforts to improve and optimize extraction techniques and that these techniques are carefully chosen based on the plant, plant part, and desired product quality. It is also important to note that the analytical composition of EOs can vary depending on the extraction technique used and that factors such as distillation duration, temperature, operating pressure, and raw plant material quality can all influence EOs yield(Souiy et al., 2022). From the techno-economic and environmental perspective, water distillation with full energy integration is the best method for rosemary oil.

Steam Distillation: Steam extraction is an established and commonly used technique for obtaining EOs from plants. Highest (93%) of essential oil extractions are done using this technique, which can take one to ten hours, depending on the material type, temperature, pressure, and extraction duration. Plant material is subjected to a steam in this extraction method without first macerating. Essential oil is released when the plant material's cells are broken down by the heat. After the steam, which is saturated with volatile chemicals, condenses, the water/oil mixture is decanted to extract the essential oil (Gavahian & Chu, 2018). One benefit of steam extraction is that essential oil is not hydrolysed or degraded because there is no direct contact between water and plant material or water and aromatic molecules. Ninety five percentage of the volatile molecules are recovered in as little as 30 min, which contain the most volatile molecules (Okoh et al., 2010). By making sure that the ambient pressure and the combined vapor pressure are equal at around 100 °C, the method enables volatile substances with boiling points between 150 and 300 °C to evaporate at a temperature that is comparable to that of water. It's also noteworthy that, depending on how challenging the EO extraction is, this method can be used under pressure (Bousbia et al., 2009).

Hydro-distillation: Hydro-distillation has established itself as the industry standard for extracting EOs from plant material, such as wood or flowers, and is frequently employed to separate high-boiling-point, non-water-soluble natural materials. The procedure entails completely submerging plant components in water and then boiling them. Since the surrounding water serves as a barrier to keep it from scorching, this technique somewhat protects the extracted oils. Condensation of the steam and essential oil vapour results in an aqueous fraction. This process has the benefit of being able to distillate the necessary material at a temperature lower than 100 °C. The benefit of this method is that it allows the necessary material to be distilled at temperatures lower than 100 °C. Okoh and colleagues examined the effects of solvent-free microwave extraction and hydro-distillation on the yield and characteristics of essential oil extracted from rosemary (*Rosmarinus officinalis L.*) examined. In hydro-distillation and solvent-free microwave extraction, the overall yields of the volatile components were 0.31 and 0.39%, respectively. More monoterpene hydrocarbons (32.95%) were obtained by hydro-distillation than via solvent-free microwave extraction (25.77%), whereas, solvent-free microwave extracted oil had more oxygenated monoterpenes (28.6%) than hydro-distilled oil (26.98%) (Okoh et al., 2010).

Cold pressing: It is one of the oldest extraction techniques. This technique describes any physical procedure that releases the oil by breaking the essential oil glands in the cuticles and peel. A watery emulsion is produced as a result of this procedure, and the essential oil is then separated by centrifugation (Bousbia et al., 2009). Cold-pressed citrus oils were produced industrially by hand until the turn of the twentieth century (Stratakos & Koidis, 2016). This method is more environmentally friendly and uses less energy than other methods of extracting oil. Particularly in the oil production process from oilseeds, it is utilized to extract oil from a variety of matrices. Using the cold press process to produce oils at low temperatures can result in high-quality oils. It is solvent-free and has an eco-friendly application. The cold-press extraction method, in other words, does not use heat or chemical extraction. Low productivity is one of this technique's drawbacks. The difficulty of extracting the same quality product is another drawback of this method (Cakaloglu et al., 2018).

Solvent extraction method: The multicomponent solid used in solvent extraction dissolves the target material in the medium. This extraction method's basic idea is to remove the oil from the environment by dissolving it in a solvent. It is exceedingly challenging to achieve equilibrium in this process since diffusion is slow in solids (Pereira & Meireles, 2007). The structure and amount of solvent to dissolve oil, the ease with which the solution can be removed from the environment (lower boiling points are preferred), temperature (diffusion rate increases as temperature increases), and the contact of oil with solvent (oil is typically localized between cells, so size reduction pre-treatment is required, increasing the surface area and the contact area as well) are the parameters affecting solvent extraction. Solvent extraction has the following benefits: low cost, easy equipment uses, great efficiency, and no need to filter the extracted oil. However, there are few drawbacks such as high-temperature applications and too much poisonous solution to discharge into the environment. A study conducted in which the cold press technique, solvent extraction, and supercritical carbon dioxide extraction is used to analyse the chemical makeup of the oil extracted from flaxseed. This indicates that the cold press method had the lowest efficiency, supercritical CO_2 extraction came in second, and solvent extraction using hexane had the highest oil yield. But after looking at the quality standards, it was determined that the cold press process produced oil of higher quality than the other methods (Pradhan et al., 2010).

3.2 Modern Extraction Methods

Supercritical carbon dioxide extraction: This method has been considered as a substitute medium for the extraction of EOs. Due to its mild critical conditions, carbon dioxide is the most often utilized supercritical fluid (Jimenez-Carmona et al., 1999; Senorans et al., 2000). Conventional techniques, such as steam distillation and solvent extraction, have drawbacks such as lengthy preparation times and high levels of organic solvents (Deng et al., 2005). Additionally, there may be hazardous solvent residue in the extract, low extraction efficiency, degradation of unsaturated compounds, and losses of some volatile chemicals (Glišić et al., 2007; Jimenez-Carmona et al., 1999). CO_2 becomes a liquid under high pressure, making it an extremely safe and inert medium for removing aromatic compounds from raw materials. Under normal atmospheric pressure and temperature, the liquid CO_2 simply turns back into a gas and evaporates, leaving no solvent residue in the final product. Even though the components of EOs were highly soluble in supercritical CO_2, the extraction rates with pure CO_2 were comparatively slow around 80% recovery after 90 min. Nonetheless, the combined techniques of a 15 min dynamic extraction with pure CO_2 and a 15 min static extraction with methylene chloride as a modifier produced significant recoveries. After four hours of operation, the extraction efficacy was comparable to hydro-distillation (Hawthorne et al., 1993a). A study demonstrated that supercritical fluid extraction is cost-effective than steam distillation technique. The latter's higher energy use and lower yield are the primary causes of this (Pereira & Meireles, 2007).

Microwave-assisted extraction: A modern technology for biological material extraction that has been seen as a significant alternative to other extraction methods due to its many benefits, which include volumetric heating, controlled heating, selectivity, and a reduction in extraction time and solvents. Numerous studies have demonstrated the effectiveness of this technique in the extraction of various chemicals, including pigments, EOs, perfumes, antioxidants, and other organic compounds found in food, plants, and animal tissues. This approach offers additional advantages such as more efficient heating, quicker energy transfer, smaller equipment, quicker warming initiation, and higher yields in addition to the decrease of time, solvent usage, and energy consumption (Farhat et al., 2009). In comparison to the conventional steam distillation method for extracting EOs from herbs and spices, the use of Microwave- assisted extraction in the isolation of herbal EOs is an intriguing option that offers more efficacy. In various studies, EOs were extracted from herbs using microwave extraction methods. In the 1980s, the use of microwave-assisted extraction to extract EOs was first documented. Few more advanced techniques, including solvent-free microwave extraction, microwave accelerated steam distillation, vacuum microwave hydro distillation, microwave hydro distillation, compressed air microwave distillation, and microwave by hydro diffusion and gravity, have been developed as a result of advancements in microwave-assisted extraction (Farhat et al., 2009; Périno-Issartier et al., 2011).

4 Properties of EOs and Their Underlying Mechanism

EOs are widely recognized for their potent antimicrobial properties and the presence of diverse therapeutic constituents. They are commonly used not only for their flavor but also for their therapeutic and aromatic qualities in a broad range of products, including food, pharmaceuticals, and cosmetics. Notably, EOs cannot be replaced by synthetic alternatives. Their biological activities are well-documented, encompassing insecticidal, antifungal, antibacterial, antiviral, antiparasitic, antimycotic, and antioxidant effects (Hou et al., 2022; Raikwar et al., 2025; Souiy et al., 2021).

Chemically, antioxidant activity refers to a substance's ability, even at trace levels, to inhibit the oxidation of readily oxidizable compounds like polyunsaturated lipids. Oxidation proceeds via a radical-chain reaction known as peroxidation, where oxygen is incorporated into organic molecules, forming hydroperoxides, epoxides, and other oxygenated derivatives. Alkyl peroxyl radicals (ROO) are the key propagators of this chain reaction. Thus, the capacity of a particular molecule to capture ROO radicals is closely linked to its antioxidant activity. Regarding the chemical mechanism of peroxidation inhibition, the constituents of EOs can be categorized into three primary groups: phenols, which function as radical trapping agents and highly oxidizable compounds, which improve the termination. Several EOs are frequently used to prevent a variety of chronic diseases and are essential in lowering oxidative stress. Originating from the EOs of *Matricaria chamomilla*, or c*hamomile, chamazulene* is a bicyclic sesquiterpene that is well-known for its health benefits. The amount of reactive oxygen species (ROS) in bovine aortic endothelial cells-1 were shown to be well balanced by chamazulene in one investigation. Exposure to high glucose and H_2O_2 caused these cells ROS levels to rise, but *chamazulene* treatment helped them return to equilibrium, demonstrating the compound's potential as an antioxidant.

EOs and their constituents have been demonstrated to have antiviral properties by disrupting the viral envelope, dissolving the capsid, or preventing the virus from attaching to host cell receptor. Numerous studies have demonstrated that EOs from Australian tea tree, oregano, *Eucalyptus caesia Benth*, *Mentha suaveolens* and *Star anise* have antiviral properties against HSV-1 because they bind to viral particles and prevent the virus from attaching itself to host cells. EOs have become a fascinating field of study due to their potential as natural antibacterial agents in the face of increasing antibiotic resistance. With their broad-spectrum antibacterial qualities, many EOs can fight off a variety of bacteria, viruses, and fungus. For example, tea tree oil has shown strong antibacterial properties against a variety of bacterial and fungal species (Carson et al., 2006) Strong antibacterial and antifungal qualities are another well-known benefit of oregano oil (Force et al., 2000). Understanding the processes underlying these antimicrobial actions and investigating the possible application of EOs in creating substitute therapies to fight infectious diseases are the main goals of current study. It is crucial to remember that additional research is necessary to confirm their efficacy and safety for application in medical settings.

Inflammation is a complicated defensive reaction to damaging external stimuli or endogenous signals, and the immune system is essential in removing the underlying cause and protecting tissue and cell integrity, which ultimately leads to homeostasis. Geranium essential oil possesses anti-inflammatory, antiseptic, antifungal, and antibacterial properties, it is frequently used in conventional medicine to treat inflammatory skin disorders. Bergapten is a furocoumarin that is found in *Cnidium monnieri*, primary constituent of bergamot essential oil. Bergapten's medicinal activity has led to its usage in the treatment of skin conditions such vitiligo, dermatitis, mycosis, and psoriasis. According to current experimental studies, it has anti-inflammatory and anti-allergic properties and can prevent osteoporosis.

5 Emerging Trends and Opportunities Specifically in Health Sector

In this section, we examine the many uses of EOs in healthcare, from their use in antimicrobial treatments and pain management to evidence-based procedures in conventional and alternative medicine.

For ages, people from many countries and traditions have utilized EOs for therapeutic purposes. The effectiveness of these aromatic miracles in conventional and alternative medicine practices has been clarified by scientific research in recent years. Aromatherapy, a supplementary therapy that uses the aromatic qualities of EOs to support healing and wellbeing, is a great illustration of their efficacy. Numerous studies have shown how effective aromatherapy is at treating stress-related illnesses, anxiety, and depression. Additionally, in order to improve patient comfort and lessen the need for traditional drugs, aromatherapy has been implemented in healthcare settings, including hospitals and palliative care institutions. More research into the medicinal potential of EOs is encouraged by the steadily increasing body of evidence proving their usage in traditional medicine.

EOs have become a supplemental method of treating pain and discomfort, which is still a major problem in healthcare. Some EOs have analgesic and anti-inflammatory qualities that make them intriguing substitutes for traditional painkillers. Peppermint oil, for instance, has been shown to reduce migraines and headaches (Göbel et al., 1994). Although lavender oil has been demonstrated to lessen pregnant women's labour pains. EOs like frankincense and chamomile have been used in palliative care to ease the symptoms and comfort people with terminal illnesses and chronic pain. Essential oil-based complementary therapies, like massage and aromatherapy, have grown in popularity because of their capacity to lessen pain, promote relaxation, and enhance general wellbeing (Nasiri & Mahmodi, 2018).

6 Challenges in Addressing the Sustainable Applications of EOs

There are several significant challenges to solve in order to ensure the safe and efficient use of EOs with sustainable use. Among these challenges are (a). *Low stability and volatility:* EOs are ignited and exposed to changes in temperature, light, oxygen and humidity. This instability can reduce shelf life and efficacy, especially in drug and pharmaceutical applications, leading to rapid degradation and loss of efficacy, (b). *Variation in composition:* Various plant species, growth environments, seasons of harvest and extraction methods—all have a major impact on the composition of EOs. It is difficult to ensure a certain level of quality of this variety of products. This can complicate normalization of EOs and lead to unequal outcomes of treatment, (c). *Insolubility in water:* EOs are not soluble in water, because they are hydrophobic, which limits their use in the compositions containing water. Since water solubility is generally necessary for adequate absorption, this limits their internal potential in oral or injection, (d). *Low bioavailability:* Low bioavailability of EOs limits their effectiveness and absorption in the human body. Since chemicals in EOs often experience problems with passing through biological barriers, they are less likely to reach target areas in fairly large quantities to have the therapeutic effect. Several other factors including cytotoxicity, safety, efficacy etc. are represented in Fig. 3. To sum up, researchers, manufacturers, legislators, and practitioners must work together to address these challenges and guarantee the continued safety, efficacy, and sustainability of EOs in industrial and medicinal uses (Aljaafari et al., 2021).

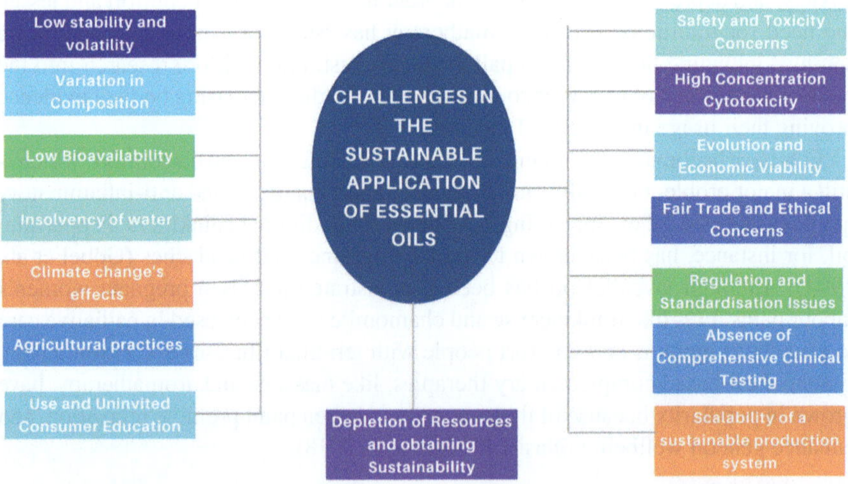

Fig. 3 Figure representing various challenges in sustainable applications of essential oils

7 EOs Based Nano-Formulations and Their Applications

One of the major properties of EOs is that they are lipophilic, which enables them to penetrate the cell membrane and impact the cellular leakage, transportation of ions, and ultimate cell death, especially in the cases of gram-positive bacteria. Due to such nature, there is a higher scope for the research being conducted to demonstrate various properties of the EOs in combating diseases and influencing treatments (Reyes-Jurado et al., 2020; Wińska et al., 2019).

Medicinal and aromatic plants have an important role in biological, pharmacological, and chemical activities (Abdoul-Latif et al., 2021). Throughout the centuries, they have been used in various treatments. Research on plant-derived modern drugs to derive these therapeutic compounds is ongoing (Al-Radadi, 2022). Such compounds could be produced for safer and more effective drugs. Further studies are required to know about the molecular mechanism of EOs in cancer treatment (Majdalawieh et al., 2017).

Conventional chemotherapeutic drugs are 50% plant-based, either chemically modified or half-directly derived compounds. Paclitaxel, from *Taxus brevifolia*, disrupts mitosis by targeting tubulin, treating pancreatic, breast, and ovarian cancers (Amin et al., 2009). Patchoulol oxide was made from the EOs component patchoulol by laboratory synthesis because natural sources were scarce. These results demonstrate the potential of EOs and plant chemicals in cancer treatment. Table 1 lists the compounds extracted from EOs from plant species and their activity on cancer cells.

Nanotechnology along with the herbal activity of EOs has evolved and led to numerous breakthroughs in target drug delivery, improvement of bioavailability, and limiting the systemic toxicity in cancer treatment leading to a great impact and developing a potentially efficient clinical approach. The EOs have many hurdles to overcome to be successfully administered to the patients, like poor water solubility, instability, and vaporization. With the use of nanotechnology-based approaches, EOs could be used as targeted drugs in cancer treatment, which could be selectively accumulated or released in a controlled manner at the tumor site. The approaches may include nano-emulsion, solid-lipid nanoparticles, polymeric nanoparticles, liposomes, or more that could enhance the potential and stability of EO within the target site as it could be more sustainable, safe and a better approach for cancer treatment.

Solid lipid nanoparticles (SLNs) serve as effective drug delivery systems due to their hydrophobic core, sustained release, and biocompatibility. *Zataria multiflora* SLN (ZMSLN) was formulated to enhance the therapeutic potential of *Zataria multiflora* essential oil (ZMEO). Key parameters like encapsulation efficiency and particle size were analysed, and cytotoxicity was assessed via MTT assay against MDA-MB-468 and A-375 cell lines. Previous studies confirm that EO-loaded SLNs improve efficacy. Similarly, *Croton argyrophyllus* and *Eugenia caryophyllata* showed high entrapment and bioactivity. ZMSLN showed superior anticancer effects due to enhanced cellular uptake, outperforming bulk ZMEO, which showed a dose-dependent response. Similar trends were observed with *Citrus* EO-loaded

Table 1 Table representing bioactive compounds from essential oils from various sources and their activity on cancer cells and mechanisms involved

Species	Compounds	Observations	Mechanism	References
Thymus fallax	Carvacrol, p-cymene, thymol, γ-terpinene	Cytotoxic only to cancerous cells	Antioxidant activity	Çetinus et al. (2013)
Boswellia sacra	α-pinene, α-thujene, β-pinene, myrcene, boswellic acid	Cytotoxic to cancerous cells	Antiproliferative	Suhail et al. (2011)
Amomum tsaoko	1,8-cineole, ρ-propylbenzaldehyde, geraniol, geranial, α-terpineol, α-phellandrene, neral, β-pinene	Cytotoxic to some cancer cells, like HepG2 Hela and Bel-7402, less cytotoxicity towards HL-7702 and HUVEC Non cytotoxicity towards some	Antiproliferative	Yang et al. (2010)
Lippia alba	Geranial, neral, geraniol, *trans*-β-caryophyllene, 6-methyl-5-hepten-2-one, limonene, linalool	Cytotoxic to cancerous cells	Citral-dependent cytotoxicity	Mesa-Arango et al. (2009)
Boswellia sp.	Duva-3,9,13-trien-1,5alpha-diol-1-acetate, octyl acetate, o-methyl anisole, naphthalene decahydro-1,1,4a-trimethyl-6-methylene-5-(3-methyl-2-pentenyl), thunbergol	Cytotoxic to cancerous cells	Antiproliferative	Frank et al. (2009)
Casearia sylvestris	Bicyclogermacrene, β-caryophyllene, spathulenol, α-humulene, α-pinene	Cytotoxic to HeLa, A549 and HT-29 with less cytotoxicity to Vero and macrophages	Cytotoxic	Silva et al. (2008)
Zanthoxylum rhoifolum Lam	ß-caryophyllene, α-humulene, α-pinene, myrcene, linalool	Cytotoxic to HeLa, A549, and HT-29	Cytotoxicity	Silva et al. (2007)

(continued)

Table 1 (continued)

Species	Compounds	Observations	Mechanism	References
Commiphora gileadensis	Sabinene, ß-caryophyllene, germacrene D, *α*- pinene	Cytotoxic to cancerous cells but efficiency varies with concentrations of EOs in dilution	Antiproliferative	Amiel et al. (2012)
Juniperus excelsa	α-Pinene, Cedrol, L-verbenol, D-verbenol, Eucarvone, p-Cymene, Limonene, α-Muurolene, 3-Carene, β-Myrcene, Terpinolene	Moderately cytotoxic against A2780 ovarian cancer cell line	Cytotoxic	Topçu et al. (2005)
Aniba rosaeodora	Rosewood essential oil, Linalool	Cytotoxicity to cancer cells A431 and HaCaT and minor cytotoxic to normal cells HEK001 & NHEK	Cytotoxic	Sœur et al. (2011)
Ocimum kilimandscharicum	11,8-cineole, limonene	High selectivity and potent anticancer activity against human ovarian cancer cells	Anti-inflammatory (inhibits leukocyte migration), antioxidant (DPPH scavenging), anticancer (cytotoxicity)	De Lima et al. (2014)
Ruta chalepensis L	*β*-pinene; *α*-pinene (Undecanone, Octyl Acetate)	Highly cytotoxic in HEK293, NCI-N87; moderate in K562, MIA-Paca2, U2OS, U87-MG, JIMT-T1, MRC-5, A549, PC3; least in RT4, HCT116, A2780	Apoptosis	Abdoul-Latif et al. (2022)

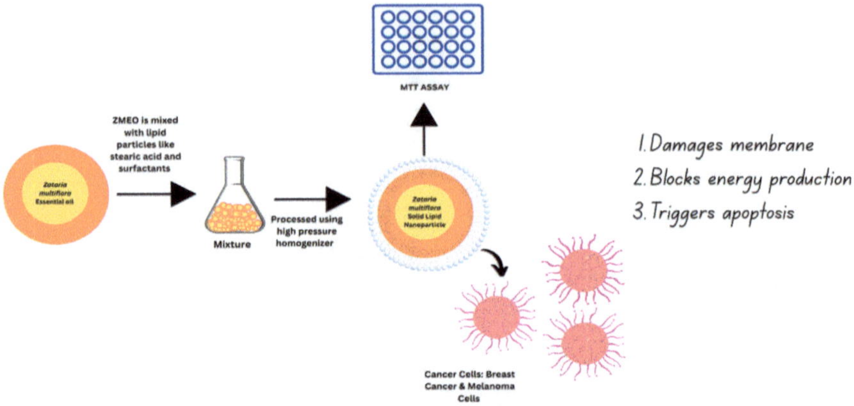

Fig. 4 Preparation of *Zataria multiflora* Solid Lipid Nanoparticle and its action on cancer cells

chitosan nanoparticles. Figure 4 illustrates ZMSLN synthesis and action on cancer cells.

Cellular and molecular interactions control the complex biological process of wound healing. Disruptions in this process can lead to chronic wounds such as surgical wounds, ulcers, and pressure sores. These wounds are often caused by systemic disorders such as starvation, infection, or inflammation (Ayuk et al., 2016; Guo & DiPietro, 2010). EOs, such as eucalyptus and rosemary have potent antibacterial properties even against types of bacteria that are resistant to many drugs (Ait-Ouazzou et al., 2011; Davis & Perez, 2009; Mayaud et al., 2008). However, because of their instability, their efficacy is increased when they are encapsulated in nano carriers (Drulis-Kawa & Dorotkiewicz-Jach, 2010). Studies have created solid lipid nanoparticles (SLNs) and nanostructured lipid carriers (NLCs) using stabilisers, including lecithin, olive/sesame oil, and cocoa butter. These were made by ultrasonography and high-shear homogenisation, and their physicochemical characteristics, bio-adhesion, cytocompatibility, and potential for fibroblast-mediated wound healing were assessed (Moss et al., 2003; Saporito et al., 2017; Takahashi et al., 2004).

Wound infections, often caused by methicillin resistant *Staphylococcus aureus* (MRSA), *Pseudomonas aeruginosa, Candida albicans,* and *E. coli*, impair healing due to biofilm formation and antibiotic resistance (Dabiri et al., 2016; Gefen, 2019; Järbrink et al., 2017). Silver nanoparticles (AgNPs) are perfect for advanced wound dressings since they have minimal cytotoxicity, broad-spectrum antibacterial qualities, and the capacity to promote tissue regeneration; they are ideal for advanced wound dressings (Aljghami et al., 2019). EOs improve immunity and healing: clove oil fights germs and fungus, niaouli oil strengthens the immune system, and mandarin oil promotes skin regeneration (Chouhan et al., 2017). Clove, niaouli, and mandarin EOs were used to functionalise three different types of wound dressings, whereas AgNPs were the sole ingredient in one kind. To improve adherence, sodium alginate was added to the dressings and allowed to dry before use. Figure 5 shows the synthesis of silver nanoparticles and modifies wound dressings demonstrating the

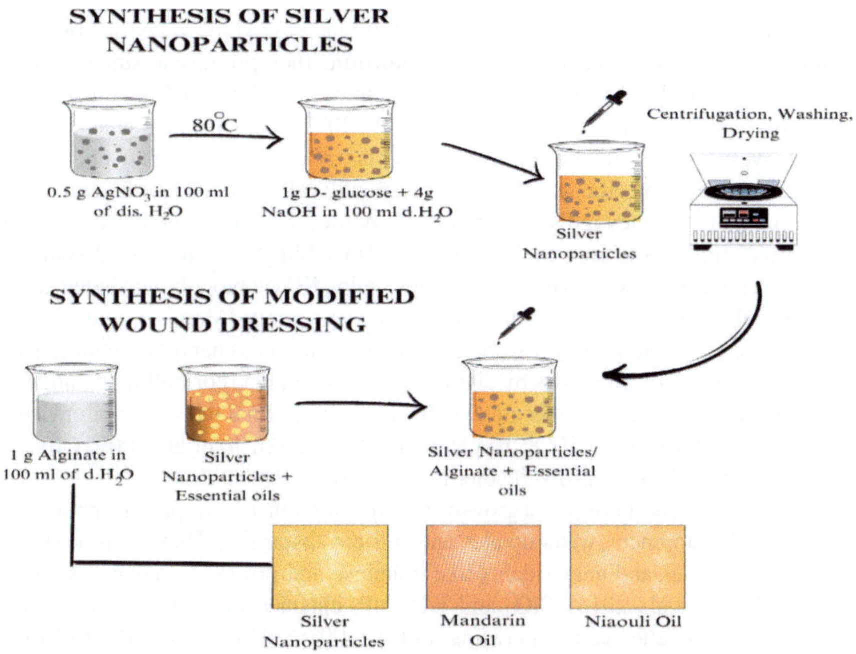

Fig. 5 Synthesis of silver nanoparticles and modified wound dressing model using essential oils

composite process by AgNPs, sodium alginate, and EOs when combined together can provide a useful method of bioactive, infection-resistant wound care.

Silver sulfadiazine (SSD), an antibacterial medication, is still widely used to treat burn wound infections due to its broad-spectrum efficacy and low toxicity (Heyneman et al., 2016; Khattak et al., 2021). The disadvantages of typical SSD creams include limited skin penetration, low solubility, and the formation of pseudo-eschar, which may delay healing (Luan et al., 2012). Water-in-oil (w/o) nano-emulsions are one example of a nanotechnology-based formulation designed to enhance SSD delivery while addressing these constraints. In addition to increasing skin permeability and medication absorption, these nano-formulations enable regulated and prolonged release. The nano-emulsion's stability is enhanced by surfactants such as Span-80, Triton X-100, and Tween-80, as well as polymers such as HPMCP, Methocel, and CAP. Argan oil contains anti-inflammatory qualities and is a biocompatible oil that moisturises the skin (Irshad et al., 2022; Menni et al., 2020; Mofazzal Jahromi et al., 2018; Siddiqui et al., 2020, 2022). These SSD-loaded nano-emulsions shown potent antibacterial action against burn wound pathogens such as *Staphylococcus aureus, Escherichia coli, Klebsiella pneumoniae,* and *Candida albicans* (de Sousa et al., 2019; Klifto et al., 2022; Shao et al., 2015). The formulations increased drug penetration while decreasing systemic toxicity, minimising irritation from traditional cream bases, boosting biocompatibility with skin tissues, and encouraging better therapeutic benefits. SSD nano-emulsions exhibit increased bioavailability, biocompatibility, and robust antibacterial activity, indicating their promise as superior alternatives to current burn wound treatments, despite limitations such as formulation scalability and long-term stability (Fatima et al., 2022; Nithianandam et al., 2020).

The overuse of synthetic agrochemicals has risen environmental and health concerns, encouraging the adoption of eco-friendly alternatives like essential oils (EOs) and nanoparticles (NPs) in agriculture. While EOs offer benefits such as low toxicity and biodegradability, their volatility and instability limit their use. Nanotechnology overcomes these drawbacks by encapsulating EOs in biopolymer-based nanocarriers, enhancing stability, controlled release, and safety (Menossi et al., 2021). Solid NPs such as alumina, silica, and nano-clays function as nano-herbicides, nano pesticides, and nano-fungicides by disrupting insect cuticles, controlling pathogens, and facilitating targeted agrochemical delivery with minimal resistance or environmental harm (Barik et al., 2008; Bin Hussein et al., 2005; Jo et al., 2009; Oliveira et al., 2015; Sarlak et al., 2014; Stadler et al., 2010).

Nanoparticles also boost plant growth and stress tolerance by improving photosynthesis, root development, water uptake, and nutrient absorption. Under abiotic stress, NPs activate genes and antioxidants like trehalose and proline to reduce oxidative damage (Rejeb et al., 2014). SiO_2 and TiO_2 NPs enhance water use efficiency and protect photosynthetic machinery (Mingyu et al., 2008; Siddiqui et al., 2014). Cerium oxide NPs (nanoceria) mimic antioxidant enzymes via redox cycling between Ce^{3+}/Ce^{4+}, enhancing stress resistance and reducing reactive oxygen species. In precision agriculture, nano-fertilizers further aid nutrient use efficiency and sustainability (Collin et al., 2014; Liu & Shi, 2019; Milenković et al., 2019; Pulido-Reyes et al., 2015; Saxena et al., 2016; Wakeel et al., 2020; Wang et al., 2012, 2012).

8 Overcoming Hurdles for Sustainable Application of EOs

Nanotechnology has revolutionized biomedical science through its application in targeted drug delivery, gene therapy, cancer treatment, and diagnostics. Due to their unique physicochemical properties, especially their ultra-small size, large surface area-to-volume ratio, and tunable surface functionalities nanoparticles (NPs) are capable of crossing biological barriers such as capillaries and cellular membranes. This improves the precision and effectiveness of treatment programs by enabling medications to accumulate at the site of action more efficiently (Parveen et al., 2016). NPs enhance the solubility, bioavailability, and retention length of medications to enable longer and controlled release patterns that are less affected by patient-specific characteristics such as fed or fasted states.

By preventing off-target effects and allowing for lower dosages of active medications, nanoparticles lessen systemic toxicity and the likelihood of negative side effects and drug overdoses. Crucially, a large number of nanocarriers have minimal cytotoxicity and are biodegradable, which makes them suitable for application in biomedicine. Silver nanoparticles (AgNPs), zinc oxide nanoparticles (ZnO–NPs), and mesoporous silica nanoparticles (MSNs) are common nanomaterial utilised in drug delivery; these materials have shown encouraging outcomes in antibacterial, anticancer, and immune-modulatory applications (Gul et al., 2021; Kalpana & Rajeswari, 2018).

By strengthening selective targeting, extending systemic circulation duration, and boosting biocompatibility, advanced delivery methods including PEGylation-the coating of nanoparticles with polyethylene glycol (PEG) further improve nanoparticle performance. PEGylated nanoparticles, for example, have been used to deliver antibiotics straight to infection sites, lowering microbial resistance and enhancing the effectiveness of treatment (Suk et al., 2016). Furthermore, nanoparticles have shown promise in tissue regeneration and dental surgery. By minimising polymerisation shrinkage and lowering the amount of synthetic resin needed, their inclusion into resin-based composites enhances patient safety and clinical results.

By tackling major issues including volatility, poor solubility, and instability under environmental stress, nanotechnology is revolutionising the use of EOs in tandem with medicinal advancements. EO-loaded nanosystems especially chitosan-based carriers, solid lipid nanoparticles (SLNs), nano emulsions, and nanofibers, improve EO bioactivity, extend shelf life, and allow for controlled release. Ultrasonication and high-pressure homogenisation are frequently employed to create EO nano emulsions with better stability and dispersion. For instance, citrus EO nano emulsions (34 nm) show improved storage stability, while ultrasonication increases the antibacterial action of *Zataria multiflora* oil (Gahruie et al., 2017; Zhiwei et al., 2022).

Biodegradable food preservation films that contain EOs have been successfully integrated. While cinnamon oil in potato starch films improves antifungal, antioxidant, and heat resistant qualities, clove oil permits a controlled release of antimicrobial chemicals in pullulan-gelatin films (Shen et al., 2021). EO may be encapsulated into nanofibers for sophisticated food and medical packaging due to electrospinning. For instance, peppermint oil embedded in polycaprolactone nanofibers has been produced for antibacterial cutaneous patches, while cumin oil combined with ZnO–NPs has been used to manufacture packaging films that prolong the shelf life of cheese (Shanbehzadeh et al., 2022).

Furthermore, EO encapsulation in SLNs and chitosan-based systems provides multifunctional platforms for pharmaceutical and food applications. Lime oil in chitosan nanoparticles shows potent activity against foodborne pathogens, and eucalyptus oil in SLNs demonstrates improved antimicrobial efficacy. Furthermore, in cancer therapy models, SLN-encapsulated cumin oil has notable cytotoxic effects and high encapsulation efficiency. Advanced carriers, including smectite clay and mesoporous silica nanoparticles, also preserve EOs bioactivity over time. For example, when enclosed in silica nanoparticles, bergamot oil demonstrates long-lasting antidepressant action, and cinnamon oil embedded in clay particles demonstrates persistent antifungal activity against *Candida albicans* (Saad et al., 2024; Tavasoli et al., 2022; Yang et al., 2021).

New encapsulating techniques such layer-by-layer assembly, solvent displacement, and ionic gelation are enhancing EOs delivery systems by facilitating controlled release, decreased volatility, and increased biocompatibility. Green synthesis techniques that employ natural polymers as environmentally benign nano-carriers, such as chitosan, alginate, and pectin, further promote sustainability. These promote safe and scalable nanotechnology applications by lowering the need for organic solvents and adhering to green chemistry principles (Hasheminejad et al., 2019).

The future of integrative, sustainable treatments is demonstrated by the combination of nanotechnology with plant-based bioactive, such as EOs. In addition to improving medication and essential oil delivery methods, these developments pave the way for low-toxicity therapeutic treatments, personalised medicine, and natural preservative systems. By fusing cutting-edge material science with substances found in nature, scientists are opening the door to multipurpose nano-carriers that satisfy environmental and health requirements.

References

Abd El-Hamid, M. I., El-Tarabili, R. M., Bahnass, M. M., Alshahrani, M. A., Saif, A., Alwutayd, K. M., Safhi, F. A., Mansour, A. T., Alblwi, N. A. N., Ghoneim, M. M., Elmaaty, A. A., Al-harthi, H. F., & Bendary, M. M. (2023). Partnering essential oils with antibiotics: Proven therapies against bovine Staphylococcus aureus mastitis. *Frontiers in Cellular and Infection Microbiology, 13*, 1265027.

Abdoul-Latif, F. M., Oumaskour, K., Boujaber, N., Ainane, A., Mohamed, J., & Ainane, T. (2021). Formulations of a cosmetic product for hair care based on extract of the microalga Isochrysis galbana: In vivo and in vitro activities. *Journal of Analytical Sciences and Applied Biotechnology, 3*(1), Article 1.

Abdoul-Latif, F. M., Elmi, A., Merito, A., Nour, M., Risler, A., Ainane, A., Bignon, J., & Ainane, T. (2022). Essential oil of Ruta chalepensis L. from Djibouti: Chemical analysis and modeling of in vitro anticancer profiling. *Separations, 9*(12), Article 12.

Ait-Ouazzou, A., Lorán, S., Bakkali, M., Laglaoui, A., Rota, C., Herrera, A., Pagán, R., & Conchello, P. (2011). Chemical composition and antimicrobial activity of essential oils of Thymus algeriensis, Eucalyptus globulus and Rosmarinus officinalis from Morocco. *Journal of the Science of Food and Agriculture, 91*(14), 2643–2651.

Aljaafari, M. N., AlAli, A. O., Baqais, L., Alqubaisy, M., AlAli, M., Molouki, A., Ong-Abdullah, J., Abushelaibi, A., Lai, K.-S., & Lim, S.-H.E. (2021). An overview of the potential therapeutic applications of essential oils. *Molecules, 26*(3), 3.

Aljghami, M. E., Saboor, S., & Amini-Nik, S. (2019). Emerging innovative wound dressings. *Annals of Biomedical Engineering, 47*(3), 659–675.

Al-Radadi, N. S. (2022). Biogenic proficient synthesis of (Au–NPs) via aqueous extract of Red Dragon Pulp and seed oil: Characterization, antioxidant, cytotoxic properties, anti-diabetic anti-inflammatory, anti-Alzheimer and their anti-proliferative potential against cancer cell lines. *Saudi Journal of Biological Sciences, 29*(4), 2836–2855.

Amiel, E., Ofir, R., Dudai, N., Soloway, E., Rabinsky, T., & Rachmilevitch, S. (2012). β-Caryophyllene, a compound isolated from the biblical balm of gilead (Commiphora gileadensis), is a selective apoptosis inducer for tumor cell lines. *Evidence-Based Complementary and Alternative Medicine: Ecam, 2012*, Article 872394.

Amin, A., Gali-Muhtasib, H., Ocker, M., & Schneider-Stock, R. (2009). Overview of major classes of plant-derived anticancer drugs. *International Journal of Biomedical Science: IJBS, 5*(1), 1–11.

Ayuk, S. M., Abrahamse, H., & Houreld, N. N. (2016). The role of matrix metalloproteinases in diabetic wound healing in relation to Photobiomodulation. *Journal of Diabetes Research, 2016*(1), 2897656.

Barik, T. K., Sahu, B. & Swain, V. (2008). *Nanosilica—From medicine to pest control | Parasitology Research.* https://doi.org/10.1007/s00436-008-0975-7

Bin Hussein, M. Z., Yahaya, A. H., Zainal, Z., & Kian, L. H. (2005). Nanocomposite-based controlled release formulation of an herbicide, 2,4-dichlorophenoxyacetate incapsulated in zinc–aluminium-layered double hydroxide. *Science and Technology of Advanced Materials, 6*(8), 956–962.

Bousbia, N., Vian, M. A., Ferhat, M. A., Meklati, B. Y., & Chemat, F. (2009). A new process for extraction of essential oil from Citrus peels: Microwave hydrodiffusion and gravity. *Journal of Food Engineering, 90*(3), 409–413.

Brah, A. S., Armah, F. A., Obuah, C., Akwetey, S. A., & Adokoh, C. K. (2023). Toxicity and therapeutic applications of citrus essential oils (CEOs): A review. *International Journal of Food Properties, 26*(1), 301–326.

Cakaloglu, B., Ozyurt, V. H., & Otles, S. (2018). Cold press in oil extraction. A review. *Ukrainian Food Journal, 7*(4), 640–654.

Carson, C. F., Hammer, K. A., & Riley, T. V. (2006). *Melaleuca alternifolia* (Tea Tree) Oil: A review of antimicrobial and other medicinal properties. *Clinical Microbiology Reviews, 19*(1), 50–62.

Çetinus, E., Temiz, T., Ergül, M., Altun, A., Çetinus, Ş., & Kaya, T. (2013). Thyme essential oil inhibits proliferation of DLD-1 colorectal cancer cells through antioxidant effect. *Cumhuriyet Medical Journal, 35*(1), 14–24.

Chauhan, B. S., & Dahiya, P. (2016). Evaluation of In -vitro antimicrobial potential and phytochemical analysis of Spruce, Cajeput and Jamrosa essential oil against clinical isolates. *International Journal of Green Pharmacy, 10*(1), 27–32.

Chouhan, S., Sharma, K., & Guleria, S. (2017). Antimicrobial activity of some essential oils—present status and future perspectives. *Medicines, 4*(3), Article 3.

Collin, B., Auffan, M., Johnson, A. C., Kaur, I., Keller, A. A., Lazareva, A., Lead, J. R., Ma, X., Merrifield, R. C., Svendsen, C., White, J. C., & Unrine, J. M. (2014). Environmental release, fate and ecotoxicological effects of manufactured ceria nanomaterials. *Environmental Science: Nano, 1*(6), 533–548.

Dabiri, G., Damstetter, E., & Phillips, T. (2016). Choosing a wound dressing based on common wound characteristics. *Advances in Wound Care, 5*(1), 32–41.

Davis, S. C., & Perez, R. (2009). Cosmeceuticals and natural products: Wound healing. *Clinics in Dermatology, 27*(5), 502–506.

De Lima, V. T., Vieira, M. C., Kassuya, C. A. L., Cardoso, C. A. L., Alves, J. M., Foglio, M. A., De Carvalho, J. E., Formagio, A. S. N. (2014). Chemical composition and free radical-scavenging, anticancer and anti-inflammatory activities of the essential oil from Ocimum kilimandscharicum. *Phytomedicine: International Journal of Phytotherapy and Phytopharmacology, 21*(11), 1298–1302.

De Sousa, D. P., Damasceno, R. O. S., Amorati, R., Elshabrawy, H. A., De Castro, R. D., Bezerra, D. P., Nunes, V. R. V., Gomes, R. C., & Lima, T. C. (2023). Essential oils: Chemistry and pharmacological activities. *Biomolecules, 13*(7), 1144.

de Sousa, J. K. T., Haddad, J. P. A., de Oliveira, A. C., Vieira, C. D., & dos Santos, S. G. (2019). In vitro activity of antimicrobial-impregnated catheters against biofilms formed by KPC-producing Klebsiella pneumoniae. *Journal of Applied Microbiology, 127*(4), 1018–1027.

Deng, C., Yao, N., Wang, A., & Zhang, X. (2005). Determination of essential oil in a traditional Chinese medicine, Fructus amomi by pressurized hot water extraction followed by liquid-phase microextraction and gas chromatography–mass spectrometry. *Analytica Chimica Acta, 536*(1–2), 237–244.

Drulis-Kawa, Z., & Dorotkiewicz-Jach, A. (2010). Liposomes as delivery systems for antibiotics. *International Journal of Pharmaceutics, 387*(1–2), 187–198.

Farhat, A., Ginies, C., Romdhane, M., & Chemat, F. (2009). Eco-friendly and cleaner process for isolation of essential oil using microwave energy: Experimental and theoretical study. *Journal of Chromatography A, 1216*(26), 5077–5085.

Fatima, Q. U. A., Ahmed, N., Siddiqui, B., Rehman, A. U., Haq, I. U., Khan, G. M., & Elaissari, A. (2022). Enhanced antimicrobial activity of silver sulfadiazine cosmetotherapeutic nanolotion for burn infections. *Cosmetics, 9*(5), Article 5.

Force, M., Sparks, W. S., & Ronzio, R. A. (2000). Inhibition of enteric parasites by emulsified oil of oreganoin vivo. *Phytotherapy Research, 14*(3), 213–214.

Frank, M. B., Yang, Q., Osban, J., Azzarello, J. T., Saban, M. R., Saban, R., Ashley, R. A., Welter, J. C., Fung, K.-M., & Lin, H.-K. (2009). Frankincense oil derived from Boswellia carteri induces tumor cell specific cytotoxicity. *BMC Complementary and Alternative Medicine, 9*, 6. https://doi.org/10.1186/1472-6882-9-6

Gahruie, H., Ziaee, E., Eskandari, M. H., & Hosseini, S. M. H. (2017). Characterization of basil seed gum-based edible films incorporated with *Zataria multiflora* essential oil nanoemulsion. *Carbohydrate Polymers, 166*, 93–103.

Gavahian, M., & Chu, Y.-H. (2018). Ohmic accelerated steam distillation of essential oil from lavender in comparison with conventional steam distillation. *Innovative Food Science and Emerging Technologies, 50*, 34–41.

Gefen, A. (2019). How medical engineering has changed our understanding of chronic wounds and future prospects. *Medical Engineering and Physics, 72*, 13–18.

Glišić, S. B., Mišić, D. R., Stamenić, M. D., Zizovic, I. T., Ašanin, R. M., & Skala, D. U. (2007). Supercritical carbon dioxide extraction of carrot fruit essential oil: Chemical composition and antimicrobial activity. *Food Chemistry, 105*(1), 346–352.

Göbel, H., Schmidt, G., & Soyka, D. (1994). Effect of peppermint and eucalyptus oil preparations on neurophysiological and experimental algesimetric headache parameters. *Cephalalgia, 14*(3), 228–234.

Gul, A. R., Shaheen, F., Rafique, R., Bal, J., Waseem, S., & Park, T. J. (2021). Grass-mediated biogenic synthesis of silver nanoparticles and their drug delivery evaluation: A biocompatible anti-cancer therapy. *Chemical Engineering Journal, 407*, Article 127202.

Guo, S., & DiPietro, L. A. (2010). Factors affecting wound healing. *Journal of Dental Research, 89*(3), 219–229. https://doi.org/10.1177/0022034509359125

Hasheminejad, N., Khodaiyan, F., & Safari, M. (2019). Improving the antifungal activity of clove essential oil encapsulated by chitosan nanoparticles. *Food Chemistry, 275*, 113–122. https://doi.org/10.1016/j.foodchem.2018.09.085

Hawthorne, S. B., Rickkola, M.-L., Screnius, K., Holm, Y., Hiltunen, R., & Hartonen, K. (1993a). Comparison of hydrodistillation and supercritical fluid extraction for the determination of essential oils in aromatic plants. *Journal of Chromatography A, 634*(2), 297–308.

Hawthorne, S. B., Rickkola, M.-L., Screnius, K., Holm, Y., Hiltunen, R., & Hartonen, K. (1993b). Comparison of hydrodistillation and supercritical fluid extraction for the determination of essential oils in aromatic plants. *Journal of Chromatography A, 634*(2), 297–308. https://doi.org/10.1016/0021-9673(93)83017-M

Heyneman, A., Hoeksema, H., Vandekerckhove, D., Pirayesh, A., & Monstrey, S. (2016). The role of silver sulphadiazine in the conservative treatment of partial thickness burn wounds: A systematic review. *Burns, 42*(7), 1377–1386.

Hou, T., Sana, S. S., Li, H., Xing, Y., Nanda, A., Netala, V. R., & Zhang, Z. (2022). Essential oils and its antibacterial, antifungal and antioxidant activity applications: A review. *Food Bioscience, 47*, Article 101716.

Irshad, S., Siddiqui, B., ur. Rehman, A., Farooq, R. K., & Ahmed, N. (2022). Recent trends and development in targeted delivery of therapeutics through enzyme responsive intelligent nanoplatform. *International Journal of Polymeric Materials and Polymeric Biomaterials, 71*(6), 403–413.

Järbrink, K., Ni, G., Sönnergren, H., Schmidtchen, A., Pang, C., Bajpai, R., & Car, J. (2017). The humanistic and economic burden of chronic wounds: A protocol for a systematic review. *Systematic Reviews, 6*(1), 15.

Jimenez-Carmona, M. M., Ubera, J. L., & De Castro, M. L. (1999). Comparison of continuous subcritical water extraction and hydrodistillation of marjoram essential oil. *Journal of Chromatography A, 855*(2), 625–632.

Jo, Y.-K., Kim, B. H., & Jung, G. (2009). Antifungal activity of silver ions and nanoparticles on phytopathogenic fungi. *Plant Disease, 93*(10), 1037–1043. https://doi.org/10.1094/PDIS-93-10-1037

Kalpana, V. N., & Devi Rajeswari, V. (2018). A review on green synthesis, biomedical applications, and toxicity studies of ZnO NPs. *Bioinorganic Chemistry and Applications, 2018*(1), 3569758. https://doi.org/10.1155/2018/3569758

Keivanfar, L., Nateghi, L., Rashidi, L., Pourahmad, R., & Rashidi Nodeh, H. (2023). Comparing two different extraction techniques on chemical composition and antioxidant property of three essential oils of Ferulago contracta, Rosmarinus officinalis and Lavendula sublepoidota. *Journal of Food Measurement and Characterization, 17*(4), 3579–3591. https://doi.org/10.1007/s11694-023-01859-y

Khattak, S., Qin, X.-T., Huang, L.-H., Xie, Y.-Y., Jia, S.-R., & Zhong, C. (2021). Preparation and characterization of antibacterial bacterial cellulose/chitosan hydrogels impregnated with silver sulfadiazine. *International Journal of Biological Macromolecules, 189*, 483–493. https://doi.org/10.1016/j.ijbiomac.2021.08.157

Klifto, K. M., Gurno, C. F., Seal, S. M., & Hultman, C. S. (2022). Factors associated with mortality following burns complicated by necrotizing skin and soft tissue infections: A systematic review and meta-analysis of individual participant data. *Journal of Burn Care and Research, 43*(1), 163–188.

Kokina, M., Salević, A., Kalušević, A., Lević, S., Pantić, M., Pljevljakušić, D., Šavikin, K., Shamtsyan, M., Nikšić, M., & Nedović, V. (2019). Characterization, antioxidant and antibacterial activity of essential oils and their encapsulation into biodegradable material followed by freeze drying. *Food Technology and Biotechnology, 57*(2), 282–289.

Liu, Y., & Shi, J. (2019). Antioxidative nanomaterials and biomedical applications. *Nano Today, 27*, 146–177. https://doi.org/10.1016/j.nantod.2019.05.008

Luan, J., Wu, J., Zheng, Y., Song, W., Wang, G., Guo, J., & Ding, X. (2012). Impregnation of silver sulfadiazine into bacterial cellulose for antimicrobial and biocompatible wound dressing. *Biomedical Materials, 7*(6), Article 065006.

Majdalawieh, A. F., Fayyad, M. W., & Nasrallah, G. K. (2017). Anti-cancer properties and mechanisms of action of thymoquinone, the major active ingredient of Nigella sativa. *Critical Reviews in Food Science and Nutrition, 57*(18), 3911–3928.

Mayaud, L., Carricajo, A., Zhiri, A., & Aubert, G. (2008). Comparison of bacteriostatic and bactericidal activity of 13 essential oils against strains with varying sensitivity to antibiotics. *Letters in Applied Microbiology, 47*(3), 167–173.

Menni, H. B., Belarbi, M., Menni, D. B., Bendiab, H., Kherraf, Y., Ksouri, R., Djebli, N., & Visioli, F. (2020). Anti-inflammatory activity of argan oil and its minor components. *International Journal of Food Sciences and Nutrition, 71*(3), 307–314.

Menossi, M., Ollier, R. P., Casalongué, C. A., & Alvarez, V. A. (2021). Essential oil-loaded bionanomaterials for sustainable agricultural applications. *Journal of Chemical Technology and Biotechnology, 96*(8), 2109–2122.

Mesa-Arango, A. C., Montiel-Ramos, J., Zapata, B., Durán, C., Betancur-Galvis, L., & Stashenko, E. (2009). Citral and carvone chemotypes from the essential oils of Colombian Lippia alba (Mill.) N.E. Brown: Composition, cytotoxicity and antifungal activity. *Memorias Do Instituto Oswaldo Cruz, 104*(6), 878–884.

Milenković, I., Mitrović, A., Algarra, M., Lázaro-Martínez, J. M., Rodríguez-Castellón, E., Maksimović, V., Spasić, S. Z., Beškoski, V. P., & Radotić, K. (2019). Interaction of carbohydrate coated cerium-oxide nanoparticles with wheat and pea: Stress induction potential and effect on development. *Plants (Basel, Switzerland), 8*(11), 478. https://doi.org/10.3390/plants8110478

Mingyu, S., Chao, L., Chunxiang, Q., Lei, Z., Liang, C., Hao, H., Xiaoqing, L., Xiao, W., & Fashui, H. (2008). Nano-anatase relieves the inhibition of electron transport caused by linolenic acid in chloroplasts of spinach. *Biological Trace Element Research, 122*(1), 73–81.

Mofazzal Jahromi, M. A., Sahandi Zangabad, P., Moosavi Basri, S. M., Sahandi Zangabad, K., Ghamarypour, A., Aref, A. R., Karimi, M., & Hamblin, M. R. (2018). Nanomedicine and advanced technologies for burns: Preventing infection and facilitating wound healing. *Advanced Drug Delivery Reviews, 123*, 33–64.

Moss, M., Cook, J., Wesnes, K., & Duckett, P. (2003). Aromas of rosemary and lavender essential oils differentially affect cognition and mood in healthy adults. *The International Journal of Neuroscience, 113*(1), 15–38.

Nasiri, A., & Mahmodi, M. A. (2018). Aromatherapy massage with lavender essential oil and the prevention of disability in ADL in patients with osteoarthritis of the knee: A randomized controlled clinical trial. *Complementary Therapies in Clinical Practice, 30*, 116–121.

Nithianandam, P., Das, S., & Park, Y. C. (2020). Effect of surfactant-keratin hydrolysate interactions on the hydration properties of a stratum corneum substitute. *Langmuir, 36*(10), 2543–2552.

Okoh, O. O., Sadimenko, A. P., & Afolayan, A. J. (2010). Comparative evaluation of the antibacterial activities of the essential oils of Rosmarinus officinalis L. obtained by hydrodistillation and solvent free microwave extraction methods. *Food Chemistry, 120*(1), 308–312.

Oliveira, H. C., Stolf-Moreira, R., Martinez, C. B. R., Grillo, R., de Jesus, M. B., & Fraceto, L. F. (2015). Nanoencapsulation enhances the post-emergence herbicidal activity of atrazine against mustard plants. *PLoS ONE, 10*(7), Article e0132971.

Parveen, K., Banse, V., & Ledwani, L. (2016). Green synthesis of nanoparticles: Their advantages and disadvantages. *AIP Conference Proceedings, 1724*(1), Article 020048.

Pereira, C. G., & Meireles, M. A. A. (2007). Economic analysis of rosemary, fennel and anise essential oils obtained by supercritical fluid extraction. *Flavour and Fragrance Journal, 22*(5), 407–413. https://doi.org/10.1002/ffj.1813

Périno-Issartier, S., Zill-e-Huma, Abert-Vian, M., & Chemat, F. (2011). Solvent free microwave-assisted extraction of antioxidants from sea buckthorn (Hippophae rhamnoides) food by-products. *Food and Bioprocess Technology*, *4*(6), 1020–1028.

Pezantes-Orellana, C., German Bermúdez, F., Matías De La Cruz, C., Montalvo, J. L., & Orellana-Manzano, A. (2024). Essential oils: A systematic review on revolutionizing health, nutrition, and omics for optimal well-being. *Frontiers in Medicine, 11*, 1337785. https://doi.org/10.3389/fmed.2024.1337785

Pradhan, R. C., Meda, V., Rout, P. K., Naik, S., & Dalai, A. K. (2010). Supercritical CO_2 extraction of fatty oil from flaxseed and comparison with screw press expression and solvent extraction processes. *Journal of Food Engineering, 98*(4), 393–397.

Pulido-Reyes, G., Rodea-Palomares, I., Das, S., Sakthivel, T. S., Leganes, F., Rosal, R., Seal, S., & Fernández-Piñas, F. (2015). Untangling the biological effects of cerium oxide nanoparticles: The role of surface valence states. *Scientific Reports, 5*, 15613.

Purkayastha, S., Narain, R., & Dahiya, P. (2012). Evaluation of antimicrobial and phytochemical screening of Fennel, Juniper and Kalonji essential oils against multi drug resistant clinical isolates. *Asian Pacific Journal of Tropical Biomedicine, 2*(3), S1625–S1629.

Raikwar, G., Kumar, D., Mohan, S., & Dahiya, P. (2024). Synergistic potential of essential oils with antibiotics for antimicrobial resistance with emphasis on mechanism of action: A review. *Biocatalysis and Agricultural Biotechnology, 61*, Article 103384. https://doi.org/10.1016/j.bcab.2024.103384

Raikwar, G., Kumar, D., Mohan, S., & Dahiya, P. (2025). Exploring synergistic effects of *Piper betle* and *Anethum graveolens* essential oils with antibiotics against Methicillin-resistant *Staphylococcus aureus*: Insights from In silico targeting of PBP 2a. *Microbial Pathogenesis, 203*, Article 107484.

Rejeb, I. B., Pastor, V., & Mauch-Mani, B. (2014). Plant responses to simultaneous biotic and abiotic stress: Molecular mechanisms. *Plants*, *3*(4), Article 4.

Reyes-Jurado, F., Navarro-Cruz, A. R., Ochoa-Velasco, C. E., Palou, E., López-Malo, A., & Ávila-Sosa, R. (2020). Essential oils in vapor phase as alternative antimicrobials: A review. *Critical Reviews in Food Science and Nutrition, 60*(10), 1641–1650.

Saad, H., Ayed, A., Srasra, M., Mezni, M., Essid, R., Jallouli, S., Tabbene, O., & Srasra, E. (2024). Smectite clay nanoparticles as a sustained release system for cinnamom essential oil targeting candida albicans. *Chemistry Africa, 7*(6), 3185–3200.

Santos, C., de Araújo Gonçalves, M., de Macedo, L. F., Torres, A. H. F., Marena, G. D., Chorilli, M., & Trovatti, E. (2023). Green nanotechnology for the development of nanoparticles based on alginate associated with essential and vegetable oils for application in fruits and seeds protection. *International Journal of Biological Macromolecules, 232*, Article 123351. https://doi.org/10.1016/j.ijbiomac.2023.123351

Saporito, F., Sandri, G., Bonferoni, M. C., Rossi, S., Boselli, C., Icaro Cornaglia, A., Mannucci, B., Grisoli, P., Vigani, B., & Ferrari, F. (2017). Essential oil-loaded lipid nanoparticles for wound healing. *International Journal of Nanomedicine, 13*, 175–186.

Sarlak, N., Taherifar, A., & Salehi, F. (2014). Synthesis of nanopesticides by encapsulating pesticide nanoparticles using functionalized carbon nanotubes and application of new nanocomposite for plant disease treatment. *Journal of Agricultural and Food Chemistry, 62*(21), 4833–4838.

Saxena, R., Tomar, R. S., & Kumar, M. (2016). Exploring nanobiotechnology to mitigate abiotic stress in crop plants. *Journal of Pharmaceutical Sciences, 8*.

Senorans, F. J., Ibanez, E., Cavero, S., Tabera, J., & Reglero, G. (2000). Liquid chromatographic–mass spectrometric analysis of supercritical-fluid extracts of rosemary plants. *Journal of Chromatography A, 870*(1–2), 491–499.

Shanbehzadeh, F., Saei-Dehkordi, S. S., & Semnani, D. (2022). Fabrication and characterization of electrospun nanofibrous mats of polycaprolactone/gelatin containing ZnO nanoparticles and cumin essential oil and their anti-staphylococcal potency in white cheese. *Food Bioscience, 49*, Article 101904.

Shao, W., Liu, H., Liu, X., Wang, S., Wu, J., Zhang, R., Min, H., & Huang, M. (2015). Development of silver sulfadiazine loaded bacterial cellulose/sodium alginate composite films with enhanced antibacterial property. *Carbohydrate Polymers, 132*, 351–358.

Shen, Y., Ni, Z.-J., Thakur, K., Zhang, J.-G., Hu, F., & Wei, Z.-J. (2021). Preparation and characterization of clove essential oil loaded nanoemulsion and pickering emulsion activated pullulan-gelatin based edible film. *International Journal of Biological Macromolecules, 181*, 528–539.

Siddiqui, M. H., Al-Whaibi, M. H., Faisal, M., & Al Sahli, A. A. (2014). Nano-silicon dioxide mitigates the adverse effects of salt stress on Cucurbita pepo L. *Environmental Toxicology and Chemistry, 33*(11), 2429–2437.

Siddiqui, B., Rehman, A. U., Haq, I. U., Ahmad, N. M., & Ahmed, N. (2020). Development, optimisation, and evaluation of nanoencapsulated diacerein emulgel for potential use in osteoarthritis. *Journal of Microencapsulation, 37*(8), 595–608.

Siddiqui, B., Rehman, A. U., Haq, I. U., Al-Dossary, A. A., Elaissari, A., & Ahmed, N. (2022). Exploiting recent trends for the synthesis and surface functionalization of mesoporous silica nanoparticles towards biomedical applications. *International Journal of Pharmaceutics: X, 4*, 100116.

Silva, S. L. D., Figueiredo, P. M., & Yano, T. (2007). Cytotoxic evaluation of essential oil from Zanthoxylum rhoifolium Lam. Leaves. *Acta Amazonica, 37*, 281–286. https://doi.org/10.1590/S0044-59672007000200015

Silva, S. L. D., Chaar, J. D. S., Figueiredo, P. D. M. S., & Yano, T. (2008). Cytotoxic evaluation of essential oil from Casearia sylvestris Sw on human cancer cells and erythrocytes. *Acta Amazonica, 38*, 107–112. https://doi.org/10.1590/S0044-59672008000100012

Sœur, J., Marrot, L., Perez, P., Iraqui, I., Kienda, G., Dardalhon, M., Meunier, J.-R., Averbeck, D., & Huang, M.-E. (2011). Selective cytotoxicity of Aniba rosaeodora essential oil towards epidermoid cancer cells through induction of apoptosis. *Mutation Research, 718*(1–2), 24–32. https://doi.org/10.1016/j.mrgentox.2010.10.009

Souiy, Z., Elaissi, A., Jlassi, I., Sghair, W., Allouch, N., Mastouri, M., & Krifia, B. (2021). Application of simplex-centroid design methodologies to optimize the anti-bacterial and anti-candidal activity of the mixture of Menthapulegium, Pituranthos Chloranthus and Thymus Algeriensis essential oils. *Medicinal and Aromatic Plants, 10*, 365.

Souiy, Z., Zakhama, N., Cheraief, I., & Hammami, M. (2022). Nutritional, physical, microbial, and sensory characteristics of gluten-and sugar-free cereal bar enriched with spirulina and flavored with neroli essential oil. *LWT, 169*, Article 113955.

Souiy, Z., Al-Shaikh, T. M., Alghamdi, O. A., & Krifi, B. (2023). ptimization of extraction yield, chemical composition, antioxidant, and antimicrobial activities of Thymus algeriensis Boiss. and Reut essential oils. *Biocatalysis and Agricultural Biotechnology, 47*, 102593.

Stadler, T., Buteler, M., & Weaver, D. K. (2010). Novel use of nanostructured alumina as an insecticide. *Pest Management Science, 66*(6), 577–579.

Stratakos, A. C., & Koidis, A. (2016). Methods for extracting essential oils. *Essential Oils in Food Preservation, Flavor and Safety*, 31–38.

Suhail, M. M., Wu, W., Cao, A., Mondalek, F. G., Fung, K.-M., Shih, P.-T., Fang, Y.-T., Woolley, C., Young, G., & Lin, H.-K. (2011). Boswellia sacra essential oil induces tumor cell-specific apoptosis and suppresses tumor aggressiveness in cultured human breast cancer cells. *BMC Complementary and Alternative Medicine, 11*(1), 129.

Suk, J. S., Xu, Q., Kim, N., Hanes, J., & Ensign, L. M. (2016). PEGylation as a strategy for improving nanoparticle-based drug and gene delivery. *Advanced Drug Delivery Reviews, 99*, 28–51. https://doi.org/10.1016/j.addr.2015.09.012

Takahashi, T., Kokubo, R., & Sakaino, M. (2004). Antimicrobial activities of eucalyptus leaf extracts and flavonoids from Eucalyptus maculata. *Letters in Applied Microbiology, 39*(1), 60–64. https://doi.org/10.1111/j.1472-765X.2004.01538.x

Tavasoli, M., Tatari, M., Samadi Kazemi, M., & Taghizadeh, S. F. (2022). In vitro cytotoxicity of Cuminum cyminum essential oil loaded SLN nanoparticle. *Nanomedicine Journal, 9*(3), 525–260. https://doi.org/10.22038/nmj.2022.63943.1668

Topçu, G., Gören, A. C., Bilsel, G., Bilsel, M., Çakmak, O., Schilling, J., & Kingston, D. G. I. (2005). Cytotoxic activity and essential oil composition of leaves and berries of Juniperus excelsa. *Pharmaceutical Biology.*

Vogt, T. (2010). Phenylpropanoid biosynthesis. *Molecular Plant, 3*(1), 2–20.

Wakeel, A., Xu, M., & Gan, Y. (2020). Chromium-induced reactive oxygen species accumulation by altering the enzymatic antioxidant system and associated cytotoxic, genotoxic, ultrastructural, and photosynthetic changes in plants. *International Journal of Molecular Sciences, 21*(3), 728.

Wang, Q., Ma, X., Zhang, W., Pei, H., & Chen, Y. (2012). The impact of cerium oxide nanoparticles on tomato (Solanum lycopersicum L.) and its implications for food safety. *Metallomics: Integrated Biometal Science, 4*(10), 1105–1112.

Wang, Z., Xie, X., Zhao, J., Liu, X., Feng, W., White, J. C., & Xing, B. (2012). Xylem- and phloem-based transport of CuO nanoparticles in maize (Zea mays L.). *Environmental Science and Technology, 46*(8), 4434–4441.

Wińska, K., Mączka, W., Łyczko, J., Grabarczyk, M., Czubaszek, A., & Szumny, A. (2019). Essential oils as antimicrobial agents—myth or real alternative? *Molecules, 24*(11), Article 11.

Yang, Y., Yue, Y., Runwei, Y., & Guolin, Z. (2010). Cytotoxic, apoptotic and antioxidant activity of the essential oil of Amomum tsao-ko. *Bioresource Technology, 101*(11), 4205–4211. https://doi.org/10.1016/j.biortech.2009.12.131

Yang, X.-H., Peng, H., Hao, Q.-L., Wang, J.-Z., Lu, Z.-G., Niu, Y.-W., Xiao, Z.-B., & Zhang, X. (2021). Nano-aromatic drugs based on mesoporous silica nanoparticles and bergamot essential oil for anti-depression. *Journal of Biomedical Nanotechnology, 17*(6), 1242–1248. https://doi.org/10.1166/jbn.2021.3100

Zaidi, S., & Dahiya, P. (2015). In vitro antimicrobial activity, phytochemical analysis and total phenolic content of essential oil from *Mentha spicata and Mentha piperita. International Food Research Journal, 22*(6), 2440–2445.

Zhiwei, K., Shan, C., Yan, Z., Sadeeq, U., & Hao, L. (2022). Rational construction of citrus essential oil nanoemulsion with robust stability and high antimicrobial activity based on combination of emulsifiers. *Innovative Food Science and Emerging Technologies, 80.*